Animal Biochemistry: From Theory to Applications

Animal Biochemistry: From Theory to Applications

Edited by **Mia Steers**

SYRAWOOD
PUBLISHING HOUSE
New York

Published by Syrawood Publishing House,
750 Third Avenue, 9th Floor,
New York, NY 10017, USA
www.syrawoodpublishinghouse.com

Animal Biochemistry: From Theory to Applications
Edited by Mia Steers

International Standard Book Number: 978-1-68286-274-2 (Hardback)

Printed in China.

Contents

Preface

Every book is a source of knowledge and this one is no exception. The idea that led to the conceptualization of this book was the fact that the world is advancing rapidly; which makes it crucial to document the progress in every field. I am aware that a lot of data is already available, yet, there is a lot more to learn. Hence, I accepted the responsibility of editing this book and contributing my knowledge to the community.

The understanding of animal biochemistry is essential for disciplines like animal husbandry and veterinary science. This book is a valuable compilation of topics, ranging from the basic to the most complex advancements in animal biochemistry. Included herein are insightful topics like metabolism, animal nutrition, feed processing technology, etc. which will enable the readers to have an in-depth understanding of this subject. It has been written and edited by internationally renowned scholars and the aim of this book is to present researches that have transformed this discipline and aided its advancement. For the students of zoology, veterinary sciences and similar disciplines, this text will prove to be an ideal resource material.

While editing this book, I had multiple visions for it. Then I finally narrowed down to make every chapter a sole standing text explaining a particular topic, so that they can be used independently. However, the umbrella subject sinews them into a common theme. This makes the book a unique platform of knowledge.

I would like to give the major credit of this book to the experts from every corner of the world, who took the time to share their expertise with us. Also, I owe the completion of this book to the never-ending support of my family, who supported me throughout the project.

Editor

1

Improving efficiency of sow productivity: nutrition and health

Sung Woo Kim*, Alexandra C Weaver, Yan Bin Shen and Yan Zhao

Abstract

This reviews research focused to understand the nutrient requirement and balance to meet the needs of fetal growth, mammary growth, and milk production. This summary will handle how feeding strategies can be adjusted according to the nutrient needs for a sow to enhance productivity and health. Most research data used in this summary are based on the studies conducted by the authors between 1996 and 2013. Nutrient requirements of sows are affected by stage of gestation and parity of sows. Dietary antioxidant concentrations need to be re-evaluated for its sufficiency in sow diets especially to prevent excessive oxidative stress during late gestation and lactation. When feeding sows, consideration of phase feeding of gestating sows and parity feeding of lactating sows could enhances production longevity and health of sows. Use of selected nutrients and additives seems to help productivity and health of sows.

Keywords: Colostrum, Gestation, Lactation, Milk, Nutrition, Pig

Introduction

Productivity of sows has been changed dramatically during the last decades. Continuous genetic selection led to high prolificacy of sows and production of high lean progeny. As consequences, modern sows produce larger litters than before [1] and each of offspring is leaner and grows faster [2]. A sow currently gives birth to 10 to 16 piglets at birth producing 25 to 30 pigs per year as a litter size has been increased by 3 pigs during 40 years [3]. Recent comparison shows that a porcine fetus is 40% heavier than 40 years ago (Figure 1). However, selection of pigs for high leanness also resulted in high lean type sows possessing a low appetite [4,5].

A sow, therefore, need to produce an ample amount of milk to meet the demands by her large and fast growing litter. In fact, between 1935 and 2010, milk yield has been increased by 4 folds (Figure 2). This suggests that the porcine mammary gland has adapted to support the increased demand for milk production as well. All these improvement with a sow and her litter warrants continuous updates on the nutritional management program. Without proper nutritional supports, sows will face severe catabolic condition. Severe maternal catabolic condition impairs the growth and survival of the litter.

There has been research focused to evaluate the nutrient requirement to meet the accurate needs for the growth of fetus, mammary gland and milk production. This will review how feeding strategies can be adjusted according to the nutrient needs to enhance productivity and health of a sow. Most research data used in this review are based on the studies conducted by the authors between 1996 and 2013.

Current challenges

Conventional feeding program for gestating sows does not provide sufficient proteins and minerals during late gestation causing catabolic condition to sows. Typical corn soybean meal based diets are formulated to provide 8 to 11 g true ileal digestible (TID) Lys daily to sows during the entire gestational period. A recent study [10] shows that conventional feeding program would significantly underfeed Lys during late gestation as requirements of TID Lys increase from 6.8 g/d to 15.3 g/d during late gestation (Figure 3). This increase in Lys requirement is due to dramatic changes in fetal tissue gain from 0.25 to 4.63 g CP/d/fetus [2] and mammary tissue gain from 0.41 to 3.41 g CP/d/gland [11] from early to late gestation (Figures 4 and 5).

* Correspondence: sungwoo_kim@ncsu.edu
Department of Animal Science, North Carolina State University, North Carolina, Raleigh 27695, USA

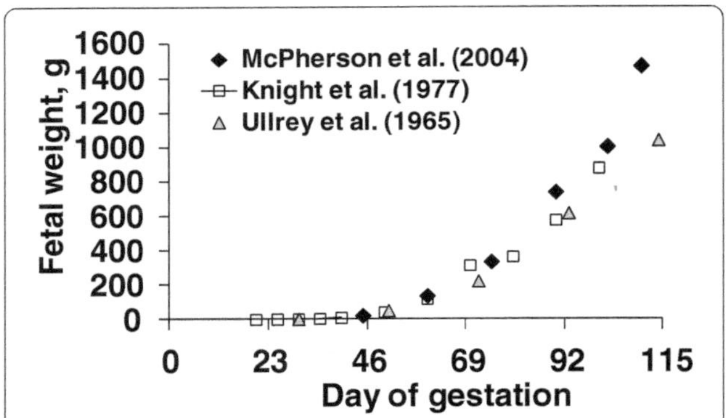

Figure 1 Growth patterns of porcine fetus. Adapted from [2,6,7].

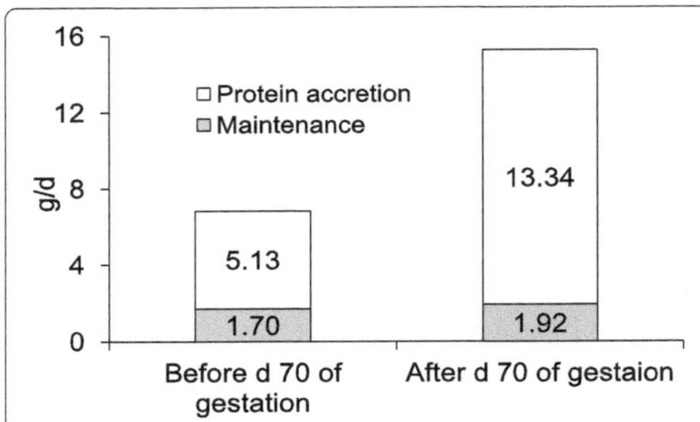

Figure 3 Requirements of true ileal digestible Lys of sows during early (d 0 to 70 of gestation) and late gestation (from d 70 of gestation to farrowing). Adapted from [10].

There are several evidences supporting that sows are not providing sufficient nutrients to fetal and mammary growth during late gestation. It has been shown that weight variations expressed as a coefficient of variation (%) among the weights of fetuses in each litter were smaller on d 45 of gestation than those on later than d 60 of gestation [10]. This indicates that fetal growth retardation occurs mainly from d 60 of gestation (Figure 6). Interestingly, fetal weight linearly decreased depending on their location on uterine horn (heaviest toward the utero-tubal junction and the lightest toward the cervix) on d 102 and 112 of gestation whereas there were no correlations between fetal weight and fetal location on d 30 and 60 of gestation (Figure 7). Limited nutrient supply from sows to support the growth of fetuses increased fetal weight variation during late gestation suggesting that current feeding programs for gestating sows is sub-optimal for fetal growth especially during late gestation.

Limited nutrient supply to fetus is also contributed from limited blood flow through placenta. Fetal weight variation could be reduced when blood flow was enhanced by dietary supplementation of arginine which can contribute nitric oxide in endothelial cells lining the blood vessels causing vasodilation [12-15].

Oxidative stress of sows

Catabolic condition increases the production of reactive oxygen species (ROS) causing increased oxidative stress [16,17]. Oxidative damage is a strong indicator of health status and wellbeing of animals in relation to aging, stress, nutritional status, and disease. Increased oxidative stress is responsible for impaired milk production, reproductive performance, and finally longevity of sows [18-20]. Impaired ability to produce milk directly affects the health and growth of nursing piglets, and may also have a long-term effect on health and growth throughout pigs' life.

Recently, it was demonstrated that sows are under severe catabolic status during late gestation causing increased oxidative stress [17]. Plasma α-tocopherol and retinoid concentrations were lower by 56% and 57% at d 110 of pregnancy as compared with d 30 of pregnancy, respectively (Figure 8).

Lymphocyte DNA damage was analyzed by alkaline single cell gel electrophoresis (comet assay) showing increased endogenous DNA damage during late gestation by 125% compared with d 30 of pregnancy (Figure 9). These results indicate that sows undergo increased

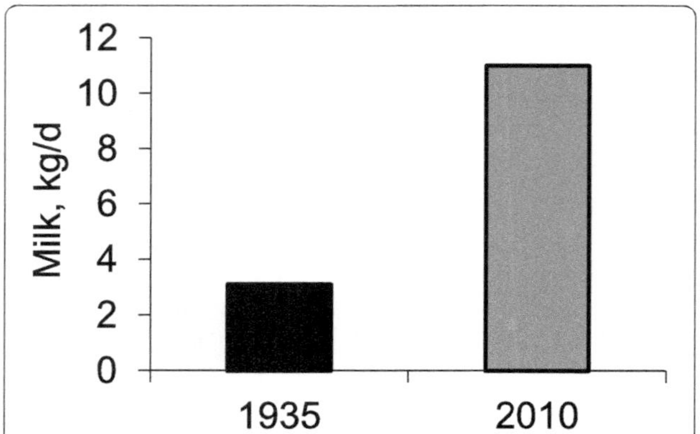

Figure 2 Milk yield of sows in 1935 and 2010. Adapted from [8,9].

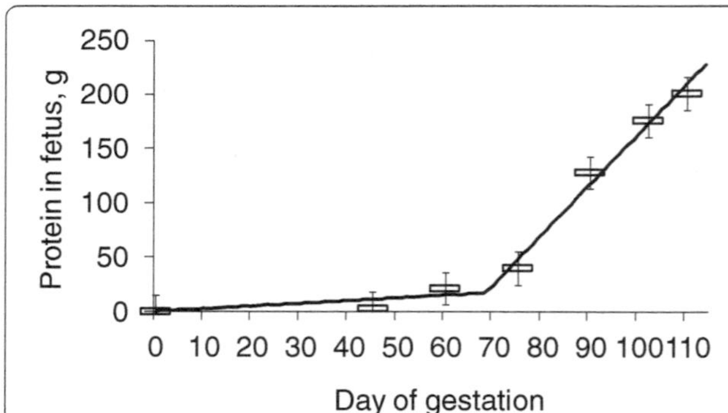

Figure 4 Protein content in a fetus during gestation. 0.25 g CP gain per day until d 70 of gestation and 4.63 g CP gain per day from d 70 of gestation. Adapted from [2].

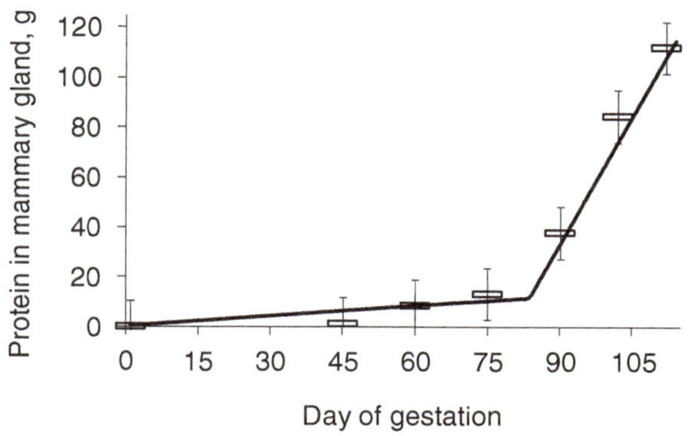

Figure 5 Protein content in a mammary gland during gestation. 0.14 g CP gain per day until d 80 of gestation and 3.41 g CP gain per day from d 70 of gestation Adapted from [11].

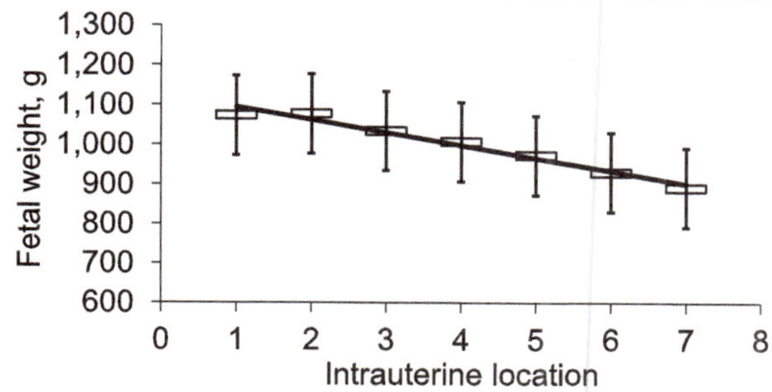

Figure 7 Litter weight variations among fetuses on d 102 of gestation. Intra-uterine location 1 indicates cranial (i.e., toward the utero-tubal junction) extremity and 7 indicates cervical (i.e., toward the cervix) extremity. Fetal weight linearly decreased as intra-uterine location changed from 1 to 7: Y = −32.1 x + 1,127.3, $R^2 = 0.97$, $P < 0.001$ where Y is fetal weight and x is intra-uterine location (1 to 7). Adapted from [10].

damage to immune cells due to increased oxidative stress to DNA and also undergo increased systemic oxidative damage due to reduced antioxidative capacity. It seems that oxidative stress to sows increases when sows are under environmental stress such as heat stress and social stress. Sows under heat stress environment showed increased oxidative stress by increased lipid peroxidation, protein oxidation, and oxidative DNA damage compared with the sows under comfort thermal neutral zone. Sows in a gestation stall may have increased oxidative stress compared with sows in a group pen environment. Increased oxidative damages in sows during late pregnancy may negatively affect the growth and health of fetuses as well as postpartum growth of piglets [20].

Hot and humid summer climate causes heat stress reducing reproductive performance and longevity of sows [21,22]. Studies showed that heat stress diminished the secretion of FSH and LH, and delayed puberty in gilts [18,21]. Heat stress may affect early development of embryos [23] causing small litter size, increased number of stillborn, and reduced birth weights [22,24,25]. Studies have shown that lactating sows exposed to high temperatures had reduced feed intake and milk production [26,27]. Hyperthermia from heat stress stimulates reactive oxygen species production causing oxidative damages [28]. A study was recently conducted to investigate the effects of hyperthermal conditions on oxidative status, and reproductive performance of sows during gestation and lactation [19]. A group of sows was under moderate ambient temperature environment (CON) and the others were under high ambient temperature environment (HT).

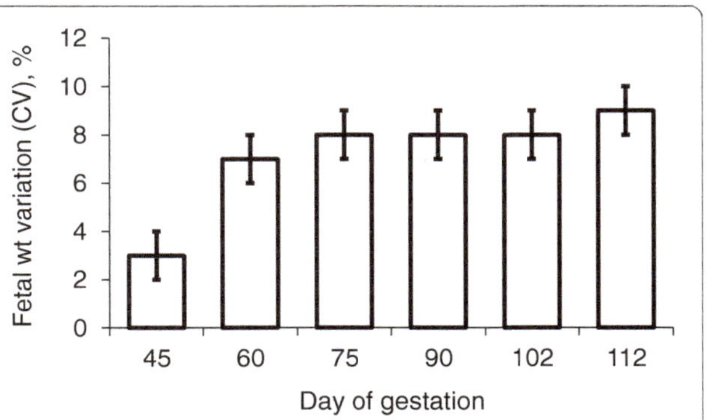

Figure 6 Weight variations among fetuses within a litter on different days of gestation. Litter weight variation was expressed as coefficient of variation [CV (%)] for each litter on different days of gestation. Adapted from [10].

Sows in HT had decreased ($P < 0.05$) number of piglets born alive and piglets per litter on d 18 of lactation. Litter weight at birth in HT tended to be smaller ($P = 0.050$) compared with those in CON. Litter weight on d 18 of lactation and litter weight gain in HT were smaller ($P < 0.05$) than those in CON. The protein carbonyl concentration in HT was greater than CON on d 90 and 109 of gestation, and d1 and 18 of lactation. Sows in HT had a greater concentration of malonedialdehyde (MDA) on d 90 and 109 of gestation, and d 1 of lactation than sows in CON.

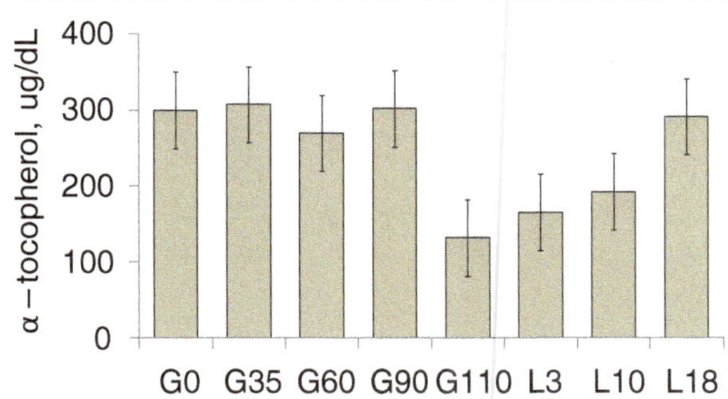

Figure 8 Plasma α -tocopherol concentration during gestation and lactation. Adapted from [17].

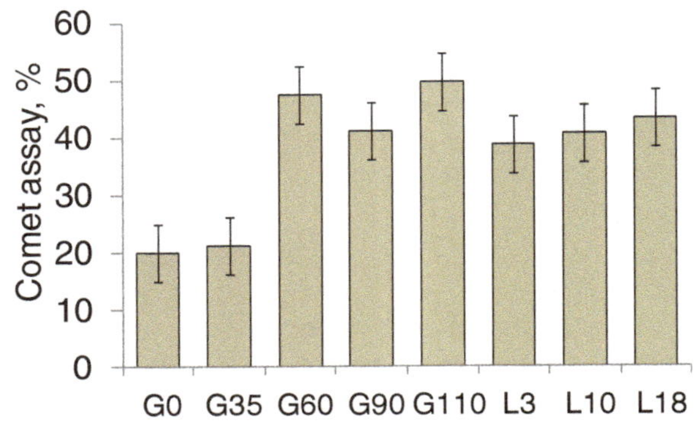

Figure 9 Lymphocyte endogenous DNA damage (comet assay) during gestation and lactation. Adapted from [17].

These data indicate that sows in HT have increased lipid and protein damage during late gestation and lactation compared with sows in CON. If comparing oxidative markers between different gestating and lactating days within each treatment, this study showed that sows under heat stress had greater plasma concentrations of 8-OHdG and protein carbonyl on d 109 of gestation than the other days, which indicating that sows under heat stress environment have increased DNA and protein damage during late gestation. Litter weight gain and litter size were negatively correlated ($P < 0.05$) with oxidative stress to sows as indicated by increased plasma concentrations of MDA, protein carbonyls, and 8-OHdG.

Decreased antioxidant capacity during late gestation and lactation can increase oxidative damage by increased production of free radicals when an animal is under high ambient temperature environment [28,29]. This indicates that oxidative stress is one of major stress responses caused by heat stress. This could lead to increased protein turnover due to increased oxidative damage to cellular proteins, increased cell death due to increased peroxidation of membrane lipids and increased DNA mutation and breakdown which can together interfere fetal development, mammary gland development, and milk production as shown as reduced number of piglets born alive, and reduced litter weight gain [19]. Increased oxidative stress during the period of embryonic implantation may cause increased embryonic death which can be related to a reduced litter size for sows under heat stress [19]. Increased oxidative damage to lipid, protein, and DNA was one of the major contributing factors for reduced reproductive performance of sows under a heat stress environment [19].

Gestation crate has widely been used in order to control individual energy intake. However, concerns about animal wellbeing enforce the removal of gestation crates. However, group housing of sows under controlled feed allowance could also potentially increase aggressive behavior between sows, health risks of low dominance sows, occurrence of stereotypic behaviors, and possible reduction in reproductive performance. It has been shown that social and behavioral stresses are associated with physical markers for oxidative stress.

A recent study [20] determined if reproductive performance and oxidative stress status of sows would be affected by different gestational housing systems. Sows were housed either in groups of 3 per pen (PEN) or individual gestational crates (CON) on d 35 of gestation. Reproductive performance of sows housed in gestational pens tended to be inferior to sows housed in gestational crates as indicated by total piglets born per litter and litter weight at born. However, oxidative stress status was not affected by gestational housing indicating that the effects of gestational housing on reproductive performance of sows may not be directly related to oxidative stress status. Oxidative damages to protein and DNA were further increased during late gestation and lactation regardless of gestational housing.

The effect of social rank of gestating sows house in group on their oxidative stress status, immune status, and reproductive performance was also investigated. The social rank of sows within a pen was determined by observing their aggressive behavior for a 4-d period after mixing. Sows within a pen were classified into high-, middle-, and low-ranking groups (HR, MR, and LR) according to their percentage of winning interactions. Sows in LR showed greater ($P < 0.05$) litter size and litter weight than sows in HR even though their BW was inferior ($P < 0.05$) to sows in HR. However, sows in LR has decreased ($P < 0.05$) farrowing rate and increased mortality. Sows in LR had higher ($P < 0.05$) DNA damage compared with HR during late gestation and lactation, which could be a major reason to their poor farrowing rate and increased mortality. The study concluded that the reproductive performance was related to oxidative status of sows regardless which rank they were in.

Milk production and porcine mammary gland

Strategy to improve milk production should consider enhancing mammary gland growth during gestation and lactation as milk synthesis occurs in a mammary epithelial cell and the number of mammary epithelial cells determines milk production [30]. Nutritional status as well as various factors affect mammary gland growth and therefore milk production. Nutritional management of gestating and lactating sows should consider increased protein and amino acid needs during late gestation and during lactation. Age of sows, litter size, and health status should also be considered in determining nutrient needs for mammary gland growth and milk production.

The growth mammary gland is affected by the anatomical location on a sow. Mammary glands in middle part of the body (typically 4th and 5th pairs of mammary glands) grow faster during gestation and bigger in size at farrowing compared with mammary glands in anterior (1st, 2nd, and 3rd pairs) and posterior (6th, 7th, and 8th pairs) location on a sow [11]. This may be because there is more physical space for mammary glands to grow in middle part of the body whereas anterior and posterior locations which are hindered by legs. It is also speculated that blood supply starts from middle location which is typically 3rd mammary glands and then extends to front and back of the body and thus mammary glands in middle location have greater chance to obtain nutrients compared with those in other locations.

During lactation, however, anterior mammary glands grew faster than others [31]. This may be because anterior mammary glands have a greater preference by piglets during lactation. Growth of suckling piglets was greater when they suckled the first 5 pairs of mammary glands compared with posterior mammary glands (Figure 10). Posterior mammary glands had greater variation in their sizes and milk production whereas anterior and middle mammary glands were more uniform in size and milk production [31].

An increase in litter size directly increase the number of functional mammary glands and thus a sow needs to have increased nutrient supply not just to produce more milk but also to support the growth of these mammary glands that are additionally needed to support the increased litter. Milk yield is more than 50% greater when litter size increased from 6 to 12 [32]. This may result from an increased number of active mammary glands, a crucial component of milk production [33,34]. Sows need an additional 1.0 g lysine per day to account for mammary gland growth for each pig added to a litter [35].

[36] demonstrated that increase in litter size causes decrease in the size of individual suckled mammary glands indicating potential decrease in milk production

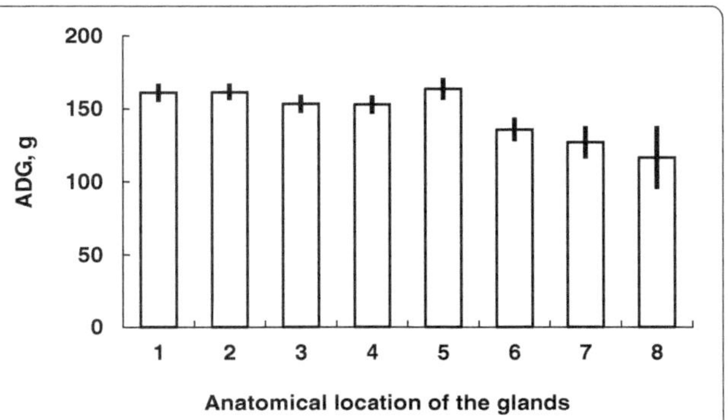

Figure 10 Growth of piglets depending on anatomical locations of suckling mammary glands [31]. Average daily gain (ADG) of piglets suckling first 5 pairs of mammary glands was greater ($P < 0.05$) than that of others.

from each mammary gland. However, mammary glands of a sow with a large litter may be more efficient at producing milk because the reduction in individual pig weight gain was only 73% of the decline in mammary gland growth rate observed in response to increased litter size [36]. Recently, [37] demonstrated that milk lactose content linearly increases, whereas protein content tended to linearly increases when a litter size increased from 8 to 14 piglets indicating changes in litter size affect both milk yield and quality.

Mammary tissue growth continues in suckled glands during lactation in sows [38]. Wet weight of individual suckled mammary gland increases by 55% and total gland DNA doubles during 4 week of lactation. This growth requires 6 g/d true ileal digestible lysine as there is 1 g/d lysine increase in mammary tissue and 5 g/d lysine used for maintenance of mammary tissue [35,39]. Interestingly, the amount of amino acids taken up by the mammary gland does not change with advancing stage of lactation [39] although the number of mammary epithelial cells increases [35] with advancing stages of lactation, indicating a diminished rate of amino acids transport by these cells.

The growth of suckled mammary glands is affected by maternal nutrition. It was shown that mammary growth is affected by both protein and energy intake and it was maximized when sows are provided with 55 g true ileal digestible lysine and 16.9 Mcal metabolizable energy daily during lactation [38]. When considering nutrient recommendation of swine by National Research Council (NRC, 1998), a sow requires 49 g/d lysine. Comparing with [35,36], the difference of 6 g/d lysine (49 vs. 55 g/d) could be accounted for by the need of mammary tissue growth during lactation which was not taken into consideration when NRC recommendation was established in 1998.

Nutritional management of sows

During gestation, sows undergo dramatic changes with fetal growth and mammary gland growth [2] investigated growth of porcine fetuses and determined their nutrient needs. Growth of fetuses was fairly limited until d 70 of gestation (0.25 g protein increase/d) whereas it was significantly increased (19 folds) to 4.63 g protein increase/d after d 70 of gestation [2,10]. This dramatic increase in growth during late gestation includes growth of heart, liver, intestines, and placenta [2]. Growth of fetuses reported from [2] is about 40% greater than previous report [6,7] indicating improvement of growth potential of fetuses during the last 40 years [11] investigated growth of porcine mammary glands during gestation. It was interesting to observe that growth was not significant until d 80 of gestation (0.41 g protein increase/gland/d) whereas it was significantly increased

Table 1 Requirements of true ileal digestible amino acid and ideal dietary amino acid ratios for sows during gestation

	Amino acid, true ileal digestible							
	Lys	Thr	Val	Leu	Ile	Phe	Arg	His
D 0 to 60 of gestation								
Amount, g/d	5.57	4.42	3.62	4.92	3.26	2.79	4.97	2.00
Ratio relative to Lys,%	100.0	79.4	65.0	88.3	58.6	50.1	89.3	35.9
D 60 to 114 of gestation								
Amount, g/d	8.78	6.25	5.83	8.36	4.87	4.54	8.59	3.12
Ratio relative to Lys,%	100.0	71.2	66.4	95.3	55.5	51.8	97.9	35.5

Adapted from [10].

(24 folds) to 3.41 g protein increase/gland/d after d 80 of gestation [10,11]. Considering this growth patterns, protein requirements should be greater in late gestation compared with early gestation.

In the case of primiparous sows used in this research, it can be estimated that requirements of true ileal digestible (TID) Lys are 6.83 g/d until d 70 of gestation and 15.26 g/d from d 70 of gestation with 2.2 fold difference in the amount [10,40]. Difference in Lys requirement, as an example, is due to dramatic differences in protein gain of fetuses and mammary glands depending on state of gestation. When considering ideal protein concept, Leu and Arg have increased importance during late gestation whereas Thr has increased importance during early gestation [41]. Different growth patterns of fetuses and mammary glands with their unique amino acid compositions contributed to changes in ideal amino acid patterns (Table 1).

Differences in nutrient requirements based on stages of gestation can create difficulties in feeding practice as it is not feasible to provide two diets if a gestation barn has only one feed line. Top dressing during late gestation can be an alternative way of providing needed nutrients during late gestation [42]. One concern with application of phase feeding during gestation is not to change diets from low to high protein concentration in one day which can cause metabolic stress to pregnant sows. It would be helpful to gradually increase protein or change diets during several days.

During lactation, most sows are under severe catabolic conditions due to produce massive amount of milk with limited nutrient intake [43]. It can be estimated that sows produce 60 g milk/kg body weight which is even greater than dairy cow (50 g milk/kg body weight). Extended catabolic condition during lactation can be a cause of increased oxidative stress [17] negatively affecting longevity and productivity of sows. If voluntary feed intake is a limiting factor leading to the catabolic condition, providing a diet with highly utilizable nutrients would be important.

In swine, the number of mammary epithelial cells is highly correlated to milk production [31]. A study showed that porcine mammary gland continue growing after farrowing and the number of mammary epithelial cells can almost doubled by 3 to 4 week of lactation [35].

Table 2 Ideal amino acid patterns and the order of limiting amino acids for lactating sows

Estimated 21-d weight loss (kg)	75 to 80	33 to 45	12 to 15	6 to 8	0	7 to 0
Level of tissue mobilization (%)	50	40	20	5	0	NRC (1998)
Ideal AA pattern (% of Lys)						
Lys	100	100	100	100	100	100
Thr	75	69	63	60	59	62
Val	78	78	78	77	77	85
Leu	128	123	118	115	115	114
Ile	60	59	59	59	59	56
Arg	22	38	59	69	72	56
Order of limiting amino acids						
First	Thr	Lys	Lys	Lys	Lys	Lys
Second	Lys	Thr	Thr	Val	Val	Val
Third	Val	Val	Val	Thr	Thr	Thr

Adapted from [10].

Lactating mammary glands have high maintenance requirements for branched chain amino acids [44]. Sows use significant amount of nutrients to support the growth of mammary glands during lactation. As an example, it was calculated that 6 g TID Lys/d is needed to support the growth of mammary gland [35]. A follow-up study further demonstrated that growth of lactating mammary glands was maximized when a primiparous sow consumed 55 g TID Lys and 17 Mcal ME/d [36] which is greater than nutrient requirements by NRC (1998) indicating that it is need to consider the amount of nutrients needed for mammary gland growth when feeding lactating sows.

Young sows with small voluntary feed intake require different quality of proteins compared with old sows with good voluntary feed intake. This is because proteins mobilized from maternal tissues has different amino acid profiles from dietary protein and thus contribute to amino acid balance for milk synthesis. The data show that ideal amino acid pattern, therefore, is different depending on parity of sows (Table 2). Sows with significant loss of body protein need more dietary Thr whereas sows without body protein loss need more dietary Val relative to Lys [10].

Similar to what it was discussed in feeding gestating sows, differences in nutrient requirements based on parity of sows can create difficulties in feeding practice as it is not feasible to provide multiple diets if a farrowing barn has only one feed line. Top dressing to primiparous sows during lactation can be an alternative way of balancing needed nutrients of sows with different parities. Or grouping sows based on their parity can be another practical way of feeding strategy if a farm has enough number of sows to do parity feeding.

Feed quality seems to be another important factor affecting the productivity of sows. Mycotoxin can cause reduction in voluntary feed intake, systemic inflammation, and increased oxidative stress even at low to moderate level contamination [45,46]. The role of functional nutrients or feed additives was further investigated for their effects on reproductive performance and health status of sows suggesting beneficial effects of selected functional nutrients. Recent publications suggest that these beneficial nutrients can include (1) functional amino acids such as Arg by enhancing blood flow and fetal growth [14,15,47] and Trp by reducing oxidative and behavioral stress [48,49], (2) omega-3-fatty acids by reducing systemic inflammation [50,51], and (3) yeast cell contents by enhancing milk production [9,42,52].

Conclusion

As discusses so far, nutrient requirements of sows are affected by stage of gestation and parity of sows. Dietary antioxidant concentrations need to be re-evaluated for its sufficiency in sow diets especially to prevent excessive oxidative stress during late gestation and lactation. When feeding sows, consideration of phase feeding of gestating sows and parity feeding of lactating sows could enhances production longevity and health of sows. Use of selected nutrients and additives seems to help productivity and health of sows.

Competing interests

The authors declare that they have no competing interests.

Authors' contributions

SWK, ACW, YBS, and YZ contributed to this review paper by collecting data and writing the text. All authors read and approved the final manuscript.

References

1. Kim SW: *Mammary Gland Growth and Nutrient Mobilization in Lactating Sows: A Dynamic Model to Describe Nutrient Flow*, Ph.D. Thesis. University of Illinois at Urbana-Champaign; 1999.
2. McPherson RL, Ji F, Wu G, Blanton JR Jr, Kim SW: **Growth and compositional changes of fetal tissues in pigs.** *J Anim Sci* 2004, **82:**2534–2540.
3. NASS: *Agricultural statistics book by year. National Agricultural Statistics Service U.S.* Washington, DC: Dept. of Agriculture, Government Printing Office, Superintendent of Documents; 2011.
4. Kanis E: **Effect of food intake capacity on genotype by feeding regimen interactions in growing pigs.** *Anim Prod* 1990, **50:**343–351.
5. Kim SW, Wu G, Baker DH: **Amino acid nutrition of breeding sows during gestation and lactation.** *Pig News Info* 2005, **26:**89N–99N. CABI.
6. Ullrey DE, Sprague JL, Becker DE, Miller ER: **Growth of the swine fetus.** *J Anim Sci* 1965, **24:**711–717.
7. Knight JW, Bazer FW, Thatcher WW, Franke DE: **Conceptus development in intact and unilaterally hysterectomized-ovariectomized gilts: interrelations among hormonal status, placental development, fetal fluids and fetal growth.** *J Anim Sci* 1977, **44:**620–636.
8. Hughes EH, Hart GH: **Production and composition of sow's milk.** *J Nutr* 1935, **9:**311–322.
9. Shen YB, Carroll JA, Yoon I, Mateo RD, Kim SW: **Effects of supplementing Saccharomyces cerevisiae fermentation product in sow diets on performance of sows and nursing piglets.** *J Anim Sci* 2011, **89:**2462–2471.
10. Kim SW, Hurley WL, Wu G, Ji F: **Ideal amino acid balance for sows during gestation and lactation.** *J Anim Sci* 2009, **87:**E123–E132.
11. Ji F, Hurley WL, Kim SW: **Characterization of mammary gland development in pregnant gilts.** *J Anim Sci* 2006, **84:**579–587.
12. Campos PH, Silva BA, Donzele JL, Oliveira RF, Knol EF: **Effects of sow nutrition during gestation on within-litter birth weight variation: a review.** *Animal* 2012, **6:**797–806.
13. Gao K, Jiang Z, Lin Y, Zheng C, Zhou G, Chen F, Yang L, Wu G: **Dietary L-arginine supplementation enhances placental growth and reproductive performance in sows.** *Amino Acids* 2012, **42:**2207–2014.
14. Mateo RD, Wu G, Bazer FW, Park JC, Shinzato I, Kim SW: **Dietary L-arginine supplementation enhances the reproductive performance of gilts.** *J Nutr* 2007, **137:**652–656.
15. Mateo RD, Wu G, Moon HK, Carroll JA, Kim SW: **Effects of dietary arginine supplementation during gestation and lactation on the performance of lactating primiparous sows and nursing piglets.** *J Anim Sci* 2008, **86:**827–835.
16. Bernardi F, Constantino L, Machado R, Petronilho F, Dal-Pizzol F: **Plasma nitric oxide, endothelin-1, arginase and superoxide dismutase in pre-eclamptic women.** *J Obstet Gynaecol Res* 2008, **34:**957–963.
17. Berchieri-Ronchi CB, Kim SW, Zhao Y, Correa CR, Yeum K-J, Ferreira ALA: **Oxidative stress status of high prolific sows during pregnancy and lactation.** *Animal* 2011, **5:**1774–1779.
18. Flowers B, Day BN: **Alterations in gonadotropin secretion and ovarian function in prepubertal gilts by elevated environmental temperature.** *Biol Reprod* 1990, **42:**465–471.

19. Zhao Y, Flowers WL, Saraiva A, Yeum K-J, Kim SW: **Effect of heat stress on oxidative stress status and reproductive performance of sows.** *J Anim Sci* 2011, **89**(2):108.
20. Zhao Y, Flowers WL, Saraiva A, Yeum K-J, Kim SW: **Effect of social ranks and gestation housing systems on oxidative stress status, reproductive performance, and immune status of sows.** *J Anim Sci.* in press.
21. Flowers B, Cantley TC, Martin MJ, Day BN: **Effect of elevated ambient temperatures on puberty in gilts.** *J Anim Sci* 1989, **67:**779–784.
22. Johnston LJ, Ellis M, Libal GW, Mayrose VB, Weldon WC: **Effect of room temperature and dietary amino acid concentration on performance of lactating sows.** *J Anim Sci* 1999, **77:**1638–1644.
23. Edwards RL, Omtvedt IT, Tuesman EJ, Stephens DF, Mahoney GWA: **Reproductive performance of gilts following heat stress prior to breeding and in early gestation.** *J Anim Sci* 1968, **27:**1634–1637.
24. Omtvedt I, Nelson T, Ronnie RE, Edwards L, Stephens DF: **Influence of heat stress during early, mid and late gestation.** *J Anim Sci* 1971, **32:**312–317.
25. Renaudeau D, Noblet J: **Effects of exposure to high ambient temperature and dietary protein level on sow milk production and performance of piglets.** *J Anim Sci* 2001, **79:**1540–1548.
26. Schoenherr WD, Stably TS, Cromwell GL: **The effects of dietary fat and fiber addition on energy and nitrogen digestibility in lactating, primiparous sows housed in a warm or hot environment.** *J Anim Sci* 1989, **67:**473–481.
27. Black JL, Mullan BP, Lorschy ML, Giles LR: **Lactation in the sow during heat stress.** *Livest Prod Sci* 1993, **35:**153–170.
28. Ozawa M, Hirabayashi M, Kanai Y: **Developmental competence and oxidative state of mouse zygotes heat-stressed maternally or in vitro.** *Reproduction* 2002, **124:**683–689.
29. Matsuzuka T, Ozawa M, Nakamura A, Ushitani A, Hirabayashi M, Kanai Y: **Effects of heat stress on the redox status in the oviduct and early embryonic development in mice.** *J Reprod Dev* 2005, **51:**281–287.
30. Kim SW: **Sow milk.** In *Milk and Dairy Products in Human Nutrition.* Wiley-Blackwell; 2013:614–626. ISBN 978-0-470-67418-5.
31. Kim SW, Hurley WL, Han IK, Easter RA: **Growth of nursing pigs related to the characteristics of nursed mammary glands.** *J Anim Sci* 2000, **78:**1313–1318.
32. King RH, Toner MS, Dove H: **Pattern of milk production in sows.** In *Manipulation of Pig Production II.* Edited by Batterham S. Attwood: Australasian Pig Science Association; 1989:98.
33. Knight CH, Peaker M: **Mammary development and regression during lactation in goat in relation to milk secretion.** *Q J Exp Physiol* 1984, **69:**331–338.
34. Knight CH, Docherty AH, Peaker M: **Milk yield in rats in relation to activity and size of the mammary secretory cell population.** *J Dairy Res* 1984, **51:**29–36.
35. Kim SW, Hurley WL, Han IK, Easter RA: **Changes in tissue composition associated with mammary gland growth during lactation in the sow.** *J Anim Sci* 1999, **77:**2510–2516.
36. Kim SW, Hurley WL, Han IK, Stein HH, Easter RA: **Effect of nutrient intake on mammary gland growth in lactating sows.** *J Anim Sci* 1999, **77:**3304–3315.
37. Voilqué G, Zhao Y, Kim SW: **Composition of porcine colostrum and milk as affected by various production environments.** *J Anim Sci* 2012, **90**(2):32.
38. Kim SW, Osaka I, Hurley WL, Easter RA: **Mammary gland growth as affected by litter size in lactating sows: impact on lysine requirement.** *J Anim Sci* 1999, **77:**3316–3321.
39. Trottier NL, Shipley CF, Easter RA: **Plasma amino acid uptake by the mammary gland of the lactating sow.** *J Anim Sci* 1997, **75:**1266–1278.
40. Ji F, Wu G, Blanton JR Jr, Kim SW: **Changes in weight and composition in various tissues of pregnant gilts and their nutritional implication.** *J Anim Sci* 2005, **83:**366–375.
41. Kim SW, Chaytor AC, Shen Y, Voilque G: *Application of ideal protein and amino acid requirements for gestating sows.* Estância de São Pedro, SP, Brasil: IV Congresso Latino Americano de Nutrição Animal; 2010:201–204.
42. Kim SW, Brandherm M, Newton B, Cook D, Yoon I, Fitzner G: **Dietary supplementation of yeast culture to sow diets for litter performance.** *Canadian J Anim Sci* 2010, **90:**229–232.
43. Kim SW, Easter RA: **Amino acid utilization for reproduction in sows.** In *Amino Acids in Animal Nutrition.* Edited by D'Mello JPF. CABI Publishing; 2003:203–222. ISBN 085199654X.
44. Li P, Knabe DA, Kim SW, Lynch CJ, Hutson SM, Wu G: **Lactating porcine mammary tissue catabolizes branched-chain amino acids for glutamine and aspartate synthesis.** *J Nutr* 2009, **139:**1502–1509.
45. Chaytor AC, See MT, Hansen JA, de Souza ALP, Middleton TF, Kim SW: **Effects of chronic exposure of diets with low levels of aflatoxin and deoxynivalenol on growth and immune status of pigs.** *J Anim Sci* 2011, **89:**124–135.
46. Chaytor AC, Kim SW: **Effects of aflatoxin and deoxynivalenol on oxidative damage in pigs.** *J Anim Sci* 2012, **90**(2):13. Abstr.
47. Kim SW, Wu G: **Regulatory role for amino acids in mammary gland growth and milk synthesis.** *Amino Acids* 2009, **37:**89–95.
48. Shen YB, Voilque G, Odle J, Kim SW: **Effects of elevating Trp intake on growth and hypothalamic serotonin secretion and stress hormone secretion in nursery pigs.** *J Anim Sci* 2012, **90:**2264–2275.
49. Shen YB, Voilque G, Odle J, Kim SW: **Dietary L-tryptophan supplementation with reduced large neutral amino acids enhances feed efficiency and reduces stress hormone secretion of nursery pigs under social-mixing stress.** *J Nutr* 2012, **142:**1540–1546.
50. Kim SW, Mateo RD, Yin Y-L, Wu G: **Functional amino acids and fatty acids for enhancing production performance of sows and piglets.** *Asian-Australasian J Anim Sci* 2006, **20:**295–306.
51. Mateo RD, Carroll JA, Hyun Y, Smith S, Kim SW: **Effect of dietary supplementation of omega-3 fatty acids and high protein on reproductive outcome of primiparous sows for two parities.** *J Anim Sci* 2009, **87:**948–959.
52. Kim SW, Brandherm M, Freeland M, Newton B, Cook D, Yoon I: **Effects of yeast culture supplementation to gestation and lactation diets on growth of nursing piglets.** *Asian-Australasian J Anim Sci* 2008, **21:**1011–1014.

2

Subcutaneous fatty acid composition of steers finished as weanlings or yearlings with and without growth promotants

Cletos Mapiye[1], Tyler D Turner[1], John A Basarab[2], Vern S Baron[1], Jennifer L Aalhus[1] and Michael ER Dugan[1*]

Abstract

Background: The current study evaluated the subcutaneous fatty acid (FA) composition of calf- and yearling-fed steers with or without growth promoting implants. Crossbred steers (n = 112; 267 ± 5.0 kg) of the same contemporary group were allocated to one of four production system and implant strategy based treatments in a completely randomized design with a 2 × 2 factorial arrangement of treatments.

Results: There were no interactions ($P > 0.05$) between production systems and growth promoting implants for the total and individual subcutaneous FA. Yearling as opposed to calf finishing reduced ($P < 0.05$) subcutaneous proportions of C20:3*n*-6, *trans (t)*12-18:1, C14:0, several minor *cis*-monounsaturated FA (*c*-MUFA; *c*9-14:1, *c*11-16:1, *c*11-18:1, *c*12-18:1, *c*13-18:1, *c*9-20:1 and *c*11-20:1), and increased ($P < 0$ .05) subcutaneous proportions of *t*11*c*15-18:2, total and individual branched-chain FA. Subcutaneous fat from steers implanted with growth promotants had higher ($P < 0.05$) proportions of total polyunsaturated FA (PUFA), total *n*-6 PUFA, C18:2*n*-6 and individual *t*-18:1 isomers (*t*6 to *t*10) compared to non-implanted steers.

Conclusions: Overall, current findings show that production systems and growth promotants led to only minor differences in subcutaneous FA composition of beef steers.

Keywords: Age at feedlot entry, Beef, Fatty acids, Growth implant, Steers

Background

In North America, beef cattle are typically finished using calf or yearling production systems. In calf production systems, cattle enter the feedlot immediately after weaning at 6–8 mo of age and are fed a high-energy finishing diet *ad libitum* until slaughter at 12–14 mo of age to take advantage of the faster growth of large-framed cattle [1,2]. On the contrary, in yearling production systems cattle graze pasture and/or crop residues post-weaning prior to entering a feedlot at 15–18 mo of age to allow for growth of frame in small- to medium-framed cattle and takes advantage of lower priced forages and subsequent compensatory growth in the feedlot. Although age at feedlot entry has been reported to consistently affect fat deposition [3,4], little is known about its effect on fatty acid (FA) composition. It has been suggested that cattle entering a feedlot as yearlings may have a healthier FA profile from a consumer's perspective due to forage-based diets compared to those entering the feedlot as weanlings [5,6].

Growth promoting implants characterised as being estrogenic (*e.g.,* estradiol benzoate and estradiol) or androgenic (*e.g.,* trenbolone acetate and progesterone) are used extensively in calf and yearling production systems to increase growth rates, carcass yield [3,7] and to reduce carcass fatness [8,9]. Regarding FA composition, studies with estrogen- or androgen-implanted steers showed small increases in the proportions of saturated FA (SFA) and decreases in proportions of monounsaturated FA (MUFA) [10,11]. In contrast, other studies with estrogen-implanted bull calves [12] or steers [13] found reduced SFA and increased MUFA. In the same studies, Hozler et al. [12] and Ibrahim et al. [13] reported a decrease and an increase in the proportions of polyunsaturated FA (PUFA), respectively. In general, the available information on the effects of growth promoting implants on beef FA composition are inconclusive.

* Correspondence: mike.dugan@agr.gc.ca
[1]Agriculture and Agri-Food Canada, Lacombe Research Centre, 6000 C & E Trail, Lacombe, Alberta T4L 1 W1, Canada
Full list of author information is available at the end of the article

In addition, the interactive effects of implant strategies with age entering the feedlot have not been investigated. The objective of the present study was, therefore, to determine the FA composition of subcutaneous adipose tissue from beef steers finished as weanlings or yearlings with and without growth promotants. More specifically the objectives were to determine the extent reduced adiposity can influence concentrations of rumen derived FA such as conjugated linoleic acid (CLA), *trans* (*t*)-18:1 isomers and branched-chain FA (BCFA). This report is part of a detailed study which also investigated greenhouse gas emissions [14], economic profitability [15], carcass merit traits [4], beef quality attributes [16] and beef texture [17].

Materials and methods

Animal management and treatments

Cattle used in this study were cared for under the guidelines provided by the Canadian Council on Animal Care [18] and the experimental procedures were approved by the Lacombe Research Centre Animal Care Committee. One hundred and twelve crossbred Hereford-Aberdeen Angus (n = 64) or Charolais-Red Angus (n = 48) steers born in March and April 2008 at the Lacombe Research Centre were used in the current study. The management of the cow-calf herd has been previously described by Basarab et al. [19]. Calves were weaned at an average age of 182 d. At weaning, calves were assigned to one of four production system and implant strategy based treatments in a completely randomized design (CRD) with a 2 × 2 factorial arrangement of treatments. There were four pens per treatment (seven steers per pen). Each breed cross was equally represented across treatments.

Post-weaning management of calf- and yearling-finished steers were detailed by López-Campos et al. [15]. In summary, following weaning, calf-finished steers (n = 56; 268 ± 5.4 kg; 191 ± 3 d) were adjusted from a high forage-based diet to a high-grain finishing diet over 42 d and subsequently finished on a high-grain diet containing 81.4% barley grain-based concentrate, 8.9% barley silage and 7.9% grass silage on DM basis for 86 d. After weaning, 56 steer calves (266 ± 4.6 kg, 193 ± 3 d), assigned to the yearling production system, rotationally grazed alfalfa (*Medicago sativa* L.)/meadow brome grass (*Bromus riparius* Rehm.) pasture (fall pasture) for 52 d. Thereafter, a grower diet (on DM basis) containing 43.1% barley silage, 41.1% grass hay and 15.8% rolled barley:oat (60:40) grain mix was fed for 192 d prior to grazing alfalfa/ meadow brome grass pasture (summer pasture) for 90 d. Yearling steers were then placed into a feedlot pen and allowed 21–23 d to adapt to the high-grain diet before finished on a high-grain diet (on DM basis) comprised of 79.0% barley grain-based concentrate and 21.0% barley silage for a period of 86 d. Half of the calf-finished steers (n = 28) were implanted with 20 mg estradiol benzoate and 200 mg progesterone (Synovex-S) at weaning and 120 mg trenbolone acetate and 24 mg estradiol (Revalor-S) 90 d before slaughter. Yearling-finished steers (n = 28) were implanted with Synovex-S at weaning and re-implanted with Synovex-S 83 d after weaning (second time), 71 d after the second implant, 86 d after the third implant and finally implanted with Revalor-S 90 d before slaughter.

Feed analyses

Feed samples for the fall pasture were collected twice, initially when the cattle went onto pasture and then when the cattle came off pasture. Feed samples for the summer grazing period (June to August 2009) were collected twice per mo, once early and once late, from each of three paddocks where the animals were grazing in a particular month. Finishing feed samples of the total mixed ration for the steers were collected weekly, pooled monthly and analyzed for nutrient and fatty acid composition. Feed analysis procedures of the experimental finishing diets fed to weanling and yearling steers are detailed by López-Campos et al. [15] and Girard et al. [16]. Fatty acid methyl esters (FAME) from the finishing total mixed ration were prepared as described by Sukhija and Palmquist [20] and analyzed using the chromatographic conditions reported Dugan et al. [21].

Animal slaughter and sample collection

For both production systems, weanling and yearling steers were targeted to be harvested in four groups of 14 animals at a constant backfat end point of 8–10 mm between the 12th and 13th rib over the right *longissimus thoracis* muscle of each animal which corresponded to 11–14 and 19–23 mo of age, respectively. Backfat thickness was measured on the first and last day of feed intake by a certified ultrasound technician using an Aloka 500 V diagnostic real-time ultrasound with a 17 cm 3.5 Mhz linear array transducer (Overseas Monitor Corporation Ltd., Richmond, B.C., Canada) following procedures of Brethour [22]. At one to two wk intervals steers were trucked 3 km for slaughter at the Lacombe Research Centre abattoir such that there were seven implanted and seven non-implanted steers within each slaughter group. At slaughter, final live weights were recorded and animals were stunned, exsanguinated and dressed in a commercial manner. At approximately, 20 min *post-mortem*, during evisceration, a cube of subcutaneous fat (5 cm × 5 cm × the thickness of subcutaneous fat) was collected from the posterior end of the 12th rib and stored at −80°C for subsequent FA analysis.

Subcutaneous fatty acid analysis

Subcutaneous fat samples (50 mg) were freeze-dried and directly methylated with sodium methoxide according to Cruz-Hernandez et al. [23]. Internal standard (1 mL of 1 mg 23:0 methyl ester/ml toluene; EMD Chemicals Inc. Darmstadt, Germany) was added before the addition of the methylating reagent. The majority of FAME were analysed with gas chromatography (GC) using a 175°C temperature program as described by Dugan et al. [21].

For the identification of FAME by GC, the reference standard no. 601 from Nu-Check Prep Inc, Elysian, MN, USA was used. Branched-chain FAME were identified using a GC reference standard BC-Mix1 purchased previously from Applied Science (State College, PA, USA). *Trans*-18:1 isomers and other PUFA biohydrogenation intermediates, not included in the standard mixtures, were identified by their retention times and elution order [21,23,24]. The FAME were quantified using chromatographic peak area and internal standard (23:0 methyl esters)-based calculations (mg FAME = FAME peak area × relative response factor × mg internal standard added / internal standard peak area). The FAME concentrations were reported as percentage of total FA identified.

Statistical analysis

Fatty acid data were analyzed using PROC MIXED procedures [25] as a CRD with a 2 × 2 factorial arrangement of treatments. The model fitted production system (calf-fed, yearling-fed), implant strategy (implant, no implant) and their interaction as the main effects and the random effects of pen nested within production system × implant strategy interaction. Initial body weight was included as a covariate. Treatment means were determined using the LSMEANS and PDIFF options and separated using the LSD test. Significance was declared at $P < 0.05$.

Results and discussion

Nutritional composition of the experimental diets and steers subcutaneous fat thickness

Overall, the nutritional composition of the finisher diets were similar (Table 1), except that calcium was slighlty higher for the yearling-fed steers compared to calf- fed steers. Total FA content and FA composition of the finishing diets were similar (Table 1) with the diet fed to yearling-finished steers having slightly more linoleic acid (C18:2*n*-6, LA) than that fed to calf-finished steers. Alpha-linolenic acid (C18:3*n*-3, ALA) was the dominant FA in fall and summer pasture grazed by yearling-finished steers (Table 1). End ultrasound subcutaneous thickness of yearling-fed steers (10.1 ± 0.30 mm) was slightly higher than that of calf-fed steers (8.4 ± 0.30 mm).

Effects of production system on fatty acid composition

There were no significant ($P > 0.05$) interactions between production system and growth implant strategy for total

Table 1 Nutritional and fatty acid composition (% of total fatty acids) of the diets fed to calf- and yearling-finished steers

	Calf-fed steers	Yearling-fed steers			
Variable	Finishing diet	Fall pasture	Grower diet	Summer pasture	Finishing diet
Nutrient, % DM					
Crude protein	13.2 ± 1.09	9.40 ± 2.83	12.6 ± 1.52	13.4 ± 2.51	12.6 ± 0.49
ADF	15.2 ± 3.68	36.8 ± 4.54	35.3 ± 2.32	34.4 ± 2.71	16.7 ± 1.22
NDF	27.6 ± 4.45	55.6 ± 3.86	55.1 ± 1.81	56.7 ± 2.68	28.1 ± 1.88
Calcium	0.6 ± 0.19	0.7 ± 0.13	0.7 ± 0.07	0.8 ± 0.21	0.9 ± 0.27
Phosphorus	0.4 ± 0.09	0.3 ± 0.03	0.3 ± 0.03	0.3 ± 0.14	0.5 ± 0.06
TDN, %	75.2 ± 1.72^{x}	57.0 ± 5.91^{y}	59.0 ± 1.08^{y}	60.2 ± 3.53^{y}	74.5 ± 0.57^{z}
ME^{2}, MJ/kg DM	11.4 ± 0.26^{x}	8.61 ± 0.89^{y}	8.91 ± 0.16^{y}	9.1 ± 0.53^{y}	12.6 ± 0.49^{z}
Fatty acid, % total fatty acids					
C16:0	18.7 ± 1.40	24.0 ±1.32	22.1 ± 0.71	28.0 ± 3.82	18.8 ± 0.41
C18:0	2.15 ± 0.17	2.51 ± 0.06	2.07 ± 0.22	2.73 ± 0.56	1.75 ± 0.09
*c*9-18:1	20.8 ± 1.88	9.67 ± 6.32	14.1 ± 5.37	4.67 ± 2.12	18.5 ±0.27
*c*11-18:1	4.37 ± 0.61	2.23 ± 0.60	2.62 ± 0.86	1.39 ± 0.54	4.00 ± 0.26
C18:2*n*-6	43.3 ± 0.87	23.7 ± 8.40	28.8 ± 2.39	15.2 ± 0.21	46.5 ± 0.71
C18:3*n*-3	9.20 ± 0.72	30.1 ± 10.3	26.1 ± 8.09	38.7 ± 8.69	9.00 ± 0.56
Total fatty acids (mg/g DM)	2.30 ± 0.69	1.00 ± 0.69	1.25 ± 0.69	0.96 ± 0.69	2.34 ± 0.69

[x]Equations used to calculate total digestible nutrients (TDN) and metabolizable energy (ME) for the finishing diet fed to calf-fed steers: %TDN = 82.299 - (ADF, % × 0.467); ME, MJ/kg DM = ((%TDN/100) × 4.4 × 0.82) x 4.184 MJ/Mcal); [y]Equations used to calculate TDN and ME for the pasture, grower's diet and summer pasture fed to yearling steers: %TDN = 82.299 - (ADF, % × 0.467); ME, MJ/kg DM = ((%TDN/100) × 4.4 × 0.82) × 4.184 MJ/Mcal); [z]Equations used to calculate TDN and ME for the finishing diet fed to yearling steers: %TDN = 82.299 - (ADF, % × 0.467); ME, MJ/kg DM = ((%TDN/100) × 4.4 × 0.82) x 4.184 MJ/Mcal).

and individual FA in subcutaneous fat, and therefore only main effects were reported. Subcutaneous fat from yearling steers had greater ($P > 0.05$) total FA content than calf-finished steers (Table 2) and may relate to the maturity of the depot [26]. Subcutaneous fat from calf- and yearling-finished steers had similar ($P > 0.05$) proportions of total PUFA. The proportions of total and individual *n*-6 PUFA were not affected by production system except for C20:3*n*-6 which was slightly higher ($P < 0.05$) in calf-finished steers than in yearling-finished steers (Table 2). The effect of production systems on C20:3*n*-6 were, however, small and would be of limited biological importance. Alpha-linolenic acid, the only *n*-3 PUFA identified, was not influenced ($P > 0.05$) by production system (Table 2).

Yearling-finished steers had greater ($P < 0.05$) subcutaneous proportions of *t*11,*c*15-18:2 (3-fold), one of the major non-conjugated 18:2 biohydrogenation products (*i.e.*, atypical dienes, AD) and lower ($P < 0.05$) proportions of *t*9,*c*12-18:2 (3.5-fold) compared to their contemporaries finished as weanlings (Table 2). These results could be related to differences in FA composition of the grower and finishing diets between the two production systems. The higher *t*11,*c*15-18:2 observed for the yearling-finished steers was likely derived from ALA obtained from pasture grazing while the elevated proportions of *t*9,*c*12-18:2 observed for the calf-finished steers may have been derived from higher LA proportions obtained from the high-grain finishing diet for these steers. The finding of increased *t*11,*c*15-18:2 in yearling steers demonstrates for the first time its persistence post-grazing; as such it may be a suitable long-term marker of forage consumption, and points towards its limited metabolism in subcutaneous adipose tissue. During rumen biohydrogenation, ALA yields conjugated linolenic acids, chiefly *c*9, *t*11,*c*15-18:3, which is in turn sequentially hydrogenated to yield AD isomers, chiefly *t*11,*c*15-18:2 through the activities of isomerase and reductase enzymes [27]. Overall, high proportions of a given FA in the tissues could be indicative of its limited metabolism or reflect slower rate of metabolism [28].

Neither total nor individual CLA isomers were affected by production system ($P > 0.05$; Table 2). With the exception of *t*12-18:1, production system had no influence ($P > 0.05$) on *t*-18:1 isomers. The subcutaneous proportions of *t*12-18:1 were greater ($P < 0.05$) in calf-finished steers than in yearling-finished steers but the reason for this is not immediately apparent. Calf *vs.* yearling finishing increased ($P < 0.05$) subcutaneous proportions of several individual *cis (c)*-MUFA isomers (*c*9-14:1, *c*11-16:1, *c*11-18:1, *c*12-18:1, *c*13-18:1, *c*9-20:1 and *c*11-20:1). These findings could be attributed to age-dependent differences in gene expression and catalytic activity of stearoyl-CoA desaturase [29,30]. The increase in *c*9-14:1, *c*11-16:1 and *c*13-18:1 reported for the weanling steers could be partly explained by the higher proportions of 14:0 (myristic acid) observed for these steers. Myristic acid is desatured to *c*9-14:1 by Δ9-desaturase, which is then elongated to *c*11-16:1 and *c*13-18:1, respectively. The proportions of *c*9-17:1 were greater ($P < 0.05$) for yearling steers than for weanling steers and the explanation for this is not immediately clear given that the proportions of C17:0, its derivative, were similar across production systems. The relative benefit or risk to human health of consuming the individual *c*-MUFA isomers remains to be elucidated, and thus recommendations to either enrich or deplete these isomers should be reserved until their effects are known.

Subcutaneous fat from yearling-finished steers as opposed to calf-finished steers had elevated ($P < 0.05$) proportions of total and individual (BCFA). The observed variability in BCFA proportions found between calf- and yearling-finished steers may also be partially attributed to the carryover effects of grower diets and backgrounding feeding regimes [31,32]. In the current study, yearling-finished steers as opposed to calf-finished steers entered the feedlot after a forage-based backgrounding phase. Overall, forage feeding has been reported to increase the percentage of BCFA in beef compared to concentrate feeding [6,33]. How grower diets and backgounding feeding regimes influence subsequent feedlot performance and FA composition of concentrate-finished steers merit further investigation. Improved understanding of mechanisms involved in enriching BCFA in meat would also be of interest given their potential to reduce cancer [34] and necrotizing enterocolitis [35] in humans.

Calf-finished steers had greater ($P < 0.05$) subcutaneous proportions of C14:0 (myristic acid) and smaller ($P < 0.05$) proportions of C15:0 compared to steers finished as yearlings. The observation that calf-finished steers had higher proportions of C14:0 than yearling-finished steers agrees with previous findings [36]. This may also in part be attributed to differences in total FA observed in the current study. Myristic acid is hypercholesterolemic and atherogenic [37] and thus, it is considered a less desirable component of the human diet. There were no differences ($P > 0.05$) in the proportions of total SFA and several major long-chain SFA (C16:0, C17:0, C18:0, C19:0 and C20:0) between calf- and yearling-finished steers.

Effects of growth implants on fatty acid composition

Total FA content in adipose tissue in implanted and non-implanted steers was similar ($P > 0.05$; Table 2). Implanted steers had slightly higher ($P < 0.05$) subcutaneous proportions of total PUFA, total *n*-6 PUFA and LA, than steers without growth implants (Table 2). The subcutaneous

Table 2 Effect of production system and growth implant strategy on fatty acid composition (% of total fatty acids) from subcutaneous fat of feedlot steers

	Production system		Implant strategy			*P*-value	
Variable	Calf-finished	Yearling-finished	No	Yes	s.e.m	PS	IS
Total fatty acids (g/g of tissue)	0.90[b]	0.93[a]	0.92	0.90	0.01	0.01	0.91
∑ PUFA	1.62	1.69	1.55[b]	1.77[a]	0.07	0.54	0.01
∑ *n*-6	1.39	1.42	1.30[b]	1.51[a]	0.06	0.74	<0.001
C18:2*n*-6	1.32	1.36	1.24[b]	1.45[a]	0.06	0.67	<0.001
C20:2*n*-6	0.02	0.02	0.02	0.02	0.002	0.71	0.49
C20:3*n*-6	0.05[a]	0.04[b]	0.05	0.04	0.004	0.05	0.89
n-3							
C18:3*n*-3	0.23	0.27	0.25	0.26	0.01	0.09	0.26
∑ AD	0.61	0.64	0.62	0.63	0.02	0.35	0.71
*c*9,*t*13-/*t*8,*c*13-18:2	0.20	0.22	0.21	0.20	0.01	0.28	0.31
*t*8,*c*12-/*c*9,*t*12-18:2	0.11	0.12	0.11	0.12	0.01	0.37	0.11
*t*9,*c*12-18:2	0.07[a]	0.02[b]	0.05	0.05	0.01	<0.001	0.95
*t*11,*c*15-18:2	0.04[b]	0.12[a]	0.07	0.09	0.01	<0.001	0.09
*c*9,*c*15-18:2	0.19	0.18	0.19	0.18	0.01	0.56	0.35
∑ CLA	0.54	0.49	0.51	0.52	0.02	0.14	0.63
*t*7,*c*9-/*c*9,*t*11-18:2	0.48	0.41	0.44	0.45	0.02	0.12	0.76
*c*10,*t*12-18:2	0.02	0.02	0.02	0.02	0.01	0.11	0.73
*t*9,*t*11-18:2	0.04	0.05	0.04	0.05	0.01	0.21	0.35
∑ *t*18:1	3.29	3.19	3.00	3.46	0.36	0.49	0.61
*t*6/*t*7/*t*8-18:1	0.25	0.25	0.23[b]	0.28[a]	0.01	0.98	<0.001
*t*9-18:1	0.27	0.27	0.26[b]	0.28[a]	0.01	0.72	0.01
*t*10-18:1	1.57	1.65	1.43[b]	1.79[a]	0.12	0.67	0.01
*t*11-18:1	0.67	0.54	0.60	0.61	0.06	0.23	0.92
*t*12-18:1	0.14[a]	0.10[b]	0.12	0.12	0.01	0.01	0.46
*t*13/*t*14-18:1	0.27	0.27	0.26	0.27	0.01	0.70	0.29
*t*16/*c*14-18:1	0.12	0.10	0.11	0.11	0.01	0.31	0.93
∑ MUFA	51.4	51.9	52.1	51.2	0.52	0.63	0.14
∑ *c*-MUFA	50.0	50.0	11.2	10.4	0.05	0.50	0.85
*c*9-14:1	1.70[a]	1.32[b]	1.59[a]	1.43[b]	0.06	<0.001	0.05
*c*7-16:1	0.15	0.15	0.15	0.15	0.01	0.54	0.45
*c*9-16:1	5.24	4.71	5.16	4.80	0.17	0.09	0.15
*c*11-16:1	0.37[a]	0.29[b]	0.35	0.32	0.02	<0.001	0.07
*c*9-17:1	0.51[b]	1.31[a]	0.91	0.910	0.13	0.02	0.99
*c*9-18:1	39.0	39.5	39.6	38.9	0.39	0.43	0.16
*c*11-18:1	1.79	1.76	1.84	1.71	0.06	0.79	0.13
*c*12-18:1	0.10[a]	0.08[b]	0.09	0.09	0.01	0.03	0.34
*c*13-18:1	0.68[a]	0.58[b]	0.66	0.60	0.03	0.04	0.15
*c*16-18:1	0.06	0.05	0.05	0.05	0.01	0.26	0.68
*c*9-20:1	0.09[a]	0.02[b]	0.06	0.05	0.01	0.01	0.18
*c*11-20:1	0.32[a]	0.26[b]	0.31[a]	0.27[b]	0.01	0.02	0.01

Table 2 Effect of production system and growth implant strategy on fatty acid composition (% of total fatty acids) from subcutaneous fat of feedlot steers *(Continued)*

∑ BCFA	1.14[b]	1.45[a]	1.28	1.32	0.01	0.05	0.96
iso-15:0	0.11[b]	0.16[a]	0.13	0.13	0.01	0.01	0.84
anteiso-15:0	0.15[b]	0.19[a]	0.17	0.18	0.01	<0.001	0.19
iso-16:0	0.14[b]	0.22[a]	0.18	0.18	0.01	<0.001	0.83
iso-17:0	0.32[b]	0.35[a]	0.33	0.34	0.01	0.05	0.27
anteiso-17:0	0.56[b]	0.66[a]	0.59[b]	0.63[a]	0.01	<0.001	0.01
iso-18:0	0.12[b]	0.17[a]	0.15	0.14	0.01	<0.001	0.46
∑ SFA	42.2	41.9	41.7	42.4	0.52	0.75	0.23
C14:0	3.85[a]	3.39[b]	3.65	3.57	0.08	0.01	0.43
C15:0	0.47[b]	0.65[a]	0.56	0.56	0.01	<0.001	0.94
C16:0	27.5	26.7	27.2	26.9	0.31	0.13	0.54
C17:0	1.02	1.09	0.99	1.12	0.14	0.81	0.11
C18:0	9.34	10.1	9.23	10.2	0.35	0.20	0.09
C19:0	0.25	0.24	0.24	0.25	0.01	0.65	0.38
C20:0	0.04	0.04	0.04	0.05	0.01	0.84	0.73

[a,b,c]Means with different superscripts for a particular FA profile are significantly different ($P < 0.05$); s.e.m, standard error of mean; PS, production system effect; IS, Implant strategy effect; *c*, *cis*; *t*, *trans*; ∑ PUFA, sum of polyunsaturated FA = C18:3*n*-3 + C18:2*n*-6 + C20:2*n*-6 + C20:3*n*-6; ∑ *n*-6, sum of omega-6 FA = C18:2*n*-6 + C20:2*n*-6 + 20:3*n*-6; ∑ AD, total atypical dienes = *c*9,*t*13-/*t*8,*c*13 + *t*8,*c*12-/*c*9,*t*12 + *t*9,*c*12 + *t*11,*c*15 + *c*9,*c*15; ∑ CLA, conjugated linoleic acid = *c*7,*t*9-/*c*9,*t*11- + *c*10,*t*12-18:2 + *t*9,*t*11-18:2; ∑ *t*-18:1, sum of *trans*-18 :1 isomers = *t*6-*t*8-18:1 + *t*9-18:1 + *t*10-18:1 + *t*11-18:1 + *t*12-18:1 + *t*13/*t*14-18:1 + *t*16-/*c*14-18:1; ∑ MUFA, sum of monounsaturated FA = ∑ *t*-18:1 + ∑ *t*-MUFA; ∑ *c*-MUFA, sum of *cis*- monounsaturated FA = *c*9-14:1 + *c*7-16:1 + *c*9-16:1 + *c*11-16:1 + *c*9-17:1 + *c*11-18:1 + *c*12-18:1 + *c*13-18:1 + *c*16-18:1 + *c*9-20:1 + *c*11-20:1; ∑ BCFA, sum of branched chain FA = *iso*-15:0 + *anteiso*-15:0 + *iso*-16:0 + *anteiso*-17:0 + *iso*-18:0; ∑ SFA, sum of saturated FA = C14:0 + C15:0 + C16:0 + C17:0 + C18:0 + C19:0 + 20:0.

proportions of other individual *n*-6 PUFA (C20:2*n*-6 and C20:3*n*-6) were similar ($P > 0.05$) across treatments. In support of the current findings, Ibrahim et al. [13] reported that fat from bull calves implanted with growth promotants had increased total PUFA and total *n*-6 PUFA. These findings could be partly related to the numerical differences in total FA content between implanted and non-implanted steers. Overall, the proportions of PUFA increases with decreasing fat content due to less dilution by *de novo* synthesised FA [38]. Growth implants had no effect on ALA, the only *n*-3 PUFA identified.

Neither total nor individual proportions of AD and CLA were influenced by growth implants ($P > 0.05$). For *t*-18:1, a pattern of isomers was found related to the implant strategy (Table 2). Implantation with growth promotants increased ($P < 0.05$) subcutaneous proportions of *t*-18:1 isomers with double bonds from carbon 6 to 10 compared to non-implantation. Growth promotants did not, however, influence ($P < 0.05$) total *t*-18:1. In agreement with the current study, implantation with estradiol benzoate and progesterone [11] or trenbolone acetate and estradiol [11,39] increased the proportions of some *t*-18:1 isomers (*t*6-*t*10) in the subcutaneous fat of concentrate-finished steers. These findings may relate to decreased dilution of *t*-18:1 isomers in subcutaneous fat by *de novo* synthesised FA. *Trans*-18:1 isomers other than *t*11-18:1 have been associated with unhealthy changes in blood lipid profiles in animal models [40] and are considered as undesirable components of the human diet. It is, however, not certain if the small differences (< 0.5%) in *t*-18:1 attributed to the use of growth promotants in the current study would be enough to impact human health.

Total and individual *c*-MUFA were not affected by growth implants except for *c*9-14:1 and *c*11-20:1. Steers implanted with growth promotants had low ($P < 0.05$) *c*9-14:1 and *c*11-20:1 compared to non-implanted steers. The observation that growth implants decreased the proportions of *c*9-14:1 agrees with earlier findings [39]. Proportions of total and individual BCFA were not affected by growth promotants except for *anteiso*-17:0 which was greater ($P < 0.05$) in steers implanted with growth promotants than steers without growth promotants (Table 2), but the differences were small (0.04%). Implantation had no effect on the proportions of total and several individual SFA ($P > 0.05$), however, the proportions of C18:0 tended to be greater ($P = 0.09$) in implanted than non-implanted steers. The finding that growth implants tended to increase the proportions of C18:0 may be related high proportions of PUFA reported for the implanted in the current study. Generally, high PUFA proportions inhibit Δ9 desaturase responsible for converting SFA to their respective MUFA [41].

Conclusions

Yearling as opposed to a calf production system yielded lower subcutaneous proportions of individual *c*-MUFA and higher subcutaneous proportions of *t*11,*c*15-18:2, individual and total BCFA. Growth promotants increased subcutaneous proportions of total PUFA, total *n*-6 PUFA, LA and individual *t*-18:1 isomers (*t*6-*t*10), likely relating to reduced dilution by *de novo* synthesised FA. Overall, changes in subcutaneous FA composition of finished beef steers due to production systems and growth implants were limited with an average difference of 0.15% in FA composition which may not be sufficient to result in differences in human health.

Competing interests

The authors declare that there are no competing interests in relation to this manuscript.

Authors' contributions

CM, TDT and MERD participated in the analyses of feed and subcutaneous fatty acids, statistical analysis of data and drafted the manuscript. JAB, JLA and VSB conceived the study, acquired funds, conducted the live animal study and participated in its design and write-up of the manuscript. All authors read and approved the final manuscript.

Acknowledgements

The authors gratefully acknowledge funding support from Alberta Livestock and Meat Agency Ltd., Alberta Agriculture and Rural Development (ARD), Alberta Environment, Agriculture and Agri-Food Canada Matching Initiatives Program, Elanco Animal Health and Pioneer Hybrid International, and the in-kind contribution in animals, facilities and people received from Agriculture and Agri-Food Canada (AAFC), Lacombe Research Centre, AB, Canada. Drs C. Mapiye and T. Turner gratefully acknowledge the receipt of NSERC fellowships funded through Alberta Meat and Livestock Agency (ALMA). We also wish to thank the significant contributions of David C. Rolland, Wayne Robertson (retired AAFC Meat Quality Biologist), Cathy Bryant and Sheldon Johnston of ARD, and Stan Landry, Ivy Larsen, Adele Ohama and Dave Young of AAFC for technical support. Special thanks are extended to Cletus Sehn, Ken Grimson, and their staff at the Lacombe Research Centre Beef Unit of AAFC for animal care, animal management and sample collection. The slaughter and processing of the cattle by Chuck Pimm and his staff is gratefully acknowledged.

Author details

[1]Agriculture and Agri-Food Canada, Lacombe Research Centre, 6000 C & E Trail, Lacombe, Alberta T4L 1 W1, Canada. [2]Alberta Agriculture and Rural Development, Lacombe Research Centre, 6000 C & E Trail, Lacombe, Alberta T4L 1 W1, Canada.

References

1. Brooks MA, Choi CW, Lunt DK, Miller RK, Choi CB, Smith SB: **Case Study: Carcass and meat characteristics and M. longissimus thoracis histology of beef from calf-fed and yearling-fed Angus steers.** *Prof Anim Scient* 2011, **27:**385–393.
2. Smith SB, Chapman AA, Lunt DK, Harris JJ, Savell JW: **Adiposity of calf- And yearling-fed Brangus steers raised to constant-age and constant-body weight endpoints.** *J Anim Sci* 2007, **85:**1136–1143.
3. Barham B, Beck P, Apple J, Whitworth W, Miller M, Gadberry S: **Effect of age entering feedlot and implant regimen on beef cattle performance, carcass characteristics, and sensory evaluation.** *Prof Anim Scient* 2012, **28:**20–31.
4. López-Campos Ó, Basarab JA, Baron VS, Aalhus JL, Juárez M: **Reduced age at slaughter in youthful beef cattle: Effects on carcass merit traits.** *Can J Anim Sci* 2012, **92:**449–463.
5. Purchas RW, Knight TW, Busboom JR: **The effect of production system and age on concentrations of fatty acids in intramuscular fat of the longissimus and triceps brachii muscles of Angus-cross heifers.** *Meat Sci* 2005, **70:**597–603.
6. Humada MJ, Serrano E, Sañudo C, Rolland DC, Dugan MER: **Production system and slaughter age effects on intramuscular fatty acids from young Tudanca bulls.** *Meat Sci* 2012, **90:**678–685.
7. Cleale RM, Amodie D, Bechtol DT, Drouillard JS, Edmonds JD, Edmonds M, Hunsaker BD, Kraft LA, Lawrence TE, Rulli RD, Waite AR: **Effects of estradiol benzoate and trenbolone acetate, alone or in combination at dose levels present in Synovex Choice, on performance by feedlot heifers.** *J Anim Sci* 2013, **91:**970–977.
8. Bruns KW, Pritchard RH, Boggs DL: **The effect of stage of growth and implant exposure on performance and carcass composition in steers.** *J Anim Sci* 2005, **83:**108–116.
9. McPhee MJ, Oltjen JW, Famula TR, Sainz RD: **Meta-analysis of factors affecting carcass characteristics of feedlot steers.** *J Anim Sci* 2006, **84:**3143–3154.
10. Duckett SK, Wagner DG, Owens FN, Dolezal HG, Gill DR: **Effect of anabolic implants on beef intramuscular lipid content.** *J Anim Sci* 1999, **77:**1100–1104.
11. Fritsche S, Rumsey TS, Yurawecz MP, Ku Y, Fritsche J: **Influence of growth promoting implants on fatty acid composition including conjugated linoleic acid isomers in beef fat.** *Europ Food Res Techn* 2001, **212:**621–629.
12. Holzer Z, Aharoni Y, Brosh A, Orlov A, Veenhuizen JJ, Kasser TR: **The effects of long-term administration of recombinant bovine somatotropin (Posilac) and Synovex on performance, plasma hormone and amino acid concentration, and muscle and subcutaneous fat fatty acid composition in Holstein-Friesian bull calves.** *J Anim Sci* 1999, **77:**1422–1430.
13. Ibrahim RM, Marchello JA, Duff GC: **Effects of implanting beef steers with zeranol on fatty acid composition of subcutaneous and iIntramuscular fat.** *Prof Anim Scient* 2006, **22:**301–306.
14. Basarab J, Baron V, López-Campos Ó, Aalhus J, Haugen-Kozyra K, Okine E: **Greenhouse gas emissions from calf- and yearling-fed beef production systems, with and without the use of growth promotants.** *Animals* 2012, **2:**195–220.
15. López-Campos Ó, Aalhus JL, Okine EK, Baron VS, Basarab JA: **Effects of calf- and yearling-fed beef production systems and growth promotants on production and profitability.** *Can J Anim Sci* 2013, **93:**171–184.
16. Girard I, Aalhus JL, Basarab JA, Larsen IL, Bruce HL: **Modification of beef quality through steer age at slaughter, breed cross and growth promotants.** *Can J Anim Sci* 2012, **92:**175–188.
17. Juárez M, Basarab JA, Baron VS, Valera M, Larsen IL, Aalhus JL: **Quantifying the relative contribution of ante- and post-mortem factors to the variability in beef texture.** *Animal* 2012, **6:**1878–1887.
18. Canadian Council on Animal Care: **Guide to the Care and Use of Experimental Animals. Vol. 1.** 2nd edition. Edited by Olfert ED, Cross BM, McWilliams AA. Ottawa, Ontario, Canada: Canadian Council on Animal Care; 1993.
19. Basarab JA, McCartney D, Okine EK, Baron VS: **Relationships between progeny residual feed intake and dam productivity traits.** *Can J Anim Sci* 2007, **87:**489–502.
20. Sukhija PS, Palmquist DL: **Rapid method for determination of total fatty acid content and composition of feedstuffs and feces.** *J Agric Food Chem* 1988, **36:**1202–1206.
21. Dugan MER, Kramer JKG, Robertson WM, Meadus WJ, Aldai N, Rolland DC: **Comparing subcutaneous adipose tissue in beef and muskox with emphasis on trans 18:1 and conjugated linoleic acids.** *Lipids* 2007, **42:**509–518.
22. Brethour JR: **The repeatability and accuracy of ultrasound in measuring backfat of cattle.** *J Anim Sci* 1992, **70:**1039–1044.
23. Cruz-Hernandez C, Deng Z, Zhou J, Hill AR, Yurawecz MP, Delmonte P, Mossoba MM, Dugan MER, Kramer JKG: **Methods for analysis of conjugated linoleic acids and trans-18:1 isomers in dairy fats by using a combination of gas chromatography, silver-ion thin-layer chromatography/gas chromatography, and silver-ion liquid chromatography.** *J AOAC International* 2004, **87:**545–562.
24. Kramer JKG, Hernandez M, Cruz-Hernandez C, Kraft J, Dugan MER: **Combining results of two GC separations partly achieves determination of all cis and trans 16:1, 18:1, 18:2 and 18:3 except CLA isomers of milk fat as demonstrated using ag-ion SPE fractionation.** *Lipids* 2008, **43:**259–273.
25. SAS: *SAS user's guide: Statistics. SAS for windows. Release 9.2.* Cary NC: SAS Institute Inc; 2009.
26. Wood JD, Enser M, Fisher AV, Nute GR, Sheard PR, Richardson RI, Hughes SI, Whittington FM: **Fat deposition, fatty acid composition and meat quality: A review.** *Meat Sci* 2008, **78:**343–358.

27. Lee YJ, Jenkins TC: **Biohydrogenation of linolenic acid to stearic acid by the rumen microbial population yields multiple intermediate conjugated diene isomers.** *J Nutr* 2011, **141**:1445–1450.
28. Kramer JKG, Sehat N, Dugan MER, Mossoba MM, Yurawecz MP, Roach JAG, Eulitz K, Aalhus JL, Schaefer AL, Ku Y: **Distributions of conjugated linoleic acid (CLA) isomers in tissue lipid classes of pigs fed a commercial CLA mixture determined by gas chromatography and silver ion-high-performance liquid chromatography.** *Lipids* 1998, **33**:549–558.
29. Duckett SK, Pratt SL, Pavan E: **Corn oil or corn grain supplementation to steers grazing endophyte-free tall fescue. II. Effects on subcutaneous fatty acid content and lipogenic gene expression.** *J Anim Sci* 2009, **87**:1120–1128.
30. Chung KY, Lunt DK, Kawachi H, Yano H, Smith SB: **Lipogenesis and stearoyl-CoA desaturase gene expression and enzyme activity in adipose tissue of short- and long-fed Angus and Wagyu steers fed corn- or hay-based diets.** *J Anim Sci* 2007, **85**:380–387.
31. Pordomingo AJ, García TP, Volpi Lagreca G: **Effect of feeding treatment during the backgrounding phase of beef production from pasture on: II. Longissimus muscle proximate composition, cholesterol and fatty acids.** *Meat Sci* 2012, **90**:947–955.
32. Reuter RR, Beck PA: **Southern section interdisciplinary beef cattle symposium: Carryover effects of stocker cattle systems on feedlot performance and carcass characteristics.** *J Anim Sci* 2013, **91**:508–515.
33. Manner W, Maxwell RJ, Williams JE: **Effects of dietary regimen and tissue site on bovine fatty acid profiles.** *J Anim Sci* 1984, **59**:109–121.
34. Wongtangtintharn S, Oku H, Iwasaki H, Toda T: **Effect of branched-chain fatty acids on fatty acid biosynthesis of human breast cancer cells.** *J Nutr Sci Vitaminol* 2004, **50**:137–143.
35. Ran-Ressler RR, Khailova L, Arganbright KM, Adkins-Rieck CK, Jouni ZE, Koren O, Ley RE, Brenna JT, Dvorak B: **Branched chain fatty acids reduce the incidence of necrotizing enterocolitis and alter gastrointestinal microbial ecology in a neonatal rat model.** *PLoS One* 2011, **6**(12):e29032. doi:10.1371/journal.pone.0029032.
36. Duckett S, Wagner D, Dolezal H, Yates L, Gill D, King M: *Effect of animal age on the fatty acid and cholesterol content of ribeye steaks. Animal Science Research report.* Oklahoma, USA: Oklahoma Agricultural Experiment Station; 1994:40-43, http://beefextension.com/research_reports/1994rr/94-8.pdf.
37. Salter AM: **Dietary fatty acids and cardiovascular disease.** *Animal* 2013, **7**:163–171.
38. Wood JD: **Fat deposition and the quality of fat tissue in meat animals.** In *Fats in animal nutrition.* Edited by Wiseman J. London: Butterworths; 407:435–1984.
39. Faucitano L, Chouinard PY, Fortin J, Mandell IB, Lafrenière C, Girard CL, Berthiaume R: **Comparison of alternative beef production systems based on forage finishing or grain-forage diets with or without growth promotants: 2. Meat quality, fatty acid composition, and overall palatability.** *J Anim Sci* 2008, **86**:1678–1689.
40. Roy A, Chardigny JM, Bauchart D, Ferlay D, Lorenz S, Durand D, Gruffat D, Faulconnier Y, Sébédio JL, Chilliard Y: **Butters rich either in trans-10-C18:1 or in trans-11-C18:1 plus cis-9, trans-11 CLA differentially affect plasma lipids and aortic fatty streak in experimental atherosclerosis in rabbits.** *Animal* 2007, **1**:467–476.
41. Smith SB, Gill CA, Lunt DK, Brooks MA: **Regulation of fat and fatty acid composition in beef cattle.** *Asian-Aust J Anim Sci* 2009, **22**:1225–1233.

3

Effects of Chinese herbal medicine on plasma glucose, protein and energy metabolism in sheep

Xi Liang, Kyota Yamazaki, Mohammad Kamruzzaman, Xue Bi, Arvinda Panthee and Hiroaki Sano*

Abstract

Background: The use of antibiotics in animal diets is facing negative feedback due to the hidden danger of drug residues to human health. Traditional Chinese herbal medicine has been used to replace antibiotics in the past two decades and played an increasingly important role in livestock production. The present study was carried out to assess the feeding effects of a traditional nourishing Chinese herbal medicine mixture on kinetics of plasma glucose, protein and energy metabolism in sheep. Ruminal fermentation characteristics were also determined.

Methods: Four sheep were fed on either mixed hay (MH-diet) or MH-diet supplemented with 2% of Chinese herbal medicine (mixture of Astragalus root, Angelica root and Atractylodes rhizome; CHM-diet) over two 35-day periods using a crossover design. The turnover rate of plasma glucose was measured with an isotope dilution method using [U-^{13}C]glucose. The rates of plasma leucine turnover and leucine oxidation, whole body protein synthesis (WBPS) and metabolic heat production were measured using the [1-^{13}C]leucine dilution and open circuit calorimetry.

Results: Body weight gain of sheep was higher ($P = 0.03$) for CHM-diet than for MH-diet. Rumen pH was lower ($P = 0.02$), concentration of rumen total volatile fatty acid tended to be higher ($P = 0.05$) and acetate was higher ($P = 0.04$) for CHM-diet than for MH-diet. Turnover rates of plasma glucose and leucine did not differ between diets. Oxidation rate of leucine tended to be higher ($P = 0.06$) for CHM-diet than for MH-diet, but the WBPS did not differ between diets. Metabolic heat production tended to be greater ($P = 0.05$) for CHM-diet than for MH-diet.

Conclusions: The sheep fed on CHM-diet had a higher body weight gain and showed positive impacts on rumen fermentation and energy metabolism without resulting in any adverse response. Therefore, these results suggested that the Chinese herbal medicine mixture should be considered as a potential feed additive for sheep.

Keywords: Chinese herbal medicine, Glucose metabolism, Heat production, Protein metabolism, Ruminal fermentation characteristics, Sheep

Background

Antibiotics have been widely used as feed additive in livestock production for more than 50 yr [1]. They have played a very important role in helping animals to prevent diseases and enhance productivity. However, their use in animal diets has also brought along the hidden danger of drug residues to human health [2]. At present, the use of antibiotic growth promoters in animal industry is restricted in the European Union. It is likely that the restrictions on the use of antibiotics in animal husbandry will spread to the rest of the world due to the increasing public concern on human health. Therefore, replacing antibiotics with alternative feed additives has become the specific research interest to animal scientists.

Chinese herbs as the traditional medicine have been widely used in East Asian countries. For thousands of years, they have made great contributions to the maintenance of human health. Most of the Chinese herbal medicine comes from the different parts of perennial herbs, such as the leaves, roots and stems. It has been well-known that Chinese herbal medicine contains the bioactive components which have anti-bacterial activity, anti-inflammatory properties and immune enhancing effects [3]. Because of the natural origin, Chinese herbal medicine will not cause excessive drug residue or toxicity and thus it can be considered as a safe and

* Correspondence: sano@iwate-u.ac.jp
Faculty of Agriculture, Iwate University, 3-18-8 Ueda, Morioka 020-8550, Iwate, Japan

suitable substitute for antibiotics in animal feeding. In recent years, a lot of Chinese herbal medicine has already been reported to promote growth and boost immune system in pigs, chickens and other animals [4-6].

It was considered that such numerous Chinese herbal medicine provide a great potential for practical application and some of them could be used as alternative feed additive for ruminants. Until now, little information is available regarding the performance of Chinese herbal medicine on nutrients and energy metabolism in ruminants. We hypothesized that Chinese herbal medicine might be beneficial to nutrients and energy metabolism in ruminants due to its bioactive properties. Therefore, the present study was carried out to assess the effects of feeding a traditional nourishing Chinese herbal medicine mixture on kinetics of plasma glucose, protein and energy metabolism in sheep.

Methods

Animals, diets and management

The handling of animals, including cannulation and blood collection, was reviewed and approved by the Animal Care Committee of Iwate University. All experimental procedures were performed without any noticeable stress to the animals.

According to the classical Chinese pharmacopoeia, a traditional nourishing Chinese herbal medicine mixture was utilized in this study. Three herbs were mixed in proportion as 55% of Astragalus root (*Astragalus membranaceus*), 27% of Angelica root (*Angelica sinensis*) and 18% of Atractylodes rhizome (*Atractylodes lancea*). The herbs are known to be rich in polysaccharides (Astragalus root) and essential oils (Angelica root and Atractylodes rhizome). In humans, the Chinese herbal medicine mixture is commonly used as a health regulator to remove tiredness and comfort stress by inducing hematopoiesis.

Four crossbred (Corriedale × Suffolk) shorn wethers, aged 7 mo on average, weighing 29 ± 2 kg, were used in the experiment. The sheep were assigned to two dietary treatments, including either mixed hay (MH-diet) of orchardgrass (*Dactylis glomerata*) and reed canarygrass (*Phalaris arundinacea*) offered at maintenance level [7] or MH-diet supplemented with 2% of Chinese herbal medicine mixture (CHM-diet). The chemical compositions of mixed hay and Chinese herbal medicine on air dry matter basis are shown in Table 1. The sheep received mixed hay at 67 g/kgBW$^{0.75}$/d for both dietary treatments. The experiment was performed using a crossover design over two 35-day periods. The sheep were housed in individual pens in an animal room during the first four weeks of the experiment for adaptation. Then animals were moved to metabolic cages in a controlled environment house at an air temperature of 23 ± 1°C, with lighting present from 08:00 h to 22:00 h. Two sheep were fed on the MH-diet during the first period and then fed on the CHM-diet during the second period. The other two sheep were subjected to the dietary treatments in the reverse order. The animals were given either diet twice a day at 08:30 h and 20:30 h, and they commonly consumed all the diets which were fed within 1 h. Water was available *ad libitum*. The sheep were weighted at the start of experiment, on day 15 and 29, and at the end of each dietary treatment.

Table 1 Chemical composition of experimental diets

Items*	MH	CHM
Dry matter, %	88.3	91.2
CP, %	11.9	10.8
NDF, %	65.1	47.2
ME, kcal/g	1.79	N/A

*Values are presented on air dry matter basis.
MH = mixed hay of orchardgrass and reed canarygrass; CHM = Chinese herbal medicine mixture; CP = crude protein; NDF = neutral detergent fiber; ME = metabolizable energy; N/A = not available.

Experimental procedures

On day 33 of each dietary treatment, rumen fluid (30 ml) was collected through a stomach tube before feeding, 3 and 6 h after feeding. After determination of pH values of rumen fluid with a pH meter (HM-20E, Toa Electronics Ltd., Japan), the rumen fluid was centrifuged at 8,000 × *g* for 10 min at 4°C (RS-18IV, Tomy, Japan). Then 1 ml of supernatant was mixed with 1 ml of 0.1 N HCl. The prepared samples and the residuals of rumen fluid were stored at −30°C for the determinations of rumen ammonia (NH_3) and volatile fatty acid (VFA), respectively.

Two sheep were subjected to the experimental procedures to determine plasma glucose, protein and energy metabolism on day 34 and 35, respectively. The other two sheep were determined in the reverse order. The turnover rate of plasma glucose was measured with an isotope dilution method using [U-^{13}C]glucose. The rates of plasma leucine turnover and leucine oxidation, whole body protein synthesis and metabolic heat production were measured using the [1-^{13}C]leucine dilution and open circuit calorimetry. Two catheters, one for isotope infusion and the other for blood collection were inserted into the left and right jugular veins on the morning of day 34 of each dietary treatment. The catheters were filled with sterile solution of 3.8% trisodium citrate.

At 12:00 h on the day of [U-^{13}C]glucose dilution, 3.0 μmol/kgBW$^{0.75}$ of [U-^{13}C]glucose (D-glucose-^{13}C, 99 atom% excess ^{13}C; Cambridge Isotope Laboratories, USA) dissolved in saline solution (0.9% sodium chloride) was injected into the jugular catheter for infusion as a priming dose. After the priming injection, [U-^{13}C]glucose was then continuously infused by multichannel peristaltic pumps (AC-2120, Atto, Japan) at rate of 3.0 μmol/kg$^{0.75}$/h through the same catheters for 4 h. Blood samples were taken from the sampling catheter immediately before (12 ml) and at 30-min intervals (6 ml) during the last 120 min of isotope infusion. The collected samples were immediately transferred into centrifuge tubes containing heparin sodium and were stored in crushed ice. Plasma was separated from blood samples by means of centrifugation at 10,000 × *g* for 10 min at 4°C and then stored at −30°C for further analysis.

At 11:30 h on the day of [1-^{13}C]leucine dilution, the sheep were fitted with a clear head chamber (approximately 0.2 m^3) to collect the samples of exhaled gas. After collecting gaseous samples in the pre-infusion period, 7.2 μmol/kgBW$^{0.75}$ of [1-^{13}C]leucine (L-leucine-1-^{13}C, 99 atom% excess ^{13}C; Cambridge Isotope Laboratories, USA) and 3.5 μmol/kgBW$^{0.75}$ of $NaH^{13}CO_3$ (sodium bicarbonate-^{13}C, 99.2 atom% excess ^{13}C; Cambridge Isotope Laboratories, USA) dissolved in saline solution were injected into the infusion catheter as a priming dose. Then, [1-^{13}C]leucine was continuously infused by multichannel peristaltic pumps at rates of 7.2 μmol/kg$^{0.75}$/h through the same catheters for 6 h. Blood samples were taken from the sampling catheter immediately before (12 ml) and at 30-min intervals (6 ml) during the last 120 min of isotope infusion. The catheters were removed at the end of isotope infusions on day 35 of each dietary treatment.

Oxygen (O_2) consumption and carbon dioxide (CO_2) production were monitored continuously for 30 min at the end of [1-^{13}C]leucine infusion using a metabolic monitor (Coast Electronics, UK). An aliquot of exhaled CO_2 was collected in 4 ml of 1 N NaOH for 30 min immediately before and three times during the last 90 min of [1-^{13}C]leucine infusion to measure the enrichment of exhaled CO_2 derived from leucine oxidation. The CO_2 samples were stored at −30°C for further analysis.

Chemical analysis

Analysis of chemical components of the diets was performed according to Association of Official Analytical Chemists [8]. Nitrogen contents in diets were analyzed using the Foss Kjeltec System (Kjeltec 2300, Foss, Sweden). The NDF contents in diets were measured according to Van Soest *et al.* [9] using the Foss Analytical FiberCap™ system (Foss, Sweden).

Concentrations of rumen NH_3 were determined with a colorimetric method [10] using a spectrophotometer (V-630, JASCO, Japan). Concentrations of rumen VFA were measured using a gas chromatography (HP-5890, Hewlett Packard, USA) after steam distillation.

Concentrations of plasma free amino acids and urea were determined using an automatic amino acid analyzer (JLC-500/V, JEOL, Japan). Concentrations of plasma non-esterified fatty acid (NEFA) were enzymatically determined using a diagnostic kit (NEFA C, Wako Pure Chemicals, Japan).

Plasma glucose was derivatized to glucose aldonitrile pentaacetate according to the procedure of Tesrng and Kalhan [11] with slight modifications by Fujita *et al.* [12]. Enrichments of plasma [U-^{13}C]glucose were measured by gas chromatography mass spectrometry (QP-2010, Shimadzu, Japan). Concentrations of plasma glucose were enzymatically determined with a glucose oxidase method of Huggett and Nixon [13].

Concentrations of plasma α-ketoisocaproic acid (α-KIC) and enrichments of plasma α-[1-^{13}C]KIC were measured by gas chromatography mass spectrometry according to the procedures of Rocchiccioli *et al.* [14] and Calder and Smith [15]. The exhaled CO_2 captured in 1 N NaOH solutions was released by adding 6 N H_2SO_4 into the samples in rubber capped vials under a vacuum condition, and the isotopic abundance of exhaled $^{13}CO_2$ was determined using a gas chromatography-combustion-isotope ratio mass spectrometric system (DELTA^{plus}, Thermo Electron, USA).

Calculations

The turnover rates of plasma glucose and leucine (GluTR and LeuTR, respectively) as well as leucine oxidation rate (LeuOX) were calculated using the equations given by Wolfe [16]:

$$\mathrm{TR} = I \times (1/E\text{-}1)$$

Where, *I* represents the infusion rates of [U-^{13}C]glucose and [1-^{13}C]leucine, and *E* represents the plasma isotopic enrichments of [U-^{13}C]glucose and α-[1-^{13}C]KIC during the steady state, respectively. In the present study, the enrichment of plasma α-[1-^{13}C]KIC was used to calculate LeuTR instead of plasma [1-^{13}C]leucine, because it is a more useful indicator to determine leucine metabolism [17].

$$\mathrm{LeuOX} = E_{CO2}/E_{KIC}/0.81 \times V_{CO2}$$

Where, Eco_2 is the isotopic enrichment of exhaled $^{13}CO_2$ and Vco_2 is the CO_2 production rate. The recovery fraction of exhaled CO_2 production in the animal body was estimated to be 0.81 [18]. Whole body protein synthesis (WBPS) was calculated using the following equation as described by Wolfe *et al.* [19]:

$$\mathrm{WBPS} = (LeuTR\text{–}LeuOX)/\textit{leucine concentration in carcass protein}$$

Leucine concentration in carcass protein (66 g/kg) was used as described by Harris *et al.* [20].

Heat production (HP) was calculated based on the Brouwer's equation [21] using O_2 consumption and CO_2 production as described by Young *et al.* [22]:

$$HP = 3.866 \times V_{O2} + 1.2 \times V_{CO2}$$

Where, Vo_2 is the O_2 consumption rate and Vco_2 is the CO_2 production rate.

Statistical analysis

Statistical analysis was conducted according to the MIXED procedure of SAS [23]. The analysis of variance was used to test the effects of period and diet, and sheep were the random effect. Results were considered significant at $P < 0.05$ level, and tendency was at $0.05 \le P < 0.10$.

Results

Mean values with standard error of the mean (SEM) were given. Body weight gain of sheep and the ruminal fermentation parameters are shown in Table 2. The sheep fed on CHM-diet had a greater ($P = 0.03$) daily weight gain than those fed on MH-diet. Rumen pH was lower ($P = 0.02$) for CHM-diet than for MH-diet. Concentration of rumen NH_3 remained similar between diets. Concentration of rumen total VFA tended to be higher ($P = 0.05$) and acetate was higher ($P = 0.04$) for CHM-diet than for MH-diet.

Plasma free amino acids, urea and NEFA determined in the pre-infusion period of isotope dilution method are presented in Table 3. Concentration of plasma aspartic acid was higher ($P = 0.02$) and those of threonine, leucine, phenylalanine, alanine, proline and total amino acid tended to be higher ($P < 0.10$) for CHM-diet than for MH-diet. Concentrations of plasma urea and NEFA did not differ between diets.

Table 3 Effects of Chinese herbal medicine on concentrations of plasma metabolites at pre-infusion period in sheep

Items	MH-diet	CHM-diet	SEM	*P*-value
Essential amino acids, μmol/L				
Threonine	258	282	20	0.07
Valine	268	315	17	0.13
Methionine	25	27	2	0.52
Iso-leucine	89	109	7	0.21
Leucine	129	156	9	0.09
Phenylalanine	61	68	4	0.06
Histidine	58	64	3	0.44
Lysine	112	130	9	0.16
Non essential amino acids, μmol/L				
Aspartic acid	7.2	8.1	1.1	0.02
Serine	141	159	6	0.30
Asparagine	52	60	3	0.24
Glutamic acid	78	90	4	0.15
Glutamine	311	349	17	0.32
Glycine	467	501	25	0.39
Alanine	190	207	9	0.08
Tyrosine	90	95	7	0.37
Tryptophan	32	47	4	0.34
Arginine	141	149	11	0.19
Proline	107	124	8	0.09
Total amino acid	2616	2940	268	0.07
Urea, mmol/L	6.5	7.2	0.8	0.10
NEFA, mEq/L	0.16	0.14	0.03	0.12

MH-diet = mixed hay of orchardgrass and reed canarygrass; CHM-diet = MH-diet supplemented with 2% of Chinese herbal medicine mixture; SEM = standard error of the mean; NEFA = non-esterified fatty acid.

Table 2 Effects of Chinese herbal medicine on body weight gain and ruminal fermentation characteristics in sheep

Items	MH-diet	CHM-diet	SEM	*P*-value
BW gain, g/day	16	25	6	0.03
pH	6.99	6.84	0.04	0.02
NH_3, mmol/L	10.7	11.2	0.6	0.65
Total VFA, mmol/L	86.2	92.8	3.2	0.05
Acetate, mmol/L	60.5	64.5	2.1	0.04
Propionate, mmol/L	17.2	19.1	0.9	0.35
Iso-butyrate, mmol/L	0.6	0.7	0.01	0.15
Butyrate, mmol/L	6.5	7.2	0.3	0.34
Iso-valerate, mmol/L	0.8	0.9	0.02	0.21
Valerate, mmol/L	0.5	0.5	0.01	0.12

MH-diet = mixed hay of orchardgrass and reed canarygrass; CHM-diet = MH-diet supplemented with 2% of Chinese herbal medicine mixture; SEM = standard error of the mean; BW = body weight; VFA = volatile fatty acids.

Concentration of plasma glucose and enrichment of plasma [U-^{13}C]glucose remained constant during the last 2 h of isotope infusion for each dietary treatment (Figure 1). Concentration of plasma glucose and GluTR did not differ between diets (Table 4).

Concentration of plasma α-KIC and enrichments of plasma α-[1-^{13}C]KIC and exhaled $^{13}CO_2$ were stable during the latter period of [1-^{13}C]leucine infusion (Figure 2). Concentration of plasma α-KIC remained similar for both dietary treatments. Plasma LeuTR did not differ between diets. The LeuOX tended to be greater ($P = 0.06$) for CHM-diet than for MH-diet, but the WBPS did not differ between diets. Metabolic heat production tended to be greater ($P = 0.05$) for CHM-diet than for MH-diet (Table 4).

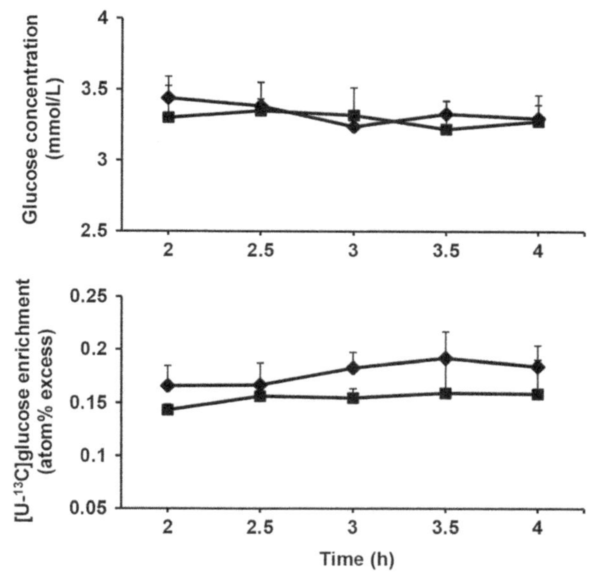

Figure 1 Time course changes of plasma glucose concentration and [U-^{13}C]glucose enrichment. Time course of changes in plasma glucose concentration and [U-^{13}C]glucose enrichment during the last 2 h of [U-^{13}C]glucose infusion in sheep fed two different diets (◆ = MH-diet, ■ = CHM-diet).

Discussion

Ruminal fermentation characteristics

In ruminants, the rumen is a large fermentation chamber which plays a very important role in feed digestion. The VFA is produced as the digestive product of dietary carbohydrates through microbial fermentation in the rumen. In the present study, concentration of rumen total VFA tended to be higher and acetate was higher in sheep fed on CHM-diet than on MH-diet. These results indicated that ruminal fermentation conditions might be improved and the fermentation of dietary carbohydrates was modified by Chinese herbal medicine. Angelica root and Atractylodes rhizome contain considerable amounts of essential oils which have been well recognized as the main bioactive components of the two herbs [24,25]. Both *in vitro* and *in vivo* studies have shown that essential oils could favorably manipulate rumen fermentation by changing VFA production. Castillejos *et al.* [26,27] reported that essential oils could increase total VFA production and acetate proportion in continuous culture fermenters. The feeding effects of essential oils in growing lambs were studied by Chaves *et al.* [28], who observed the significantly higher concentration of rumen total VFA in the experimental groups than in the control group. Similar results can also be found in dairy cows [29]. These research findings may clarify the observation of our current experiment, where the changes in rumen VFA concentration might be largely due to the effect of essential oils in CHM-diet.

The NH_3 is produced as the product of dietary nitrogenous substances through microbial fermentation in the rumen. In the present study, rumen NH_3 concentration remained similar between diets. Castillejos *et al.* [30] reported that essential oils had no influence on NH_3 concentration of rumen fluid in a long term *in vitro* incubations. Similar response on rumen NH_3 concentration was also observed in cows by Yang *et al.* [31]. In addition, various reports have verified that rumen NH_3 concentration was influenced by dietary CP intake [32,33]. Because the addition of Chinese herbal medicine was small (only 2%) in our current experiment, the lack of change in rumen NH_3 concentration might also be partly related to a comparable CP intake between dietary treatments.

Rumen pH values were within the normal range for both dietary treatments. The lower rumen pH for CHM-diet might be associated with the higher rumen VFA concentration, which agrees with Salman *et al.* [34] who demonstrated that rumen pH values were contrary to total VFA concentration in goats.

Table 4 Effects of Chinese herbal medicine on plasma glucose, protein and energy metabolism in sheep

Items	MH-diet	CHM-diet	SEM	*P*-value
Glucose concentration, mmol/L	3.36	3.33	0.15	0.84
GluTR, mmol/kg$^{0.75}$/h	1.45	1.56	0.04	0.17
α-KIC concentration, μmol/L	9.5	8.9	1.3	0.42
LeuTR, μmol/kg$^{0.75}$/h	398	422	23	0.16
LeuOX, μmol/kg$^{0.75}$/h	102	160	11	0.06
WBPS, g/kgBW$^{0.75}$/d	14.1	12.5	1.4	0.15
HP, kcal/kg$^{0.75}$/h	3.94	4.25	0.07	0.05

MH-diet = mixed hay of orchardgrass and reed canarygrass; CHM-diet = MH-diet supplemented with 2% of Chinese herbal medicine mixture; SEM = standard error of the mean; GluTR = turnover rate of plasma glucose; α-KIC = α-ketoisocaproic acid; LeuTR = turnover rate of plasma leucine; LeuOX = oxidation rate of leucine; WBPS = whole body protein synthesis; HP = heat production.

Plasma NEFA and free amino acids

Plasma NEFA is an important indicator of nutritional status in farm animals, which can directly indicate the adverse responses to such as fasting, negative energy balance or stress [35-37]. In our present study, concentration of plasma NEFA did not differ between diets and the values were presented within normal range for sheep. It indicated that the sheep fed on CHM-diet had comparable nutritional status with those fed on MH-diet without causing any adverse response or stress.

El-Shafei *et al.* [38] supplemented Astragalus root to broiler chick diets, and reported that the total protein in serum significantly increased probably due to a hormonal regulating effect of Astragalus root on protein metabolism. In association with the current experiment, although the level of total protein in blood was not

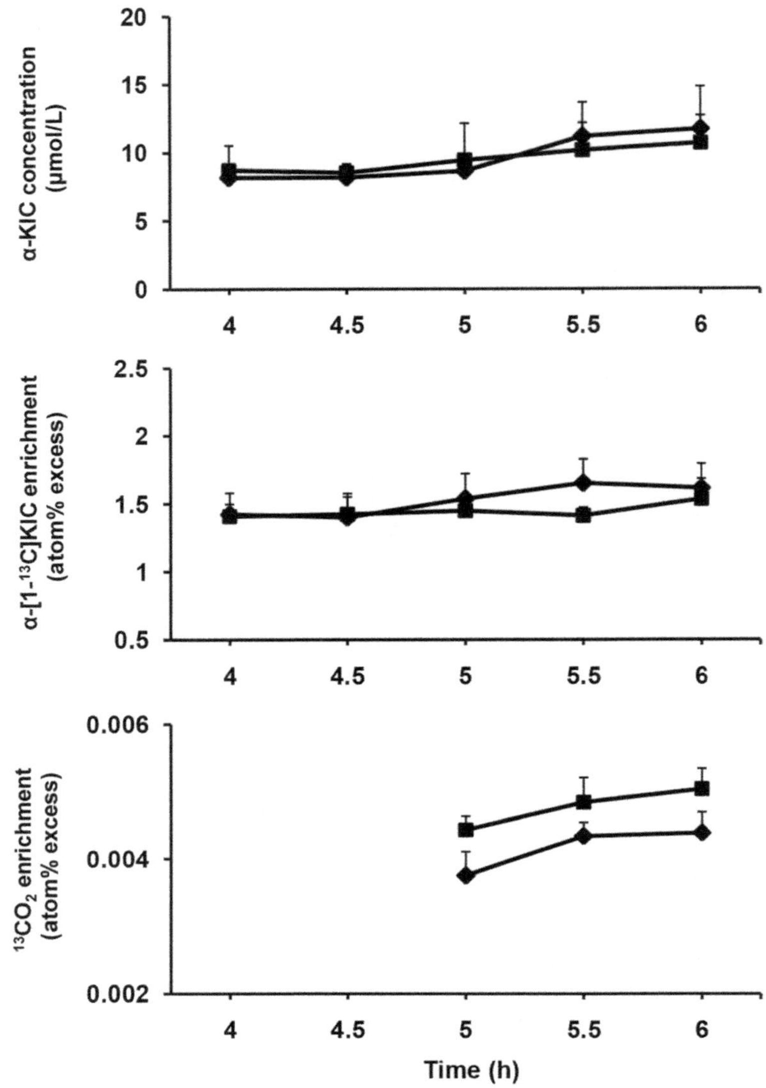

Figure 2 Time course changes of plasma α-KIC concentration and enrichments of plasma α-[1-^{13}C]KIC and exhaled $^{13}CO_2$. Time course of changes in plasma α-ketoisocaproic acid (α-KIC) concentration and enrichments of plasma α-[1-^{13}C]KIC and exhaled $^{13}CO_2$ during the latter period of [1-^{13}C]leucine infusion in sheep fed two different diets (◆ = MH-diet, ■ = CHM-diet).

determined, the trend of higher concentrations of certain plasma amino acids and total amino acid might be related to the bioactivity of herbs in CHM-diet.

Plasma glucose and protein kinetics

Based on the knowledge from human research, it has been well-known that the Chinese herbal medicine mixture is able to induce hematopoiesis by accelerating the generation, growth and maturity of blood cells [39]. Although it was expected that plasma glucose metabolism would be enhanced by the Chinese herbal medicine in sheep due to its bioactive properties on hematopoiesis, no positive impact was found on either plasma glucose concentration or plasma GluTR in the current experiment. In ruminants, because the dietary carbohydrates are fermented to VFA by microbes in the rumen, most of the glucose must be supplied through gluconeogenesis. Therefore, the supply of glucose precursor is considered as the major factor to influence glucose metabolism [40]. In our present study, propionate, the major glucogenic substrate produced in the rumen, did not differ between diets. It indicated that glucose precursors from the rumen were similar and thus the gluconeogenesis might be comparable between dietary treatments. Al-Mamun *et al.* [41] reported that plasma glucose concentration and turnover rate did not differ in sheep fed on plantain herb (a perennial herb having anti-bacterial and anti-inflammatory properties) or mixed hay, which is in accordance with our present findings.

Plasma LeuTR and the WBPS did not differ between diets. Nevertheless, it was found that higher numerical values of plasma LeuTR resulted in lower numerical values of the WBPS for CHM-diet than for MH-diet. This may be largely due to the calculation method as we used plasma LeuTR and LeuOX to calculate the WBPS, where the differences in LeuOX were greater than those in LeuTR. Similar data can also be found in a previous report of Sano *et al.* [42]. So far, little information is available regarding the performance of Chinese herbal medicine on plasma amino acid kinetics and protein synthesis in ruminants. The data obtained from human research is also limited. Al-Mamun *et al.* [43] reported that plasma LeuTR decreased but the WBPS remained unchanged in sheep fed on plantain herb compared with fed on mixed hay, which partly agrees with our present findings. Li *et al.* [44] used [^{15}N]glycine as the tracer to determine the effect of a Chinese herbal medicine mixture (Astragalus root and Angelica root, same as the present study) on protein metabolism in nephrotic patients, and observed that Chinese herbal medicine could improve the disorder of protein metabolism and increase the level of serum protein by improving the net rate of protein synthesis. However, their findings are inconsistent with the results of our current experiment in sheep.

Taken together, the sheep fed on CHM-diet did not enhance plasma GluTR, LeuTR and WBPS as expected under the conditions of the present study. In the treatment of human diseases, Chinese herbal medicine is normally taken as syrup of its extract. The herbs are processed by boiling in order to remove foreign substances and reduce toxic contents as well as increase therapeutic effects [39]. During the feeding periods of our current experiment, the Chinese herbal medicine was directly given to the animals without any treatment, thus the lack of changes in GluTR, LeuTR and WBPS between dietary treatments might be partly related to the factor that we did not process the Chinese herbal medicine, so that it was not as effective as expected.

Heat production

Until now, to our knowledge, the effect of Chinese herbal medicine on heat production has not been investigated either in humans or in animals. Heat is defined as the released energy that is produced from the oxidation process of the nutritive substances such as carbohydrate, protein and fat in the body. In our present study, metabolic heat production tended to be greater for CHM-diet than for MH-diet. The trend of increased heat production indicated that Chinese herbal medicine might play a role in accelerating nutrients oxidation in sheep. It is in good accordance with the result of LeuOX, which also tended to be greater for CHM-diet than for MH-diet. Furthermore, the sheep fed on CHM-diet had a higher body weight gain than those fed on MH-diet. Because energy is essential to animals in almost all life activities including maintenance, growth and production, the trend of increased heat production also demonstrated that the CHM-diet could provide more available energy for growth than MH-diet in sheep.

Conclusions

Although no significant effect on plasma glucose and protein metabolism was found, the sheep fed on CHM-diet had a higher body weight gain and showed positive impacts on rumen fermentation and energy metabolism without resulting in any adverse response than those fed on MH-diet. Therefore, these results suggested that the Chinese herbal medicine mixture should be considered as a potential feed additive for sheep.

Competing interests

The authors declare that they have no competing interests.

Authors' contributions

XL, as the lead author, was in charge of all research work, including designing the protocol, carrying out the experiment and writing the manuscript. KY, MK, XB and AP participated in the lab work to perform the chemical analysis of all samples, and provided kind suggestions for the manuscript. HS, as the supervisor to the lead author, was involved in the design and execution of the study, and approved the final manuscript. All authors read and approved the final manuscript.

Acknowledgements

The authors are grateful to Jianjun Liu, Harbin University of Commerce, China, for providing Chinese herbal medicine for the experiment.

References

1. Dibner JJ, Richards JD: **Antibiotics growth promoter in agriculture: history and mode of action.** *Poult Sci* 2005, **84:**634–643.
2. Barton MD: **Antibiotic use in animal feed and its impact on human health.** *Nutri Res Rev* 2000, **13:**279–299.
3. Huang KC: *The Pharmacology of Chinese Herbs.* 2nd edition. Boca Raton: CRC Press; 1998.
4. Kong X, Hu Y, Rui R, Wang D, Li X: **Effects of Chinese herbal medicinal ingredients on peripheral lymphocyte proliferation and serum antibody titer after vaccination in chicken.** *Int Immunopharmacolo* 2004, **4:**975–982.
5. Lien TF, Horng YM, Wu CP: **Feasibility of replacing antibiotic feed promoters with the Chinese traditional herbal medicine Bazhen in weaned piglets.** *Livest Prod Sci* 2007, **107:**97–102.
6. Wang J, Zhou H: **Comparison of the effects of Chinese herbs, probiotics and prebiotics with those of antibiotics in diets on the performance of meat ducks.** *J Anim Feed Sci* 2007, **16:**96–103.
7. National Research Council (NRC): *Nutrient requirement of sheep.* 6th edition. Washington, DC: National Academy Press; 1985.
8. AOAC: *Official Methods of Analysis.* 16th edition. Arlington, VA: Association of Official Analytical Chemists; 1995.
9. Van Soest PJ, Robertson JB, Lewis BA: **Methods for dietary fiber, neutral detergent fiber and non-starch polysaccharides in relation to animal nutrition.** *J Dairy Sci* 1991, **74:**3583–3597.

10. Weatherburn MW: **Phenol-hypochlorite reaction for determination of ammonia.** *Anal Chem* 1967, **39**:971–974.
11. Tserng KY, Kalhan SC: **Estimation of glucose carbon recycling and glucose turnover with [U-^{13}C]glucose.** *Am J Physiol* 1983, **245**:E476–E482.
12. Fujita T, Majima H, Itoh T, Sano H: **Combined effect of salinomycin and feeding on whole body glucose kinetics in sheep fed a high-concentrate diet.** *Reprod Nutr Dev* 2006, **46**:503–514.
13. Huggett AG, Nixon DA: **Enzymatic determination of blood glucose.** *Biochem J* 1957, **66**:12.
14. Rocchiccioli F, Leroux JP, Cartier P: **Quantitation of 2-ketoacids in biological fluids by gas chromatography chemical ionization mass spectrometry of o-trimethylsilyl-quinoxalinol derivatives.** *Biomed Mass Spectrom* 1981, **8**:160–164.
15. Calder AG, Smith A: **Stable isotope ratio analysis of leucine and ketoisocaproic acid in blood plasma by gas chromatography/mass spectrometry. Use of tertiary butyldimenthylsilyl derivatives.** *Rapid Commun Mass Spectrom* 1988, **2**:14–16.
16. Wolfe RR: *Tracers in metabolic research: radioisotope and stable isotope/mass spectrometry methods.* New York: Alan R. Liss; 1984.
17. Magni F, Arnoldi L, Galati G, Kienle MG: **Simultaneous determination of plasma levels of α-ketoisocaproic acid and leucine and evaluation of α-[1-^{13}C]ketoisocaproic acid and [1-^{13}C]leucine enrichment by gas chromatography–mass spectrometry.** *Anal Biochem* 1994, **220**:308–314.
18. Allsop JR, Wolfe RR, Burke JF: **Tracer priming the bicarbonate pool.** *J Appl Physiol* 1978, **45**:137–139.
19. Wolfe RR, Goodenough RD, Wolfe MH, Royle GT, Nadel ER: **Isotopic analysis of leucine and urea metabolism in exercising humans.** *J Appl Physiol* 1982, **52**:458–466.
20. Harris PM, Skene PA, Buchan V, Milne E, Calder AG, Anderson SE, Connell A, Lobley GE: **Effect of food intake on hind-limb and whole-body protein metabolism in young growing sheep: chronic studies based on arterio-venous techniques.** *Brit J Nutr* 1992, **68**:389–407.
21. Brouwer E: In *Energy Metabolism.* Edited by Blaxter KL. London: Academic Press; 1965:302–304.
22. Young BA, Kerrigan B, Christopherson RJ: **A versatile respiratory pattern analyzer for studies of energy metabolism of livestock.** *Can J Anim Sci* 1975, **55**:17–22.
23. SAS (Statistical Analysis System): *SAS/STAT® Software: Changes and enhancements through release. Version 6–11.* Cary, NC: SAS Institute; 1996.
24. Wu MJ, Sun XJ, Dai YH, Guo FQ, Huang LF, Liang YZ: **Determination of constituents of essential oil from *Angelica sinensis* by gas chromatography–mass spectrometry.** *J Cent South Univ* 2005, **12**:430–436.
25. Zhou J, Fang L, Wang X, Zhang J, Guo LP, Huang LQ: **Comparison of the volatile compounds of crude and processed Atractylodis rhizome analyzed by GC-MS.** *Afr J Pharm Pharmacol* 2012, **6**:2155–2160.
26. Castillejos L, Calsamiglia S, Ferret A, Losa R: **Effects of a specific blend of essential oil compounds and the type of diet of on rumen microbial fermentation and nutrient flow from continuous culture systems.** *Anim Feed Sci Tech* 2005, **119**:29–41.
27. Castillejos L, Calsamiglia S, Ferret A, Losa R: **Effects of dose and adaptation time of a specific blend of essential oil compounds on rumen fermentation.** *Anim Feed Sci Tech* 2007, **132**:186–201.
28. Chaves AV, Stanford K, Gibson LL, McAllister TA, Benchaar C: **Effects of carvacrol and cinnamaldehyde on intake, rumen fermentation, growth performance, and carcass characteristics of growing lambs.** *Anim Feed Sci Tech* 2008, **145**:396–408.
29. Soltan MAE, Shewita RS, Al-Sultan SI: **Influence of essential oils supplementation on digestion, rumen fermentation, rumen microbial populations and productive performance of dairy cows.** *Asian J Anim Sci* 2009, **3**:1–12.
30. Castillejos L, Calsamiglia S, Ferret A: **Effect of essential oil active compounds on rumen microbial fermentation and nutrient flow in *in vitro* systems.** *J Dairy Sci* 2006, **89**:2649–2658.
31. Yang WZ, Benchaar C, Ametaj BN, Chaves AV, He ML, McAllister TA: **Effects of garlic and juniper berry essential oils on ruminal fermentation and on the site and extent of digestion in lactating cows.** *J Dairy Sci* 2007, **90**:5671–5681.
32. Freeman AS, Galyean ML, Caton JS: **Effects of supplemental protein percentage and feeding level on intake, ruminal fermentation, and digesta passage in beef steers fed prairie hay.** *J Anim Sci* 1992, **70**:1562–1572.
33. Alam MK, Ogata Y, Sako Y, Al-Mamun M, Sano H: **Intermediary metabolism of plasma acetic acid, glucose and protein in sheep fed a rice straw-based diet.** *Asian-Aust J Anim Sci* 2010, **23**:1333–1339.
34. Salman FM, El-Kadi RI, Abdel-Rahman H, Ahmed SM, Mohammad MI, Shoukry MM: **Biologically treated sugar beet pulp as a supplement in goat rations.** *Int J Agric Biol* 2008, **10**:412–416.
35. Sticker LS, Thompson DLJ, Bunting LD, Fernandez JM, DePew CL: **Dietary protein and (or) energy restriction in mares: plasma glucose, insulin, nonesterified fatty acid, and urea nitrogen responses to feeding, glucose, and epinephrine.** *J Anim Sci* 1995, **73**:136–144.
36. Chelikani PK, Ambrose JD, Keisler DH, Kennelly JJ: **Effect of short-term fasting on plasma concentrations of leptin and other hormones and metabolites in dairy cattle.** *Domest Anim Endocrin* 2004, **26**:33–48.
37. Hristov S, Maksimovic N, Stankovic B, Zujovic M, Pantelic V, Stanisic N, Zlatanovic Z: **The most significant stressors in intensive sheep production.** *Biotechnology in Animal Husbandry* 2012, **28**:649–658.
38. El-Shafei AA, Al-Gamal MA, Abdelrahman AS, Arafa MM: **Influence of different levels of Astragalus root powder in broiler chick diets on the physiological and biochemical changes.** *J Appl Sci Res* 2013, **9**:2104–2118.
39. Li X, Wei W: *Chinese Materia Medica: Combinations and Applications.* Potters Bar: Donica Publishing; 2002.
40. Ortigues-Marty I, Vernet J, Majdoub L: **Whole body glucose turnover in growing and non-productive adult ruminants: meta-analysis and review.** *Reprod Nutr Dev* 2003, **43**:371–383.
41. Al-Mamun M, Tanaka C, Hanai Y, Tamura Y, Sano H: **Effects of plantain (*Plantago lanceolata* L.) herb and heat exposure on plasma glucose metabolism in sheep.** *Asian-Aust J Anim Sci* 2007, **20**:894–899.
42. Sano H, Kajita M, Fujita T: **Effect of dietary protein intake on plasma leucine flux, protein synthesis, and degradation in sheep.** *Comp Biochem Physiol* 2004, **139**:163–168.
43. Al-Mamun M, Hanai Y, Tanaka C, Tamura Y, Sano H: **Responses of whole body protein synthesis and degradation to plantain herb in sheep exposed to heat.** *Arch Anim Nutr* 2008, **62**:219–229.
44. Li L, Yu H, Pan J: **A study on protein metabolism in nephrotic patients treated with Chinese herbs.** *Zhonghua Nei Ke Za Zhi* 1995, **34**:670–672.

4

Estimation of endogenous protein and amino acid ileal losses in weaned piglets by regression analysis using diets with graded levels of casein

Tércia Cesária Reis de Souza[1], Araceli Aguilera Barreyro[1] and Gerardo Mariscal-Landín[2*]

Abstract

Background: Many studies have investigated endogenous loss of proteins and amino acids (AAs) at the ileal level in growing pigs. However, only a few studies have researched this subject in piglets. Knowledge regarding AA ileal digestibility in piglets would be helpful during the formulation of diets for weaning piglets, rather than just using coefficients obtained in growing pigs. Therefore, in this study, we sought to estimate endogenous protein and AA ileal losses in piglets. Furthermore, apparent and true ileal digestibility (AID and TID) of protein and AAs from casein were measured.

Results: The average flow of protein was 20.8 g/kg of dry matter intake (DMI). Basal protein loss, as estimated by regression, was 16.9 g/kg DMI. Glutamic acid, arginine, and aspartic acid (2.2, 1.4, and 1.2 g/kg DMI, respectively) were the AAs for which greater losses were seen. The AID of protein and AAs increased as the protein level in the diet increased. A higher increment in AID was observed between diets with 80 and160 g CP/kg of feed; this finding was mainly attributable to increases in glycine and arginine (46.1% and 18%, respectively). The TID of protein was 97.8, and the TID of AAs varied from 93.9 for histidine to 100.2 for phenylalanine.

Conclusions: The basal endogenous protein loss in piglets was 16.9 g/kg DMI. Endogenous protein was rich in glutamic acid, aspartic acid, and arginine, which represented 32.7% of endogenous protein loss in weaning piglets. The TID of casein was high and varied from 93.0 for histidine to 100.2 for phenylalanine.

Keywords: Amino acid loss, Ileal digestibility, Protein loss, Weaned piglets

Background

Many studies have investigated endogenous loss of proteins and amino acids (AAs) at the ileal level in growing pigs [1-3]. However, only a few studies have researched this subject in piglets [4-7]. Mariscal-Landín and Reis de Souza [6] reported high proline endogenous losses following prolonged intake (15 d) of a low protein diet in piglets, and similar effects of low protein or protein-free diets have been reported in growing pigs [1,8,9]. Further knowledge regarding AA ileal digestibility in piglets would be helpful during the formulation of diets for weaning piglets, rather than just using coefficients obtained from data for optimal diets in growing pigs, especially for cereals and vegetable proteins [10,11]. In order to achieve this goal, it is essential to determine the endogenous loss of protein and AAs in the ileum of piglets.

Therefore, the objectives of this study were: 1) measure endogenous ileal loss of protein and AAs in weaned piglets, avoiding changes in the AA profile of endogenous protein through prolonged intake of a low protein diet; 2) measure the apparent and true ileal digestibility (AID and TID, respectively) of protein and AAs derived from casein.

Materials and methods

The experiment was performed at the experimental farm of CENID-Physiology. The care of animals used in this study was in accordance with the guidelines issued by the Mexican Official Standard for the Production, Protection and Use of Lab Animals [12] and the guidelines of the International Guiding Principles for Biomedical Research Involving Animals [13].

* Correspondence: mariscal.gerardo@inifap.gob.mx
[2]Centro Nacional de Investigación en Fisiología Animal, Instituto Nacional de Investigaciones Forestales Agrícolas y Pecuarias, Km 1, Carretera a Colón, Ajuchitlán Colón Querétaro, 76280 México
Full list of author information is available at the end of the article

Animals

Eighteen piglets (Duroc × Landrace; 6 per group), weaned at 18 ± 1 d of age and weighing 6 ± 1 kg, were used. The piglets were placed in individual metabolism cages provided with a self-feeder and a low pressure drinking nipple, in a temperature controlled room (26 ± 3°C). In order to modify the enzymatic profile of suckling animals as little as possible; piglets were fed from 18 to 20 d of age with a mixture of dry whole milk (80%) and maize starch (20%) 3 times/d, at 0800, 1300 and 1800 h, at a concentration of 30 g/kg body weight (BW). When the piglets were 21 d old, a T cannula was fit at the terminal ileum as previously described [14]. After surgery, a therapeutic treatment (penicillin 600,000 IU, streptomycin 750 mg, oxytetracycline 500 mg) was administered for 3 d, and the piglets received a casein diet containing 240 g of crude protein (CP)/kg for 8 d.

Experimental diets

From the ninth d after surgery, piglets received 1 of 3 experimental treatments (Table 1). Casein was used as the sole protein source to provide 80, 160 or 240 g of CP/kg of feed. Because the low protein diets used in this study (80 or 160 g CP/kg) did not provide sufficient levels of lysine, methionine, threonine, or tryptophan, these AAs were supplemented to fulfill NRC nutrient requirements (Table 2) [15]. All 3 experimental diets contained the same amount of lactose and maize oil. To adjust calcium and phosphorus levels, calcium carbonate and dicalcium phosphate were used. Salt and vitamin and mineral premixes were included at levels that met or exceeded the NRC nutritional requirements [15]. Chromic oxide was added at a rate of 3 g/kg of feed as an inert marker.

Table 1 Experimental diets composition (g/kg feed[1])

g CP/kg feed	80	160	240
Maize starch	677.3	604.2	523.3
Casein	89.8	180.0	270.1
Lactose	126.3	126.3	126.3
Maize oil	40.0	40.0	40.0
Calcium Phosphate			
(dicalcium)	32.1	27.9	23.8
Calcium carbonate		1.1	2.1
Salt	6.4	6.4	6.4
Vitamins[2]	3.6	3.6	3.6
Minerals[3]	1.2	1.2	1.2
Antioxidant[4]	0.2	0.2	0.2
Chromic oxide	3.0	3.0	3.0
Lysine HCl	1.7		
Tryptosine[5]	9.1	2.7	
DL Methionine	4.5	1.9	
L Threonine	4.8	1.5	
Protein, g/kg[6]	98.0	151.3	231.5
ED, MJ · kg[7]	15.1	15.1	15.1

[1] As fed basis.
[2] Vitamin premix that provided per kilogram of piglet diet: Vitamin A 10,200 IU, D 1,980 IU, E 60 IU, K 1.20 mg, Choline 967 mg, Niacin 36 mg, Pantothenic acid 17 mg, Riboflavin 7.2 mg, Vitamin B_{12} 40 µg, Thiamine 0.30 mg, Pyridoxine 0.31 mg, Biotin 0.08 mg, Folic acid 0.75 mg.
[3] Mineral premix that provided per kilogram of piglet diet: Cu 14.4 mg, I 0.96 mg, Fe 120 mg, Mn 36 mg, Se 0.3 mg, Zn 144 mg.
[4] ANTI-OX-HP, Compagnie Chimique d'Aquitaine.
[5] 15/70 Tryptophan/Lysine Blend. Produced by Indukern de México.
[6] Analyzed levels.
[7] Estimated from INRA tables, [16].

Sample collection

Experimental diets were supplied starting on d 9 after surgery, according to schedules described previously. The experimental period lasted for 7 d, that is 5 d for adaptation and 2 d for collection of ileal digesta. Feed intake was fixed at 26 g/kg BW during the experimental period to maintain feed intake at a constant level in relation to BW for all animals. The piglets always had free access to water.

Ileal digesta were collected in plastic bags (5 cm × 3 cm) containing 2 mL of a 0.2 mol/L solution of HCl to minimize further bacterial activity. Bags were attached to the barrel of the cannula with a rubber band. Ileal digesta were collected continuously over the course of 12 h when the piglets were 36 and 37 d of age (d 16 and 17 after surgery).

Table 2 Protein and amino acids composition of experimental diets (g/kg DM)

g CP/kg feed	80	160	240
Protein	105.0	162.0	247.0
Essential amino acids			
Arginine	3.4	6.2	10.2
Histidine	2.6	4.9	7.8
Lysine	13.8	14.2	19.8
Isoleucine	4.6	8.6	14.1
Leucine	8.2	15.3	24.9
Valine	5.8	10.9	17.7
Methionine	7.0	7.5	6.9
Phenylalanine	4.7	8.6	15.1
Threonine	9.8	10.8	12.0
Non essential amino acids			
Alanine	2.8	5.2	8.4
Aspartic acid	6.2	11.5	18.9
Cystine	0.6	0.9	1.4
Glutamic acid	20.0	37.7	62.1
Glycine	1.8	3.4	5.3
Proline	8.2	17.3	23.7
Serine	4.8	9.0	14.8
Tyrosine	4.4	8.1	14.2

When ileal digesta filled the bags, they were transferred to a container and frozen at −20°C until lyophilization.

Preparation of samples and chemical analysis

Ileal digesta were lyiophilized and ground in a laboratory mill by using a 0.5-mm mesh (Arthur H. Thomas Co. Philadelphia, PA). Experimental diets and ileal digesta were analyzed for dry matter (DM) and CP, according to AOAC methods 934.01 and 976.05 [17] respectively, and for chromic oxide according to the methods reported by Fenton and Fenton [18]. Sample preparation for AA analysis was performed following method 994.12 of AOAC [17], samples were hydrolysed in 6 mol/L HCl at 110°C for 24 h. Methionine and cystine were oxidized with performic acid before analysis [17]. Amino acids analysis was performed according to the method reported by Henderson et al. [19]. Tryptophan was not analyzed.

Data analysis

Endogenous protein and AA losses in the ileal digesta were estimated using the following equation [20]:

$$ILAA = AF \times (ID/IF)$$

Where *ILAA* is the ileal loss of CP or AA expressed in g/kg DMI, *AF* is the nutrient concentration in ileal digesta (g/kg DM), *ID* is the marker concentration in the diet (g/kg DM), and *IF* is the marker concentration in the ileal digesta (g/kg DM).

The AIDs of DM, protein and AAs were estimated using the following equation [20]:

$$AID = [1-[(ID \times AF)/(AD \times IF)]] \times 100$$

Where *AID* is the apparent ileal digestibility of a nutrient in the diet, *ID* is the marker concentration in the diet (g/kg DM), *AF* is the nutrient concentration in the ileal digesta (g/kg DM), *AD* is the nutrient concentration in diet (g/kg DM), and *IF* is the marker concentration in the ileal digesta (g/kg DM). The calculi excluded crystalline amino acids (L-Lys-HCl, DL-Met, L-Thr and L-Trp) because they were considered completely digested at the end of the ileum [21,22].

Basal protein and AA ileal losses and the TIDs of protein and AAs from casein were estimated by linear regression [2]:

$$y = a + bx$$

Where y is the protein or amino acid amount disappearing from the digestive tract in terms of g/kg DMI, a is the protein or amino acid basal ileal loss in terms of g/kg DMI, b is the protein or amino acid disappearing in the digestive tract, (equivalent to the TID), and x is the intake in g/kg DMI. Regression was performed using the REG procedure of the SAS software [23]. Performing regression analysis in this way allowed us to obtain a high value of the regression coefficient R^2, because almost the total amount of nutrient ingested disappeared at the ileal level. The intercept was negative because an endogenous loss was, by definition, an amount found at zero intake, and consequently, at zero intake there was no dietary nutrient absorption, only excretion. Nevertheless, the absolute value of the intercept using excreted (non-digestible) or absorbed (digestible) nutrient was the same and could be used as reported previously [2].

Statistical analysis

Homogeneity of variance for all data was tested by Levene's test using the hovtest of the SAS software [23]. Protein and AA Ileal losses and AID data were analyzed as a completely randomized design under the general model:

$$Yij = \mu + Ti + eij$$

Where Y_{ij} is the dependent variable; μ is the general mean; T_i is the treatment$_i$ 1, 2, 3; and e_{ij} is the experimental error, calculated using the GLM procedure of the SAS software [23]. The piglets were the experimental units for all analyses. Treatment means were compared using the Duncan method, an α-value of 0.05 was used to assess significance and orthogonal polynomial contrast were performed to find a linear or quadratic response [24].

Results

Protein and amino acids flow at ileal level

Protein flow was higher in the digesta of piglets fed 160 g CP/kg of feed than in the digesta of piglets fed 80 g CP/kg of feed ($P < 0.05$, Table 3). Moreover, significant differences among these groups were observed for the AAs valine ($P < 0.05$), threonine ($P < 0.05$), aspartic acid ($P < 0.05$), glutamic acid ($P < 0.05$), serine ($P < 0.01$) and histidine ($P < 0.001$). Protein flow was similar between digesta of piglets fed 160 or 240 g CP/kg of feed ($P > 0.05$).

For all 3 groups, the average protein flow was 20.8 g/kg DMI. Protein was rich in glutamic acid, aspartic acid, arginine and threonine (3.2; 1.6; 1.4 and 1.3 g/kg DMI, respectively). These amounts represented 17.6%, 8.7%, 7.9% and 7.0% of total AAs. AAs with lower flow included methionine, histidine and tyrosine (0.3, 0.5, and 0.5 g/kg DMI, respectively); these flow represented 1.6%, 3.0%, and 3.1% of total AA flow, respectively.

Basal endogenous losses

The basal endogenous protein ileal loss estimated by regression was 16.9 g/kg DMI (Table 4). Glutamic acid, arginine and aspartic acid were the AAs exhibiting larger basal endogenous ileal losses (2.2, 1.4, and 1.2 g/kg DMI, respectively). AAs with lower endogenous losses included histidine, methionine, and tyrosine (0.2, 0.4 and 0.5 g/kg

Table 3 Protein and amino acids flow at the ileum level, in terms of dry matter intake (g/kg)

g CP/kg feed	80	160	240	SEM[1]	*P* value[2]
Protein[C]	17.67[a]	22.10[b]	22.72[b]	0.77	0.05
Essential amino acids					
Arginine	1.25	1.56	1.47	0.06	NS
Histidine[B]	0.36[a]	0.52[b]	0.74[b]	0.04	0.001
Lysine	0.61	0.74	0.71	0.04	NS
Isoleucine[B]	0.65[a]	0.88[b]	0.93[b]	0.03	0.01
Leucine[E]	0.91	1.17	1.11	0.04	NS
Valine[B]	0.81[a]	1.08[b]	1.11[b]	0.04	0.05
Methionine	0.30	0.31	0.25	0.02	NS
Phenylalanine[E]	0.55	0.67	0.60	0.03	NS
Threonine[B]	1.01[a]	1.34[b]	1.43[b]	0.05	0.05
Non essential amino acids					
Alanine[D]	0.97	1.24	1.22	0.05	NS
Aspartic acid[C]	1.26[a]	1.77[b]	1.73[b]	0.08	0.05
Cystine	0.32	0.37	0.39	0.05	NS
Glutamic acid[C]	2.33[a]	3.65[b]	3.76[b]	0.21	0.05
Glycine	0.94	0.97	1.02	0.04	NS
Proline[D]	1.03	1.37	1.40	0.08	NS
Serine[A]	1.04[a]	1.49[b]	1.69[b]	0.06	0.01
Tyrosine[E]	0.49	0.61	0.55	0.03	NS

[1] Standard error of the mean.
[2ab] Means with different superscript differ at probability level show in the fifth column.
NS non significant.
[A] Linear effect ($P < 0.001$), [B] Linear effect ($P < 0.01$), [C] Linear effect ($P < 0.05$), [D] Linear effect ($P < 0.10$), [E] Quadratic effect ($P < 0.10$).

DMI, respectively). All regressions were significant ($P <$ 0.001) and had determination coefficients of at least 0.98.

AID

The AIDs of protein, arginine, leucine, isoleucine, valine, phenylalanine, methionine, aspartic acid, serine, glycine, alanine and tyrosine, increased as casein levels increased (Table 5). The AIDs of histidine, cystine and proline differed significantly between piglets fed 80 and 160 g CP/kg feed, while the AIDs of these AAs were similar between piglets fed 160 and 240 g CP/kg of feed. The AID of lysine was different between piglets consuming 80 and 240 g CP/kg, and the AID observed in piglets fed with 160 g CP/kg was similar to those of the other 2 treatments. The AID of glutamic acid was similar in piglets consuming 80 and 160 g CP/kg and lower than that in piglets consuming 240 g CP/kg. However, all amino acids and protein showed a linear increment ($P < 0.01$) in their AIDs as protein level increased in the diet.

The AAs with the greatest differences in their AIDs between piglets consuming 80 and 160 g CP/kg were glycine (46.1%), arginine (18%), valine (16.3%) and alanine (14.4%).

Table 4 Basal ileal losses (g/kg DMI) and true ileal digestibility estimated by regression[1]

	BEL[2]		TID[3]		$R^{2(5)}$
	a	S.D.[4]	*b*	S.D.	
Protein	16.95	3.69	97.8	0.3	0.99
Essential amino acids					
Arginine	1.36	0.26	99.0	3.8	0.98
Histidine	0.18	0.10	93.0	1.9	0.99
Lysine	0.60	0.16	99.4	1.1	0.99
Isoleucine	0.64	0.17	98.0	1.8	0.99
Leucine	0.98	0.19	99.5	1.1	0.99
Valine	0.81	0.18	98.4	1.5	0.99
Methionine	0.35	0.09	97.0	9.0	0.99
Phenylalanine	0.63	0.10	100.2	1.0	0.99
Threonine	0.96	0.21	96.0	2.6	0.99
Non essential amino acids					
Alanine	0.99	0.20	97.2	3.5	0.98
Aspartic acid	1.25	0.26	97.1	2.1	0.99
Cystine	0.30	0.07	95.3	4.8	0.96
Glutamic acid	2.18	0.69	97.1	1.7	0.99
Glycine	0.95	0.13	99.3	3.4	0.98
Proline	0.97	0.27	98.1	1.6	0.99
Serine	0.91	0.26	95.0	2.6	0.99
Tyrosine	0.55	0.09	100.1	1.0	0.99

[1] According to equation: $Y = a + bX$.
[2] *BEL* Basal endogenous losses (*a* = intercept, protein or amino acids endogenous losses at ileal level).
[3] *TID* True ileal digestibility (*b* = slope, proportion of protein or amino acid that disappeared in the digestive tract, per protein or amino acid unit intake).
[4] *SD* Standard deviation of the intercept (*a*) or slope (*b*).
[5] R^2 Coefficient of determination for regression straight line.

The difference in the AID of protein between piglets consuming 80 and 240 g CP/kg feed was 9.2%, which was much lower differences in the AIDs of glycine (66.2%), arginine (35%) and alanine (28.2%) between these groups. The lowest difference in AID between these groups was for Lysine (2.9%).

TID

The TID of protein was 97.8, and the average TID for AA was 97.6. The minimum TID value was obtained for histidine (93.0) and the maximum TID value was obtained for phenylalanine (100.2). The TID values for lysine and threonine were 99.4 and 96.0, respectively (Table 4).

Discussion

In this study, we estimated endogenous protein and AA ileal losses in piglets. Our data demonstrated that endogenous protein was rich in glutamic acid, aspartic acid, and arginine, which represented 32.7% of endogenous protein loss in weaning piglets. Moreover, the TID

Table 5 Apparent ileal digestibility

g CP/kg feed	80	160	240	SEM[1]	*P* value[2]
Protein[A]	82.8[a]	86.5[b]	90.4[c]	0.06	0.001
Essential amino acids					
Arginine[A]	62.3[a]	73.5[b]	84.1[c]	1.17	0.001
Histidine[A]	85.5[a]	89.0[b]	89.8[b]	0.71	0.05
Lysine[A]	93.4[a]	94.5[ab]	96.1[b]	0.35	0.05
Isoleucine[A]	85.6[a]	89.3[b]	92.6[c]	0.39	0.001
Leucine[A]	88.7[a]	92.0[b]	95.1[c]	0.35	0.001
Valine[A]	85.7[a]	90.7[b]	93.1[c]	0.39	0.001
Methionine[A]	89.7[a]	93.3[b]	96.9[c]	0.42	0.001
Phenylalanine[A]	88.0[a]	92.0[b]	95.6[c]	0.38	0.001
Threonine[A]	76.7[a]	80.2[b]	86.9[b]	0.78	0.001
Non essential amino acids					
Alanine[A]	65.5[a]	74.9[b]	84.0[c]	1.09	0.001
Aspartic acid[A]	79.5[a]	83.9[b]	90.0[c]	0.72	0.001
Cystine[A]	51.4[a]	72.3[b]	79.5[b]	4.26	0.01
Glutamic acid[A]	88.1[a]	89.8[a]	93.4[b]	0.55	0.01
Glycine[A]	47.7[a]	69.7[b]	79.3[c]	1.23	0.001
Proline[A]	87.1[a]	91.7[b]	93.6[b]	0.49	0.001
Serine[A]	78.1[a]	82.5[b]	87.5[c]	0.73	0.001
Tyrosine[A]	88.8[a]	92.1[b]	95.7[c]	0.37	0.001

[1] Standard Error of the Mean.
[2abc] Means with different superscript differ at probability level show in the fifth column.
[A] Linear effect ($P < 0.001$).

of casein was high, varying from 93.0 for histidine to 100.2 for phenylalanine.

Ileal losses

The protein and AA ileal losses found in this study were similar to those reported by Mariscal-Landín and Reis de Souza [6], except that proline losses were higher in the earlier study. This difference was likely to be attributable to the fact that we used higher dietary protein levels and that lysine, methionine, threonine and tryptophan were supplemented to fulfill NRC requirements in the current study [15]. Another reason could be the duration of the experimental period: 7 d in the current study compared to 14 d in the previous study. This shorter period was used to avoid increased proline loss caused by prolonged consumption of a low protein diet [9].

AA supplementation and the short period of consumption of the low protein diet could reduce muscle catabolism, thereby decreasing blood glutamine levels. These effects could lead to suppression of proline synthesis because enterocytes use glutamine as proline precursor [25-27]. Moreover, proline synthesis dependent of pyrroline-5-caboxylate, has been associated with increased proline loss in growing pigs fed protein-free diets [1,8,28]. Because the activity of pyrroline-5-carboxylate reductase, the enzyme responsible for proline synthesis, is greater than the activity of proline oxidase, the enzyme responsible for proline degradation [1].

The arginine losses reported here agree with the results of some previously reported studies Leibholz [29] and Mariscal-Landín and Reis de Souza [6], but disagree with the results of studies by Eklund [4,5]. These differences are likely due to differences in study design; Eklund used during 4 periods of 7 d each with the same piglets and reported values as the mean of the 4 periods. Our values represented only those obtained during the fourth wk of age, which was also observed in studies by Leibholz [29] and Mariscal-Landin and Reis de Souza [6].

Additionally, the glycine losses reported in our study agree with values reported in 4-wk-old piglets fed a protein-free diet [30] and in 4- and 5-wk-old piglets [6]. Glycine losses are associated with biliary secretion because glycine represents 95% of amino nitrogen present in biliary salts [31].

Ileal digestibility

The AID for protein from casein (86.6 on an average) and for AA were similar to those reported by Mariscal-Landín and Reis de Souza [6]. Additionally, the AIDs for glycine, arginine and alanine (47.7, 62.3 and 65.5 respectively) were the most affected by protein levels. Glycine has been reported as the least digestible AA from casein [6,32], as observed in this study, because lacteal proteins, which include casein, have a low glycine content, and glycine endogenous losses are important [30]. Both factors interact to reduce the AID of glycine. Moreover a combination of the same factors could explain the low AID of arginine because lacteal proteins have low arginine content [33] and because endogenous protein is rich in arginine [6,29].

The lack of variation in the AIDs of lysine, histidine, threonine, cystine and proline between treatments (160 and 240 g CP/kg of feed) could be due to the high levels of these AAs in protein. Jørgensen [34] fed growing pigs a casein-based diet (13%, 16%, 19%, and 22% CP) and observed differences between the AIDs of AAs only for pigs fed 13% and 16% CP. No differences were observed in groups fed 16%, 19% or 22% CP. The high TID values (greater than 93.0) were due to high casein digestibility and its low induction of endogenous secretion [35]; because of these characteristics, casein is often used to measure endogenous losses. In our study, the TIDs of AAs were similar to those reported previously [4,6].

Conclusions

The basal endogenous protein loss in piglets was 16.9 g/kg DMI; Endogenous protein was rich in glutamic acid, aspartic acid and arginine, which represented 32.7% of endogenous protein loss in weaning piglets. The TID of casein was high and varied from 93.0 for histidine to 100.2 for phenylalanine.

Abbreviations

AA: Amino acids; AID: Apparent ileal digestibility; TID: True ileal digestibility; BW: Body weight; DM: Dry matter; CP: Crude protein.

Competing interests

The authors declare that they have no competing interests.

Authors' contributions

All authors contributed equally to this research. All authors read and approved the final manuscript.

Acknowledgments

This study was funded in part by the Minister of Agriculture, Livestock and Fishing (SAGARPA) of Mexico and The National Council for Science and Technology (CONACYT) of Mexico via the financial support given to Research Project SAGARPA-CONACYT 2003-2-169.

Author details

[1]Facultad de Ciencias Naturales, Universidad Autónoma de Querétaro, Santiago de Querétaro, Av. De las Ciencias s/n Juriquilla 76000, Querétaro, Querétaro, México. [2]Centro Nacional de Investigación en Fisiología Animal, Instituto Nacional de Investigaciones Forestales Agrícolas y Pecuarias, Km 1, Carretera a Colón, Ajuchitlán Colón Querétaro, 76280 México.

References

1. Mariscal-Landín G, Sève B, Collèaux Y, LeBreton Y: **Endogenous amino nitrogen collected from pigs with end to end ileorectal anastomosis is affected by the method of estimation and altered by dietary fiber.** *J Nutr* 1995, **125:**136–146.
2. Furuya S, Kaji Y: **Estimation of the true ileal digestibility of amino acids and nitrogen from their apparent values for growing pigs.** *Anim Feed Sci Technol* 1989, **26:**271–285.
3. Hodgkinson SM, Souffrant WB, Moughan PJ: **Comparison of the enzyme-hydrolyzed casein, guanidination, and isotope dilution methods for determining ileal endogenous protein flow in the growing rat and pig.** *J Anim Sci* 2003, **81:**2525–2534.
4. Eklund M, Mosenthin R, Piepho HP, Rademacher M: **Estimates of the basal ileal endogenous losses of amino acids by regression analysis and determination of standardised ileal amino acids digestibilities from casein in newly weaned piglets.** *J Sci Food Agric* 2008, **88:**641–651.
5. Eklund M, Mosenthin R, Piepho HP, Rademacher M: **Effect of dietary protein crude level on basal ileal endogenous losses and standardized ileal digestibilities of crude protein and amino acids in newly weaned pigs.** *J Anim Physiol Anim Nutr* 2008, **92:**578–590.
6. Mariscal-Landín G, Reis de Souza TC: **Endogenous ileal losses of nitrogen and amino acids in pigs and piglets fed graded levels of casein.** *Arch Anim Nutr* 2006, **60:**454–466.
7. Apolônio LR, Donzele JL, de Oliveira RFM, Silva FCO, Lopes DC, Freitas LS: **Digestibilidade ileal de aminoácidos de alimentos utilizados em dietas pré-iniciais para leitões, determinada pelo método do sacrificio.** *R Bras Zootec* 2002, **31:**1983–1998.
8. de Lange CFM, Sauer WC, Souffrant WB: **The effect of protein status of the pig on the recovery and amino acid composition of endogenous protein in digesta collected from the distal ileum.** *J Anim Sci* 1989, **67:**755–762.
9. Hodgkinson SM, Moughan PJ, Reynolds GW: **Effect of the duration of feeding of a protein-free diet on endogenous ileal nitrogen and amino acid loss in the growing pig.** *J Sci Food Agric* 2000, **80:**1407–1412.
10. Mariscal-Landín G, Reis de Souza TC, Parra SJE, Aguilera BA, Mar BB: **Ileal digestibility of protein and amino acids from canola meal in weaned piglets and growing pigs.** *Livest Sci* 2008, **116:**53–62.
11. Mariscal-Landín G, Reis de Souza TC, Ávalos MA: **Ileal amino acids digestibility of sorghum in weaned piglets and growing pigs.** *Animal* 2010, **4:**1341–1348.
12. Diario Oficial de la Federación: *Especificaciones técnicas para la producción, cuidado y uso de los animales de laboratorio*, Norma Oficial Mexicana NOM-062-ZOO-1999. México D.F: Diario Oficial de la Federación; 2001.
13. **International guiding principles for biomedical research involving animals.** 1985:. The development of science-based guidelines for laboratory animal care - NCBI Bookshelf. [http://cioms.ch/publications/guidelines/1985_texts_of_guidelines.htm]
14. Reis de Souza TC, Mar BB, Mariscal LG: **Canulación de cerdos posdestete para pruebas de digestibilidad ileal: Desarrollo de una metodología.** *Téc Pecu Méx* 2000, **38:**143–150.
15. NRC: *Nutrient requirements of swine.* 10th edition. Washington, DC: The National Academy Press; 1998.
16. INRA: *L'alimentation des animaux monogastriques: porc, lapin, volailles.* Paris, France: Institut National de la Recherche Agronomique; 1984.
17. AOAC: *Official methods of analysis.* 17th edition. Arlington, VA: Assoc. Offic. Anal. Chem.; 2000.
18. Fenton TW, Fenton M: **An improved procedure for determination of chromic oxide in feed and feces.** *Can J Anim Sci* 1979, **59:**631–634.
19. Henderson JH, Ricker RD, Bidlingmeyer BA, Woodward C: *Rapid, accurate and reproducible HPLC analysis of amino acids. Amino acid analysis using Zorbax Eclipse AAA columns and the Agilent 1100 HPLC.* USA: Agilent technologies; 2000. Part No5980-1193E 2000:10 pag. Agilent Technologies home page at: www.agilent.com/chem/supplies.
20. Fan MZ, Sauer WC: **Determination of apparent ileal amino acid digestibility in barley and canola meal for pigs with the direct, difference, and regression methods.** *J Anim Sci* 1995, **73:**2364–2374.
21. Chung TK, Baker DH: **Apparent and true amino acid digestibility of a crystalline amino acid mixture and of casein: comparison of values obtained with ileal-cannulated pigs and cecectomized cockerels.** *J Anim Sci* 1992, **70:**3781–3790.
22. Urbaityte R, Mosenthin R, Eklund M, Piepho HP, Rademacher M: **Determination of standardized ileal crude protein and amino acid digestibilities in protein supplements for piglets.** *Animal* 2009, **3:**1696–1705.
23. SAS version 9.2: *Book statistical analysis systems institute User's guide.* 9.2nd edition. SAS Institute Inc; 2008.
24. Steel RGD, Torrie JH: *Principles and procedures of statistics. A biometrical approach.* 2nd edition. New York: McGraw-Hill; 1980.
25. Wu G, Davis PK, Flynn NE, Knabe DA, Davidson T: **Endogenous synthesis of arginine plays an important role in maintaining arginine homeostasis in postweaning growing pigs.** *J Nutr* 1997, **127:**2342–2349.
26. Dekaney CM, Wu G, Jaeger LA: **Ornithine aminotransferase messenger RNA expresion and enzymatic activity in fetal porcine intestine.** *Pediatr Res* 2001, **50:**104–109.
27. Bertolo RFP, Brunton JA, Pencharz PB, Ball RO: **Arginine, ornithine, and proline interconversion is dependent on small intestinal metabolism in neonatal pigs.** *Am J Physiol Endocrinol Metab* 2003, **284:**E915–E922.
28. Stein HH, Trottier NL, Bellaver C, Easter RA: **The effect of feeding level and physiological status on total flow and amino acid composition of endogenous protein at the distal ileum in swine.** *J Anim Sci* 1999, **77:**1180–1187.
29. Leibholz J: **The flow of endogenous nitrogen in the digestive tract of young pigs.** *Br J Nutr* 1982, **48:**509–517.
30. Darragh AJ, Moughan PJ: **The amino acid composition of human milk corrected for amino acid digestibility.** *Br J Nutr* 1998, **80:**25–34.
31. Juste C: **Apports endogènes par les secretions digestives chez le porc.** In *Les Colloques de l'INRA.* Edited by Laplace JP, Corring T, Rérat A. Paris: INRA; 1982:155–173.
32. Nyachoti CM, de Lange CFM, McBride BW, Schulze H: **Significance of endogenous gut nitrogen losses in the nutrition of growing pigs: a review.** *Can J Anim Sci* 1997, **77:**149–163.
33. Wu G, Bazer FW, Davis TA, Jaeger LA, Johnson GA, Kim SW, Knabe DA, Meininger CJ, Spencer TE, Yin Y-L: **Important roles for the arginine family of amino acids in swine nutrition and production.** *Livest Sci* 2007, **112:**8–22.
34. Jørgensen H, Gabert VM: **The ileal digestion of semipurified diets containing increasing levels of casein and the use of the regression method to estimate the endogenous flow of amino acids at the distal ileum of growing pigs.** In *Digestive physiology in pigs.* Edited by Lindberg JE, Ogle B. New York: CABI publishing; 2001:198–201.
35. Valette P, Malouin H, Corring T, Savoie L, Gueugneau AM, Berot S: **Effects of diets containing casein and rapeseed on enzyme secretion from the exocrine pancreas in the pig.** *Br J Nutr* 1992, **67:**215–222.

5

Effects of dietary tryptophan and stocking density on the performance, meat quality, and metabolic status of broilers

Bo Wang, Zhizhi Min, Jianmin Yuan*, Bingkun Zhang and Yuming Guo

Abstract

Background: Highly automated cage-rearing systems are becoming increasingly popular in China. However, a high stocking density can cause oxidative stress and decrease broiler performance. The tryptophan (TRP) derivative 5-hydroxytryptophan (5-HT) has been shown to preserve membrane fluidity in birds suffering from oxidative stress. Therefore, this experiment was conducted to determine the effects of dietary TRP supplementation on performance, breast meat quality and oxidative stress in broilers reared in cages with a high or low stocking density.

Methods: Female Arbor Acres broilers (25-d-old, n = 144) were randomly allocated to 1 of 4 treatments. The birds were fed a diet based on corn, soybean meal, cottonseed meal and corn gluten meal containing either 0.18 or 0.27% TRP and were housed with stocking densities of 11 or 15.4 birds/m^2 in a 2 × 2 factorial experiment. Broiler performance was evaluated from d 25 to 42. Eight birds from each treatment were slaughtered on d 42 and plasma and breast muscle samples were collected to measure biochemical indices.

Results: A higher stocking density tended to be associated with reduced weight gain ($P < 0.10$), and significantly increased plasma glutamic-pyruvic transaminase (GPT) activity ($P < 0.001$). Increased dietary TRP significantly reduced the activities of lactic dehydrogenase and GPT while increasing total cholesterol in the plasma ($P < 0.01$), reducing drip loss of breast muscle ($P < 0.10$) and improving feed efficiency ($P < 0.10$).

Conclusions: An increase in dietary TRP, 1.5-fold higher than the standard supplementation level, can alleviate oxidative stress as well as improve welfare and feed efficiency in broilers reared in cages with a high stocking density.

Keywords: Broiler, Meat quality, Performance, Stocking density, Tryptophan

Background

Increased attention is being paid to animal welfare and meat quality, factors that are strongly affected by stocking density [1]. A low stocking density is thought to be associated with high product quality, although it produces less meat and can therefore result in economic losses [1]. In contrast, a high stocking density can increase the quantity of broiler meat produced per unit area [2]. However, previous studies have shown that high stocking densities affect the amount of time broilers spend walking [3], decrease locomotor activity [4,5], as well as increase the temperature, moisture, and ammonia concentrations in the chicken house, and cause leg and hock problems [6-8]. In addition, a high stocking density can increase disturbances [5] and conflicts between birds [9], and lead to apparent fearfulness [8,10]. Other problems associated with a high stocking density include increased heterophil: lymphocyte ratios [11], and increased plasma concentrations of glutathione peroxidase, leading to oxidative stress [4,8,12]. Thus, high stocking densities result in reduced growth, impaired feed efficiency, and poor quality of life for broilers [4,13].

Tryptophan (TRP), the rate-limiting substrate of 5-hydroxytryptamine (5-HT), is synthesized in the brain and competes with large neutral amino acids (LNAA) to cross the blood–brain barrier [14]. TRP supplementation significantly increased 5-HT turnover in the hippocampus and archistriatum and tended to do the same in the remainder of the forebrain [14]. Previous studies have shown

* Correspondence: yuanjm@cau.edu.cn
State Key Laboratory of Animal Nutrition, College of Animal Science and Technology, China Agricultural University, Beijing 100193, China

that birds fed low-TRP diets exhibited abnormal behavior and signs of stress [15]. An increase in dietary TRP has been shown to reduce aggression, and alleviate stress in livestock and poultry [16,17], reduce feather-pecking behavior and reduce hysteria in layers [18].

Excessive TRP reduced weight gain in weaned piglets [19] and rats [20]. However, no studies have investigated the effects of stocking density in relation to different concentrations of dietary TRP especially in broilers raised using cage rearing systems. Multi-storey cages or highly automated cage-rearing systems are becoming increasingly popular in China and other countries because they enable convenient management and removal of excreta and minimize bird contact with wastes, which reduces ammonia concentrations and improves chicken welfare. However, whether TRP supplementation can reduce stress and improve the welfare of broilers and quality of meat produced under high stocking densities is unknown. In this study, the effect of TRP supplementation on performance, breast meat quality, and the oxidative stress status of broilers reared in cages under high stocking density was investigated.

Materials and methods

Bird management

This protocol used in this study was approved and conducted in accordance with the guidelines of the Animal Ethics Committee of China Agricultural University (Beijing, China). Day-old female broiler chicks (Arbor Acres Plus), obtained from a commercial hatchery (Aviagen Group, Beijing, China), were raised in the middle floor of a 3 tier set of overlap cages. Individual cages were 70 cm long × 65 cm wide × 38 cm tall, and contained two nipple drinkers. A galvanized iron trough, hanging outside of the cage, was used to feed the broilers. Excreta was collected in trays under each cage, and was removed every 2 or 3 d to decrease ammonia concentrations in the broiler house.

Broilers were raised according to the Arbor Acres Broiler Management Guide (Aviagen Group, Beijing, China). The temperature was maintained at 34°C on day 1, and was gradually reduced to 22°C by day 21. Feed was offered ad libitum and water was freely available at all times. The relative humidity was maintained at 40–65%. The experiment was performed during winter and open windows were used to achieve cross-ventilation.

Experimental design and diets

The trial used a 2 × 2 factorial arrangement, involving 2 stocking densities and 2 concentrations of TRP. The normal stocking density was 5 birds/cage (11 birds/m^2), and the high stocking density was 7 birds/cage (15.4 birds/m^2), according to Sørensen et al. [21]. The normal TRP concentration was defined as 0.18%, on the basis of NRC (1994) nutritional requirements for broilers [22], and the high TRP concentration was 0.27% (1.5-fold higher than the normal concentration). Since the effects of stocking density occur primarily during the growing period [6], the experiment commenced when the broilers were 25 days of age. From days 1 to 24, the birds were fed a commercially prepared mash diet. On day 25, the birds were individually weighed, and 144 birds with similar body weights, were randomly assigned to 24 cages. The difference in the average body weight among the cages was <25 g. From days 25 to 42, birds were fed the experimental diets in mash form (Table 1).

Measurements

Broiler performance

From days 25 to 42, mortality, culling and feed intake were recorded daily for each pen. On day 42, BW was recorded on a per cage basis to calculate the final BW and BW gain. Daily feed intake and feed efficiency were corrected for mortality.

Metabolic status

On day 42, blood samples were collected into heparinized tubes from the wing vein of 8 birds per treatment. The samples were centrifuged for 10 min at 3,000 × *g*, and plasma was collected and frozen at −40°C for subsequent analysis of lactic dehydrogenase (LDH), glutamic-pyruvic transaminase (GPT), glutamic-oxalacetic transaminase (GOT), glucose (GLU), triglyceride (TG), and total cholesterol (TC). The analyses were performed using commercial analytical kits according to the manufacturer's recommendations (Jian Cheng Bioengineering Institute, Nanjing, China).

Breast meat quality

On day 42, 8 birds closest to the average BW for each treatment were slaughtered. The left breast muscle was dissected, and breast muscle yield (without bone or skin) was measured as the ratio of 2 times the Pectoralis major and Pectoralis minor muscle mass to the final BW of the bird. The pH_{45} min post-mortem (pH_{45}) was measured using a TESTO 205 pH meter (Testo Limited, Alton, Hampshire, UK) by inserting the electrode into three different points of the right section of the Pectoralis muscle and the average values were used for statistical analysis. Crude fat was measured by ether extraction of a 20-g sample of breast meat. An additional 20 g sample was placed in a polyethylene bag and drip loss was determined after 2 d of storage at 4°C [23].

Statistical analysis

The data was analyzed to assess the interactive effects of stocking density and TRP concentrations using the General Linear Model (GLM) of SPSS 15.0 [24]. Percentage

Table 1 Composition and nutrient level of diets (air-dry basis)

Item	0.18% TRP	0.27% TRP
Ingredients, %		
Corn	64.64	64.54
Soybean meal	16.12	16.12
Cottonseed meal	6.00	6.00
Corn gluten meal	6.00	6.00
Soybean oil	3.23	3.23
Limestone	1.15	1.15
Dicalcium phosphate	1.56	1.56
NaCl	0.35	0.35
Vitamin premix[1]	0.02	0.02
Trace mineral premix[2]	0.20	0.20
DL-Met (98%)	0.04	0.04
L-Lys · HCl (78%)	0.40	0.40
L-Thr	0.06	0.06
Choline chloride (50%)	0.16	0.16
L-Trp (98%)	0.01	0.10
Nutrient levels	0.06	0.06
AME, Mcal/kg		
Crude protein,%	3.10	3.10
Calcium, %	19.00	19.00
Available phosphorus, %	0.88	0.88
Nutrient levels	0.40	0.40
Amino acid content		
Lys, %	1.00	1.00
Met, %	0.38	0.38
Thr, %	0.64	0.64
Trp, %	0.18	0.27

[1]The vitamin premix provided the following per kilogram of diet: vitamin A, 10,000 IU; vitamin D_3, 2,400 IU; vitamin E, 20 IU; vitamin K_3, 2.00 mg; thiamin, 2.00 mg; riboflavin, 6.40 mg; pyridoxin, 3.00 mg; VB_{12}, 0.02 mg; folic acid, 1.00 mg; pantothenic acid, 10.00 mg; nicotinic acid, 30.00 mg; biotin, 0.10 mg.
[2]The mineral premix provided the following per kg of diet: Cu, 8 mg; Fe, 80 mg; Zn, 60 mg; Mn, 100 mg; I, 0.35 mg.

data was converted to degrees before Analysis of Variance was conducted and was subsequently back-transformed to generate an estimate of the standard error of the mean (SEM). Significance was set at $P < 0.05$, and trends were recognized at $P < 0.10$.

Results

Performance

Only one bird died during the experiment and as a result, mortality is not shown in the tables. There were no significant differences in the initial bodyweight among the experimental groups. However, increased stocking density tended to reduce the final BW and bodyweight gain ($P < 0.10$) (Table 2). There were no significant differences in feed efficiency or feed intake. The addition of dietary TRP tended to decrease feed efficiency ($P <$ 0.10) but had no effect on BW gain or feed intake. There were no significant interactions between dietary TRP concentrations and stocking density on the final BW, BW gain, or feed intake. However, high stocking density in combination with increased TRP supplementation significantly lowered feed efficiency compared with a high stocking density with normal TRP supplementation ($P < 0.05$), while a lower stocking density in combination with increased TRP supplementation did not significantly affect feed efficiency.

Breast meat quality

Increased stocking density had no significant effect on breast muscle yield, fat content, pH_{45}, or drip loss (Table 3). Increased TRP concentration tended to reduce drip loss in the breast muscle of broilers ($P < 0.10$). There was no influence of TRP concentration on breast muscle yield, breast fat content, or pH_{45}. Interactions between TRP concentration and stocking density were not significant for fat content, pH_{45}, or drip loss. However, the high stocking density in combination with the low level of TRP supplementation tended to produce significantly higher breast muscle yield compared with the other treatments.

Metabolic status

A high stocking density tended to increase plasma GLU ($P < 0.10$), and significantly increased plasma GPT activity ($P < 0.001$). There were no significant differences in the activity of LDH and GOT or in the plasma levels of TG and TC.

Increased dietary TRP significantly increased plasma TC ($P < 0.05$), and decreased LDH and GPT activities ($P < 0.01$). There were no significant differences in GOT activity or in GLU and TG levels between the TRP treatments. Significant interactive effects were observed between TRP concentration and stocking density for plasma GLU ($P < 0.01$), LDH activity ($P < 0.05$), and GPT activity ($P < 0.001$). A high stocking density in combination with increased dietary TRP significantly increased GLU content compared with the other treatments and significantly decreased GPT activity compared with the normal TRP treatment with a high stocking density, although the level did not differ significantly from the low stocking density treatments. A high stocking density in combination with increased TRP supplementation significantly decreased LDH activity compared with the low-TRP treatment, but showed no significant differences from the low stocking density treatment.

Table 2 Effects of stocking density and tryptophan (TRP) level on broiler performance[1]

Stocking density, birds/m^2	Tryptophan Level, %	Initial Bodyweight, g	Final Bodyweight, g	Daily bodyweight Gain, g	Feed Intake, g/d	Feed efficiency
Stocking density × tryptophan level						
11.0	0.18	778	1,960	69.53	127.2	1.83^b
11.0	0.27	780	1,961	69.47	126.5	1.82^b
15.4	0.18	781	1,887	65.00	123.5	1.90^a
15.4	0.27	781	1,939	68.06	123.2	1.81^b
SEM		5.6	76.2	4.250	6.30	0.059
Stocking density						
11.0		779	1,963	69.65	126.8	1.82
15.4		781	1,913	66.59	123.4	1.85
SEM		1.1	16.4	0.900	1.28	0.012
Tryptophan level						
0.18%		780	1,924	67.24	125.3	1.86
0.27%		782	1,950	68.76	124.9	1.82
SEM		1.1	15.5	0.057	1.28	0.015
P-values						
Stocking density		0.351	0.084	0.089	0.214	0.144
TRP level		0.891	0.564	0.509	0.851	0.052
Density × TRP		0.659	0.265	0.375	0.938	0.045

[1]Each mean represents 6 replicates for each treatment with each replicate containing 5 (11.0 birds/m^2) or 7 birds (15.4 birds/m^2).
[a,b]Means in the same row with different superscripts are significantly different ($P < 0.05$).

Discussion

In this study, increased stocking density tended to reduce weight gain of broilers (Table 2), a finding that is consistent with previous studies that demonstrated that a high stocking density is associated with decreased BW [4], especially during the growing period [7]. However, Estevez [25] suggested that the recommended stocking density could be increased to approximately 18 birds/m^2 without any major adverse effects on the final BW. Ekstrand and Carpenter [26] proposed that the response to changes in stocking density depended on environmental variables. Differences in the environment provided to chickens by poultry producers might have greater effects on welfare than stocking density [27].

Clements [28] reviewed numerous articles on stocking density and animal welfare and concluded that there was no significant relationship between stocking density and behavior, leg problems, or skin damage. Under deep litter conditions, growth rate was significantly reduced when stocking density exceeded approximately 30 kg/m^2 live weight [7]. However, the effects of litter and stocking density on growth were attenuated when ventilation was increased or when broilers were raised with perforated flooring.

Increased litter temperature has been identified as a causal factor for reduced growth rate [29]. High stocking densities might contribute to reduced broiler performance because of high temperatures and reduced airflow at the bird level [26]. The reduction in BW gain at a stocking density of 15.4 birds/m^2 compared with that at a stocking density of 11 birds/m^2 might be related to the cage environment (e.g., air flow, temperature, and cage area). Cage area might affect broiler performance, which also affects locomotor activity [4,5]. Suitable stocking density when birds are raised in cages might differ from the stocking density required when birds are raised with open litter floor or perforated flooring. The suitable stocking density of broilers raised in 3-tier-overlap cages would be <15.4 birds/m^2 if simple cross ventilation were used by opening windows during the winter.

Increasing the stocking density of female birds from 11 to 15.6 birds/m^2 did not affect breast meat yield (Table 3), which is in agreement with the findings of Moreira et al. [30] who showed that densities between 10 and 16 birds/m^2 did not affect broiler meat quality characteristics (cooking loss, shear force, and pH) in Ross 308, Cobb 500, and Hybro PG commercial strains. Thomas et al. [10] also showed that stocking densities of 10, 15, and 20 birds/m^2 had no influence on carcass characteristics. However, Rilgili and Hess [31] found that increased stocking density could significantly reduce

Table 3 Effects of stocking density and tryptophan level on breast meat quality of broilers[1]

Stocking density, birds/m^2	Tryptophan level, %	Body weight, kg	Breast muscle			
			Yield, %	Fat content, %	pH$_{45}$[2]	Drip loss, %
Stocking density × tryptophan level						
11.0	0.18	1.87	14.25	0.79	6.07	9.17
11.0	0.27	1.98	14.82	0.69	6.11	8.50
15.4	0.18	1.87	15.45	0.66	6.16	11.60
15.4	0.27	1.82	14.43	0.67	6.13	8.16
SEM		0.228	1.083	0.121	0.195	2.502
Stocking density						
11.0		1.93	14.54	0.74	6.09	8.84
15.4		1.85	14.94	0.66	6.14	9.63
SEM		0.046	0.173	0.025	0.040	0.511
Tryptophan level						
0.18%		1.87	14.85	0.73	6.11	10.14
0.27%		1.90	14.62	0.68	6.12	8.33
SEM		0.046	0.173	0.025	0.040	0.511
***P* Value**						
Stocking density		0.405	0.356	0.130	0.533	0.425
TRP level		0.745	0.599	0.351	0.907	0.077
Density × TRP		0.405	0.077	0.302	0.710	0.258

[1]Each mean represents 8 birds for each treatment.
[2]Note: pH$_{45}$ represents pH value for breast meat 45 min after slaughter.

carcass quality and breast meat yield in fast-growing male birds, whereas slow-growing female birds were not affected. Simitzis et al. [4] showed that high stocking density decreased intramuscular breast fat.

The effects of stocking density on meat quality might be related to growth rate. Puron et al. [2], showed that similar masses of meat were produced when housing density exceeded 17 male or 19 female birds/m^2 which was close to the densities recommended to minimize the adverse effects of crowding on broiler welfare. The stocking densities of female broilers in the current study (11 and 15.4 birds/m^2) were lower than those used by Puron et al. [2], and slow growing female broilers were used in the current study, similar to the experiment of Rilgili and Hess [31] which might explain the lack of significant differences in breast meat quality between the 2 stocking densities.

Simsek et al. [12] compared stocking densities of 6 and 13 birds/m^2 and found that a high stocking density decreased feeding time and caused oxidative stress. Simitzis et al. [4] showed that a high stocking density was associated with decreased locomotor activity and increased indicators of physiological (heterophil: lymphocyte ratio and bursa weight), and oxidative stress. However, other researchers have shown that stocking density did not affect the major physiological stress indicators, including plasma glucose and cholesterol [32] and fecal corticosterone concentrations [8,27]. In this study, a high stocking density did not influence the levels of TG or TC or the activities of LDH and GOT in the plasma, which is in agreement with previous studies [7,33]. However, an increased stocking density significantly increased GPT activity in the plasma.

Serum GPT and GOT are commonly measured clinically as a part of diagnostic liver-function tests to determine liver health. Levels of serum GPT and GOT generally increase with muscle or liver damage [34]. The significantly increased GPT activity indicates that a high stocking density might cause oxidative lesions, which is in agreement with the findings of Simsek et al. [14]. However, whether or not increased stocking density causes oxidative stress with no change in LDH activity related to the stress level needs further research.

Previous studies have shown that the TRP derivative 5-hydroxytryptophan could help preserve membrane fluidity in animals experiencing oxidative stress [35]. Another important TRP derivative, melatonin, was found to have anti-oxidative properties [36,37]. In the current study, increased dietary TRP significantly decreased the TC and the activities of LDH and GPT in plasma

Table 4 Effects of stocking density and tryptophan (TRP) level on plasma parameters of broilers[1]

Stocking density, birds/m^2	Tryptophan Concentration, %	LDH[2], U/L	GLU, mg/dL	TG, mg/dL	TC, mg/dL	GPT, U/L	GOT, U/L
Stocking density × tryptophan level							
11.0	0.18	24,219^{ab}	239.1^{b}	39.49	106.1	2.88^{b}	32.92
11.0	0.27	22,317^{b}	228.2^{b}	51.48	121.2	3.83^{b}	31.95
15.4	0.18	28,028^{a}	232.6^{b}	48.30	104.8	9.10^{a}	32.60
15.4	0.27	19,695^{b}	256.9^{a}	58.97	121.2	5.05^{b}	32.33
SEM		5,106.0	16.66	17.956	16.67	2.595	4.575
Stocking density							
11.0		23,268	233.6	45.49	113.6	3.36	32.43
15.4		23,861	244.8	53.63	113.0	7.08	32.47
SEM		1,042.3	3.40	3.665	3.40	0.529	0.934
TRP level							
0.18%		26,123	235.9	43.89	105.4	5.99	32.76
0.27%		21,006	242.5	55.23	121.2	4.44	32.14
SEM		1,042.3	3.40	3.665	3.40	0.529	0.934
***P* value**							
Stocking density		0.742	0.053	0.272	0.925	<0.001	0.987
TRP level		0.009	0.232	0.132	0.023	0.001	0.761
Density × TRP		0.024	0.004	0.928	0.925	<0.001	0.863

[1]Each mean represents 8 birds for each treatment.
[2]*LDH*: lactic dehydrogenase; *GPT*: glutamic-pyruvic transaminase; *GOT*: glutamic-oxalacetic transaminase; *GLU*: glucose; *TG*: triglyceride; *TC*: total cholesterol.
[a,b]Means in the same row with different superscripts are significantly different ($P < 0.05$).

(Table 4). Increased plasma GPT activity under a high stocking density was moderated by the increase in dietary TRP concentration, suggesting that increased dietary TRP could help prevent liver injury resulting from a high stocking density.

The cytoplasmic enzyme LDH is widely used as a marker of organ or tissue lesions in toxicology and clinical chemistry. LDH is generally associated with cellular metabolic activity, which is inhibited or elevated under oxidative stress [38]. The significant decrease in LDH activity observed in the present study with dietary supplementation of 0.27% of TRP under the high stocking density conditions, indicates that TRP level can reduce the stress caused by a high stocking density.

Corzo et al. [15], showed that supplemental TRP increased plasma glucose in a linear manner, whereas other physiological stress variables were unaffected by dietary TRP. Plasma glucose exhibited a linear response to TRP supplementation, most likely due to adrenergic-driven glycolytic factors, unlike the gluconeogenic events associated with physiological stress [34].

Previous study showed that a high stocking density increased plasma concentrations of glutathione peroxidase, leading to oxidative stress [4,8,12]. TRP treatment significantly increased the TRP/LNAA ratio in the plasma of the chicks. Furthermore, TRP treatment increased baseline and stress-induced levels of plasma corticosterone, which inhibits glycolysis [16]. Supplementation with the high concentration of TRP significantly increased TC levels compared with low TRP concentrations (Table 4), which also indicates that increased TRP could help alleviate oxidative stress. This might explain why high concentrations of TRP tended to alter feed efficiency. This indicates that increased TRP concentration could help to improve feed efficiency under high stocking densities.

Drip loss can indicate quality deviations associated with the rate and extent of postmortem glycolysis in muscle tissue. The volume of drip loss is related to the lipid peroxide content in muscle [39]. Our finding that high TRP supplementation reduced drip loss under a high stocking density indicates that administration of TRP could lead to a reduction of lipid peroxide and glycolysis values caused by stress due to a high stocking density, thus improving meat quality.

In conclusion, increasing the stocking density from 11 to 15.4 birds/m^2 caused oxidative stress in broilers and tended to reduce their performance. Supplementation with TRP at concentrations 1.5-fold higher than the standard dose commonly supplemented could alleviative oxidative stress caused by stocking density and could improve broiler welfare, meat quality, and feed efficiency.

Abbreviations

BW: Body weight; GLU: Glucose; GOT: Glutamic-oxalacetic transaminase; GPT: Glutamic-pyruvic transaminase; LDH: Lactic dehydrogenase; LNAA: Large neutral amino acids; TC: Total cholesterol content; TG: Triglyceride; TRP: Tryptophan.

Competing interests

The authors declare that they have no competing interests.

Authors' contributions

BW and ZM carried out the experiments, JY wrote the manuscript, BZ, and YG participated in the design of the study. All authors read and approved the final manuscript.

Acknowledgments

This research was supported by the System for Poultry Production Technology, Beijing Innovation Research Team of Modern Agriculture (CARS-PSTP), and the National Key Technology R&D Program of China during the 12th five-year plan (2012BAD39B04).

References

1. Vanhonacker F, Verbeke W, Van Poucke E, Buijs S, Tuyttens FAM: **Societal concern related to stocking density, pen size and group size in farma animal production.** *Livestock Sci* 2008, **113**:123–132.
2. Puron D, Santamaria R, Segura JC, Alamilla JL: **Broiler performance at different stocking densities.** *J Appl Poult Res* 1995, **4**:55–60.
3. Febrer K, Jones TA, Donnelly CA, Dawkins MS: **Forced to crowd or choosing to cluster? Spatial distribution indicates social attraction in broiler chickens.** *Anim Behav* 2006, **72**:1291–1300.
4. Simitzis PE, Kalogeraki E, Goliomytis M, Charismiadou MA, Triantaphyllopoulos K, Ayoutanti A, Niforou K, Hager-Theodorides AL, Deligeorgis SG: **Impact of stocking density on broiler growth performance, meat characteristics, behavioural components and indicators of physiological and oxidative stress.** *Br Poult Sci* 2012, **53**:721–730.
5. Ventura BA, Siewerdt F, Estevez I: **Access to barrier perches improves behavior repertoire in broilers.** *PLoS One* 2012, **7**:e29826. doi:10.1371/journal.pone.0029826. Epub 2012 Jan 27.
6. Dozier WA 3rd, Thaxton JP, Branton SL, Morgan GW, Miles DM, Roush WB, Lott BD, Vizzier-Thaxton Y: **Stocking density effects on growth performance and processing yields of heavy broilers.** *Poult Sci* 2005, **84**:1332–1338.
7. Dozier WA 3rd, Thaxton JP, Purswell JL, Olanrewaju HA, Branton SL, Roush WB: **Stocking density effects on male broilers grown to 1.8 kilograms of body weight.** *Poult Sci* 2006, **85**:344–351.
8. Buijs S, Keeling L, Rettenbacher S, Van Poucke E, Tuyttens FAM: **Stocking density effects on broiler welfare: Identifying sensitive ranges for different indicators.** *Poult Sci* 2009, **88**:1536–1543.
9. Yakubu A, Gwaska JA, Salako AE: **Strain and placement density effects on welfare, haematological and serum biochemical indices of broilers in north central Nigeria.** *Acta Agric Slovenica* 2009, **94**:153–158.
10. Thomas DG, Ravindran V, Thomas DV, Camden BJ, Cottam YH, Morel PC, Cook CJ: **Influence of stocking density on the performance, carcass characteristics and selected welfare indicators of broiler chickens.** *N Z Vet J* 2004, **52**:76–81.
11. Puvadolpirod S, Thaxton JP: **Model of physiological stress in chickens 1. Response parameters.** *Poult Sci* 2000, **79**:363–369.
12. Simsek UG, Dalkılıc B, Ciftci M, Yuce A: **The influences of different stocking densities on some welfare indicators, lipid peroxidation (MDA) and antioxidant enzyme activities (GSH, GSH-Px, CAT) in broiler chickens.** *J Anim Vet Adv* 2009, **8**:1568–1572.
13. Cortamira NO, Seve B, Lebreton Y, Ganier P: **Effect of dietary tryptophan on muscle, liver and whole-body protein-synthesis in weaned piglets - Rrelationship to plasma-insulin.** *Br J Nutr* 1991, **66**:423–435.
14. Van Hierden YM, Koolhaas JM, Korte SM: **Chronic increase of dietary l-tryptophan decreases gentle feather pecking behaviour.** *Appl Anim Behav Sci* 2004, **89**:71–84.
15. Corzo A, Kidd MT, Thaxton JP, Kerr BJ: **Dietary tryptophan effects on growth and stress responses of male broiler chicks.** *Br Poult Sci* 2005, **46**:478–484.
16. Shea MM, Mench JA, Thomas OP: **The effect of dietary tryptophan on aggressive behavior in developing and mature broiler breeder males.** *Poult Sci* 1990, **69**:1664–1669.
17. Mench JA: **Research note: Feed restriction in broiler breeders causes a persistent elevation in corticosterone secretion that is modulated by dietary tryptophan.** *Poult Sci* 1991, **70**:2547–2550.
18. Laycock SR, Ball RO: **Alleviation of hysteria in laying hens with dietary tryptophan.** *Can J Vet Res* 1990, **54**:291–295.
19. Edmonds MS, Baker DH: **Methionine, tryptophan, threonine or leucine amino acid excesses for young pigs: Effects of excess.** *J Anim Sci* 1987, **64**:1664–1671.
20. Gross B, Ronen N, Honigman S, Livne E: **Tryptophan toxicity-time and dose response in rats.** *Adv Exp Med Biol* 1999, **467**:507–516.
21. Sørensen P, Su G, Kestin SC: **Effects of age and stocking density on leg weakness in broiler chickens.** *Poult Sci* 2000, **79**:864–870.
22. Council NR: *Nutrient Requirement of Poultry.* 9th edition. Washington, DC: National Academy Press; 1994.
23. Lundstrom K, Malmfors G: **Variation in light scatter and water-holding capacity along the porcine longissimus doris muscle.** *Meat Sci* 1985, **15**:203–214.
24. *Spss 15.0 Command Syntax Reference.* Chicago Ill: SPSS Inc; 2006.
25. Estevez I: **Density allowances for broilers: Where to set the limits?** *Poult Sci* 2007, **86**:1265–1272.
26. Ekstrand C, Carpenter TE: **Temporal aspects of foot-pad dermatitis in Swedish broilers.** *Acta Vet Scand* 1998, **39**:229–236.
27. Dawkins MS, Donnelly CA, Jones TA: **Chicken welfare is influenced more by housing conditions than by stocking density.** *Nature* 2004, **427**:342–344.
28. Clements M: **Is the poultry industry making the most of its data?** *Poult Int* 2013, **52**:6–12.
29. Feddes JJR, Emmanuel EJ, Zuidhof MJ: **Broiler performance, BW variance, feed and water intake, and carcass quality at different stocking densities.** *Poult Sci* 2002, **81**:774–779.
30. Moreira J, Mendes AA, Roca RO, Garcia EA, Naas IA, Garcia RG, Lima IC, Paz A: **Effect of stocking density on performance, carcass yield and meat quality in broilers of different commercial strains.** *R Bras Zootec* 2004, **33**:1506–1519.
31. Rilgili SF, Hess JB: **Placement density influences broiler carcass grade and meat yields.** *J Appl Poult Res* 1995, **4**:384–389.
32. Thaxton JP, Dozier WA 3rd, Branton SL, Morgan GW, Miles DW, Roush WB, Lott BD, Vizzier-Thaxton Y: **Stocking density and physiological adaptive responses of broilers.** *Poult Sci* 2006, **85**:819–824.
33. Škrbić Z, Pavlovski Z, Lukić M, Perić L, Milošević N: **The effect of stocking density on certain broiler welfare parameters.** *Biotechnol Anim Husbandry* 2009, **25**:11–21.
34. Özbey O, Esen F: **The effects of breeding system and stocking density on some blood parameters of rock partridges (Alectoris graeca).** *Poult Sci* 2007, **86**:420–422.
35. Reyes-Gonzales MC, Fuentes-Broto L, Martínez-Ballarín E, Miana-Mena FJ, Berzosa C, García-Gil FA, Aranda M, García JJ: **Effects of tryptophan and 5-hydroxytryptophan on the hepatic cell membrane rigidity due to oxidative stress.** *J Membr Biol* 2009, **231**:93–101.
36. Ghosh G, De K, Maity S, Bandyopadhyay D, Bhattacharya S, Reiter RJ, Bandyopadhyay A: **Melatonin protects against oxidative damage and restores expression of GLUT4 gene in the hyperthyroid rat heart.** *J Pineal Res* 2007, **42**:71–82.
37. Mollaoglu H, Topal T, Ozler M, Uysal B, Reiter RJ, Korkmaz A, Oter S: **Antioxidant effects of melatonin in rats during chronic exposure to hyperbar.** *J Pineal Res* 2007, **42**:50–54.
38. Das PC, Ayyappan S, Das BK, Jena JK: **Ni- trite toxicity in Indian major carps: Sublethal effect on selected enzymes in fingerlings of Catla catla, Labeo rohita, and Cirrhi- nus mrigala.** *Comp Biochem Physiol C Toxicol Pharmacol* 2004, **138**:3–10.
39. Tavárez MA, Boler DD, Bess KN, Zhao J, Yan F, Dilger AC, Mckeith FK, Killefer J: **Effect of antioxidant inclusion and oil quality on broiler performance, meat quality, and lipid oxidation.** *Poult Sci* 2011, **90**:922–930.

6

Predicting corn digestible and metabolizable energy content from its chemical composition in growing pigs

Quanfeng Li, Jianjun Zang, Dewen Liu, Xiangshu Piao, Changhua Lai and Defa Li*

Abstract

Background: The nutrient composition of corn is variable. To prevent unforeseen reductions in growth performance, grading and analytical methods are used to minimize nutrient variability between calculated and analyzed values. This experiment was carried out to define the sources of variation in the energy content of corn and to develop a practical method to accurately estimate the digestible energy (DE) and metabolisable energy (ME) content of individual corn samples for growing pigs. Twenty samples were taken from each of five provinces in China (Jilin, Hebei, Shandong, Liaoning, and Henan) to obtain a range of quality.

Results: The DE and ME contents of the 100 corn samples were measured in 35.3 ± 1.92 kg growing pigs (six pigs per corn sample). Sixty corn samples were used to build the prediction model; the remaining forty samples were used to test the suitability of these models. The chemical composition of each corn sample was determined, and the results were used to establish prediction equations for DE or ME content from chemical characteristics. The mean DE and ME content of the 100 samples were 4,053 and 3,923 kcal/kg (dry matter basis), respectively. The physical characteristics were determined, as well, and the results indicated that the bulk weight and 1,000-kernel weight were not associated with energy content. The DE and ME values could be accurately predicted from chemical characteristics. The best fit equations were as follows: DE, kcal/kg of DM = 1062.68 + (49.72 × EE) + (0.54 × GE) + (9.11 × starch), with $R^2 = 0.62$, residual standard deviation (RSD) = 48 kcal/kg, and $P < 0.01$; ME, kcal/kg of dry matter basis (DM) = 671.54 + (0.89 × DE) − (5.57 × NDF) − (191.39 × ash), with $R^2 = 0.87$, RSD = 18 kcal/kg, and $P < 0.01$.

Conclusion: This experiment confirms the large variation in the energy content of corn, describes the factors that influence this variation, and presents equations based on chemical measurements that may be used to predict the DE and ME content of individual corn samples.

Keywords: Corn, Digestible energy, Metabolizable energy, Pigs, Prediction equation

Background

Corn is the principal cereal grain used in swine diets because it is widely grown, have highly DE and ME, and is generally economical. However, variation in nutrient content of corn has the potential to greatly affect profits in pig production. For example, variation in valuable energy may translate to economically significant changes in feed conversion [1]. To prevent unforeseen reductions in growth performance, grading and analytical methods are used to minimize nutrient variability between calculated and analyzed values. In the present Chinese grading system, corn is graded based on bulk weight and damaged kernels even though other factors may affect its feeding value.

Nutrient digestibility of corn is affected by agronomic conditions, genetics, postharvest processing, storage conditions, and anti-nutritional factors [2,3]. Differences between corn samples can yield variability in available energy, nutrient digestibility and growth performance in pigs [4-6]. However, the United States department of agriculture (USDA) corn grading system and Chinese grading system are based primarily on physical characteristics, such as bulk weight. Therefore, corn is priced with

* Correspondence: Defali@public2.bta.net.cn
State Key Laboratory of Animal Nutrition, Ministry of Agriculture Feed Industry Centre, China Agricultural University, Beijing 100193, China

disregard to variations in chemical quality due to the vast scale of analysis that would be required commercially [6] and the acceptance that nutrient value of feed ingredients may be constant based on broad-based quality designations [7].

By potentially ignoring inherent variation in nutrient content and digestibility, the grading methods used to evaluate corn, such as bulk weight, may be poor estimators of feeding value [8,9]. Furthermore, prediction equations for digestible energy (DE) and metabolisable energy (ME) in feed ingredients based on chemical composition can be a useful tool in feed ingredient evaluation, but such equations are currently available only for barley [3], DDGS [10,11], wheat [12], and complete diets [13]. To our knowledge, there is a lack of peer-reviewed information regarding the combination of these techniques to predict nutrient digestibility of diverse samples of corn in pigs. The objectives of the present study were to characterize the nature of the variation in the energy content of corn and to develop a system(s) that accurately estimates the DE and ME levels in individual corn samples.

Methods

Selection and preparation of the corn samples

The corn samples were obtained from the main corn producing areas of China. Jilin, Liaoning, and northern Hebei provinces are spring corn-growing areas; seeds are planted from the end of April to May. Southern Hebei, Shandong and Henan provinces are summer corn-growing areas; seeds are planted in mid-June. To obtain a range of quality of Chinese feed corn, a total of 100 corn samples were taken, twenty from each of five provinces (Jilin, Hebei, Shandong, Liaoning, and Henan). From each location, one sample was selected to be below average and another was selected to be above average in bulk weight. Thus, the primary goal of the sample selection process was not to compare cultivars, but rather to provide a diverse array of corn samples for investigation into the nature of energy variability in corn. The chemical characteristics and physical characteristics of corn are shown in Table 1. The Institutional Animal Care and Use Committee at China Agricultural University (Beijing, China) reviewed and approved the protocols used in this study.

Experimental design

Sixty corn samples were randomly selected from the five provinces of China, every province contains twelve samples, to develop a prediction model for DE and ME that could be utilized for the formulation of diets for pigs. The remaining forty corn samples were used to test the accuracy of the DE prediction model. One hundred diets were formulated to contain 96.8% of one of each of the corn samples and 3.2% minerals and vitamins (Table 2). Corn was assumed to be the only source of energy in the diet as the slight contribution of energy from vitamin and mineral premixes was assumed to be negligible. Vitamins and minerals were supplied at levels formulated to exceed the requirements of 20 to 50 kg growing pigs as defined by NRC [14].

The total experiment consisted of five digestibility trials conducted from October 2011 to March 2012 under similar experimental conditions. We have ten metabolism rooms, and each room has twelve metabolism cages. Six workers were employed to collect feces. Each successive trial measured twenty diets. A total of six hundred crossbred barrows (Duroc × Landrace × Yorkshire) (initial BW, 35.3 ± 1.9 kg) were used according to a completely randomized design, and each diet was tested with six pigs.

Table 1 Chemical and physical characteristic characteristics of 100 corn samples

Item	Mean	SD*	CV※	Minimum	Maximum
Chemical composition, % of DM					
Crude protein	9.69	0.50	5.16	7.78	11.03
Ash	1.36	0.14	10.07	0.99	1.79
Calcium	0.03	0.01	26.67	0.01	0.03
Phosphorus	0.25	0.03	11.20	0.18	0.32
NDF	11.13	0.93	8.35	9.56	17.36
ADF	2.29	0.23	10.04	1.86	2.95
Ether extract	3.65	0.54	14.79	2.04	4.81
Starch	72.77	3.32	4.56	53.46	79.80
Gross energy, kcal/kg of DM	4,447	44.96	1.01	4,357	4,537
Physical characteristic					
Bulk weight, g/L	698.68	25.91	3.71	573.62	752.35
1,000 kernel weight, g	325.48	41.16	12.64	220.20	411.10

*SD, Standard deviation; ※CV, Coefficient of variation.

Table 2 Composition of the experimental diets (as-fed basis) fed to growing pigs for comparison of the energy digestibility between different corn samples

Ingredient	%
Corn	96.8
Antioxidant[1]	0.1
Dicalcium phosphate	1.7
Limestone	0.6
Salt	0.3
Vitamin and trace mineral premix[2]	0.5

[1]Santoquin MAX composite antioxidant, contained no less than 10% Ethoxyquin, no less than 3% Butylated Hydroxytoluene (BHT) and Citric acid, provided by Novus International, Inc.
[2]Premix provided the following per kg of complete diet for growing pigs: vitamin A, 5,512 IU; vitamin D_3, 2,200 IU; vitamin E, 30 IU; vitamin K_3, 2.2 mg; vitamin B_{12}, 27.6 μg; riboflavin, 4 mg; pantothenic acid, 14 mg; niacin, 30 mg; choline chloride, 400 mg; folacin, 0.7 mg; thiamine 1.5 mg; pyridoxine 3 mg; biotin, 44 μg; Mn, 40 mg (MnO); Fe, 75 mg ($FeSO_4 \cdot H_2O$); Zn, 75 mg (ZnO); Cu, 100 mg ($CuSO_4 \cdot 5H_2O$); I, 0.3 mg (KI); Se, 0.3 mg (Na_2SeO_3).

Pigs were individually housed in stainless steel metabolism cages (1.4 m × 0.45 m × 0.6 m), and were weighed at the beginning of each period. Pigs were adapted to the diet and the digestibility cage for more than ten days before total collection of feces and urine for five days. The crates were located in an environmentally controlled room with a temperature of 22 ± 1°C.

Feed was provided twice daily at 08:00 and 17:00 h as a mash. Water was continuously available through a nipple drinker. During a ten days period of adjustment to the metabolism crates and diets, average daily feed intake was gradually increased until it was estimated to supply 4% of the average BW determined at the initiation of each adaptation period. Feed refusals and spillage were collected daily and weighed. The collection and sample preparation of feces and urine were conducted according to the methods described by Song et al. [15]. Feces were collected as they appeared in the metabolism crates and placed in plastic bags to be stored at −20°C. Urine was collected in a bucket placed under the metabolic crate. The bucket contained 10 mL of 6 mol/L HCl for every 1,000 mL of urine. Each day, the total urine volume was measured and a 10% aliquot was filtered through gauze and the urine samples were transferred into a screw-capped tube and immediately stored at −20°C until needed for analysis. At the end of the collection period, feces were thawed, pooled by pig within period, homogenized, sub-sampled, dried for 72 h in a 65°C drying oven and ground through a 1-mm screen. For analysis, all the corn samples were ground through a 1-mm screen as well.

Chemical analyses

All chemical analysis were conducted in duplicate and repeated if the results differed by more than 5%. The ingredients used in this experiment were analyzed for dry matter (DM) [16], ether extract (EE) [17], ash [16], calcium [16], and phosphorus [16]. Kjeldahl N was determined according to the method used by Thiex et al. [18]. The content of neutral detergent fibre (NDF) and acid detergent fibre (ADF) were determined using filter bags and fiber analyzer equipment (Fiber Analyzer, Ankom Technology, Macedon, NY) following a modification of the procedure of Van Soest et al. [19]. Starch content was determined after converting starch to glucose using an enzyme assay kit (Megazym International Ireland, Wicklow, Ireland). The GE of feces, diets and corn samples were measured using an automatic adiabatic oxygen bomb calorimeter (Parr 6300 Calorimeter, Moline, IL). The GE of urine was measured by injecting 4 ml of the sample into 2 filter papers in a special crucible, and dried for 8 h in a 65°C drying oven to determine the energy. The 1,000-kernel weight (g/1,000 seeds) was measured in each sample of test corn by first cleaning it of all foreign materials and then counting 1,000 seeds.

Calculations and statistical analysis

The apparent total tract digestibility (ATTD) of GE was measured on the 100 feed samples and was later converted to reflect the digestibility of the individual corn sample. The small portion of the experimental diets that consisted of minerals and vitamins (3.2%) was assumed to have a negligible contribution to the digestibility of GE.

This experiment was a completely randomized design; the data were analyzed using the mean, correlation, GLM, and one-way ANOVA procedures of SAS (SAS Inst. Inc., NC). The individual animal and corn sample were the experimental units for analyzing the data from the digestibility trial and analysis of the chemical constituents, respectively. The relationship between physical characteristic, chemical composition, DE and ME were analyzed using the CORR procedures of SAS (1991). The linear regression equations for predicting the DE and ME value of the corn from the chemical constituents were calculated with the forward stepwise regression procedure within SAS (1991). The level of significance adopted was 5% ($P < 0.05$). The equations with the smallest RSD are presented in the results.

Results

Chemical characteristics, physical characteristics and gross energy of corn

As expected, the chemical composition and physical characteristics of corn were quite variable for some criteria (Table 1). On a dry matter basis, the concentration of CP ranged from 7.78 to 11.03% with a mean of 9.69%. Ash concentration ranged from 0.99 to 1.79% with a mean of 1.36%. The variation was particularly high within the main fiber fractions as NDF concentration in corn ranged from 9.56 to 17.36% (mean 11.13%) of DM, while values for ADF in corn ranged from 1.86 to 2.95% (mean 2.29%) of DM. Concentrations of EE and starch varied greatly as well, ranging from 2.04 to 4.81% and from 53.46 to 79.80% and averaging 3.65 and 72.77%, respectively. In contrast, GE content of the corn samples varied slightly. The bulk weight of corn ranged from 573.62 to 752.39 g/L (mean 698.68 g/L). The 1,000-kernel weight varied greatly as well, ranging from 220.20 to 411.10 g (mean 325.48 g).

Energy concentration and energy digestibility of corn

Energy concentration and the ATTD of GE of the corn are shown in Table 3. In the 100 corn samples, DE content ranged from 3,931 to 4,180 kcal/kg with a mean DE content of 4,053 kcal/kg, resulting in a 6% range in DE. The ME content ranged from 3,798 to 4,092 kcal/kg with a mean ME content of 3,923 kcal/kg, and the overall variation in ME was 294 kcal. The ratio of ME to DE calculated from 100 measured samples ranged from 95.41 to 98.13% with a mean value of 96.78%. The ATTD of GE ranged from 83.43 to 92.25 % with a mean of 90.49%.

Effect of growing region on chemical characteristics, physical characteristics and energy values

With exception to ash content and bulk weight (Table 4), the chemical characteristics, physical characteristics and energy values of corn were influenced significantly by growing region ($P < 0.01$). Among five provinces, Henan had the highest starch, NDF and energy content (GE, DE and ME) and corn had a larger 1,000-kernel weight when grown in the Liaoning (354.59 g) and Jilin (363.60 g) provinces compared with corn grown in the Henan (299.70 g) and Shandong (307.38 g) provinces. The DE content of corn grown in Liaoning (4,032.69 kcal/kg) and Jilin (4,035.85 kcal/kg) provinces were similar; however, the DE content of corn grown in Shandong (3,996.61 kcal/kg) province was lowest. Overall, growing region significantly influenced the DE content of corn ($P < 0.01$). Corns grown in the spring growing areas (Liaoning and Jilin provinces) had a significantly higher 1,000-kernel weight compared with corn grown in the summer growing areas (Henan and Shandong provinces).

Table 3 Energy concentration and ATTD of GE of the 100 corn samples

Item	Mean	Minimum	Maximum	SD	CV
DE, kcal/kg of DM	4,053	3,931	4,180	60.39	1.49
ME, kcal/kg of DM	3,923	3,798	4,092	55.42	1.41
ME/DE	96.78	95.41	98.13	1.01	1.04
ATTD of GE, %	91.15	83.27	93.07	0.97	1.06

Correlation coefficients between physical and chemical characteristics and energy values

In the 100 corn samples, fibrous compounds had a negative correlation with DE and ME content, while the correlation of EE, GE and starch with DE content was positive (Table 4). The content of EE had the highest correlation of any characteristic with DE content ($r = 0.44$; $P < 0.01$), followed by total starch ($r = 0.38$; $P < 0.01$), NDF ($r = -0.32$; $P < 0.01$) and ash ($r = -0.29$; $P < 0.05$). Correlation analyses showed that ME content of corn was positively correlated to the DE ($r = 0.95$; $P < 0.01$) and EE content ($r = 0.29$; $P < 0.01$), while ash ($r = -0.28$, $P < 0.01$) and NDF ($r = -0.27$, $P < 0.01$) had a negative correlation with ME content. The correlation of bulk weight and 1,000-kernel weight with energy content was not significant (Table 5).

Prediction equations for digestible energy and metabolizable energy

Some equations based on simple and multiple linear regression analysis were then conducted to develop prediction equations for DE content of corns based on the results of stepwise regression analysis (Table 6). According to the high correlation between DE and EE content (Table 5), the best single predictor was always the EE estimate. Prediction slightly improved when the starch content was included (Equation 2 in Table 6). Addition of NDF and GE content to the equation improved the precision of the prediction (Equations 3 and 4 in Table 6). Among the different predictors, the predictions with the lowest RSD were obtained when EE, starch, NDF, and GE were considered (Equation 4 in Table 6). The residual standard deviation (RSD) was then equivalent to 48 kcal of DM.

Equations for estimating ME content from chemical characteristics were calculated similarly. The results showed that the ME content of corn could be predicted with a reasonable degree of accuracy by measuring the DE (Equation 9 in Table 7). As with DE, the addition of NDF and ash content to the equation improved the precision of the prediction (Equations 10 and 11 in Table 7). The content of ME can also be accurately predicted from CP, EE, NDF and ash contents without DE content (RSD, 35 kcal/kg of DM) (Equation 8 in Table 7).

Comparison of DE content in corn determined by using the *in vivo* method and prediction model

To test the suitability of these models (Table 6) to predict the DE content of a normal corn sample, the DE content of 40 samples of corn was measured by both the *in vivo* method and prediction models (Equation 4 in Table 6). Our results showed that the maximum absolute difference between DE determined by the *in vivo* method and the prediction model was 104.61 kcal/kg, while the minimum absolute difference was 0.15 kcal/kg

Table 4 Effect of growing region on chemical characteristics, physical characteristics and energy values of the 100 corn samples from five provinces

Item	Liaoning	Jilin	Hebei	Henan	Shandong	SEM	*P*-value
Chemical composition, % of DM							
Dry matter	87.57[b]	88.15[a]	86.82[c]	86.49[d]	88.20[a]	0.47	0.01
Crude protein	9.87[a]	9.78[a]	9.38[b]	9.61[ab]	9.82[a]	0.47	0.01
Ether extract	3.54[b]	3.38[b]	4.05[a]	3.71[b]	3.56[b]	0.05	0.01
ADF	2.28[b]	2.22[b]	2.46[a]	2.29[b]	2.25[b]	0.22	0.01
NDF	10.95[bc]	10.73[c]	10.76[c]	11.76[a]	11.46[ab]	0.86	0.01
Ash	1.61	1.62	1.58	1.60	1.63	0.10	0.55
Starch	70.58[c]	73.20[b]	72.97[b]	75.32[a]	71.79[bc]	2.97	0.01
Physical characteristic							
Bulk weight, g/L	693.29	689.94	699.21	707.94	703.04	25.60	0.18
1,000 kernel weight, g	354.69[a]	363.60[a]	302.19[b]	299.70[b]	307.38[b]	30.92	0.01
Energy concentration, kcal/kg of DM							
Gross energy	4448.16[b]	4436.57[b]	4435.73[b]	4476.82[a]	4436.69[b]	18.23	0.01
Digestible energy	4032.85[b]	4035.85[b]	4093.45[a]	4106.23[a]	3996.61[c]	45.20	0.01
Metabolisable energy	3906.83[b]	3903.14[b]	3970.06[a]	3968.63[a]	3868.94[c]	40.61	0.01

a-d Means followed by the same letter within each row are not significantly different from each other ($P > 0.05$).

(Table 8). The mean of observed group and prediction group were 4,035.05 and 4,021.73 kcal/kg, and the difference was only 13.32 kcal/kg. Therefore, the prediction models established from 60 corn samples as described in this article can be used to predict the DE content of corn for pigs with acceptable accuracy.

Discussion

Chemical characteristics, physical characteristics and energy variation in corn

The chemical composition and concomitant nutritional value of corn is variable and dependent on variety, growing environment, drying temperature, starch structure and the presence of various anti-nutritive factors [8,20-27]. Comparing the present study with the National Research Council [28], mean GE concentration was identical, mean CP and starch concentrations were higher than the NRC (2012) values, mean EE and ADF concentrations were lower than the NRC (2012) values. Content of GE did not vary much among the 100 samples. The CV for CP, NDF, starch, and GE were within 10%, but wide variations in the content of EE (CV: 14.79%), ash (CV: 10.07%), Ca (CV: 26.67%), P (CV: 11.20%) and ADF (CV: 10.04%) were observed (Table 1). For the physical trait, the wide variation in 1,000-kernel weight also was observed (CV: 12.64%).

Table 5 Correlation coefficients between chemical characteristics, physical characteristics and energy values of the 100 corn samples

Item	DE	ME	Crude protein	Ether extract	ADF	NDF	Ash	Gross energy	Starch	Bulk weight	1,000 kernel weight
DE	1.00										
ME	0.95**	1.00									
Crude protein	−0.15	−0.06	1.00								
Ether extract	0.44**	0.29**	−0.28	1.00							
ADF	−0.05	−0.06	−0.05	0.33	1.00						
NDF	−0.32**	−0.27**	0.14	−0.02	0.68	1.00					
Ash	−0.29*	−0.28**	0.22	−0.27	−0.11	0.17	1.00				
Gross energy	0.25**	0.23	−0.06	0.14	−0.28	0.19	0.07	1.00			
Starch	0.38**	0.32*	−0.03	0.03	0.13	0.11	−0.12	0.06	1.00		
Bulk weight	0.05	0.03	0.03	0.15	−0.16	0.13	0.22	0.05	−0.13	1.00	
1,000 kernel weight	0.11	0.06	0.25	0.02	−0.06	−0.13	0.08	0.14	−0.11	0.42	1.00

*, **, $P < 0.05$, $P < 0.01$, respectively.

Table 6 Most effective prediction equations of digestible energy (kcal/kg; dry matter) based on chemical variables (% or kcal/kg; dry matter) of the corn samples

No.	Equation	R^2	RSD	*P*-value
1	DE = 3889.81 + (46.21 × Ether extract)	0.20	59	<0.01
2	DE = 3335.46 + (44.91 × Ether extract) + (7.59 × Starch)	0.33	58	<0.01
3	DE = 3507.46 + (44.01 × Ether extract) − (20.09 × NDF) + (8.38 × Starch)	0.45	56	<0.01
4	DE = 1062.68 + (49.72 × Ether extract) − (24.89 × NDF) + (0.54 × Gross energy) + (9.11 × Starch)	0.62	48	<0.01

Variety and growing environment are major sources of variation for the chemical composition of cereal grains [29]. Table 4 also indicated that growing regions influenced significantly the chemical characteristics, physical characteristics and energy values of corn. A previous study demonstrated that chemical composition, especially the NSP composition and structure, was significantly different due to variety and growing location and was negatively correlated with the DE content [30]. One reason for not observing a huge degree of variation in some criteria may be that no extreme samples were collected, such as high-oil corn, high-lysine corn, or NutriDense corn. The other reason is that the quality standards for corn were similar for all the feed companies who collected the samples for this study.

Corn samples were collected for use in this study to increase the likelihood of observing a degree of variation in DE and ME content of the 100 corn samples. Yellow Dent corn in the NRC [28] has a DE value of 3,921 kcal/kg of DM and a ME value of 3,857 kcal/kg of DM, respectively. The average DE for corn determined in the present study was 4,053 kcal/kg of DM, and ME averaged 3,923 kcal/kg of DM, respectively. The difference between the current study and previous results could be attributable to differences in particle size of corn. The mean particle size of the corn reported herein was 441 μm, well below the 600 to 700 μm level recommended for corn fed to growing swine [31,32]. Data reported by Owsley et al. [33], and Healy et al. [32] suggest that the DE of diet increases when particle size is reduced. The screen was a new one for this study, and Chinese producers prefer the small feed particles.

The amount of energy lost in urine represented from 2 to 5% of dietary DE content (mean: 3.0%). The mean ME:DE ratio therefore averaged 97%. The ME:DE ratio obtained in the current study corresponds with the data published by NRC [28], which reported Yellow Dent corn had an ME:DE ratio of 98%. Based upon the NRC [28] reported value, Noblet and Perez [13] proposed an equation (ME/DE = 100.3 - 0.21 × CP) to be applied to feed ingredients. Under practical conditions, the dietary CP content is less variable, so ME:DE has been considered almost as a constant of approximately 96.78% [34]. The mean of the ATTD of GE determined for these corn samples was 91.15%, somewhat higher than standard values, which are typically 87.74% (Yellow Dent Corn) [28]. One possible explanation for the higher ATTD of GE could be the smaller than expected particle size of the ground corn.

Factors influencing energy variation

Corn is typically sold as a commodity and is valued by bulk weight, moisture content, and absence of foreign particulates. However, in this experiment, no physical parameter measured (bulk weight and 1,000 kernel weight) was significant at $P \leq 0.15$ to predict DE or ME. The results obtained in the current study agreed with the data provided by Leeson [8] and Dale [9]. Such a result indicates that most of the observed variation in DE or ME between samples was the result of factors other than bulk weight and 1,000 kernel weight. Compared with the data of correlation coefficients (Table 5) among DE of DM values of samples in the study, EE, NDF, ash, GE and starch were 0.44, -0.32, -0.29, 0.25 and 0.38,

Table 7 Most effective prediction equations of ME (kcal/kg; DM) based on chemical variables (% or kcal/kg; DM) of the corn samples

No.	Equation	R^2	RSD	*P*-value
5	ME = 4190.19 − (19.56 × NDF)	0.22	44	<0.01
6	ME = 4664.04 − (17.74 × NDF) − (310.93 × Ash)	0.33	40	<0.01
7	ME = 4464.24 + (20.15 × Ether extract) − (17.84 × NDF) − (233.72 × Ash)	0.44	39	<0.01
8	ME = 4289.74 + (20.02 × Crude protein) + (22.47 × EE) − (18.40 × NDF) − (245.20 × Ash)	0.49	35	<0.01
9	ME = −213.69 + (1.02 × DE)	0.79	23	<0.01
10	ME = 362.67 + (0.95 × DE) − (193.59 × Ash)	0.85	19	<0.01
11	ME = 671.58 + (0.89 × DE) − (5.57 × NDF) − (191.39 × Ash)	0.87	18	<0.01

Table 8 Comparison of digestible energy contents in corn determined by using the *in vivo* method and prediction model (kcal/kg of DM)

Test	Observed	Predicted	Difference
1	4,084.05	3,996.40	87.64
2	4,108.22	4,040.57	67.65
3	4,140.16	4,101.24	38.92
4	4,023.51	4,019.85	3.66
5	4,113.16	4,047.87	65.29
6	3,950.07	4,050.91	−100.84
7	4,058.73	4,085.59	−26.85
8	4,054.97	4,062.27	−7.30
9	4,046.38	4,007.67	38.71
10	3,997.07	3,989.82	7.25
11	4,044.88	3,969.98	74.89
12	4,062.20	4,042.23	19.97
13	4,062.67	3,994.71	67.95
14	4,009.57	3,974.33	35.23
15	3,957.41	3,962.92	−5.51
16	4,036.46	4,036.31	0.15
17	3,983.54	3,970.58	12.96
18	4,011.42	3,993.93	17.49
19	3,932.03	3,951.91	−19.88
20	3,980.49	4,035.48	−54.99
21	4,065.07	4,054.47	61.22
22	4,024.33	4043.60	−19.28
23	4,025.92	4,020.54	5.38
24	4,020.02	4,024.04	−4.02
25	4,016.88	4,005.85	11.03
26	4,051.08	3,996.01	55.06
27	4,058.55	3,987.99	70.57
28	4,051.18	4,018.28	32.90
29	4,088.96	4,036.06	52.90
30	4,087.01	4,028.30	58.71
31	4,121.96	4,017.35	104.61
32	4,084.62	4,055.96	28.66
33	4,002.90	4,050.52	−47.62
34	3,979.24	4,014.76	−35.52
35	4,003.71	4,022.21	−18.50
36	4,066.90	4,022.56	44.34
37	3,934.52	4,003.62	−69.10
38	4,041.55	4,038.95	2.59
39	4,029.50	4,062.03	−32.54
40	3,991.01	4,031.85	−40.84
SD*	49.74	32.93	
RSD※*			46.75
P-value			0.16

*SD, Standard deviation; ※RSD, residual standard deviation.

respectively. Correlation coefficients and prediction equations with chemical composition clearly indicate that EE, starch and fibrous components predominantly determined the energy values. This observation is consistent with findings for barley by Fairbairn et al. [3], in which a prediction equation with an r^2 of 0.89 for DE was reported if the similar variables were used.

Ether extract and starch were important prediction estimators of DE or ME in corn. The reasons may attribute that starch and dietary fat are important ingredients in diets because of its high energy value. Interestingly, the narrow range of GE levels in the 100 corn samples using herein (4,357 to 4,537 kcal/kg) did not seem to reduce the effectiveness of GE as an accurate estimator of DE or ME in corn. The reasons for this are unclear but the first estimate, based on a large set of experimental data, seems preferable. However, the coefficient of GE was positive with energy values in the study, and similar results were found in wheat (unpublished data) in our lab. Barley, wheat and sorghum showed the same result in Batterham's study [35].

In agreement with most literature data [13,36,37], the presence of fiber in the diet of pigs reduces the ATTD of energy and nutrients. The content of NDF was another significant factor affecting the DE and ME variation in the present experiment. The prediction was improved when the different Van Soest fractions were included (Equations 3 in Table 6). This significant correlation between NDF and DE in other cereal grains has been observed by Perez et al. [38] and Fairbairn et al., [3]. Other researchers have found that the NDF content of corn was correlated with growing performance of pigs [6]. The reason may be that cellulose and lignin act as a diluent, thus lowering the energy content by displacing more digestible fractions such as starch. However, the dilution effect does not fully explain the changes in energy content. It may be that the presence of fiber can reduce the digestion of dietary nutrients. The results of the present study clearly show that ME content of corn can be accurately predicted from the DE content, because the correlation coefficient between DE and ME is approximately 0.95.

After testing the suitability of these models (Table 6) to predict the DE content of corn samples, our results showed that the difference between *vivo* method the prediction model was not significant ($P > 0.16$ in Table 8). This suggested that the accuracy of prediction models for DE was close to that obtained *in vivo*. In conclusion, the results of this study demonstrate that it is possible to estimate the DE and ME of corn from chemical variables. The best predictors for energy values were EE, starch, GE and NDF contents. Equation 4 (Table 6) and Equation 11 (Table 7) represent the best combination of accuracy and practicality when estimating DE and ME levels in corn.

Abbreviations
DE: Digestible energy; ME: Metabolizable energy; CP: Crude protein; NDF: Neutral detergent fibre; ADF: Acid detergent fibre; ATTD: Apparent total tract digestibility; DDGS: Distillers dried grains with soluble; EE: Ether extract; GE: Gross energy; DM: Dry matter; RSD: Residual standard deviation.

Competing interests
The authors declare that they have no competing interests.

Authors' contributions
QFL carried out the experiment trial, performed the statistics and drafted the manuscript. JJZ and XSP participated in design of the study. DWL participated animal trial. DFL and CHL conceived the study, and participated in its design and coordination. All authors read and approved the final manuscript.

Acknowledgements
This research was financially supported by the National Key Technology R & D Program in the 11th Five Year Plan of China (2006BAD12B01) and Special Public Sector Fund in Agriculture (200903006). Thanks to Chuanxin Shi, Zhongchao Li and Zhaoyu Liu for their practical support.

References
1. Dozier WA, Gehring CK, Corzo A, Olanrewaju HA: **Apparent metabolizable energy needs of male and female broilers from 36 to 47 days of age.** *Poult Sci* 2011, **90:**804–814.
2. Cowieson AJ: **Factors that affect the nutritional value of maize for broilers.** *Anim Feed Sci Technol* 2005, **119:**293–305.
3. Fairbairn SL, Patience JF, Classen HL, Zijlstra RT: **The energy content of barley fed to growing pigs: characterizing the nature of its variability and developing prediction equations for its estimation.** *J Anim Sci* 1999, **77:**1502–1512.
4. Spencer JD, Allee GL, Sauber TE: **Growing-finishing performance and carcass characteristics of pigs fed normal and genetically modified low-phytate corn.** *J Anim Sci* 2000, **78:**1529–1536.
5. Lampe JF, Baas TJ, Mabry JW: **Comparison of grain sources for swine diets and their effect on meat and fat quality traits.** *J Anim Sci* 2006, **84:**1022–1029.
6. Moore SM, Stalder KJ, Beitz DC, Stahl CH, Fithian WA, Bregendahl K: **The correlation of chemical and physical corn kernel traits with growth performance and carcass characteristics in pigs.** *J Anim Sci* 2008, **86:**592–601.
7. de Coca-Sinova A, Valencia DG, Jiménez-Moreno E, Lázaro R, Mateos GG: **Apparent ileal digestibility of energy, nitrogen, and amino acids of soybean meals of different origin in broilers.** *Poult Sci* 2008, **87:**2613–2623.
8. Leeson S, Yersin A, Volker L: **Nutritive values of 1992 corn crop.** *J Appl Poult Res* 1993, **2:**208–213.
9. Dale N, Jackson D: **True metabolizable energy of corn fractions.** *J Appl Poult Res* 1994, **3:**179–183.
10. Pedersen C, Boersma MG, Stein HH: **Digestibility of energy and phosphorus in 10 samples of distillers dried grains with solubles fed to growing pigs.** *J Anim Sci* 2007, **85:**1168–1176.
11. Cozannet P, Primot C, Gady C, Metayer JP, Lessire M, Skiba F, Noblet J: **Energy value of wheat distillers grains with solubles for growing pigs and adult sows.** *J Anim Sci* 2010, **88:**2382–2392.
12. Zijlstra RT, De Lange CFM, Patience JF: **Nutritional value of wheat for growing pigs: chemical composition and digestible energy content.** *Can J Anim Sci* 1999, **79:**187–194.
13. Noblet J, Perez JM: **Prediction of digestibility of nutrients and energy values of pig diets from chemical analysis.** *J Anim Sci* 1993, **71:**3389–3398.
14. NRC: *Nutrient requirements of swine 10th rev. ed.* Washington DC, USA: National Academy Press; 1998.
15. Song GL, Li DF, Piao XS, Chi F, Yang WJ: **Apparent ileal digestibility of amino acids and the digestible and metabolizable energy content of high-oil corn varieties and its effects on growth performance of pigs.** *Arch Anim Nutr* 2003, **57:**297–306.
16. AOAC: *Official methods of analysis. 17th ed.* Arlington, VA, USA: Association of Official Analytical Chemists; 2000.
17. Thiex NJ, Anderson S, Gildemeister B: **Crude fat, diethyl ester extraction, in feed, cereal grain, and forage (Randall/Soxtec/submersion method): collaborative study.** *J AOAC Int* 2003, **86:**888–898.
18. Thiex NJ, Manson H, Anderson S, Persson JA: **Determination of crude protein in animal feed, forage, grain, and oilseeds by using block digestion with copper catalyst and steam distillation into boric acid: collaborative study.** *J AOAC Int* 2002, **85:**309–317.
19. Van Soest PJ, Robertson JB, Lewis BA: **Methods for dietary fiber and non-starch polysaccharides in relation to animal nutrition.** *J Dairy Sci* 1991, **74:**3568–3597.
20. Leigh K: *The unpredictable nature of maize.* Pigs; 1994:37–39.
21. Brown I: **Complex carbohydrates and resistant starch.** *Nutr Rev* 1996, **54:**115–119.
22. Collins NE, Moran ET, Stilborn HL: **Maize hybrid and bird maturity affect apparent metabolizable energy values.** *Poult Sci* 1998, **11:**42.
23. Cromwell GL, Calvert CC, Cline TR, Crenshaw JD, Crenshaw TD, Easter RA, Ewan RC, Hamilton CR, Hill GM, Lewis AJ, Mahan DC, Miller ER, Nelssen JL, Pettigrew JE, Tribble LF, Veum TL, Yen JT: **Variability among sources and laboratories in nutrient analyses of maize and soybean meal.** *J Anim Sci* 1999, **77:**262–3273.
24. Collins NE, Moran JR: **Influence of yellow dent maize hybrids having different kernel characteristics yet similar nutrient composition on broiler production.** *J Appl Anim Res* 2001, **10:**228–235.
25. Bohlke RA, Thaler RC, Stein HH: **Calcium, phosphorus, and amino acid digestibility in low-phytate corn, normal corn, and soybean meal by growing pigs.** *J Anim Sci* 2005, **83:**2396–2403.
26. Bird AR, Vuaran M, Brown I, Topping DL: **Two high-amylose maize starches with different amounts of resistant starch vary in their effects on fermentation, tissue and digesta mass accretion, and bacterial populations in the large bowel of pigs.** *Br J Nutr* 2007, **97:**134–144.
27. Oliveira GCD, Moreira I, Furlan AC, Piano LM, Toledo JB, Peñuela Sierra LM: **Corn types with different nutritional profiles, extruded or not, on piglets (6 to 15 kg) feeding.** *Revista rasileira de Zootecnia* 2011, **40:**2462–2470.
28. NRC: *Nutrient requirements of swine 11th rev. ed.* Washington DC, USA: National Academy Press; 2012.
29. Kim JC, Simmins PH, Mullan BP, Pluske JR: **The digestible energy value of wheats for pigs, with special reference to the post-weaned animal.** *Anim Feed Sci Technol* 2005, **122:**257–287.
30. Kim JC, Mullan BP, Simmins PH, Pluske JR: **Effect of variety, growing region and growing season on digestible energy content of wheats grown in Western Australia for weaner pigs.** *Anim Sci* 2004, **78:**53–60.
31. Wondra KJ, Hancock JD, Behnke KC, Stark CR: **Effects of mill type and particle size uniformity on growth performance, nutrient digestibility, and stomach morphology in finishing pigs.** *J Anim Sci* 1995, **73:**2564–2573.
32. Healy BJ, Hancock JD, Kennedy GA, Bramel-Cox PJ, Behnke KC, Hines RH: **Optimum particle size of corn and hard and soft sorghum for nursery pigs.** *J Anim Sci* 1994, **72:**2227–2236.
33. Owsley WF, Knabe DA, Tanksley TD: **Effect of sorghum particle size on digestibility of nutrients at the terminal ileum and over the total digestive tract of growing-finishing pigs.** *J Anim Sci* 1981, **52:**557–566.
34. Morgan CA, Whittemore CT, Phillips P, Crooks P: **The prediction of the energy value of compounded pig foods from chemical analysis.** *Anim Feed Sci Technol* 1987, **17:**81–107.
35. Batterham ES, Lewis CE, Lowe RF, McMillan CJ: **Digestible energy content of cereals and wheat by-products for growing-pigs.** *Anim Prod* 1980, **31:**259–271.
36. Dierick NA, Vervaeke IJ, Decuypere JA, Henderickx HK: **Influence de la nature et du niveau des fibres brutes sur la digestibilite ileale et fecale apparente de la matiere seche, des proteines et des acides amines et sur la retention azotee chez les porcs.** *Rev Agric* 1983, **6:**1691–1711.
37. Stanogias G, Pearce GR: **The digestion of fibre by pigs, I. The effects of amount and type of fibre on apparent digestibility, nitrogen balance and rate of passage.** *Br J Nutr* 1985, **53:**513–530.
38. Perez JM, Ramoelintsalama B, Bourdon D: **Energy evaluation of barley for pigs. Prediction from analyses of fiber content.** *J Rech Porcine Fr* 1980, **12:**273–284.

7

Placental accommodations for transport and metabolism during intra-uterine crowding in pigs

Jeffrey L Vallet*, Anthony K McNeel, Jeremy R Miles and Bradley A Freking

Abstract

Litter size and birth weights are limited by uterine capacity, defined as the ability of the uterus to maintain the appropriate development of some number of conceptuses. Uterine capacity is the result of the combined effects of uterine, placental and embryo/fetal function. The number of living conceptuses that the uterus is capable of supporting is greater during early gestation compared to later gestation. Plots of log fetal weight versus log placental weight also indicate that fetal weights are less sensitive to reduced placental weight (and therefore reduced intrauterine space) in early gestation compared to late gestation. However, even in late gestation, mechanisms still exist that maintain fetal growth when the size of the placenta is reduced. One such mechanism is likely to be improved development of the folded placental-epithelial/maternal-epithelial bilayer. Fold depth, and therefore the maternal fetal interactive surface, increases as gestation advances and is greater in placenta from small fetuses. On the fetal side of the placenta, the epithelial bilayer is embedded in stromal tissue. Glycosaminoglycans are major components of stroma, including hyaluronan and heparan sulfate. Hyaluronidases and heparanases are present within placental tissues, and likely play roles in modification of stromal components to facilitate fold development. Glycosaminoglycans are polymers of forms of glucose (glucosamine, glucuronic acid, iduronic acid) suggesting that glycosaminoglycan synthesis may compete with the glucose needs of the developing fetus. Pig conceptuses are fructogenic, such that a substantial portion of glucose transferred from mother to fetus is converted to fructose. Fructose is an intermediate product in the synthesis of glucosamine from glucose, and glucosamine is linked to regulation of trophoblast cell proliferation through regulation of mTOR. These findings suggest a link between glucose, fructose, glucosamine synthesis, GAG production, and placental morphogenesis, but the details of these interactions remain unclear. In addition, recent placental epithelial transcriptome analysis identified several glucose, amino acid, lipid, vitamin, mineral and hormone transporter mechanisms within the placenta. Further elucidation of mechanisms of placental morphogenesis and solute transport could provide clues to improving nutrient transport to the pig fetus, potentially increasing litter size and piglet birth weights.

Keywords: Glycosaminoglycans, Nutrient transport, Placenta

Introduction

A recent analysis by the National Pork Board indicated that between 2007 and 2012, the number of piglets born alive per litter improved by 0.25 piglets per year, while the number of piglets weaned per litter improved by only 0.16 piglets per year [1]. This difference translates into an increase in preweaning mortality of 0.3% per year, and is a concern not only from a production standpoint, but also from an animal well-being standpoint. The increase in the number born alive is likely due to a combination of improved management and genetic selection of breeding animals. Although it was not measured in the study, the disparity between the improvement in the number of piglets born alive and the number of piglets weaned is most likely due to the well known depression in piglet birth weight associated with increased litter size [2]. Reduced birth weights are a primary factor associated with preweaning mortality [3-5], and may explain the observed increase in preweaning mortality. Finally, depression of birth weights with increasing litter size is a manifestation of limitations on conceptus development that are imposed by uterine capacity.

* Correspondence: jeff.vallet@ars.usda.gov
U.S. Department of Agriculture, Agricultural Research Service, U.S. Meat Animal Research Center, State Spur 18D, Clay Center, NE 68933, USA

Uterine capacity

Uterine capacity has been defined in a variety of ways. The number of piglets that can be supported by the uterus during gestation until farrowing if the number of potential fetuses is not limiting, is a common definition [6]. This definition includes stillborn piglets, which die during farrowing, but does not include mummified fetuses, which die at some point during gestation. It does not fully describe uterine capacity in totality because it does not include any component reflecting differences in piglet birth weights. Clearly, a pig uterus that can support ten 2 kg fetuses has greater uterine capacity than a pig uterus that can support ten 1 kg fetuses. Finally, in practice, the independence of uterine capacity from the number of potential fetuses is also problematic. The relationship between the maximum litter size that can be maintained by the uterus and the number of potential embryos is almost certain to be curvilinear. Reports suggest that at moderate intrauterine crowding, litter size reaches a peak. Significant further crowding beyond this point reduces the number of viable embryos/fetuses, probably by reducing the number of embryos able to obtain sufficient uterine space for survival due to intrauterine competition for space among embryos [7].

Studies of intrauterine crowding and uterine capacity have employed a variety of methods. The most straightforward to apply is unilateral hysterectomy-ovariectomy (UHO) [8]. In this surgical method, one ovary and one uterine horn are removed. Compensatory ovarian hypertrophy results in a normal ovulation rate, with only half the available uterine space. The intrauterine crowding that results is relatively moderate and uniform between pigs compared to that caused by superovulation [9], and the UHO surgical procedure is simpler and more reproducible than embryo transfer [10]. The surgery can be done early in life (i.e., before puberty, typically 100 to 160 days of age) and the pig can be allowed to recover, reducing the interference of the surgery with normal pregnancy physiology that occurs when uterine ligation methods are used [11]. Genetic selection also increases uterine crowding [12], but takes several generations to accomplish. Litter size in UHO females has been reported to be independent of ovulation rate, and it has been assumed that the litter size obtained is half the uterine capacity [8] for that animal. The UHO procedure has been used successfully to select pigs for uterine capacity [13,14].

Placental efficiency

There have been some approaches to placental efficiency that have not required an understanding of the underlying mechanisms, but their utility has been controversial and limited. One approach, the fetal weight to placental weight ratio, has been suggested as a broad indicator of placental efficiency [15,16]. An early report that this ratio could be used to select animals for placental efficiency, with subsequent improvements in litter size [17], was not confirmed using more rigorous selection methods over several generations [18]. Also, litter size was improved in a line selected for uterine capacity but selection did not alter the fetal weight to placental weight ratio and selection for ovulation rate did not alter litter size but did alter the fetal weight to placental weight ratio [19]. Although it makes sense that the size of a fetus supported by a given size of placenta should reflect placental efficiency, use of the ratio as a measure of placental efficiency fails to consider the ability of the fetus and placenta to adjust efficiency as the size of the placenta is reduced. In other words, if compensatory mechanisms exist to adjust placental efficiency as the size of the placenta is reduced, the fetal weight to placental weight ratio does not measure absolute placental efficiency for a given conceptus, as would be required to implement genetic selection. Instead, the fetal weight to placental weight ratio measures placental efficiency for that conceptus given the size of the placenta. If that same conceptus developed a smaller placenta, compensatory mechanisms would raise placental efficiency, which would then be reflected in the fetal weight to placental weight ratio.

Evidence of compensatory mechanisms for placental efficiency can be found by examining relationships between log fetal weight and log placental weight during gestation [20,21]. Huxley [22] proposed that the slope of log-log relationships between individual body parts or between a body part and the entire organism reflects the relative growth between the two (Figure 1). A slope of 1 indicates proportional growth, greater than 1 indicates that the y variable grows faster than the x variable, less than one indicates that the y variable grows more slowly than the x variable [23]. Extending this concept to the fetus and placenta, a slope of one indicates a fully proportional relationship, the fetus is proportionally larger if the placenta is larger. As placental size is reduced, fetal size is also reduced, and the ratio of the two is the same (placental efficiency is constant and no compensatory mechanisms are present). A slope less than one indicates fetal growth is not fully proportional to the size of the placenta. As placental size is reduced, fetal size is less affected (placental efficiency increases suggesting compensatory mechanisms are present). The slope of the relationship between log fetal weight and log placental weight increases throughout gestation, but even in late gestation is still less than one [21]. Growth of the fetus is not very dependent on placental size during early gestation, and becomes more dependent as gestation advances, but is never fully dependent on placental size. These relationships

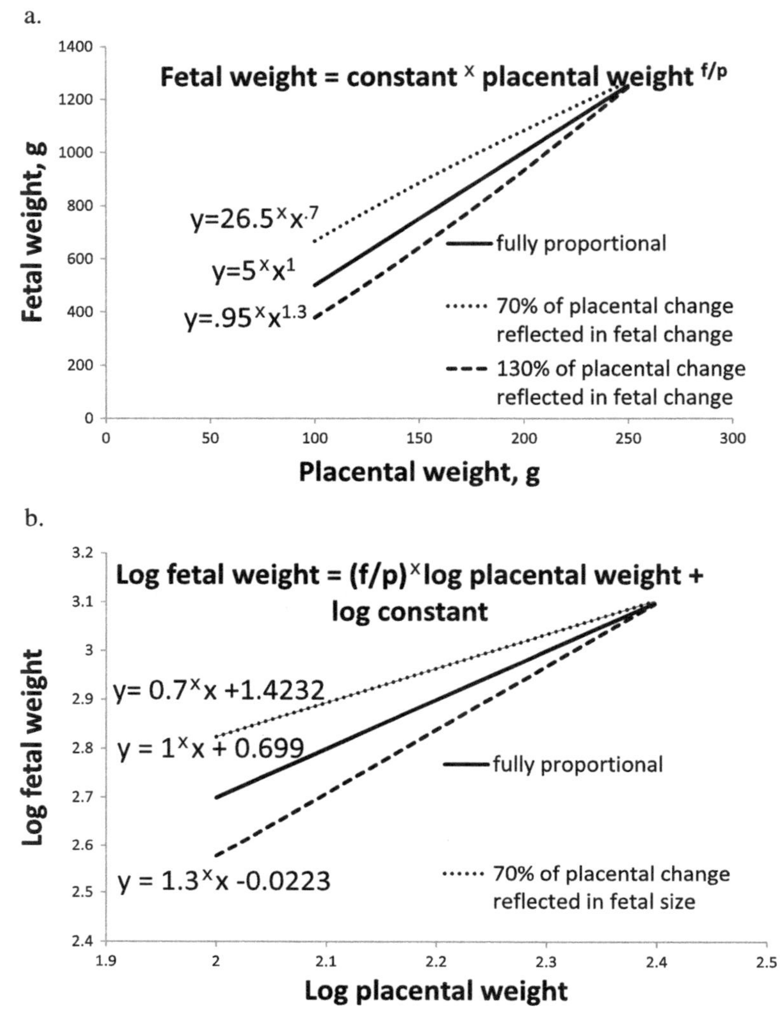

Figure 1 Possible allometric relationships between fetal weight and placental weight are illustrated. According to Huxley [22], the relationship between fetal weight and placental weight can be described with the equation fetal weight = constant × placental weight$^{f/p}$ where f and p represent growth rates of the fetus and placenta, respectively. Given this relationship, the slope of the linear relationship between log fetal weight and log placental weight is f/p, the relative growth rates of the two components. The graphs represent **(a)** untransformed and **(b)** log transformed hypothetical relationships where growth is fully proportional (i.e., growth rates are equal; solid line), only 70% of changes in placental weight are reflected in changes in fetal weight (a fetal sparing effect where fetal growth is relatively insensitive to differences in placental weight; small dashes), and 130% of changes in placental weight are reflected in changes in fetal weight (fetal growth is highly sensitive to differences in placental weight; large dashes). In pigs, even during late gestation, the slope of the log fetal weight to log placental weight relationship is less than one, indicating fetal sparing [21].

indicate that compensatory mechanisms are present, even in late gestation.

Factors affecting placental function

The pig placenta is classified as diffuse epitheliochorial [24]. This distinguishes it from other livestock species in that there are no placentomes (making it diffuse), and both the fetal and maternal epithelial cell layers are maintained throughout gestation (making it epitheliochorial). Numerous factors affect the efficiency of placental transport of nutrients. Generally, these factors fall into two broad categories, physical and nutrient specific. Physical factors include maternal and fetal blood flow (both rates and physical arrangement relative to one another), fetal and maternal interactive surface area, and the distance between maternal and fetal capillaries. These factors globally affect the transport of all nutrients. Nutrient specific factors are as varied as the nutrients transported. To fully understand placental transport, it is necessary to fully characterize the contribution of both categories.

The physical structure of the placenta divides into areolae and interareolar areas [25]. Areolae take up histotroph secreted by uterine glands [26,27]. Histotroph represents nutrient specific transport and will be dealt with later. The interareolar areas consist of a maternal endometrial epithelium layer tightly adhered to a fetal epithelium layer (trophoblast), which becomes folded beginning about day 30 to 35 of gestation (Figure 2). The folds become more elaborate and more extensive as gestation advances [20,28]. During mid gestation, the folds are relatively close to one another. During late gestation, fetal placental stroma grows into the region between the folds creating a stromal space between folds, and secondary folds develop into this space. As this is taking place, the bilayer separating maternal and fetal capillaries becomes thinner, to the point that maternal and fetal capillaries actually indent into, but do not penetrate, the epithelial cell layers. The indentation of capillaries reduces the distance between maternal and fetal capillaries to as little as 2 microns [28].

Blood flow and angiogenesis of the pig uterus and placenta have received a great deal of attention. Increased uterine blood flow in response to increased litter size has been reported, but appears to be limited [29,30]. The fetal weight to placental weight ratio has been reported to be correlated with placental blood vessel density in a comparison of Meishan and European breed pigs [15]. These results are difficult to interpret, primarily because it is not clear whether the blood vessels adjacent to the bilayer were measured preferentially or whether all placental vessels were included. Because the epithelial bilayer is the interactive surface between the dam and fetus and solute diffusion decreases rapidly with increased distance, only the capillaries directly adjacent to the bilayer are relevant to nutrient transport. Finally, placental expression of vascular endothelial growth

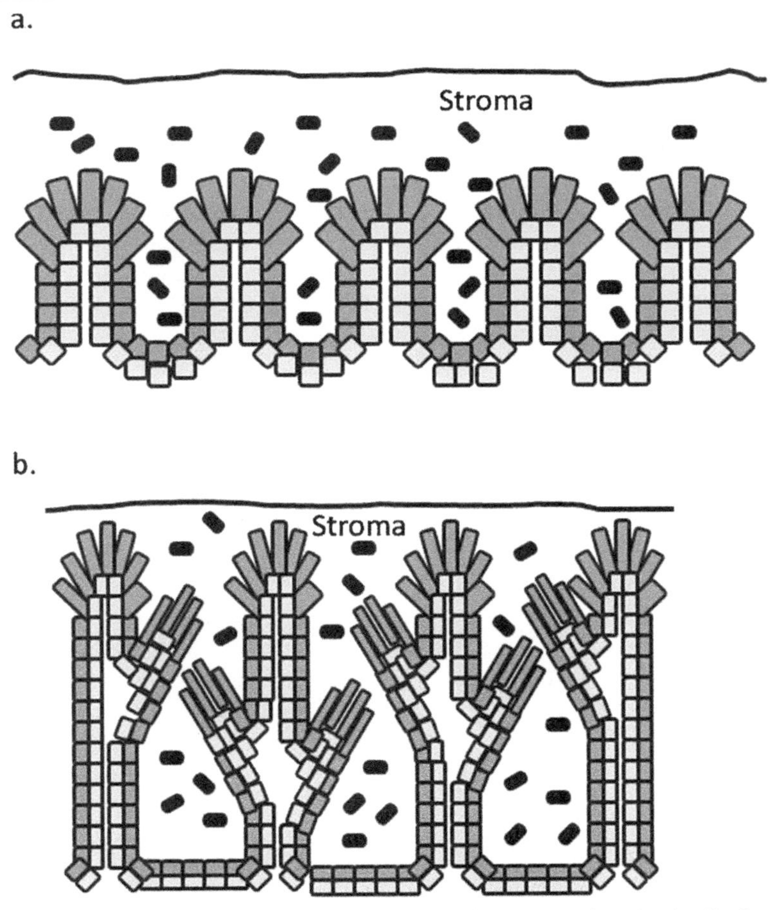

Figure 2 Schematic showing changes in the pig placental epithelial bilayer during gestation. On day 60 of pregnancy **(a)** the folded bilayer has a regular appearance and is embedded in placental stroma. The stromal covering of the folds is relatively deep. The fetal placental epithelial cells at the tops (fetus side is up in the figures) of the folds are tall columnar in appearance, the sides and bottoms are lined with cubiodal epithelia. By day 105 of gestation **(b)**, the folded bilayer is wider, more complex and has secondary folds, which increases the interacting surface area. The epithelial cells of the bilayer are thinner to reduce the distance solutes must diffuse. The stromal space between the folds is expanded to accommodate the secondary folds. The stroma covering the folds is thinner and can be absent in placenta of small fetuses. Fetal and maternal capillaries (not shown) are immediately adjacent to the epithelial cells within the folded structure and are arranged in a countercurrent fashion.

factor (VEGF), a known angiogenic growth factor, has been reported to increase with advancing gestation and is also correlated with the fetal weight to placental weight ratio [31,32]. These results suggest that fetal placental angiogenesis may contribute to placental efficiency.

Leiser and Dantzer [33] used microcast techniques to visualize maternal and fetal capillaries and concluded that they were arranged in a cross-countercurrent fashion on opposite sides of and directly adjacent to the epithelial bilayer. According to Leiser and Dantzer [33], maternal blood enters the structure at the top of the folds and exits at the bottom (fetal side of the placenta is "up"), fetal blood enters near the bottom of the folds and exits at the top. Thus, the capillary arrangement exchanges solutes between the dam and fetus within the axis perpendicular to the surface of the placenta. The distance of the interacting surface along this axis (the width of the folds) should affect efficiency of exchange, because a greater width would increase the interaction distance between the two blood supplies and facilitate exchange.

Vallet and Freking [20] measured this width throughout gestation and compared placenta associated with the largest and smallest fetuses in litters from UHO gilts. Fold width increased during late gestation, and was greater in the placenta of small fetuses compared to large fetuses. In the same experiment, the width of the stromal area between the top of the folds and the allantois was also measured. This region may represent further room for the width of the folds to expand. The width of this stromal region became progressively less with advancing gestation, and was less in the placenta of small fetuses compared to large fetuses. In placenta of some of the smallest fetuses, a stromal region above the folds was absent, suggesting that no further expansion of the folded bilayer was possible.

Given these results, we have hypothesized that expansion of the width of the folds represents a compensatory mechanism for increasing placental efficiency if the size of the placenta is restricted (e.g., during intrauterine crowding). If expansion of the folds exceeds the available placental stromal width, further compensation is not possible and fetal growth and development are impaired. If this impairment is severe enough, the fetus is lost, contributing to late gestation loss that occurs during intrauterine crowding.

If this hypothesis is true, one strategy for improving litter size would be to focus on mechanisms of placental stroma and folded bilayer development. Stromal tissues are made up of fibroblasts embedded in extracellular matrix, large constituents of which are glycosaminoglycans like hyaluronan and heparan sulfate [34,35]. Glycosaminoglycans are polymers of sugar [36,37], probably originating primarily from glucose transported from dam to fetus. Thus, adequate stromal development, and therefore ultimately placental compensatory development, is likely to be dependent on adequate glucose transport. Logically, this suggests that placental development and fetal development compete for glucose resources. This is consistent with reports in sheep where the placenta consumes 60% of incoming glucose [38]. This competition is likely to be most severe during late gestation and suggests there may be mechanisms that regulate whether glucose is directed to either fetal or placental development.

One such mechanism might be generation of fructose. Livestock conceptuses are fructogenic [39], meaning that a substantial portion of incoming glucose is converted to fructose, primarily by the placenta [40]. Studies indicate that fructose is oxidized to CO_2 at 20% the rate of glucose, so once glucose is converted to fructose it is relatively less available for oxidation [41,42]. Fructose is an intermediate in glucosamine synthesis and glucosamine is needed for glycosaminoglycan production [43]. Also of interest, conversion of fructose to glucosamine stimulates mTOR and regulates cell proliferation in porcine and human trophoblast cells, tying fructose to placental development [44,45]. We recently compared glucose and fructose concentrations in the blood of fetuses from UHO gilts during late gestation. Neither sugar was related to fetal weight despite large differences (weight range 423 to 1473 g). Glucose concentrations were positively correlated with placental weight ($P < 0.05$; weight range 73 to 413 g), but fructose concentrations were not (Vallet, unpublished observations). This suggests that the concentration of fructose is likely to be regulated within the conceptus. Whether fructose is synthesized to sequester it from oxidation and preserve its use for glycosaminoglycan synthesis, or for some other role in metabolism, requires further study.

Factors that control folded bilayer development are largely unknown. Two types of placental trophoblast cells are present within the folded bilayer, tall columnar cells at the tops of the folds and cuboidal cells at the bottom and sides of the folds. The location of the tall columnar cells at the top of the folds suggested that they might play a role in widening of the folded structure by erosion of the surrounding stroma. To begin to address the control of fold development, enzymes that degrade extracellular matrix components of the stroma, which would be needed to alter bilayer structure, were examined in the placenta throughout gestation. The cDNAs corresponding to two forms of hyaluronidase were cloned from placental tissue. Similarly, two molecular weight forms of hyaluronidase were detectable in placental tissue extracts using zymography [35], but the correspondence between the cDNAs and molecular weight forms of the proteins has not been fully clarified. Both protein forms increased with advancing gestation and were greater in placenta from small fetuses, which is consistent with a role for these enzymes in fold formation. In addition to hyaluronidase, heparanase was also cloned from placental tissues and expression was examined using in situ hybridization [46]. Heparanase mRNA was localized to trophoblast cells lining the sides and bottom of the folded bilayer, no labeling was observed in trophoblast cells at the tops of the folds. Heparanase plays a role in modifications of the basement membrane during cell movements within tissues [47]. Thus, contrary to our hypothesis, the heparanase results suggest that modifications of the folded bilayer occur along the sides and bottoms of the folds.

To gain further insight into the role of the two trophoblast cell types in fold development, we performed a transcriptomic analysis of the two cell types that had been collected from average weight fetuses on day 85 of gestation [48] using laser capture microdissection combined with high-throughput Illumina sequencing of cDNA (gestation length in the pig is 114 days). Expression levels of 7413 genes were observed by the two cell types combined. Increased expression of 434 genes were observed for tall columnar trophoblast cells compared to cuboidal trophoblast cells, while the expression of 1088 genes were greater in cuboidal trophoblast cells compared to tall columnar cells. Ingenuity Pathway Analysis was used to determine biological pathways that would be affected by these changes in transcription. Of relevance to fold development, pathways involved in cell invasion, motility and movement were increased in the cuboidal cells lining the sides and bottom of the folds compared to tall columnar cells at the tops of the folds. Heparanase sequences were observed in this analysis, and preferential transcription of heparanase in cuboidal cells compared to tall columnar cells was confirmed. These results again suggest that fold development, or at least modifications of the morphology of the epithelial bilayer, is controlled by changes in the sides and bottom of the folds.

Nutrient specific mechanisms

Most nutrients transported from the dam to the fetus have specific mechanisms that facilitate transfer. Oxygen and CO_2 cross the bilayer by simple diffusion, but hemoglobin [49] and carbonic anhydrase [50] facilitate transport, respectively. Glucose and amino acids are polar molecules that do not cross cell membranes easily, and various proteins facilitate their uptake by cells. Very little is known regarding the specifics of lipid transport. One report indicated that transport of fatty acid across the pig placenta is very poor [51]. Vitamin and mineral transport is facilitated in a variety of ways. For example, iron in the form of uteroferrin is secreted by the uterine glands as a component of histotroph, which is then taken up by placental epithelial cells within the areolae by pinocytosis [26]. Histotroph is likely to be a vehicle for the transfer of a variety of nutrients in a similar fashion. An example of a nutrient specific mechanism that facilitates vitamin transport are placental folate binding proteins [52,53], which likely mediate folate transport to the developing conceptus.

Our transcriptomic analysis of placental trophoblast cells presented an opportunity to survey genes involved in nutrient transport. To gain further information on transport mechanisms, we focused attention on solute carrier (SLC) genes expressed by the two different types of trophoblast cells. Swine genome 9.2, the version of the genome used for matching of transcriptomic sequences, included 239 SLC genes. This only included SLC genes placed in the completed sequence, and does not include all known SLC genes (e.g., notably absent are SLC2 sugar transporter genes SLC2A1, 2, 3, 5, 6, 9 and 11). In addition to this limitation, only genes with average expression between the two cell types of greater than 1.8 fragments per kilobase of exon model mapped per million mapped reads (FPKM) were considered to be present in the cells [48] (Table 1). Nevertheless, the results provide information on an extensive list of SLC genes expressed by trophoblast cells, and indicate that numerous genes for glucose, amino acid, lipid, vitamin and mineral transport proteins are expressed by placental trophoblast cells.

One final caveat regarding these results is that although SLC genes are all involved in solute transport across membranes, not all are involved in nutrient transport across the plasma membrane, which would be required for a role in transport between the dam and fetus. Many are involved in transport between organelles within cells (e.g., SLC25 genes are all mitochondrial transport genes). Some provide substrate influx into the cell, while others provide substrate efflux, and some do both depending on their cellular location (e.g., apical vs. basal) and the conditions in and around the cell. Because two epithelial cell layers separate maternal and fetal blood in the pig, all nutrients must pass four intact cell membranes to reach the fetal blood, so whether they participate in influx or efflux, they would still contribute to nutrient transport to the fetus. Despite these limitations, the results provide suggestions for future experiments that could be directed at understanding the transport of nutrients and the physiology of the pig placenta.

General

Of the 293 SLC genes in the Swine Genome build 9.2, 108 genes (37% of total) had expression values greater than 1.8 PKM. Of these, the expression of 30 (28% of expressed) were greater in short cuboidal compared to

Table 1 Expression (FPKM) of SLC genes by short cuboidal and tall columnar trophoblast epithelial cells (see Figure 2) from pig placenta collected on day 85 of gestation by laser capture microdissection (n = 4 pigs)

Gene	Substrate	Name	P value SC versus TC	Short cuboidal	Tall columnar
SLC1A3	Amino Acid	solute carrier family 1 (glial high affinity glutamate transporter), member 3	0.008342	103.518	152.46*
SLC1A5	Amino Acid	solute carrier family 1 (neutral amino acid transporter), member 5	0.005574	19.865	29.593*
SLC2A10	Glucose	solute carrier family 2 (facilitated glucose transporter), member 10	0.356081	1.774	2.437
SLC2A12	Glucose	solute carrier family 2 (facilitated glucose transporter), member 12	0.019601	93.214	124.979*
SLC2A13	Inositol	solute carrier family 2 (facilitated glucose transporter), member 13	0.012995	38.491*	26.879
SLC2A4	Glucose	solute carrier family 2 (facilitated glucose transporter), member 4	0.04111	3.39*	2.185
SLC3A1	Amino Acid	solute carrier family 3 (cystine, dibasic and neutral amino acid transporters, activator of cystine, dibasic and neutral amino acid transport), member 1	0.676755	29.255	28.814
SLC4A1AP	None	solute carrier family 4 (anion exchanger), member 1, adaptor protein	0.510953	3.625	3.356
SLC4A4	Bicarbonate	solute carrier family 4, sodium bicarbonate cotransporter, member 4	0.40478	3.657	2.806
SLC4A7	Bicarbonate	solute carrier family 4, sodium bicarbonate cotransporter, member 7	0.0625	3.787	2.531
SLC5A6	Vitamins	solute carrier family 5 (sodium-dependent vitamin transporter), member 6	2.08E-05	27.352*	7.827
SLC6A6	Amino Acid	solute carrier family 6 (neurotransmitter transporter, taurine), member 6	0.700481	3.691	4.132
SLC6A8	Creatine	solute carrier family 6 (neurotransmitter transporter, creatine), member 8	0.0925	3.97	2.598
SLC7A2	Amino Acid	solute carrier family 7 (cationic amino acid transporter, y + system), member 2	0.35208	2.194	1.559
SLC7A4	Amino Acid?	solute carrier family 7 (orphan transporter), member 4	0.612879	101.015	99.302
SLC7A7	Amino Acid	solute carrier family 7 (amino acid transporter light chain, y + L system), member 7	0.048306	3.179*	2.013
SLC7A8	Amino Acid	solute carrier family 7 (amino acid transporter light chain, L system), member 8	0.092703	6.23	4.956
SLC7A9	Amino Acid	solute carrier family 7 (glycoprotein-associated amino acid transporter light chain, bo,+ system), member 9	0.004643	2.419*	4.289
SLC9A1	Na+/H+	solute carrier family 9, subfamily A (NHE1, cation proton antiporter 1), member 1	0.462724	2.827	3.147
SLC9A3R1	Na+/H+	solute carrier family 9, subfamily A (NHE3, cation proton antiporter 3), member 3 regulator 1	4.51E-05	22.04*	10.738
SLC9A6	Na+/H+	solute carrier family 9, subfamily A (NHE6, cation proton antiporter 6), member 6	0.577227	3.025	3.173
SLC9A8	Na+/H+	solute carrier family 9, subfamily A (NHE8, cation proton antiporter 8), member 8	0.195039	5.92	4.937
SLC9A9	Na+/H+	solute carrier family 9, subfamily A (NHE9, cation proton antiporter 9), member 9	0.109026	4.589	6.648
SLC10A7	steroids	solute carrier family 10 (sodium/bile acid cotransporter family), member 7	0.226002	2.569	3.55
SLC12A6	K+/Cl-	solute carrier family 12 (potassium/chloride transporters), member 6	0.717437	6.458	6.382
SLC13A3	Dicarboxylate	solute carrier family 13 (sodium-dependent dicarboxylate transporter), member 3	0.004193	19.292*	28.657
SLC13A4	Sulfate	solute carrier family 13 (sodium/sulfate symporters), member 4	1.73E-06	91.288*	44.882
SLC14A1	Urea	solute carrier family 14 (urea transporter), member 1 (Kidd blood group)	0.579099	3.759	5.211
SLC15A1	Di/tripeptides	solute carrier family 15 (oligopeptide transporter), member 1	0.10901	2.251	1.715
SLC15A2	Di/tripeptides	solute carrier family 15 (H+/peptide transporter), member 2	6.49E-05	84.452	147.72*
SLC16A10	Amino Acid	solute carrier family 16, member 10 (aromatic amino acid transporter)	0.057011	47.966	38.846
SLC16A12	Creatine	solute carrier family 16, member 12 (monocarboxylic acid transporter 12)	5.09E-05	7.95*	2.177
SLC16A14	?	solute carrier family 16, member 14 (monocarboxylic acid transporter 14)	0.002365	9.309*	5.651
SLC16A7	Lactate	solute carrier family 16, member 7 (monocarboxylic acid transporter 2)	0.01361	3.627*	1.939
SLC16A9	Urate	solute carrier family 16, member 9 (monocarboxylic acid transporter 9)	0.348855	3.476	2.655
SLC17A5	Acid sugars	solute carrier family 17 (anion/sugar transporter), member 5	0.627423	7.041	7.408
SLC19A2	Thiamine	solute carrier family 19 (thiamine transporter), member 2	0.009728	3.455*	2.193
SLC20A1	Phosphate	solute carrier family 20 (phosphate transporter), member 1	0.003578	8.612*	4.872
SLC22A18	Organic cations	solute carrier family 22, member 18	0.12071	2.883	1.365
SLC22A23	?	solute carrier family 22, member 23	0.396248	3.315	3.06
SLC22A3	Polyamines	solute carrier family 22 (extraneuronal monoamine transporter), member 3	0.000145	3.563*	0.81

Table 1 Expression (FPKM) of SLC genes by short cuboidal and tall columnar trophoblast epithelial cells (see Figure 2) from pig placenta collected on day 85 of gestation by laser capture microdissection (n = 4 pigs) *(Continued)*

SLC25A1	Mitochondrial citrate, malate	solute carrier family 25 (mitochondrial carrier; citrate transporter), member 1	0.347209	20.091	17.097
SLC25A11	Mitochondrial oxoglutarate, malate	solute carrier family 25 (mitochondrial carrier; oxoglutarate carrier), member 11	0.108779	5.333	3.636
SLC25A12	Mitochondrial Amino Acid	solute carrier family 25 (aspartate/glutamate carrier), member 12	0.131405	7.194	5.996
SLC25A13	Mitochondrial Amino Acid	solute carrier family 25 (aspartate/glutamate carrier), member 13	0.409356	17.13	15.642
SLC25A14	Mitochondrial H+ (uncoupling)	solute carrier family 25 (mitochondrial carrier, brain), member 14	0.706134	1.957	2.11
SLC25A17	Mitochondrial CoA, FAD	solute carrier family 25 (mitochondrial carrier; peroxisomal membrane protein, 34 kDa), member 17	0.721955	7.931	8.946
SLC25A20	Mitochondrial carnitine	solute carrier family 25 (carnitine/acylcarnitine translocase), member 20	0.536736	18.412	20.059
SLC25A24	Mitochondrial ATP	solute carrier family 25 (mitochondrial carrier; phosphate carrier), member 24	0.345305	12.014	10.704
SLC25A25	Mitochondrial ATP	solute carrier family 25 (mitochondrial carrier; phosphate carrier), member 25	0.219067	1.837	2.319
SLC25A26	Mitochondrial adenosyl-methionine	solute carrier family 25 (S-adenosylmethionine carrier), member 26	0.458853	1.883	1.809
SLC25A27	Mitochondrial H+ (uncoupling)	solute carrier family 25, member 27	0.050934	3.482*	2.227
SLC25A28	Mitochondrial iron	solute carrier family 25 (mitochondrial iron transporter), member 28	0.398812	2.266	1.939
SLC25A3	Mitochondrial phosphate	solute carrier family 25 (mitochondrial carrier; phosphate carrier), member 3	0.465844	207.118	195.026
SLC25A32	Mitochondrial folate	solute carrier family 25 (mitochondrial folate carrier), member 32	0.342663	2.325	3.289
SLC25A33	Mitochondrial UTP	solute carrier family 25 (pyrimidine nucleotide carrier), member 33	0.02715	3.591*	1.889
SLC25A36	Mitochondrial UTP	solute carrier family 25 (pyrimidine nucleotide carrier), member 36	0.479823	10.419	11.013
SLC25A37	Mitochondrial iron	solute carrier family 25 (mitochondrial iron transporter), member 37	0.684822	8.405	8.581
SLC25A43	Mitochondrial ?	solute carrier family 25, member 43	0.004424	2.411	4.746*
SLC25A44	Mitochondrial ?	solute carrier family 25, member 44	0.660881	4.946	5.013
SLC25A46	Mitochondrial ?	solute carrier family 25, member 46	0.714232	3.622	4.053
SLC25A5	Mitochondrial ADP, ATP	solute carrier family 25 (mitochondrial carrier; adenine nucleotide translocator), member 5	0.225204	73.01	94.084
SLC26A2	Sulfate, chloride, hydroxyl ions	solute carrier family 26 (sulfate transporter), member 2	0.002566	18.742*	9.657
SLC27A4	Long chain fatty acids	solute carrier family 27 (fatty acid transporter), member 4	0.718866	2.528	2.328
SLC27A6	Long chain fatty acids	solute carrier family 27 (fatty acid transporter), member 6	0.001938	5.536*	1.159
SLC29A1	Nucleosides	solute carrier family 29 (nucleoside transporters), member 1	0.002403	51.871*	28.705
SLC29A3	Nucleosides	solute carrier family 29 (nucleoside transporters), member 3	0.0282	4.818*	3.052
SLC30A4	Zinc	solute carrier family 30 (zinc transporter), member 4	0.137271	2.308	1.538
SLC30A6	Zinc	solute carrier family 30 (zinc transporter), member 6	0.544857	6.784	6.425
SLC30A7	Zinc	solute carrier family 30 (zinc transporter), member 7	0.606204	10.946	11.388
SLC30A9	Zinc	solute carrier family 30 (zinc transporter), member 9	0.040343	12.638*	9.474
SLC31A2	Copper	solute carrier family 31 (copper transporters), member 2	0.692963	3.737	4.636
SLC35A1	Nucleotide sugar	solute carrier family 35 (CMP-sialic acid transporter), member A1	0.001966	8.876*	3.703

Table 1 Expression (FPKM) of SLC genes by short cuboidal and tall columnar trophoblast epithelial cells (see Figure 2) from pig placenta collected on day 85 of gestation by laser capture microdissection (n = 4 pigs) *(Continued)*

SLC35A3	Nucleotide sugar?	solute carrier family 35 (UDP-N-acetylglucosamine (UDP-GlcNAc) transporter), member A3	0.003442	19.107	33.186*
SLC35A4	Nucleotide sugar?	solute carrier family 35, member A4	0.717081	2.659	2.795
SLC35A5	Nucleotide sugar?	solute carrier family 35, member A5	0.683471	5.015	5.377
SLC35B1	Adenosine phospho-sulfate	solute carrier family 35, member B1	0.347209	14.398	13.503
SLC35B2	Adenosine phospho-sulfate	solute carrier family 35, member B2	0.53865	2.75	2.625
SLC35B3	Adenosine phospho-sulfate	solute carrier family 35, member B3	0.353606	6.322	5.239
SLC35C1	Nucleotide sugar	solute carrier family 35, member C1	0.00016	19.1*	6.88
SLC35D1	Nucleotide sugar	solute carrier family 35 (UDP-glucuronic acid/UDP-N-acetylgalactosamine dual transporter), member D1	0.688317	3.286	2.87
SLC35E1	?	solute carrier family 35, member E1	0.066081	4.953	2.826
SLC35E3	?	solute carrier family 35, member E3	0.265752	2.049	1.612
SLC35F2	?	solute carrier family 35, member F2	0.10158	3.188	2.186
SLC35F5	?	solute carrier family 35, member F5	0.289527	22.192	24.888
SLC36A1	Amino acid	solute carrier family 36 (proton/amino acid symporter), member 1	0.157349	19.844	16.848
SLC36A4	Amino acid	solute carrier family 36 (proton/amino acid symporter), member 4	0.159179	2.836	1.761
SLC37A1	Sugar phosphate	solute carrier family 37 (glycerol-3-phosphate transporter), member 1	0.000295	3.492*	1.626
SLC37A4	Sugar phosphate	solute carrier family 37 (glucose-6-phosphate transporter), member 4	0.196418	6.082	5.027
SLC38A1	Amino acid	solute carrier family 38, member 1	0.023371	11.839*	9.529
SLC38A6	Amino acid?	solute carrier family 38, member 6	0.00099	3.834	7.593*
SLC38A7	Glutamine	solute carrier family 38, member 7	0.657535	2.284	2.435
SLC38A9	Amino acid?	solute carrier family 38, member 9	0.203752	4.997	3.792
SLC39A11	Zinc	solute carrier family 39 (metal ion transporter), member 11	0.519439	18.918	20.158
SLC39A14	Zinc, iron, cadmium	solute carrier family 39 (zinc transporter), member 14	5.13E-05	19.797*	8.718
SLC39A7	Zinc	solute carrier family 39 (zinc transporter), member 7	0.644212	9.946	10.299
SLC39A8	Zinc, cadmium	solute carrier family 39 (zinc transporter), member 8	0.219082	6.948	4.375
SLC39A9	Zinc	solute carrier family 39 (zinc transporter), member 9	0.029289	12.607*	8.939
SLC44A1	Choline	solute carrier family 44, member 1	0.004045	29.547	45.151*
SLC44A2	Choline	solute carrier family 44, member 2	0.502802	22.687	21.741
SLC44A3	Choline?	solute carrier family 44, member 3	0.014683	6.273	10.268*
SLC44A4	Thiamine pyro-phosphate, choline	solute carrier family 44, member 4	0.002671	6.012*	2.58
SLC46A3	Folate?	solute carrier family 46, member 3	0.208452	4.484	3.561
SLC47A1	Organic cations (e.g. creatinine)	solute carrier family 47, member 1	0.645687	2.158	2.886
SLC47A2	Organic cations (e.g. creatinine)	solute carrier family 47, member 2	0.60417	3.179	2.282
SLCO2A1	Prostaglandins	solute carrier organic anion transporter family, member 2A1	0.000609	70.11	123.911*
SLCO2B1	Steroids, prostaglandins	solute carrier organic anion transporter family, member 2B1	0.004457	3.665*	2.195
SLCO4C1	thyroxin	solute carrier organic anion transporter family, member 4C1	1.95E-05	68.57*	36.456

*Asterisk indicates greater expression in cell type indicated ($P<0.05$).

tall columnar cells, and 10 (9% of expressed) were greater in tall columnar cells compared to short cuboidal cells. This compares to 20,461 genes in Swine Genome build 9.2; 7,413 (36% of total) genes expressed by trophoblast cells, and 1,088 genes (15% of expressed) greater in short cuboidal and 434 genes (6% of expressed) greater in tall columnar cells. Thus, transport genes do not appear to be preferentially expressed by trophoblast cells as a group compared to all genes, but preferential expression of SLC genes by short cuboidal cells appears to be greater generally than tall columnar cells, and greater than the frequency of preferential expression of all genes by short cuboidal cells. This may suggest that nutrient transport occurs preferentially through short cubiodal trophoblast cells, which is also consistent with their small size and greater surface area relative to tall columnar cells.

Sugars

Sugars are polar molecules, and their transport is mediated by specific proteins, facilitative glucose transporters (GLUT) and Na-dependent glucose transporters (SGLT), which are involved in passive and active transport of glucose, respectively [54]. SLC2 and SLC5 genes, respectively, correspond to these two types of transporters [55,56]. Sequences matching SLC2A7 were not observed. The two most highly expressed SLC2 genes were SLC2A12 and 13, corresponding to GLUT12 and proton-dependent inositol transporter, respectively. SLC2A4 and 10 (GLUT4 and 10) were also observed. SLC2A12 expression was greater in tall columnar cells compared to cuboidal cells, while the reverse was true for SLC2A13. GLUT12 and GLUT4 are insulin dependent glucose transporters [57], but levels of insulin are reported to be very low in the pig fetus [39]. However, insulin like growth factors (IGF) can also regulate GLUT4 transporters [58] (similar results are not available for GLUT12), suggesting that IGF1 and 2 may provide regulation of glucose transport to the pig conceptus via changes in GLUT4 and GLUT12. Also, the high expression and increased presence of GLUT12 in tall columnar trophoblast cells suggest that these cells may be specialized to provide a regulatory role for placental function through regulation of glucose transport.

The second highest SLC2 gene expressed by placenta, SLC2A13, corresponds to proton-dependent inositol transporter, which is an inositol-H^+ cotransporter that follows a proton gradient and therefore represents active transport for inositol [59,60]. The high expression of this active transporter suggests (1) that inositol has an important role in some aspect of conceptus development or function and (2) that transport of inositol would be encouraged by acidic conditions within the placenta. Consistent with an important role for inositol, concentrations in the fetus are unusually high (4-5 mmol/L, rivaling fructose), and are greater in small fetuses compared to large fetuses [61,62]. Phospholipids containing inositol, and inositol phosphates derived from them, are second messengers for a variety of receptor systems [63]. Inositol containing lipids also participate in cell membrane fusions required for trafficking of cell membrane components [64]. In addition, many membrane proteins are anchored to the cell membrane through glycophosphatidylinositol linkages [65]. Although these second messenger systems and other biochemical processes that require inositol are vital, they are unlikely to require transport of large amounts and high concentrations of inositol. Inositol can also be converted to glucuronic acid through the action of myo-inositol oxygenase (MIOX) [66]. Sequences matching the MIOX gene were found in trophoblast cells, and were greater in the short cuboidal cells similar to the SLC2A13 sequences. Glucuronic acid is a component of hyaluronan, but mammals like the pig apparently lack the enzymes necessary to generate UDP-glucuronate from glucuronic acid [67], instead UDP-glucuronate is generated using UDP-glucose as a substrate. Free glucuronic acid can be metabolized to CO_2 in the pentose shunt pathway, so it could be used for energy. Alternatively, glucuronate is also a substrate for the generation of ascorbic acid [68]. Ascorbic acid is one of the main antioxidants available to tissues to prevent oxidative damage from free radicals. Ascorbic acid is also required to transfer the iron contained in uteroferrin to transferrin [69]. Whether inositol is used for inositol lipids and second messengers, glucuronic acid, ascorbic acid or energy generation in conceptus development will require further study.

Amino acids

A number of SLC genes are known transporters of amino acids including members of the SLC1, SLC3, SLC6, SLC7, SLC16, SLC36 and SLC38 gene families [70-76] (Table 1). Of these, the three most highly expressed were SLC1A3, SLC7A4 and SLC16A10. The SLC1A3 gene codes for a glutamate/aspartate transport protein [74] (GLAST). In the brain, this protein is responsible for removal of glutamate, a potent neurotransmitter, from the synaptic space of glutamate releasing neurons. Glutamate is taken up by astrocytes and rapidly converted to glutamine, which is then released back to the neurons for reuptake and synthesis of glutamate. Interestingly, our results indicate that glutamine synthetase is highly expressed by both types of trophoblast cells (GLUL, short cuboidal cells 328.9 FPKM, tall columnar cells 388.2 FPKM), suggesting that some portion of glutamate transported into cells is likely to be converted to glutamine. Consistent with this, the glutamine concentration in fetal plasma was greatest of all the amino acids throughout most of gestation, and was much higher than maternal plasma levels [77,78]. As previously indicated, glutamine plays a role in glucosamine synthesis along with fructose, providing substrate for hyaluronan and other glycosamine glycans, and glucosamine synthesis is known to be tied to mTOR control of proliferation of day 12 trophoblast cells [44,45]. Thus, it seems possible that glutamine and fructose combine to provide overall regulation of placental development, through generation of glycosaminoglycans and control of proliferation of trophoblast cells.

SLC7A4 codes for the CAT-4 protein, which is related to the other members of SLC7 family of cationic amino acid transporter. Proteins encoded by the SLC7A1, 2 and 3 genes are membrane proteins that transport lysine, arginine and ornithine, and correspond to the y+ amino acid transport system. However, the CAT-4 protein apparently has no intrinsic transporter activity [76], so its function is not clear. Supplemental arginine is known to increase litter size, possibly through its role in NO synthesis [79,80]. However, because it is not clear what role the SLC7A4 gene has in cationic amino acid transport or metabolism, its role in arginine metabolism is also unclear.

The SLC16A10 gene encodes the MCT10 protein, an aromatic amino acid transporter [71]. Three of the four aromatic amino acids are nutritionally essential (histidine, tryptophan, phenylalanine), while the fourth (tyrosine) can be synthesized from phenyalanine. The high expression of this gene may be related to the fact that most of the transported amino acids cannot be synthesized by the fetus. In addition, the MCT10 protein is not Na+ or H+ coupled, so transport relies on the concentration gradient for these amino acids. The concentrations of all four amino acids in maternal and fetal plasma throughout gestation are consistent with passive transport as the mechanism of transport (fetal concentration is lower than maternal) [77]. If the high expression of the gene is related to the essential nature of the amino acids and the passivity of transport, it suggests that transport of these amino acids might be particularly sensitive to factors that impair SLC16A10 expression or mRNA translation, or factors that result in reduced concentrations of aromatic amino acids in the maternal blood (e.g. shortages in the diet).

Curiously, trophoblast cells highly express the SLC15A2 gene, which is a di/tripeptide transporter [81]. This suggests an alternative route of amino acid absorption by the placenta, via absorption of di- and tri-peptides, even though low levels of these are likely to be circulating in maternal blood. Alternatively, this transporter could function to allow transport of specific di- and tripeptides, such as carnosine and glutathione, respectively. Both provide antioxidant activity, which may be important in protecting the fetus from free radicals generated during respiration. Whether significant amino acids are transported to the placenta as di- and tri-peptides, or whether the SLC15A2 gene serves another role in conceptus metabolism will require further study.

Lipids

Of the SLC genes found to be expressed by trophoblast cells, only SLC27A4 and SLC27A6 transport lipids in the form of long chain fatty acids [82,83]. Neither of these genes is highly expressed by trophoblast cells (Table 1), which is consistent with previous reports indicating very poor transport of fatty acids by pig placenta [51]. Alternatively, lipids could be transported to the developing conceptus in the form of triglycerides associated with lipoproteins. Although they are not SLC genes, receptors for lipoprotein complexes are expressed by trophoblast cells, with the greatest being LRP6 (SC cells 10.9, TC cells 8.8 FPKM; McNeel et al., unpublished observations) [84]. Thus, it is likely that lipids reach the conceptus through two mechanisms, transport as free fatty acids, which appears likely to be poor, and as lipids associated with lipoproteins.

Vitamins

Vitamins are transported by SLC5A6 (biotin, pantothenic acid) [85], SLC19A2 (thiamine) [86], SLC25A32 (mitochondrial folate) [87], SLC44A1, 2 and 3 (choline) and SLC44A4 (thiamine) [88] and SLC46A3 (folate) [86]. The two most highly expressed are SLC5A6 and SLC44A1. SLC5A6 is a multivitamin transporter, and the high expression is likely to be related to the essential nature of these vitamins as cofactors in metabolic reactions. High expression of SLC44A1 (and SLC44A2) suggests that high levels of choline are needed for some aspect of fetal development. Choline is a component of choline containing phospholipids (phosphatidylcholine and sphingomyelin), which themselves are components of cell membranes. Also, choline is a component of the neurotransmitter acetylcholine. Finally, choline is a source of methyl groups for methylation reactions [89]. Methylations are particularly important for placental function, as many imprinted genes that influence placental function are controlled by differential methylation of gene copies originating from either the sire or the dam [90].

Minerals

Zinc is transported by SLC30A4, A6, A7, and A9, and SLC39A7, 8, 9, 11, and 14 [91-93]. The combined expression of these genes suggests that zinc is very important to the physiology and biochemistry of the fetus. Zinc participates in a variety of biochemical reactions, including antioxidant activity (superoxide dismutase) and prostaglandin synthesis [94-96]. But the most important to the fetus is likely to be its incorporation into carbonic anhydrase, which plays a role in metabolism of CO_2

[50]. Carbonic anhydrase catalyzes the formation of carbonate ion from CO_2, and is a major mechanism enabling the safe transport in serum of CO2 away from tissues after it is produced. Interestingly, a recent study indicated that supplementation of zinc during late pregnancy reduced the incidence of stillbirths in low birth weight piglets, reinforcing the importance of zinc to the developing pig fetus [97].

Other minerals are also transported by SLC genes [87,98-101], specifically iron (SLC25A28 and A37, both mitochondrial), copper (SLC31A2), sulfate (SLC13A4, SLC26 A2) and phosphate (SLC20A1, mitochondrial SLC25A3). Transport by these SLC genes likely support heme production for respiratory chain enzymes, copper containing superoxide dismutase and other copper containing proteins, sulfation of proteoglycans and organic compounds and myriad phosphorylation reactions.

Hormones

Although not nutrients, transport of hormones in and out of trophoblast cells is likely to have important effects on placental function in an autocrine and paracrine manner, and on the fetus in an endocrine manner. SLC genes [102,103] participate in transport of prostaglandins (SLCO21A1), steroids (SLC10A7, SLCO2B1) and thyroid hormones (SLCO4C1). Secretion of prostaglandins by the placenta are key regulators of the initiation of farrowing [104] and probably also participate in the control of blood flow. Placental tissues secrete large amounts of estrogen [105,106], which likely plays a role in mammary gland development during pregnancy. The role of thyroid hormones in the placenta is less clear, but the expression level of the SLCO4C1 gene suggests that further work may be warranted to understand how thyroid hormones may regulate placental function or fetal development in pigs.

Summary

The pig placenta mediates nutrient transport between the dam and the developing pig fetus. Physical characteristics and the expression of nutrient specific mechanisms of the placenta combine to determine the efficiency of transport of the various nutrients required for the development of a healthy piglet at birth. Further work is needed to understand and enhance the development of the folded epithelial bilayer of the pig placenta. Clues to important transport mechanisms for the pig placenta are found in the expression levels of various genes, but further work is needed to confirm the roles of these genes in transport and suggest dietary or management strategies that will improve the transport of specific nutrients to the developing pig fetus. It is still unclear what nutrients currently limit fetal development in the pig.

Conclusions

We are only beginning to understand the physiological mechanisms that control the morphology and nutrient transport capability of the pig placenta, but it is likely that further understanding will allow not only improvements in litter size, but also reduction of stillbirth and preweaning mortality. Inefficiencies in placental function during gestation are likely to be a primary contributor to perinatal and neonatal piglet losses, as well as the number of fully formed fetuses at farrowing.

Animal care and use

Previously unpublished experiments described in this report were approved by the USMARC Animal Care and Use committee, and conformed to FASS guidelines for use of Agricultural animals in research.

Abbreviations

GLAST: Glutamate/aspartate transport protein; GLUL: Glutamate-ammonia ligase; GLUT: Glucose transporters; IGF: Insulin like growth factors; MIOX: Myo-inositol oxygenase; FPKM: Fragments per kilobase of exon model mapped per million mapped reads; SC: Short cuboidal; SGLT: Na-dependent glucose transporters; SLC: Solute carrier; TC: Tall columnar; UHO: Unilateral hysterectomy-ovariectomy; USMARC: U.S. Meat Animal Research Center; VEGF: Vascular endothelial growth factor.

Competing interests

The authors declare that they have no competing interests.

Authors' contributions

JLV collaborated on many of the U.S. Meat Animal Research Center (USMARC) studies included in this review and drafted the manuscript. AKM performed the transcriptomic analyses. JRM led or collaborated on the glycosaminoglycan work performed at USMARC that is reviewed in the manuscript. BAF led the uterine capacity selection line studies and collaborated on fetal and placental growth studies performed at USMARC and reviewed the manuscript. All authors read and approved the final manuscript.

Acknowledgements

Mention of trade names or commercial products in this article is solely for the purpose of providing specific information and does not imply recommendation or endorsement by the USDA. The U.S. Department of Agriculture (USDA) prohibits discrimination in all its programs and activities on the basis of race, color, national origin, age, disability, and where applicable, sex, marital status, familial status, parental status, religion, sexual orientation, genetic information, political beliefs, reprisal, or because all or part of an individual's income is derived from any public assistance program. (Not all prohibited bases apply to all programs). Persons with disabilities who require alternative means for communication of program information (Braille, large print, audiotape, etc.) should contact USDA's TARGET Center at (202) 720-2600 (voice and TDD). To file a complaint of discrimination, write to USDA, Director, Office of Civil Rights, 1400 Independence Avenue, S.W., Washington, D.C. 20250-9410, or call (800) 795-3272 (voice) or (202) 720-6382 (TDD). USDA is an equal opportunity provider and employer.

References

1. Stalder KJ: **Pork industry productivity analysis.** 2013, [http://www.pork.org/filelibrary/Research/IPAfull.pdf]
2. Rootwelt V, Reksen O, Farstad W, Framstad T: **Postpartum deaths: piglet, placental, and umbilical characteristics.** *J Anim Sci* 2013, **91**:2647–2656.
3. Fahmy MH, Bernard C: **Causes of mortality in Yorkshire pigs from birth to 20 weeks of age.** *Can J Anim Sci* 1971, **51**:351–359.

4. Fahmy MH, Holtmann WB, MacIntyre TM, Moxley JE: **Evaluation of piglet mortality in 28 two-breed crosses among eight breeds of pig.** *Anim Prod* 1978, **26**:277–285.
5. Mesa H, Safranski TJ, Cammack KM, Weaber RL, Lamberson WR: **Genetic and phenotypic relationships of farrowing and weaning survival to birth and placental weights in pigs.** *J Anim Sci* 2006, **84**:32–40.
6. Vallet JL: **Fetal erythropoiesis and other factors which influence uterine capacity in swine.** *J Appl Anim Res* 2000, **17**:1–26.
7. Chen Z-Y, Dziuk PJ: **Influence of initial length of uterus per embryo and gestation stage on prenatal survival, development, and sex ratio in the pig.** *J Anim Sci* 1993, **71**:1895–1901.
8. Christenson RK, Leymaster KA, Young LD: **Justification of unilateral hysterectomy-ovariectomy as a model to evaluate uterine capacity in swine.** *J Anim Sci* 1987, **65**:738–744.
9. Longenecker DE, Day BN: **Fertility level of sows superovulated at post-weaning estrus.** *J Anim Sci* 1968, **27**:709–711.
10. Bazer FW, Robison OW, Clawson AJ, Ulberg LC: **Uterine capacity at two stages of gestation in gilts following embryo superinduction.** *J Anim Sci* 1969, **29**:30–34.
11. Wu MC, Hentzel MD, Dziuk PJ: **Relationships between uterine length and number of fetuses and prenatal mortality in pigs.** *J Anim Sci* 1987, **65**:762–770.
12. Johnson RK, Nielsen MK, Casey DS: **Responses in ovulation rate, embryonal survival, and litter traits in swine to 14 generations of selection to increase litter size.** *J Anim Sci* 1999, **77**:541–557.
13. Leymaster KA, Christenson RK: **Direct and correlated responses to selection for ovulation rate or uterine capacity in swine [abstract].** *J Anim Sci* 2000, **78**(Suppl. 1):68 (Abstr # 284).
14. Freking BA, Leymaster KA, Vallet JL, Christenson RK: **Number of fetuses and conceptus growth throughout gestation in lines of pigs selected for ovulation rate or uterine capacity.** *J Anim Sci* 2007, **85**:2093–2103.
15. Biensen NJ, Wilson ME, Ford SP: **The impacts of uterine environment and fetal genotype on conceptus size and placental vascularity during late gestation in pigs.** *J Anim Sci* 1999, **77**:954–959.
16. Wilson ME, Ford SP: **Comparative aspects of placental efficiency.** *Reprod Suppl* 2001, **58**:223–232.
17. Wilson ME, Biensen NJ, Ford SP: **Novel insight into the control of litter size in pigs, using placental efficiency as a selection tool.** *J Anim Sci* 1999, **77**:1654–1658.
18. Mesa H, Safranski TJ, Fischer KA, Cammack KM, Lamberson WR: **Selection for placental efficiency in swine: genetic parameters and trends.** *J Anim Sci* 2005, **83**:983–991.
19. Vallet JL, Leymaster KA, Cassady JP, Christenson RK: **Are hematocrit and placental efficiency selection tools for uterine capacity in swine? [abstract].** *J Anim Sci* 2001, **79**(Suppl. 2):89 (Abstr # 257).
20. Vallet JL, Freking BA: **Differences in placental structure during gestation associated with large and small pig fetuses.** *J Anim Sci* 2007, **85**:3267–3275.
21. Vallet JL, McNeel AK, Johnson G, Bazer FW, Triennial Reproduction Symposium: **Limitations in uterine and conceptus physiology that lead to fetal losses.** *J Anim Sci* 2013, **91**:3030–3040.
22. Huxley JS: *Problems of Relative Growth.* New York: The Dial Press; 1932.
23. Vallet JL, Freking BA: **Changes in fetal organ weights during gestation after selection for ovulation rate and uterine capacity in swine.** *J Anim Sci* 2006, **84**:2338–2345.
24. MacDonald AA, Bosma AA: **Notes on placentation in the Suina.** *Placenta* 1985, **6**:83–91.
25. Leiser R, Dantzer V: **Initial vascularisation in the pig placenta: II. Demonstration of gland and areola-gland subunits by histology and corrosion casts.** *Anat Rec* 1994, **238**:326–334.
26. Raub TJ, Bazer FW, Roberts RM: **Localization of the iron transport glycoprotein, uteroferrin, in the porcine endometrium and placenta by using immunocolloidal gold.** *Anat Embryol* 1985, 171:253–258.
27. Roberts RM, Bazer FW: **The functions of uterine secretions.** *J Reprod Fertil* 1988, **82**:875–892.
28. Friess AE, Sinowatz F, Skolek-Winnisch R, Träutner W: **The placenta of the pig. I. Finestructural changes of the placental barrier during pregnancy.** *Anat Embryol* 1980, **158**:179–191.
29. Ford SP, Reynolds LP, Ferrell CL: **Blood flow, steroid secretion and nutrient uptake of the gravid uterus during the periparturient period in sows.** *J Anim Sci* 1984, **59**:1085–1091.
30. Pére M-C, Etienne M: **Uterine blood flow in sows: effects of pregnancy stage and litter size.** *Reprod Nutr Dev* 2000, **40**:369–382.
31. Vonnahme KA, Wilson ME, Ford SP: **Relationship between placental vascular endothelial growth factor expression and placental/endometrial vascularity in the pig.** *Biol Reprod* 2001, **64**:1821–1825.
32. Vonnahme KA, Ford SP: **Differential expression of the vascular endothelial growth factor-receptor system in the gravid uterus of Yorkshire and Meishan pigs.** *Biol Reprod* 2004, **71**:163–169.
33. Leiser R, Dantzer V: **Structural and functional aspects of porcine placental microvasculature.** *Anat Embryol* 1988, **177**:409–419.
34. Steele VS, Froseth JA: **Effect of gestational age on the biochemical composition of porcine placental glycosaminoglycans.** *Proc Soc Exp Biol Med* 1980, **165**:480–485.
35. Vallet JL, Miles JR, Freking BA: **Effect of fetal size on fetal placental hyaluronan and hyaluronoglucosaminidases throughout gestation in the pig.** *Anim Reprod Sci* 2010, **118**:297–309.
36. Laurent TC, Fraser JRE: **Hyaluronan.** *FASEB J* 1992, **6**:2397–2404.
37. Heldin P: **Importance of hyaluronan biosynthesis and degradation in cell differentiation and tumor formation.** *Braz J Med Biol Res* 2003, **36**:967–973.
38. Bell AW, Hay WW Jr, Ehrhardt RA: **Placental transport of nutrients and its implications for fetal growth.** *J Reprod Fertil Suppl* 1999, **54**:401–410.
39. Père MC: **Maternal and fetal blood levels of glucose, lactate, fructose, and insulin in the conscious pig.** *J Anim Sci* 1995, **73**:2994–2999.
40. White CE, Piper EL, Noland PR: **Conversion of glucose to fructose in the fetal pig.** *J Anim Sci* 1979, **48**:585–590.
41. Huggett ASG, Pelc SR: **The utilization of foetal blood fructose in the sheep.** *J Physiol* 1964, **171**(Suppl):42P–43P.
42. Meznarich HK, Hay WW Jr, Sparks JW, Meschia G, Battaglia FC: **Fructose disposal and oxidation rates in the ovine fetus.** *Q J Exp Physiol* 1987, **72**:617–625.
43. Moussian B: **The role of GlcNAc in formation and function of extracellular matrices.** *Comp Biochem Physiol B Biochem Mol Biol* 2008, **149**:215–226.
44. Wen HY, Abbasi S, Kellems RE, Xia Y: **mTOR: a placental growth signaling sensor.** *Placenta* 2005, **26**(Suppl A):S63–S69.
45. Kim J, Song G, Wu G, Bazer FW: **Functional roles of fructose.** *Proc Natl Acad Sci U S A* 2012, **109**:E1619–E1628.
46. Miles JR, Vallet JL, Freking BA, Nonneman DJ: **Molecular cloning and characterisation of heparanase mRNA in porcine placenta throughout gestation.** *Reprod Fertil Dev* 2009, **21**:757–772.
47. Fux L, Ilan N, Sanderson RD, Vlodavsky I: **Heparanase: busy at the cell surface.** *Trends Biochem Sci* 2009, **34**:511–519.
48. McNeel AK, Chen C, Schroeder S, Sonstegard T, Dawson H, Vallet JL: **Application of RNA-seq transcriptomic analysis to reproductive physiology of the pig: Insights into differential trophoblast function within the late gestation porcine placenta.** In *Control of Pig Reproduction IX.* Edited by Rodriquez-Martinez H, Soede NM, Flowers WL. Leicestershire: Context Products Ltd; 2013:175–183.
49. Kim HD, Duhm J: **Postnatal decrease in the oxygen affinity of pig blood induced by red cell 2,3-DPG.** *Am J Physiol* 1974, **226**:1001–1006.
50. Geers C, Gros G: **Carbon dioxide transport and carbonic anhydrase in blood and muscle.** *Physiol Rev* 2000, **80**:681–715.
51. Thulin AJ, Allee GL, Harmon DL, Davis DL: **Utero-placental transfer of octanoic, palmitic and linoleic acids during late gestation in gilts.** *J Anim Sci* 1989, **67**:738–745.
52. Kim JG, Vallet JL: **Secreted and placental membrane forms of folate-binding protein occur sequentially during pregnancy in swine.** *Biol Reprod* 2004, **71**:1214–1219.
53. Kim JG, Vallet JL: **Placental expression of the membrane form of folate binding protein during pregnancy in swine.** *Theriogenology* 2007, **67**:1279–1284.
54. Olson AL, Pessin JE: **Structure, function, and regulation of the mammalian facilitative glucose transporter gene family.** *Annu Rev Nutr* 1996, **16**:235–256.
55. Zhao F-Q, Keating AF: **Functional properties and genomics of glucose transporters.** *Curr Genomics* 2007, **8**:113–128.
56. Wright EM: **Glucose transport families SLC5 and SLC50.** *Mol Aspects Med* 2013, **34**:183–196.
57. Stuart CA, Howell MEA, Zhang Y, Yin D: **Insulin-stimulated translocation of glucose transporter (GLUT) 12 parallels that of GLUT4 in normal muscle.** *J Clin Endocrinol Metab* 2009, **94**:3535–3542.
58. Montessuit C, Rosenblatt-Velin N, Papageorgiou I, Campos L, Pellieux C, Palma T, Lerch R: **Regulation of glucose transporter expression in cardiac myocytes: p38 MAPK is a strong inducer of GLUT4.** *Cardiovasc Res* 2004, **64**:94–104.

59. Di Daniel E, Mok MHS, Mead E, Mutinelli C, Zambello E, Caberlotto LL, Pell TJ, Langmead CJ, Shah AJ, Duddy G, Kew JNC, Maycox PR: **Evaluation of expression and function of the H^+/*myo*-inositol transporter HMIT.** *BMC Cell Biology* 2009, **10**:54.
60. Fu H, Li B, Hertz L, Peng L: **Contributions in astrocytes of SMIT1/2 and HMIT to *myo*-inositol uptake at different concentrations and pH.** *Neurochem Int* 2012, **61**:187–194.
61. Nissen PM, Nebel C, Oksbjerg N, Bertram HC: **Metabolomics reveals relationship between plasma inositols and birth weight: Possible markers for fetal programming of type 2 diabetes.** *J Biomed Biotechnol* 2011, **2011**:Article ID 378268.
62. Metges CC, Görs S, Lang IS, Hammon HM, Brüssow K-P, Weitzel JM, Nürnberg G, Rehfeldt C, Otten W: **Low and high dietary protein: carbohydrate ratios during pregnancy affect materno-fetal glucose metabolism in pigs.** *J Nutr* 2014, **144**:155–163.
63. Decrock E, De Bock M, Wang N, Gadicherla AK, Bol M, Delvaeye T, Vandenabeele P, Vinken M, Bultynck G, Krysko DV, Leybaert L: **IP3, a small molecule with a powerful message.** *Biochim Biophys Acta* 1833, **2013**:1772–1786.
64. Michell RH: **Inositol lipids: from an archaeal origin to phosphatidylinositol 3,5-bisphosphate faults in human disease.** *FEBS J* 2013, **280**:6281–6294.
65. Ikezawa H: **Glycosylphosphatidylinositol (GPI)-anchored proteins.** *Biol Pharm Bull* 2002, **25**:409–417.
66. Thorsell A-G, Persson C, Voevodskaya N, Busam RD, Hammarström M, Gräslund S, Gräslund A, Hallberg BM: **Structural and biophysical characterization of human *myo*-inositol oxygenase.** *J Biol Chem* 2008, **283**:15209–15216.
67. Gangl R, Behmüller R, Tenhaken R: **Molecular cloning of a novel glucuronokinase/putative pyrophosphorylase from Zebrafish acting in an UDP-glucuronic acid salvage pathway.** *PLoS One* 2014, **9**:e89690.
68. Linster CL, Van Schaftingen E: **Vitamin C. Biosynthesis, recycling and degradation in mammals.** *FEBS J* 2007, **274**:1–22.
69. Buhi WC, Ducsay CA, Bazer FW, Roberts RM: **Iron transfer between the purple phosphatase uteroferrin and transferrin and its possible role in iron metabolism of the fetal pig.** *J Biol Chem* 1982, **257**:1712–1723.
70. Fotiadis D, Kanai Y, Palacín M: **The SLC3 and SLC7 families of amino acid transporters.** *Mol Aspects Med* 2013, **34**:139–158.
71. Halestrap AP: **The SLC16 gene family - Structure, role and regulation in health and disease.** *Mol Aspects Med* 2013, **34**:337–349.
72. Pramod AB, Foster J, Carvelli L, Henry LK: **SLC6 transporters: structure, function, regulation, disease association and therapeutics.** *Mol Aspects Med* 2013, **34**:197–219.
73. Hägglund MGA, Sreedharan S, Nilsson VCO, Shaik JHA, Almkvist IM, Bäcklin S, Wrange Ö, Fredriksson R: **Identification of SLC38A7 (SNAT7) protein as a glutamine transporter expressed in neurons.** *J Biol Chem* 2011, **286**:20500–20511.
74. Kanai Y, Hediger M: **The glutamate/neutral amino acid transporter family SLC1: molecular, physiological and pharmacological aspects.** *Pflugers Arch - Eur J Physiol* 2004, **447**:469–479.
75. Thwaites DT, Anderson CMH: **The SLC36 family of proton-coupled amino acid transporters and their potential role in drug transport.** *Br J Pharmacol* 2011, **164**:1802–1816.
76. Wolf S, Janzen A, Vékony N, Martiné U, Strand D, Closs EI: **Expression of solute carrier 7A4 (SLC7A4) in the plasma membrane is not sufficient to mediate amino acid transport activity.** *Biochem J* 2002, **364**:767–775.
77. Wu G, Bazer FW, Tou W: **Developmental changes of free amino acid concentrations in fetal fluids of pigs.** *J Nutr* 1995, **125**:2859–2868.
78. Self JT, Spencer TE, Johnson GA, Hu J, Bazer FW, Wu G: **Glutamine synthesis in the developing porcine placenta.** *Biol Reprod* 2004, **70**:1444–1451.
79. Mateo RD, Wu G, Carroll JA, Shinzato I, Kim SW: **Dietary L-arginine supplementation improves pregnancy outcome in gilts [abstract].** *J Anim Sci* 2006, **84**(Suppl. 2):49.
80. Wu G, Bazer FW, Burghardt RC, Johnson GA, Kim SW, Li XL, Satterfield MC, Spencer TE: **Impacts of amino acid nutrition on pregnancy outcome in pigs: Mechanisms and implications for swine production.** *J Anim Sci* 2010, **88**:E195–E204.
81. Bahadduri PM, D'Souza VM, Pinsonneault JK, Sadée W, Bao S, Knoell DL, Swaan PW: **Functional characterization of the peptide transporter PEPT2 in primary cultures of human upper airway epithelium.** *Am J Respir Cell Mol Biol* 2005, **32**:319–325.
82. Stahl A: **A current review of fatty acid transport proteins (SLC27).** *Pflugers Arch - Eur J Physiol* 2004, **447**:722–727.
83. Gimeno RE: **Fatty acid transport proteins.** *Curr Opin Lipidol* 2007, **18**:271–276.
84. Joiner DM, Ke J, Zhong Z, Xu HE, Williams BO: **LRP5 and LRP6 in development and disease.** *Trends Endocrinol Metab* 2013, **24**:31–39.
85. de Carvalho FD, Quick M: **Surprising substrate versatility in SLC5A6: Na^+-coupled I^- transport by the human Na^+/multivitamin transporter (hSMVT).** *J Biol Chem* 2011, **286**:131–137.
86. Zhao R, Goldman ID: **Folate and thiamine transporters mediated by facilitative carriers (SLC19A1-3 and SLC46A1) and folate receptors.** *Mol Aspects Med* 2013, **34**:373–385.
87. Gutiérrez-Aguilar M, Baines CP: **Physiological and pathological roles of mitochondrial SLC25 carriers.** *Biochem J* 2013, **454**:371–386.
88. Traiffort E, O'Regan S, Ruat M: **The choline transporter-like family SLC44: properties and roles in human diseases.** *Mol Aspects Med* 2013, **34**:646–654.
89. Zeisel SH: **Metabolic crosstalk between choline/1-carbon metabolism and energy homeostasis.** *Clin Chem Lab Med* 2013, **51**:467–475.
90. Renfree MB, Suzuki S, Kaneko-Ishino T: **The origin and evolution of genomic imprinting and viviparity in mammals.** *Phil Trans R Soc B* 2013, **368**:20120151.
91. Huang L, Kirschke CP, Zhang Y, Yu YY: **The *ZIP7* gene (*Slc39a7*) encodes a zinc transporter involved in zinc homeostasis of the Golgi apparatus.** *J Biol Chem* 2005, **280**:15456–15463.
92. Yu Y, Wu A, Zhang Z, Yan G, Zhang F, Zhang L, Shen X, Hu R, Zhang Y, Zhang K, Wang F: **Characterization of the GufA subfamily member SLC39A11/Zip11 as a zinc transporter.** *J Nutr Biochem* 2013, **24**:1697–1708.
93. Huang L, Tepaamorndech S: **The SLC30 family of zinc transporters - A review of current understanding of their biological and pathophysiological roles.** *Mol Aspects Med* 2013, **34**:548–560.
94. Richards MP: **Zinc, copper, and iron metabolism during porcine fetal development.** *Biol Trace Elem Res* 1999, **69**:27–44.
95. Favier AE: **The role of zinc in reproduction. Hormonal mechanisms.** *Biol Trace Elem Res* 1992, **32**:363–382.
96. Mistry HD, Williams PJ: **The importance of antioxidant micronutrients in pregnancy.** *Oxid Med Cell Longev* 2011, **2011**:Article 841749.
97. Vallet JL, Rempel LA, Miles JR, Webel SK: **Effect of essential fatty acid and zinc supplementation during pregnancy on birth intervals, neonatal piglet brain myelination, stillbirth, and preweaning mortality.** *J Anim Sci* 2014, **92**:2422–2432.
98. Pajor AM: **Molecular properties of the SLC13 family of dicarboxylate and sulfate transporters.** *Pflugers Arch - Eur J Physiol* 2006, **451**:597–605.
99. Forster IC, Hernando N, Biber J, Murer H: **Phosphate transporters of the SLC20 and SLC34 families.** *Mol Aspects Med* 2013, **34**:386–395.
100. Kim H, Wu X, Lee J: **SLC31 (CTR) family of copper transporters in health and disease.** *Mol Aspects Med* 2013, **34**:561–570.
101. Ohana E, Shcheynikov N, Park M, Muallem S: **Solute carrier family 26 member a2 (Slc26a2) protein functions as an electroneutral $SO_4^{2-}/OH^-/Cl^-$ exchanger regulated by extracellular Cl^-.** *J Biol Chem* 2012, **287**:5122–5132.
102. Hagenbuch B, Stieger B: **The *SLCO* (former *SLC21*) superfamily of transporters.** *Mol Aspects Med* 2013, **34**:396–412.
103. da Silva TC, Polli JE, Swaan PW: **The solute carrier family 10 (SLC10): beyond bile acid transport.** *Mol Aspects Med* 2013, **34**:252–269.
104. First NL, Bosc MJ: **Proposed mechanisms controlling parturition and the induction of parturition in swine.** *J Anim Sci* 1979, **48**:1407–1421.
105. Kensinger RS, Collier RJ, Bazer FW, Kraeling RR: **Effect of number of conceptuses on maternal hormone concentrations in the pig.** *J Anim Sci* 1986, **62**:1666–1674.
106. Robertson HA, Dwyer RJ, King GJ: **Oestrogens in fetal and maternal fluids throughout pregnancy in the pig and comparisons with the ewe and cow.** *J Endocrinol* 1985, **106**:355–360.

8

Dual potential of microalgae as a sustainable biofuel feedstock and animal feed

Krystal K Lum, Jonggun Kim and Xin Gen Lei*

Abstract

The rise in global population has led to explorations of alternative sources of energy and food. Because corn and soybean are staple food crops for humans, their common use as the main source of dietary energy and protein for food-producing animals directly competes with their allocation for human consumption. Alternatively, de-fatted marine microalgal biomass generated from the potential biofuel production may be a viable replacement of corn and soybean meal due to their high levels of protein, relatively well-balanced amino acid profiles, and rich contents of minerals and vitamins, along with unique bioactive compounds. Although the full-fatted (intact) microalgae represent the main source of omega-3 (n-3) polyunsaturated fatty acids including docohexaenoic acid (DHA) and eicosapentaenoic acid (EPA), the de-fatted microalgal biomass may still contain good amounts of these components for enriching DHA/EPA in eggs, meats, and milk. This review is written to highlight the necessity and potential of using the de-fatted microalgal biomass as a new generation of animal feed in helping address the global energy, food, and environmental issues. Nutritional feasibility and limitation of the biomass as the new feed ingredient for simple-stomached species are elaborated. Potential applications of the biomass for generating value-added animal products are also explored.

Keywords: Algae, Biofuel, Feed, Health, Nutrition

Introduction

The projected rise of global population from the current 7 billion to over 9 billion within the next several decades [1,2] garners urgent needs for renewable energy and alternative foods. As global supplies of petroleum continue to decline, renewable nature fuels including biofuel are being explored. Biofuels are defined as the energy derived from raw biological materials, and show promise in harnessing adequate energy and reducing greenhouse gas emissions associated with fossil fuels. As corn and soybean are staple food crops for humans as well as the two main conventional feedstuffs that provide energy and protein for food-producing animals, the projected global population rise, along with the expansion of animal production, presents a serious threat to nutrition security for both humans and animals. However, the increasing use of these feedstocks for biodiesel and bioethanol production has driven up their global prices. From 2007 to 2011, the worldwide production of bioethanol nearly doubled from 49.6 to 84.6 billion liters [3]. In parallel, the price of corn was increased from $163 to $291 per metric ton over the four years [4]. In 2011, the United States produced approximately 12.4 billion bushels of corn, and 38% of which was used to produce bioethanol or to generate other co-products [3]. Apparently, current allocations of corn and(or) soybean for the biofuel and animal productions are unsustainable. Alternatives to these ingredients are required to maintain a harmonious infrastructure among the fuel, food, and feed industries.

Marine microalgae bear attractive properties for biofuel production [5-8]. Subsequently, the fat-extracted microalgal biomass derived from the biofuel production may be a promising carbon-neutral animal feed supplement [9-11]. In fact, our laboratory has recently demonstrated that the de-fatted biomass of *Staurospira sp.* with 19% crude protein replaced 7.5% of corn and soybean meal without affecting the growth performance or health status of broiler chickens [12]. Notably, the dual application of microalgae as a new source of biofuel and animal feed will help alleviate the greenhouse effects associated with current energy and food production.

* Correspondence: xl20@cornell.edu
Department of Animal Science, Cornell University, Ithaca, NY 14850, USA

Microalgae for biofuel production

Marine microalgae sequester carbon dioxide (CO_2) through photosynthesis, and may be used to produce biogas including methane and hydrogen via anaerobic processing [13-15]. While certain species of microalgae were recognized in the 1940s to yield high amounts of cellular lipids under selective growth pressures, it was not until the 1950's when algae were viewed as a potential energy source, and were tested for methane gas production via anaerobic digestion of their cell carbohydrates [16,17]. The flexibility and(or) adaptivity of microalgal species to water and cultural conditions allows us to spare fresh water and arable land for crop production [18]. The land use efficiency of microalgae for biofuel production, grown with 30% oil content by weight, was 130 and 338 times greater than the conventional biodiesel feedstock soybean and corn, respectively [6].

While optimal growth conditions for microalgae are species-specific, photoautotrophic cultivation of these single cell species at large scales for biofuel and co-products depends on the technical and economic feasibility. At the present time, the photoautotrophic production of microalgae is marginally cost-effective only for generating value-added co-products or feed additives used in aquaculture [19,20]. In such productions, microalgae are grown in the presence of light within constructions such as open raceway ponds. To extract the lipids, microalgae are first de-watered. The concentrated biomass is subsequently processed to optimize the solvent extraction through cell disruption, particle size reduction, and drying [21]. The remaining microalgae skeleton after lipid extraction is the so-called de-fatted microalgal biomass to be used as an animal feed. Without the presumed feed application, the commercial microalgae cultivation and processing for biofuel production [22,23] remains largely cost-ineffective. Therefore, the feed application of the de-fatted biomass would not only create a new source of animal feed to mitigate the current competition with human food supply, but also help make the biofuel production of microalgae economically feasible.

Microalgae for animal nutrition

Feasibility of microalgae as animal feed

Although the use of whole microalgae in animal diets has long been studied, dating as far back as in the 1950's, only recent literature has shown attempts to supplement lipid-extracted microalgae in animal diets [12,24]. Initially, researchers sought out methods to culture algae in ponds, and the developments of which were quickly followed by studies on algae supplementation into animal diets as a protein source [25,26]. Different sources of cultivated algae were effective in maintaining animal growth performance, and in some cases improving daily body weight gain. Sewage-grown *Chlorella* and *Scenedesmus* sp. gained attention as potential nutrient sources due to their high crude protein and carotenoid contents [27-30]. In 1952, Combs reported that supplementing 10% *Chlorella* sp. into a diet deficient in riboflavin and vitamin A improved feed efficiency and growth of chicks [27]. Later work showed no adverse response of growth by chicks to diets containing 20% sewage-grown, aluminum-free *Chlorella* and *Scenedesmus* sp., compared with those fed a corn-soybean meal based diet [28].

During the following decades, pond and tap water-grown algae were found to sustain fish growth in aquaculture, and 6-10% sewage-grown algae incorporated in a barley-based diet for growing-finishing pigs maintained their growth rate and feed conversion efficiency [31,32]. Other algal variants, such as the blue-green algae *Spirulina* sp., were also investigated not only for their effects on overall growth performance, but also on organ health and reproductive characteristics of animals. Ross and Dominy supplemented diets for broiler chickens with up to 20% blue-green algae, and found the 3 wk-old broilers experienced depressed growth when algae inclusion levels were higher than 10% [33]. At the inclusion rate of 12%, broilers showed slower growth in comparison with those fed 0, 1.5, 3.0, or 6.0% algae. Toyomizu et al. [34] fed broilers with up to 8% *Spirulina* sp. and reported no differences in their body weight, liver weight, abdominal fat, or kidney weight at 16 d of age. When early-weaned piglets were fed a diet with *S. maxima* replacing up to 12% of the total protein from skim milk, apparent digestibilities of the diets were reduced, but the piglet growth remained insignificantly different from that of the control pigs [35]. Fevrier and Seve [36] supplemented *S. maxima* into the diets of sows, and showed that the algae-fed sows had lower net weight gains with more and heavier piglets at the first reproductive cycle. By the end of the second reproductive cycle, the growth and litter characteristics were similar between the algae-fed and control groups, although the algae-fed sows produced a higher culling rate of piglets. Another study showed that feeding weanling pigs with a mixture of *S. maxima* and *A. platensis*, or *Chlorella* sp. in replacing 33% of soybean meal did not affect their body weight gain, feed efficiency, or incidence of diarrhea compared with pigs fed a control diet. Apparent toxicity or gastro-intestinal lesion was absent in the algae-fed pigs [35].

In contrast, the de-fatted biomass of microalgal species, derived from the biofuel production research, has only recently been shown feasibility in replacing corn and soybean meal in diets for poultry, swine, and cattle. Studies supplementing the de-fatted biomass from *Staurospira* sp. to replace 7.5% corn and soybean meal in diets for weanling pigs did not affect their overall growth performance or plasma biochemical indicators [37,38]. However, the pigs were incapable of tolerating a 15% replacement of corn and soybean meal [38]. Factors causing this intolerance might include amino acid imbalances, disruption of the acid:base balance, high ash content of the algal biomass, and(or) an overall reduction in the buffering capacity within the gastrointestinal tract of the weanling pigs. Austic et al. [12] replaced 7.5% of soybean meal with the de-fatted biomass of the *Staurosira* sp. in the diets of broilers, and showed decreased body weight gain and feed efficiency during the first three weeks of experiment. During the following three weeks, these differences were no longer seen. In the same study, broiler chicks fed a diet containing essential amino acids (Met, Lys, Ile, Thr, Trp, and Val) co-supplemented with 7.5% of the de-fatted biomass did not show growth performance differences from the control group. A diet with 10% lipid-extracted *Nannochloropsis oculata* meal was well tolerated by adolescent male rabbits, which showed similar final body weight, serum urea nitrogen, blood glucose, and organ histology in comparison with rabbits fed a control diet [39,40]. Finishing wethers maintained similar growth performance and carcass characteristics (*longissimus* muscle area, dressing percentage, marbling score, hot carcass weight, and subcutaneous adipose depth) when fed up to 20% de-fatted algal biomass on a dry matter basis as a protein replacement, in comparison with wethers fed a control diet [24]. In a study that examined effects of supplementing dietary dry matter with de-fatted *Lithothamnium calcareum* meal in Holstein cows, up to 1% algal meal in the diet mediated venous acid–base balance after acidosis induction, but did not improve the tract digestibility or growth performance of the cows [41].

Nutritional values of microalgae

While the nutritional profiles of microalgae vary considerably with the species used, a large majority are characterized by protein, carbohydrate, and lipid contents that are comparable, if not superior, to conventional feedstuffs (Table 1). Dietary soybean meal typically contains up to 48% crude protein, and lays claim to a relatively well balanced amino acid profile. The diversity of microalgae makes certain species amenable to cultivation for diet-specific needs of humans and animals. A commonly cultivated algae species for human consumption: *S. maxima*, contains high levels of vitamin B_1, vitamin B_2, and β-carotene, and up to 71% crude protein with sufficient concentrations of all essential amino acids except for the sulfur-containing ones (Table 2) [5,42]. Since protein is considered to be the most expensive nutrient in animal feed [43], developing natural alternatives to soybean meal may be cost competitive.

Table 1 Nutrient composition of conventional feedstuffs and various algae (% dry matter)[a]

Source	Crude protein	Carbohydrates	Lipids
Soybean	37	30	20
Corn	10	85	4
Wheat	14	84	2
Anabaena cylindrical	43-56	25-30	4-7
Arthrospira maxima	60-71	13-16	6-7
Chlorella vulgaris	51-58	12-17	14-22
Spirogyra sp.	6-20	33-64	11-21
Synechococcus sp.	73	15	11

[a]Adapted from [11,44-46].

Among all dietary amino acids, lysine and methionine are the first and second limiting amino acids. Many microalgal species contain relatively high amounts of lysine, but as previously described, are somewhat deficient in the sulfur-containing amino acids: cysteine and methionine [5]. To maximize amino acid utilization by animals, diets are typically formulated by mixing different feedstuffs to balance amino acid profiles and(or) by supplementing synthetic amino acids to meet their nutrient requirements. Austic et al. [12] reported that the decreased growth performance of broilers fed the de-fatted *Staurospira* sp. biomass in the first three weeks was prevented by the supplementation of essential amino acids. In a laying hen study, a 7.5% replacement of corn and soybean meal with the de-fatted microalgal biomass did not negatively affect hen production or health parameters, yet significantly increased the redness and decreased both the lightness and yellowness of the egg yolks [47]. However, inclusions of the biomass at 10% reduced feed intake, plasma uric acid, and egg albumen weight. Up to a third of soybean meal was successfully replaced in the diets of weanling pigs by algae biomass from the cyanobacteria *S. maxima*, *Arthrospira platensis*, and *Chlorella* sp. [35]. In another study conducted on laying hens, an inclusion of 10% *Porphyridium* sp. red algal biomass did not affect their body weight, egg production rate, or egg weight, but lowered egg yolk cholesterol levels by 24% [48].

Microalgae for value-added animal products

The research on microalgal biomass supplementation in food-producing animals has opened a new gateway to improve human health. A study conducted in the last decade showed not only 10% enhancement of growth but also potential in producing iodine-rich pork for human consumption when pigs were fed algae containing naturally high iodine [49]. Likewise, microalgae may be cultivated for their advantageous fatty acid profile (Table 3), with notable enrichment in omega-3 (n-3) polyunsaturated fatty acids (PUFA) including arachidonic acid, docohexaenoic acid (DHA), eicosapentaenoic acid (EPA), and γ–linoleic acid [21]. The enriched concentrations of these omega-3 fatty acids by a variety of microalgal species represent a largely untapped natural resource with multiple health

Table 2 Amino acid profile of conventional protein sources and various algae (g/100 protein)[a]

Source	Ala	Arg	Asp	Cys	Glu	Gly	His	Ile	Leu	Lys	Met	Phe	Pro	Ser	Thr	Try	Tyr	Val
Egg	-	6.2	11.0	2.3	12.6	4.2	2.4	6.6	8.8	5.3	3.2	5.8	4.2	6.9	5.0	1.7	4.2	7.2
Soybean	5.0	7.4	1.3	1.9	19.0	4.5	2.6	5.3	7.7	6.4	1.3	5.0	5.3	5.8	4.0	1.4	3.7	5.3
Dried Whole Milk	0.9	0.9	2.0	0.2	5.5	0.5	0.7	1.6	2.6	2.1	0.6	1.3	2.5	1.4	1.2	0.4	1.2	1.7
C. vulgaris	9.4	6.9	9.3	-	13.7	6.3	2.0	3.2	9.5	6.4	1.3	5.5	5.0	5.8	5.3	-	2.8	7.0
D. bardawil	7.3	7.3	10.4	1.2	12.7	5.5	1.8	4.2	11.0	7.0	2.3	5.8	3.3	4.6	5.4	0.7	3.7	5.8
S. platensis	9.5	7.3	11.8	0.9	10.3	5.7	2.2	6.7	9.8	4.8	2.5	5.3	4.2	5.1	6.2	0.3	5.3	7.1
Aphanizomenon flos-aquae	4.7	3.8	4.7	0.2	7.8	2.9	0.9	2.9	5.2	3.2	0.7	2.5	2.9	2.9	3.3	0.7	-	3.2

[a]Based on reference [11].

implications for both humans and animals. Among many fat sources, fish products represent the major source of n-3 fatty acids [50]. However, marine fish species are incapable of synthesizing n-3 fatty acids by themselves; they may obtain n-3 fatty acids by consuming microalgae or other algae-consuming fish [51]. Most notably, dietary intakes of n-3 fatty acids are associated with decreased risks of chronic diseases [52]. However, the average fish consumption in the U.S. has not changed since 1983, at approximately 6.5 kg per year [53]. To increase the public consumption of n-3 fatty acids, studies have been conducted with full-fatted or de-fatted microalgae to generate n-3 fatty acid-fortified animal products such as milk, meats, and eggs. Apparently, the profile of PUFA in the de-fatted microalgal biomass is affected by the lipid extraction procedure. While a practical solvent extraction can remove 95% of lipids from oilseeds, it is ineffective to extract microalgal lipid due to the relatively high moisture content and cellular elasticity [21]. Grima et al. [54] reported that the lipid extraction yields of dried algal biomass range from 50 to 93% depending on the polarity of solvent systems. In fact, a higher proportion of long chain polyunsaturated fatty acids in the de-fatted biomass may bolster its appeal as an n-3 fatty acids source for developing value-added animal products.

Table 3 Generalized fatty acid profiles (%) of oil extracts from *Spirulina maxima* (SP), *Chlorella vulgaris* (Cv), *Scenedesmus obliquus* (Sc), *Dunaliella tertiolecta* (Dt), *Nannochloropsis* sp, (Nanno), and *Neochloris oleabundans* (Neo)[a]

Fatty acid	Sp	Cv	Sc	Dt	Nanno	Neo
14:0	0.34	3.07	1.48	0.47	7.16	0.43
16:0	40.16	25.07	21.78	17.70	23.35	19.35
16:1	9.19	5.25	5.95	0.88	26.87	1.85
16:2	N.D.	N.D.	3.96	3.03	0.39	1.74
16:3	0.42	1.27	0.68	1.24	0.48	0.96
16:4	0.16	4.06	0.43	10.56	N.D.	7.24
18:0	1.18	0.63	0.45	N.D.	0.45	0.98
18:1	5.43	12.64	17.93	4.87	13.20	20.29
18:2	17.89	7.19	21.74	12.37	1.21	12.99
18:3	18.32	19.05	3.76	30.19	N.D.	17.43
18:4	0.08	N.D.	0.21	N.D.	N.D.	2.10
20:0	0.06	0.09	N.D.	N.D.	N.D.	N.D.
20:1	N.D.	0.93	N.D.	N.D.	N.D.	N.D.
20:2	0.48	N.D.	N.D.	N.D.	N.D.	N.D.
20:3	N.D.	0.83	N.D.	N.D.	N.D.	N.D.
20:4	N.D.	0.23	N.D.	N.D.	2.74	N.D.
20:5	N.D.	0.46	N.D.	N.D.	14.31	N.D.
SFA	41.74	28.86	23.71	18.17	30.96	20.76
MUFA	14.62	18.82	23.88	5.75	40.07	22.14
PUFA	37.35	33.09	30.78	57.39	19.13	42.46

[a]Based on reference [55].

Microalgal biomass or oil may be supplied in the feed of ruminants to manipulate their milk fatty acid composition. The content of n-3 fatty acids in milk, especially DHA, has been shown to increase with the inclusion of algal biomass or oil, without affecting the milk yield [56-58]. However, a study with similar dietary algae incorporation reported a decrease in milk fat content [59]. Many studies in ruminants have focused on producing n-3 fatty acid-fortified milk. In such an attempt, Stamey et al. [51] supplied either algal biomass or oil to dairy cows. The overall milk fat and yield were not affected by either supplement, but the milk DHA (C22:6) content was increased by both supplementations. Glover et al. [57] reported that feeding microalgae and fresh forage decreased total milk fat content, but elevated its DHA concentration. A parallel study was conducted with sheep to determine if algae and (or) their co-supplementation with sunflower oil in the diet could enhance the nutrient profile of milk [59]. While the milk yield was unaffected by the dietary treatments, the milk fat content was decreased and the milk DHA concentration was increased as dietary algae concentration rose. Franklin et al. [58] determined that feeding dairy cows with either algae protected against ruminal biohydrogenation or unprotected algae elevated the milk DHA concentration without affecting the milk fat content.

Collectively, the supplementation of microalgae to the diets of ruminants increased the concentration of n-3 PUFA in milk, with a mixed effect on the milk fat content.

Likewise, fatty acid compositions in meats and eggs are also affected by dietary supplementations of microalgae. In lambs and horses, dietary microalgae increased the n-3 fatty acid content in meat and blood, respectively [60,61]. In pigs, dietary microalgae increased DHA concentrations in the loin and subcutaneous fat [62]. However, this increment did not manifest as a dose-dependent response [63]. To enrich eggs with n-3 fatty acids, flaxseed is usually supplemented in diets of laying hens [64]. Alternatively, dietary algae have been shown to increase n-3 and decrease n-6 fatty acids in eggs [64,65]. Supplementing flaxseed at 10% or higher usually decreases egg production and the body weight of hens. Moreover, such high flaxseed supplementations may induce fatty acid oxidation, resulting in off-odor eggs [66]. Herber and Van Elswyk [65] supplemented up to 4.8% of golden marine algae into the diet of hens, and reported increased levels of DHA with a concomitant reduction of n-6 fatty acids. The inclusion of algae in the diet affected neither the egg production nor egg quality in the 56 wk-old hens. When up to 4.3% of fermented *Schizochytrum* sp. was included in the diets of hens, their egg production and feed conversion ratio were elevated [67]. The 0.86% and 4.3% supplementation of the microalgae elevated the DHA content to 134 and 220 mg/egg, respectively.

Special considerations of microalgae

Despite their high nutritional values and health implications, caution must be taken to avoid possible toxicity of the microalgal biomass [8]. Among over 200,000 existing algal species, approximately 35,000 species are characterized [68] without apparent toxicological property [5]. However, certain species contain biogenic toxins including purines and non-biogenic toxins such as heavy metals. Some algal species are known to rapidly accumulate heavy metals at concentrations higher than their surroundings [5], while others generate pathological metabolites that cause neuro-degenerative disorders [69]. Such species require establishment and evaluation prior to their commercialization as feed supplements. Currently, no official regulations for heavy metals in microalgal products exist [5].

In spite of the relatively high crude protein contents, many microalgae species may still show limited biological values for the proteinaceous biomass due to the presence of non-protein nitrogen that consists of nucleic acids, nitrogen-containing cell walls, and amines. As single cell proteins, microalgae are photosynthetic unicellular eukaryotes containing nucleic acids that may represent approximately 10% of the crude protein [5]. Ruminal bacteria are effective in utilizing non-protein nitrogen to synthesize protein [70]. Simple-stomached animals lack ruminal microbes and thus are incapable of utilizing the relatively high levels of non-protein nitrogen in the microalgal biomass. Because the individual amino acid content and digestibility are important factors for formulating diets to support the maximum growth of animals without excreting excess nitrogen, it is necessary to optimize dietary amino acid profile and digestibility when the microalgal biomass is used to replace conventional feedstuffs.

Notably, the microalgal cell wall is largely indigestible by simple-stomached animals, yet contains immunostimulating properties that may provide added value to the biomass [71,72]. Approaches to improve the absorption of microalgal nutrients include employing enzymatic treatments to enhance cell wall digestibility [71], as well as developing genetically-altered microalgae that lack cell walls [73]. Microalgal solubility is high at alkaline conditions [74]. Comprising approximately 10% cellular dry matter [71], the cell wall of *Chlorella* sp. contains high levels of β-1,3-glucan, a bio-active immunostimulating free-radical scavenger [75]. Other species, such as those of cyanobacterial origin, have well-recognized anti-tumor, anti-bacterial, and anti-viral properties [76]. Microalgal cell wall-derived sulfated polysaccharides and carrageenans show strong T and B cell mitogenic effects [77]. Animals fed the whole *Spirulina* sp. samples accumulated greater degrees of cytotoxic lymphocytes that were critical to innate immunity, and had elevated phagocytic activity and antibody production [78].

Conclusions

The use of microalgal biomass in animal feed will not only improve human and animal food security, but also facilitate cost-effective biofuel production and reduces greenhouse gas production of agriculture [6,79-81]. Recent estimates indicate that 30% of the global algal production is used by the animal feed industry [5], amounting to a fast-growing $300 million in retail value [82]. Pragmatic species-specific large-scale algae refinery techniques must be devised to reduce the cost of the biomass production. It is necessary to determine limiting factors of the microalgal biomass that hinder its digestion and utilization by animals. Novel technology should be explored to improve microalgal nutrient utilization, and their long term nutritional and metabolic effects should be assessed. In addition, the production of DHA from microalgal biomass alone is a rapidly growing market. The last estimate is that approximately 300 tons is produced to generate $1.5 billion market value each year [72,82]. However, the tremendous potential of using the microalgal biomass in producing DHA/EPA enriched eggs, meats, and milk for improving human health is yet to be fully explored.

Abbreviations
DHA: Docosahexaenoic acid; EPA: Eicosapentaenoic acid; PUFA: Poly unsaturated fatty acid; SFA: Saturated fatty acid.

Competing interests
The authors declare that they have no competing interests.

Authors' contributions
KL, JK, and XGL wrote the review. All authors read and approved the final manuscript.

Acknowledgement
Research on this subject in the authors' laboratory has been supported in part by a USDA/DOE Biomass R&D Initiative grant.

References
1. United States Census Bureau: **International Database: World population information.** http://www.census.gov/population/international/data/idb/informationGateway.php.
2. Population Reference Bureau: **2012 World Population Data Sheet.** http://www.prb.org/pdf12/2012-population-data-sheet_eng.pdf.
3. Accelerating Industry Innovation: **2012 Ethanol Industry Outlook.** http://ethanolrfa.3cdn.net/d4ad995ffb7ae8fbfe_1vm62ypzd.pdf.
4. **Monthly commodity futures price chart.** http://futures.tradingcharts.com/chart/CN/M.
5. Becker W: **Microalgae in human and animal nutrition.** In *Handbook of Microalgal Culture.* Edited by Richmond A. Oxford, UK: Blackwell Publishing Ltd; 2004:312–351.
6. Chisti Y: **Biodiesel from microalgae.** *Biotechnol Adv* 2007, **25:**294–306.
7. Alzate ME, Muñoz R, Rogalla F, Fdz-Polanco F, SI P -E: **Biochemical methane potential of microalgae biomass after lipid extraction.** *Chem Eng J.* in press.
8. Kovač DJ, Simeunović JB, Babić OB, Mišan AČ, Milovanović IL: **Algae in food and feed.** *Food Feed Res* 2013, **40:**21–32.
9. Brune DE, Lundquist TJ, Benemann JR: **Microalgal biomass for greenhouse gas reductions: potential for replacement of fossil fuels and animal feeds.** *J Environ Eng* 2009, **135**(11):1136.
10. Shields RJ, Lupatsch I: **Algae for aquaculture and animal feeds.** *J Anim Sci* 2012, **21:**23–37.
11. Becker EW: *Microalgae: Biotechnology and Microbiology.* Cambridge: Cambridge University Press; 1994.
12. Austic RE, Mustafa A, Jung B, Gatrell S, Lei XG: **Potential and limitation of a new defatted diatom microalgal biomass in replacing soybean meal and corn in diets for broiler chickens.** *J Agric Food Chem* 2013, **61:**7341–7348.
13. Mostafa S: **Microalgal biotechnology: prospects and applications.** In *Plant Science.* Edited by Nabin Kumar D. INTech. http://www.intechopen.com/books/plant-science/microalgal-biotechnology-prospects-and-applications.
14. Li Y, Horsman M, Wu N, Lan CQ, Dubois-Calero N: **Biofuels from microalgae.** *Biotechnol Prog* 2008, **24**(4):815–820.
15. Hughes AD, Kelly MS, Black KD, Stanley MS: **Biogas from microalgae: is it time to revisit the idea?** *Biotechnol Biofuel* 2012, **5:**86.
16. Meier RL: **Biological cycles in the transformation of solar energy into useful fuels.** *J Anim Sci* 1955, **14:**179–183.
17. Oswald WJ, Golueke CG: **Biological transformation of solar energy.** *Adv Appl Microbiol* 1960, **2:**223–262.
18. Mata TM, Martins AA, Caetano NS: **Microalgae for biodiesel production and other applications: a review.** *Renew Sust Energ Rev* 2010, **14:**217–232.
19. Borowitzka M: **Microalgae for aquaculture: opportunities and constraints.** *J Appl Physiol* 1997, **9:**393–401.
20. Slade R, Bauen A: **Micro-algae cultivation for biofuels: Cost, energy balance, environmental impacts and future prospects.** *Biomass Bioenerg* 2013, **53:**29–38.
21. U.S. Department of Energy: *National algal biofuels technology roadmap: 9–10 December 2008.* Maryland; 2010. http://www1.eere.energy.gov/biomass/pdfs/algal_biofuels_roadmap.pdf.
22. Chisti Y: **Constraints to commercialization of algal fuels.** *J Biotechnol* 2013, **167:**201–214.
23. Saharan BS, Sharma D, Sahu R, Sahin O, Warren A: **Towards algal biofuel production: a concept of green bio energy development.** *Innovat Rom Food Biotechnol* 2013, **12:**1–21.
24. Dib MD, Engle TE, Han H, Roman-Muniz IN, Archibeque SL: **Effects of algal meal supplementation to finishing wethers on performance and carcass characteristics [abstract].** *J Anim Sci* 2012, **90**(Suppl 3):148.
25. Gotaas HB, Oswald WJ, Ludwig HF: **Photosynthetic reclamation of organic wastes.** *Sci Monthly* 1954, **79:**368–378.
26. Oswald WJ, Gotaas HB: **Photosynthesis in sewage treatment.** *Trans Am Soc Civ Eng* 1957, **122:**73–105.
27. Combs GF: **Algae (chlorella) as a source of nutrients for the chick.** *Science* 1952, **116:**453–454.
28. Grau CR, Klein NW: **Sewage-grown algae as a feedstuff for chicks.** *Poult Sci* 1957, **36:**1046–1051.
29. Hundley JM, Ing RB: **Algae as sources of lysine and threonine in supplementing wheat and bread diets.** *Science* 1956, **124:**536–537.
30. Fink H, Herold E: **Biological value of the protein of unicellular algae. On the protein value of unicellular algae and of young green leaves of higher plants/A contribution on alimentary liver necrosis of the rat and the treatment of protein deficiency in the tropics.** *Naturwissenschaften* 1955, **42:**516–517.
31. Ahmad MR: **Observations on the effect of feeding *Labeo rohita*(Ham.) with microcystis aeruginosa *Mycrocystis aeroginosa*.** *Hydrobiologia* 1967, **29:**388–388.
32. Hintz HF, Heitman H: **Sewage-grown algae as a protein supplement for swine.** *Anim Prod* 1967, **9:**135–140.
33. Ross E, Dominy W: **The nutritional value of dehydrated, blue-green algae (spirulina plantensis) for poultry.** *Poult Sci* 1990, **69:**794–800.
34. Toyomizu M, Sato K, Taroda H, Kato T, Akiba Y: **Effects of dietary spirulina on meat colour in muscle of broiler chickens.** *Br Poult Sci* 2001, **42:**197–202.
35. Yap TN, Wu JF, Pond WG, Krook L: **Feasibility of feeding spirulina-maxima, arthrospira-platensis or chlorella sp to pigs weaned to a dry diet at 4 to 8 days of age.** *Nutr Rep Int* 1982, **25:**543–552.
36. Fevrier C, Seve B: **Incorporation of a spiruline (spirulina maxima) in swine food.** *Ann Nutr Aliment* 1975, **29:**625–650.
37. Isaacs R, Roneker KR, Huntley M, Lei XG: **A partial replacement of soybean meal by whole or defatted algal meal in diet for weanling pigs does not affect their plasma biochemical indicators [abstract].** *J Anim Sci* 2011, **89**(Suppl 1):723.
38. Lum KK, Roneker KR, Lei XG: **Effects of various replacements of corn and soy by defatted microalgal meal on growth performance and biochemical status of weanling pigs [abatract].** *J Anim Sci* 2012, **90**(Suppl 3):701.
39. Howe BA, Roman-Muniz IN, Willson BD, Archibeque SL: **Evaluation of algae meal from Nannochloropsis oculata as a protein source for non-ruminant animals [abstract].** *J Anim Sci* 2010, **88**(Suppl 2):101–102.
40. Howe BA, Roman-Muniz IN, Willson BD, Archibeque SL: **Nannochloropsis oculata meal did not alter nutrient usage and had no adverse health effects when fed to rabbits as a protein source [abstract].** *J Anim Sci* 2011, **89**(Suppl. 2):724.
41. Lopex NM, Pereira RAN, Pereira MN: **Intake, milk yield, and blood acid–base balance of cows in response to marine algae meal [abstract].** *J Anim Sci* 2013, **91**(Suppl. 2):29.
42. Clement G, Giddey C, Menzi R: **Amino acid composition and nutritive valve of the alga Spirulina maxima.** *J Sci Food Agric* 1967, **18:**497–501.
43. Rezaei R, Wang W, Wu Z, Dai Z, Wang J, Wu G: **Biochemical and physiological bases for utilization of dietary amino acids by young Pigs.** *J Anim Sci Biotechnol* 2013, **4:**7.
44. Aaronson S, Dubinsky Z: **Mass production of microalgae.** *Experientia* 1982, **38:**36–40.
45. Fabregas J, Herrero C: **Marine microalgae as a potential source of single cell protein (SCP).** *Appl Microbiol Biotechnol* 1985, **23:**110–113.
46. Miles R, Chapman R: **The benefits of fish meal in aquaculture diets.** 2009. https://edis.ifas.ufl.edu/pdffiles/FA/FA12200.pdf.
47. Leng XJ, Hsu HN, Austic RE, Lei XG: **Defatted algae biomass may replace one-third of soybean meal in diets for laying hens [abstract].** *J Anim Sci* 2012, **90**(Suppl 3):701.
48. Ginzberg A, Cohen M, Sod-Moriah U, Shany S, Rosenshtrauch A, Arad S: **Chickens fed with biomass of the red microalga Porphyridium sp have reduced blood cholesterol level and modified fatty acid composition in egg yolk.** *J Appl Phycol* 2000, **12:**325–330.

49. He ML, Hollwich W, Rambeck WA: **Supplementation of algae to the diet of pigs: a new possibility to improve the iodine content in the meat.** *J Anim Physiol Anim Nutr (Berl)* 2002, **86:**97–104.
50. Daviglus ML, Stamler J, Orencia AJ, Dyer AR, Liu K, Greenland P, Walsh MK, Morris D, Shekelle RB: **Fish consumption and the 30-year risk of fatal myocardial infarction.** *N Engl J Med* 1997, **336:**1046–1053.
51. Stamey JA, Shepherd DM, de Veth MJ, Corl BA: **Use of algae or algal oil rich in n-3 fatty acids as a feed supplement for dairy cattle.** *J Dairy Sci* 2012, **95:**5269–5275.
52. Albert CM, Hennekens CH, O'Donnell CJ, Ajani UA, Carey VJ, Willett WC, Ruskin JN, Manson JE: **Fish consumption and risk of sudden cardiac death.** *JAMA* 1998, **279:**23–28.
53. United States Department of Agriculture: **Food Availability (Per Capita) Data System.** http://www.ers.usda.gov/data-products/food-availability-(per-capita)-data-system.aspx#.Ua6jU0DrzVU.
54. Grima EM, González MJI, Giménez AG: **Solvent extraction for microalgae lipids.** In *Algae for biofuels and energy.* Edited by Borowitzka MA, Moheimani NR. New York, London: Springer; 2013:187–205.
55. Gouveia L, Oliveira AC: **Microalgae as a raw material for biofuels production.** *J Ind Microbiol Biotechnol* 2009, **36:**269–274.
56. Papadopoulos G, Goulas C, Apostolaki E, Abril R: **Effects of dietary supplements of algae, containing polyunsaturated fatty acids, on milk yield and the composition of milk products in dairy ewes.** *J Dairy Res* 2002, **69:**357–365.
57. Glover KE, Budge S, Rose M, Rupasinghe HP, Maclaren L, Green-Johnson J, Fredeen AH: **Effect of feeding fresh forage and marine algae on the fatty acid composition and oxidation of milk and butter.** *J Dairy Sci* 2012, **95:**2797–2809.
58. Franklin ST, Martin KR, Baer RJ, Schingoethe DJ, Hippen AR: **Dietary marine algae (Schizochytrium sp.) increases concentrations of conjugated linoleic, docosahexaenoic and transvaccenic acids in milk of dairy cows.** *J Nutr* 1999, **129:**2048–2054.
59. Toral PG, Hervas G, Gomez-Cortes P, Frutos P, Juarez M, de la Fuente MA: **Milk fatty acid profile and dairy sheep performance in response to diet supplementation with sunflower oil plus incremental levels of marine algae.** *J Dairy Sci* 2010, **93:**1655–1667.
60. Cooper SL, Sinclair LA, Wilkinson RG, Hallett KG, Enser M, Wood JD: **Manipulation of the n-3 polyunsaturated fatty acid content of muscle and adipose tissue in lambs.** *J Anim Sci* 2004, **82:**1461–1470.
61. Hess TM, Rexford JK, Hansen DK, Harris M, Schauermann N, Ross T, Engle TE, Allen KG, Mulligan CM: **Effects of two different dietary sources of long chain omega-3, highly unsaturated fatty acids on incorporation into the plasma, red blood cell, and skeletal muscle in horses.** *J Anim Sci* 2012, **90:**3023–3031.
62. Sardi L, Martelli G, Lambertini L, Parisini P, Mordenti A: **Effects of a dietary supplement of DHA-rich marine algae on Italian heavy pig production parameters.** *Livest Sci* 2006, **103:**95–103.
63. Marriott NG, Garrett JE, Sims MD, Abril JR: **Performance characteristics and fatty acid composition of pigs fed a diet with docosahexaenoic acid.** *J Muscle Foods* 2002, **13:**265–277.
64. Sim JS, Qi G: **Designing poultry products using flaxseed.** In *Flaxseed in Human Nutrition.* Edited by Thompson LU, Cunnane LU. Champaign, IL: American Oil Chemists' Society Press; 1995:315–333.
65. Herber SM, Van Elswyk ME: **Dietary marine algae promotes efficient deposition of n-3 fatty acids for the production of enriched shell eggs.** *Poult Sci* 1996, **75:**1501–1507.
66. Leeson S, Caston L, MacLaurin T: **Organoleptic evaluation of eggs produced by laying hens fed diets containing graded levels of flaxseed and vitamin E.** *Poult Sci* 1998, **77:**1436–1440.
67. Abril JR, Barclay WR, Abril PG: **Safe use of microalgae (DHA GOLD) in laying hen feed for the production of DHA-enriched eggs.** In *Egg nutrition and biotechnology.* Edited by Sim JS, Nakai S, Guenter W. New York: CABI Publishing; 1999:197–202.
68. Munir N, Sharif N, Shagufta N, Saleem F, Manzoor F: **Harvesting and processing of microalgae biomass fractions for biodiesel production (a review).** *Sci Tech and Dev* 2013, **32:**235–243.
69. Jin DQ, Lim CS, Sung JY, Choi HG, Ha I, Han JS: **Ulva conglobata, a marine algae, has neuroprotective and anti-inflammatory effects in murine hippocampal and microglial cells.** *Neurosci Lett* 2006, **402:**154–158.
70. McNaught ML, Owen EC: **The utilization of non-protein nitrogen in the bovine rumen. 8. The nutritive value of the proteins of preparations of dried rumen bacteria, rumen protozoa and brewer's yeast for rats.** *Biochem J* 1954, **56:**151–156.
71. Becker EW: **Micro-algae as a source of protein.** *Biotechnol Adv* 2007, **25:**207–210.
72. Spolaore P, Joannis-Cassan C, Duran E, Isambert A: **Commercial applications of microalgae.** *J Biosci Bioeng* 2006, **101:**87–96.
73. Davies DR: **Single cell protein and the exploitation of mutant algae lacking cell walls.** *Nature* 1971, **233:**143–144.
74. Pulz O, Gross W: **Valuable products from biotechnology of microalgae.** *Appl Microbiol Biotechnol* 2004, **65:**635–648.
75. Iwamoto H: **Industrial production of microalgal cell-mass and secondary products - major industrial species: *Chlorella*.** In *Handbook of microalgal culture: biotechnology and applied phycology.* Edited by Richmond A. New Jersey: Willey-Blackwell Publishing; 2004:255–263.
76. Singh S, Kate BN, Banerjee UC: **Bioactive compounds form cyanobacteria and microalgae: an overview.** *Crit Rev Biotechnol* 2005, **25:**73–95.
77. Schaeffer DJ, Grylov VS: **Anti-HIV activity of extracts and compounds form algae and cyanobacteria.** *Ectoxicol Environ Saf* 2000, **45:**208–227.
78. Luescher-Mattli M: **Algae, a possible source for new drugs in the treatment of HIV and other viral diseases.** *Curr Med Chem Anti Infect Agents* 2003, **7:**219–225.
79. Vijffels RA, Barbosa M: **An outlook of microalgal biofuels.** *Science* 2010, **329:**796–799.
80. Darzins A, Pienkos P, Edye L: **Current status and potential for algal biofuels production.** http://www.fao.org/uploads/media/1008_IEA_Bioenergy_-_Current_status_and_potential_for_algal_biofuels_production.pdf.
81. Singh J, Gu S: **Commercialization potential of microalgae for biofuels production.** *Renew Sust Energ Rev* 2010, **14:**2596–2610.
82. Guil-Guerrero JL, Navarro-Juarez R, Lopez-Martinez JC, Campra-Madrid P, Rebolloso-Fuentes MM: **Functional properties of the biomass of three microalgal species.** *J Food Eng* 2004, **65:**511–517.

9

Fiber effects in nutrition and gut health in pigs

Jan Erik Lindberg

Abstract

Dietary fiber is associated with impaired nutrient utilization and reduced net energy values. However, fiber has to be included in the diet to maintain normal physiological functions in the digestive tract. Moreover, the negative impact of dietary fiber will be determined by the fiber properties and may differ considerably between fiber sources. Various techniques can be applied to enhance nutritional value and utilization of available feed resources. In addition, the extent of fiber utilization is affected by the age of the pig and the pig breed. The use of potential prebiotic effects of dietary fiber is an attractive way to stimulate gut health and thereby minimize the use of anti-microbial growth promoters. Inclusion of soluble non-starch polysaccharides (NSP) in the diet can stimulate the growth of commensal gut microbes. Inclusion of NSP from chicory results in changes in gut micro-environment and gut morphology of pigs, while growth performance remains unaffected and digestibility was only marginally reduced. The fermentation products and pH in digesta responded to diet type and were correlated with shifts in the microbiota. Interestingly, fiber intake will have an impact on the expression of intestinal epithelial heat-shock proteins in the pig. Heat-shock proteins have an important physiological role in the gut and carry out crucial housekeeping functions in order to maintain the mucosal barrier integrity. Thus, there are increasing evidence showing that fiber can have prebiotic effects in pigs due to interactions with the gut micro-environment and the gut associated immune system.

Keywords: Breed, Gut health, Fiber, Pigs, Prebiotics, Utilization

Introduction

On a worldwide basis corn and soybean meal are the main staples in the diet for pigs and poultry, providing most of the energy and nutrients needed. It is argued that although other cereals, such as wheat, and by-products, such as rice bran and distiller's grains, are used as alternative feedstuffs in part of the world the quantities available are not sufficient to replace corn and soybean meal in the global pig and poultry industry [1]. This may be correct and may apply to the industrialized livestock production, but may not be applicable to countries where a major part of the livestock production relies on smallholder farmers.

With more focus on small-scale family farming, in order to improve food security and minimize the negative impact on the environment and the climate, there are opportunities for more diversified composition of the diet with respect to feedstuffs included. Potential feed resources used for animals in many countries in Asia derive primarily from the vegetable foods and agro-industry co-products, such as cassava leaves, sweet potato vines, water spinach, rice bran, cassava residue, brewer's grain and tofu residue. They represent under-utilized feedstuffs, most having high fiber content, which may impose limitations in their use in diets for mono-gastric animals, in particular young animals, due to their bulky nature and a limited capacity to ferment fiber [2]. Thus, in order to better utilize available fiber-rich feedstuffs in the diet, their chemical and physical characteristics has to be described and taken into account in feed formulation.

Dietary fiber has an important role in pig and poultry diets and a minimum level of dietary fiber has to be included to maintain normal physiological function in the digestive tract [3]. A major concern when including fiber in diets for mono-gastric animals is that high dietary fiber content is associated with decreased nutrient utilization and low net energy values [4]. However, the negative impact of dietary fiber on nutrient utilization and net energy value will be determined by the fiber properties and may differ considerably between fiber

Correspondence: jan.erik.lindberg@slu.se
Department of Animal Nutrition and Management, Swedish University of Agricultural Sciences, P.O. Box 7024, Uppsala SE-75007, Sweden

sources. Moreover, dietary fiber may have other positive effects such as to stimulate gut health, increase the satiety, affect behavior and overall improve animal well-being [2,3,5,6]. Despite the obvious need for dietary fiber in the diet it is not included in the nutrient requirement tables [7].

For a long time antibiotics have been used as growth promoters on a regularized basis to control the problems with commonly occurring enteric diseases, such as post-weaning diarrhea and swine dysentery [6]. However, this is not a sustainable production system and the regular use of antibiotics in animal feed will promote bacterial resistance that may result in less efficient antibiotic treatments for human and animal diseases [8,9]. In addition, misusing antibiotics as feed additives for animal production can result in high antibiotic residues in animal products. Since 2006 antimicrobial growth promoters (AMGP) have been banned within EU and the goal is minimal use of antibiotics in food production. A similar development can be expected to take place also in other parts of the world.

In order to minimize the use of AMGP in animal production it is important to find ways to obtain a good animal health with other means. One possible way is changing diet composition and to use various dietary interventions. A huge amount of research has been performed in the area that aims to change the gut microbial composition, such as supplementation of probiotics, prebiotics, organic acids and dietary fiber [6]. It is believed that these types of dietary interventions have the potential to improve gut health.

Dietary fiber

The carbohydrate fraction can be divided according to glycosidic linkages into sugars, oligosaccharides and two broad classes of polysaccharides, starch and non-starch polysaccharides. Non-starch polysaccharides (NSP) together with lignin, has been defined as the dietary fibre (DF) fraction in feedstuffs and food, and can be used as a collective measure of their fibre content [10-12]. However, as non-digestible oligosaccharides and resistant starch have similar physiological effects in the body as NSP and lignin, although not being part of the cell wall structure, the definition should be extended to include these constituents [13].

The amount and composition of dietary fibre vary widely between and within feedstuff [12,14,15]. As a result, the dietary fibre content and its properties in a typical diet for any animal varies between production systems and countries due to availability of feed resources. A change in chemical composition of the plant material and their co-products not only depends on the botanical origin of the plants and the type of processing applied, but also on the tissue type and the maturity of the plant at harvesting time. The proportion of middle lamella, and primary and secondary cell wall in the plant material, together with the properties of the amorphous matrix connecting the cell walls, will determine both utilization and properties. The amorphous matrix usually shows considerable variation from tissue to tissue within plant and between plants. In monocotyledonous plants, such as cereals, the main cell wall NSP are arabinoxylans, cellulose and β-glucan [12]. In contrast, the amorphous matrix in dicotyledonous plants can differ markedly from that in monocotyledonous plants due to different tissue types, exemplified by the huge difference in content and properties of pectic substances [16,17].

The physicochemical properties of dietary fibre are dependent on the polysaccharides that make up the cell wall and their intermolecular association which determine their solubility [17]. The major physicochemical properties that have been considered in animal nutrition are cation exchange capacity, hydration properties, viscosity and organic compound adsorptive properties. Characteristic for insoluble dietary fiber is that they increase rate of passage and faecal bulk whereas soluble dietary fiber increases the viscosity and hydration properties [18]. Hydration properties of the feed is very important for effective digestion to occur in the animal and can be characterized by the swelling capacity, solubility, water holding capacity (WHC) and water binding capacity (WBC). Recent studies showed a strong correlation between content of soluble non-cellulosic polysaccharides (S-NCP) and WHC in plant material and agro-industry co-products collected in Vietnam [15]. This could be due to the occurrence of more gaps within the cell matrix that can retain excess water in feed ingredients which are high in S-NCP. Similarly, it has been shown that the S-NCP fraction in co-products from vegetable food and agro-industries was linearly related to selected hydration properties, such as swelling and WBC [14] (Figure 1).

Nutritional effects

Improving feed value

Various techniques, such as pelleting, reduction of particle size and supplementation with exogenous enzymes can be applied to enhance nutritional value and utilization of available feed resources [19-21]. This may be most relevant for locally available feed resources to overcome the shortcomings of poor nutritive value [6,22] due to high fibre content and of other components in the feed affecting the physicochemical properties.

The positive effects of enzyme addition is due to disruption of intact cell walls and release of entrapped nutrients or due to the changes in the physical properties of non-starch polysaccharides, such as viscosity and

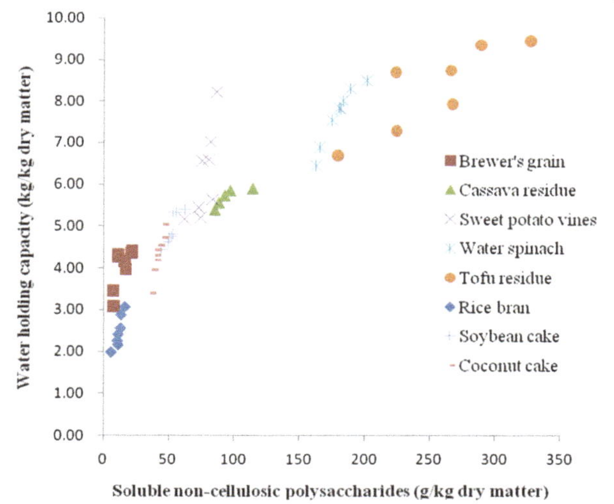

Figure 1 Correlation between water holding capacity (WHC, kg/kg DM) and soluble non-starch polysaccharides (S-NCP, g/kg DM) in selected fibre-rich plant sources and agro-industry co-products (WHC = 3.50 + 0.0214 S-NCP, R^2 = 0.82, *P* < 0.001) (Ngoc et al. [15]).

WHC, and/or changes in the composition and content of bacteria in the small and large intestine [23-25].

Reducing particle size has been reported to improve the feed efficiency and nutrient digestion of weaned pigs [19,21]. There appears to be a particle size by age of pig interaction in digestibility, resulting in a larger response in young as compared to older pigs [19]. This can be explained by the development of the digestive and absorptive capacity in the small intestine and increased colonization of carbohydrate-degrading microbiota in the large intestine with increasing age. There are also indications of a particle size by enzyme interaction for the total tract apparent digestibility of dietary components in pigs [21].

Reduction in particle size (1 versus 3 mm) increased the total tract apparent digestibility of dietary components and the average daily gain (ADG) of Landrace x Yorkshire pigs in the post-weaning period, but not in the growing period [26]. Moreover, addition of a multi-enzyme mixture (mixture of α-amylase, β-glucanase, cellulase and protease) improved the total tract apparent digestibility of dietary components and growth performance in the post-weaning period [26]. However, there was an interaction between particle size and multi-enzyme supplementation on the total tract apparent digestibility of crude protein (CP) and neutral detergent fibre (NDF) in the post-weaning period, such that multi-enzyme supplementation increased the total tract apparent digestibility of CP and NDF in the larger particle size diet (80% vs. 75%, and 58% vs. 51%, respectively), while there was no changes in a small particle size diet (80% vs. 78%, and 59% vs. 55%, respectively). However, in the growing period multi-enzyme supplementation had no positive effect on the performance and the total tract apparent digestibility of dietary components, with the exception of CP and NDF (Table 1).

Fiber utilization

Impact of age

Adult pigs have a more developed and larger GI tract, a lower feed intake per kg body weight, a slower digesta transit time and a higher cellulolytic activity than young pigs. This resulted in greater capacity of sows to digest fibrous components compared to young pigs [27] and it was shown that sows digest a larger part of the NSP in the small intestine than growing pigs [28]. They also showed that sows have a higher capacity to digest insoluble NSP, whereas the difference in digestibility of soluble NSP between growing pigs and sows were only marginal. It was suggested that because of the increased capacity to digest fibrous feedstuff by increased age and body weight, at least two different energy values, one for growing-finishing pigs and one for sows should be used for most feed ingredients in pig diets [4]. This has been implemented in the INRA net energy system for pigs (www.evapig.com).

In weaned pigs, age affected the total tract apparent digestibility of all dietary components except for NDF, with higher values at five than at three weeks post-weaning [29].

Impact of pig breed

Digestive capacity and potential body protein deposition are two important traits associated with pig performance. However, in commercial pig breeding research has

Table 1 Impact of fiber source, particle size and enzyme treatment on dry matter intake (DMI), average daily gain (ADG) and feed conversion ratio (FCR) in post-weaned and growing pigs[1]

Items	Post-weaned pigs			Growing pigs		
	DMI	ADG	FCR	DMI	ADG	FCR
Fiber source						
Cassava root meal	559	422^a	1.33^a	1289	597^a	2.16^a
Sweet potato vines	550	385^b	1.44^b	1271	540^b	2.36^b
Particle size						
Small (1 mm)	556	416^a	1.34	1273	572	2.24
Large (3 mm)	553	391^b	1.42	1286	565	2.29
Enzyme addition[#]						
-	554	387^b	1.44^b	1265	557	2.28
+	555	420^a	1.33^a	1295	580	2.24
SEM	28	12	0.07	41	17	0.08

[1]Data from Ngoc et al. [26]. Different letters within column indicates significant differences between treatments. SEM = standard error of the mean.
[#]Multi-enzyme mixture (α-amylase, β-glucanase, cellulase and protease).

mainly focused on factors associated with body lipid and protein deposition [30,31], while less attention has been directed to breed differences in the digestive capacity [32]. Thus, body protein and lipid mass and their deposition rates are key variables used in growth models to predict compositional changes in carcass muscle and fat tissue mass over time [30,33]. This priority can be explained by the increasing demand of consumers for meat quality in general and for lean meat in particular. It is also reflected in an increasing use of exotic breeds such as the Landrace, Yorkshire, Duroc and Hampshire for cross-breeding of native pigs, mainly in urban and peri-urban areas, in several countries in Asia. Crossbred Large White × Yorkshire (LY) pigs show higher growth rate and better feed conversion than native Mong Cai (MC) pigs, when fed the same daily amount of DM and CP [34,35]. This is due to greater potential for lean tissue accretion in LY than in MC pigs, as reflected in higher nitrogen retention [34,35].

However, indigenous breeds, such as the MC breed, may have better characteristics with regards to reproduction and are adapted to the local climate. Therefore, in Vietnam pure-bred MC sows are still commonly mated with boars of an exotic breed, usually Yorkshire or Landrace, using artificial insemination, and the off-springs are fattened in small-scale semi-intensive or intensive systems. Moreover, it has been shown that indigenous pig breeds may have a higher capacity to digest fibre than exotic pig breeds genetically improved to support high growth performance [22,34-36]. The main explanation to the improved digestibility in indigenous pigs is a longer MRT of digesta [37]. This will contribute to a more efficient digestion due to longer contact between digesta, digestive enzymes and absorptive surfaces [32], and between digesta and the gut microbiota [38]. A longer MRT in the indigenous MC pig can be explained by a larger GIT (in terms of diameter), reflected in more gut content, as compared to LY pigs. In accordance, it was reported that differences in gut content could explain the longer MRT in Iberian pigs than in Landrace pigs [39]. Earlier studies on indigenous and exotic pig breeds support the contention of greater length of the GIT in the indigenous pigs [36,40] (Figure 2).

In addition, recent studies show that there are interactions between breed and diet on LAB count and in the concentration of propionic acid in the ileum. Increased fibre level in the diet in combination with high soluble fiber content had a greater impact on the LAB count and the concentration of propionic acid in the ileum of MC pigs than of LY pigs [41]. This suggests differences in the gut microbiota activity and/or composition between MC and LY pigs.

Interestingly, there was also a breed effect on the small intestinal morphology in the ileum, while there were no breed related differences in the duodenum and jejunum.

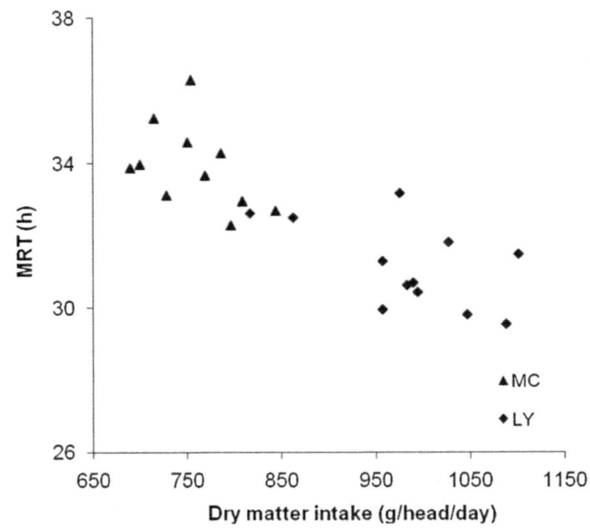

Figure 2 Impact of dry matter intake (g/head/day) on mean retention time (MRT; h) of digesta in growing Mong Cai (MC) and Landrace x Yorkshire (LY) pigs (Ngoc et al. [37]).

Mong Cai pigs had shorter villi, smaller villus width and greater crypt density in the ileum than LY pigs [41]. These differences in gut morphology can be explained by differences in digesta transit time and gut microbial activity [42]. A major part of the digested nutrients are absorbed in the proximal small intestine. This together with a rapid digesta passage and low microbial activity, results in less exposure to digesta components compared with the situation in the more distal small intestine. The MC pigs had a longer total tract mean retention time of solids than the LY pigs [37] allowing for longer contact between digestion products and absorptive surfaces [32].

Based on existing data, it appears reasonable to assume that different breeds can have different digestive capacity and will as a result show different response to the same diets in terms of nutrient utilization and performance.

Effects on gut health

Gut microenvironment

The maintenance of gut health is complex and relies on a delicate balance between the diet, the commensal microflora and the mucosa, including the digestive epithelium and the mucus overlying the epithelium [43]. The diet has a great impact on gut health, and it can provide either beneficial or harmful input [2,6,44]. The diet should be composed to create a balance between the gut, the microbiota and the gut environment and prevent disturbances in the gut. Dietary fibre interacts both with the mucosa and the microbiota and consequently has an important role in the control of gut health [2,6,43].

The complex microbial community in the gastrointestinal (GI) tract consists of different groups of microbes including bacteria, archaea, ciliate and flagellate protozoa, anaerobic phycomycete fungi and bacteriophages. Bacteria are the most abundant and studied microbes in this community. They are provided with substrates from the diet as well as components deriving from the host such as mucopolysaccharides, mucins, epithelial cells and enzymes [45]. With the introduction of molecular techniques to indentify the microbiota it has become apparent that only a minority of the GI microbes have been isolated by culture based methods [46] and consequently the knowledge we have today most likely needs to be revised in the future (Figure 3).

Prebiotic effects

The concept of prebiotics was initially defined as "non-digestible food ingredient that beneficially affects the host by selectively stimulating the growth and/or activity of one or a limited number of bacteria in the colon, and thus improves host health" [47]. However, the weakness with this definition is that almost every food oligosaccharide and polysaccharide (including dietary fibre) can be claimed to have prebiotic activity. Later, it was proposed that to be allowed to classify a compound as a prebiotic it should scientifically be demonstrated that it resists host digestion, absorption and adsorption processes, is fermented by the microbiota colonising the GI system and selectively stimulates the growth and/or the activity of one or a limited number of bacteria within the GI system [48].

Dietary fiber (DF) is a feed component that has major influence in this regard. Dietary fibre components are not digested by endogenous digestive enzymes, and consequently are the main substrates for bacterial fermentation in the distal part of the gut. The main products of fermentation are short chain organic acids (OA), predominantly lactate, acetate, propionate and butyrate. The OA have been suggested to develop the growth of the digestive tract, by stimulating epithelium cell proliferation [43]. In an acidic environment, OA can inhibit the growth of enteric bacterial pathogens, such as *Salmonella*, *E. coli* and *Clostridium* species [49-51]. Studies in pigs have shown that the various types of plant carbohydrates behave differently in the GIT depending on their structural characteristics. Inclusion of soluble NSP in the diet can stimulate the growth of commensal gut microbes, leading to increased production of OA, and a lower pH in the large intestine [2]. Insoluble NSP reduce the transit time and provide substrate that is slowly degradable by the microbiota in the distal large intestine [38], and modulate gut morphology by increasing villus length [52].

Feeding guar gum, a soluble and viscous NSP, increases the proliferation of enterotoxigenic *E. coli* [53], whereas feeding insoluble NSP reduces the occurrence of haemolytic *E. coli*, and reduces the severity of post weaning colibacillosis [54]. However, it was shown that soluble NSP per se is not detrimental to piglet health [55]. Instead it was stated that soluble NSP that does not increase the digesta viscosity may beneficially affect gut

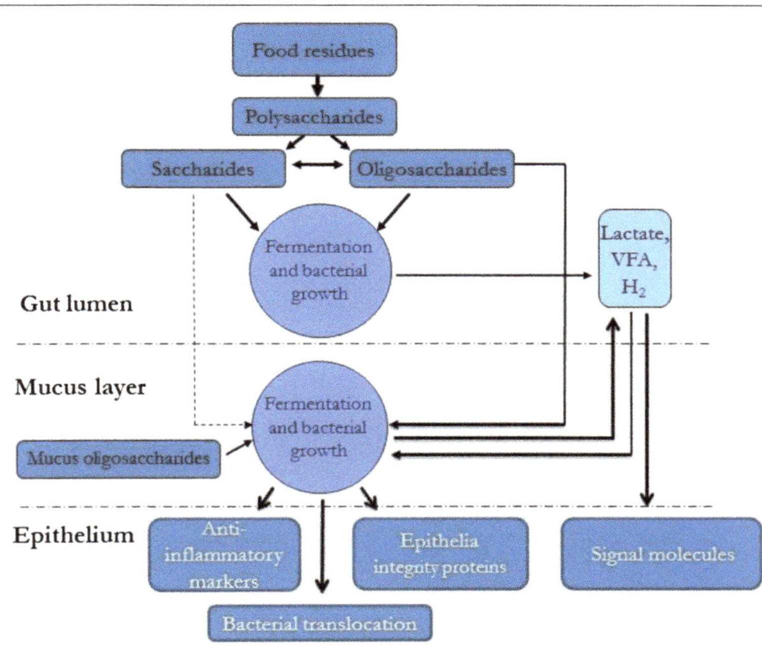

Figure 3 Interactions between dietary fiber, gut environment, gut microbiota and host response with implications on gut health.

health by increasing the lactobacilli:coliform ratio and decrease the occurrence of weaning diarrhoea.

The impact of DF source on gut microbiota composition and gut micro-environment was nicely demonstrated in a recent study on chicory. The inclusion of chicory (*Cichorium intybus* L.) forage and root in a cereal-based diet (wheat and barley) results in changes in gut micro-environment and gut morphology of pigs [56], while growth performance was unaffected and digestibility was only marginally reduced by chicory inclusion [29,57]. Within diet type, these changes followed a similar pattern in the small and large intestine. However, the dietary responses were different with inclusion of chicory root compared with chicory forage. This could be related to the chemical composition of the dietary fiber fraction [57], where the root is characterized by high content of inulin-type fructan and oligofructose, while the forage is characterized by high content of pectin. The fermentation products and pH in digesta responded to diet type and were correlated with shifts in the microbiota, showing that chicory influences the intestinal micro-environment of pigs [56]. In ileum, inclusion of chicory roots (inulin-type fructan and oligofructose) was linked with lactic acid concentration in digesta and relative abundance of LAB. In colon, inclusion of chicory forage (pectin) was associated with relative abundance of butyrate-producing bacteria and colonic acetate concentration.

Interestingly, diet fiber intake will have an impact on the expression of intestinal epithelial heat-shock proteins (HSP) that have an important physiological role in the gut. The HSP carry out crucial housekeeping functions in order to maintain the mucosal barrier integrity. In a recent study [58] we found a positive correlation between ileal HSP27 expression and daily total uronic acid intake ($r = 0.364$, $P = 0.05$). This was reinforced when ileal HSP27 and daily soluble uronic acid intake was correlated ($r = 0.390$, $P = 0.048$). Furthermore, HSP27 expression in the ileal mucosa was correlated with mucosa-associated *Megasphaera elsdenii* (TRF275) ($r = 0.553$, $P = 0.021$), which was also positively correlated with daily total uronic acid intake ($r = 0.523$, $P = 0.011$). However, daily fructan intake was not correlated with HSP27 expression ($P > 0.05$).

Conclusions

The impact of dietary fiber on pig nutrition and health is determined by the fiber properties and may differ considerably between fiber sources. Moreover, the utilization of plant fiber can be improved by using various techniques and may be improved by breeding for enhanced fiber utilization. There are increasing evidence showing that fiber can have prebiotic effects in pigs due to interactions with the gut micro-environment and the gut associated immune system. This property can be exploited and used as a means to stimulate gut health and thereby minimize the use of anti-microbial growth promoters. In addition, fiber in the diet will increase the satiety, affect behavior and overall improve animal well-being.

Competing interests

The author declares that there are no competing interests.

References

1. Leeson S: **Future considerations in poultry nutrition.** *Poult Sci* 2012, **91**:1281–1285.
2. Bach Knudsen KE, Hedemann MS, Laerke HN: **The role of carbohydrates in intestinal health of pigs.** *Anim Feed Sci Technol* 2012, **173**:41–53.
3. Wenk C: **The role of dietary fibre in the digestive physiology of the pig.** *Anim Feed Sci Technol* 2001, **90**:21–33.
4. Noblet J, Le Goff GI: **Effect of dietary fibre on the energy value of feeds for pigs.** *Anim Feed Sci Technol* 2001, **90**:35–52.
5. de Leeuw JA, Bolhuis JE, Bosch G, Gerrits WJJ: **Effects of dietary fibre on behaviour and satiety in pigs.** *Proc Nutr Soc* 2008, **67**:334–342.
6. de Lange CFM, Pluske J, Gong J, Nyachoti CM: **Strategic use of feed ingredients and feed additives to stimulate gut health and development in young pigs.** *Livest Sci* 2010, **134**:124–134.
7. NRC: **Nutrient Requirements of Swine.** In *National Research Council.* 11th edition. Washington, USA: National Academy Press; 2012.
8. Bates J, Jordens Z, Selkon J: **Evidence for an animal origin of vancomycin-resistant enterococci.** *Lancet* 1993, **342**(8869):490–491.
9. Levy SB: **Microbial resistance to antibiotics. An evolving and persistent problem.** *Lancet* 1982, **2**(8289):83–88.
10. Trowell H: **Ischemic-heart disease and dietary fiber.** *Am J Clin Nutr* 1972, **25**:926.
11. Theander O, Åman P, Westerlund E, Graham H: **Enzymatic/chemical analysis of dietary fiber.** *J AOAC Int* 1994, **77**:703–709.
12. Bach Knudsen KE: **Carbohydrate and lignin contents of plant materials used in animal feeding.** *Anim Feed Sci Technol* 1997, **67**:319–338.
13. Champ M, Langkilde AM, Brouns F, Kettlitz B, Collet YL: **Advances in dietary fibre characterisation. 1. Definition of dietary fibre, physiological relevance, health benefits and analytical aspects.** *Nutr Res Rev* 2003, **16**:71–82.
14. Serena A, Bach Knudsen KE: **Chemical and physicochemical characterisation of co-products from vegetable food and agro industries.** *Anim Feed Sci Technol* 2007, **139**:109–124.
15. Ngoc TTB, Len NT, Lindberg JE: **Chemical characterization and water holding capacity of fibre-rich feedstuffs used for pigs in Vietnam.** *Asian-Austr J Animal Sci* 2012, **25**:861–868.
16. Bacic A, Harris PJ, Stone BA: **Structure and Function of Plant Cell Walls.** In *The Biochemistry of Plants: Carbohydrates.* San Diego, California, USA: Academic Press Inc; 1988:297–372.
17. McDougall GJ, Morrison IM, Stewart D, Hillman JR: **Plant cell walls as dietary fibre: Range, structure, processing and function.** *J Sci Food Agric* 1996, **70**:133–150.
18. Bach Knudsen KE: **The nutritional significance of "dietary fibre" analysis.** *Anim Feed Sci Technol* 2001, **90**:3–20.
19. Kim JC, Mullan BP, Pluske JR: **A comparison of waxy versus non-waxy wheats in diets for weaner pigs: effects of particle size, enzyme supplementation, and collection day on total tract apparent digestibility and pig performance.** *Anim Feed Sci Technol* 2005, **120**:51–65.
20. Brufau J, Francesh M, Perez-Vendrell AM: **The use of enzyme to improve cereal diets for animal feeding.** *J Sci Food Agric* 2006, **86**:1705–1713.
21. Mavromichalis I, Hancock JD, Senne BW, Gugle TL, Kennedy GA, Hines RH, Wyatt CL: **Enzyme supplementation and particle size of wheat in diets for nursery and finishing pigs.** *J Anim Sci* 2000, **78**:3086–3095.
22. Len NT, Ngoc TB, Ogle B, Lindberg JE: **Ileal and total tract digestibility in local (Mong Cai) and exotic (Landrace x Yorkshire) piglets fed low and high fibre diets, with or without enzyme supplementation.** *Livest Sci* 2009, **126**:73–79.

23. Bedford MR, Classen HL: **Reduction of intestinal viscosity through manipulation of dietary rye and pentosanase concentration is effected through changes in the carbohydrate composition of the intestinal aqueous phase and results in improved growth rate and food conversion efficiency of broiler chicks.** *J Nutr* 1992, **122:**560–569.
24. Jensen MS, Bach Knudsen KE, Inborr J, Jakobsen K: **Effect of beta-glucanase supplementation on pancreatic enzyme activity and nutrient digestibility in piglets fed diets based on hulled and hulless barley varieties.** *Anim Feed Sci Technol* 1998, **72:**329–345.
25. O'Connell JM, Sweeney T, Callan JJ, O'Doherty JV: **The effect of cereal type and exogenous enzyme supplementation in pig diets on nutrient digestibility, intestinal microflora, volatile fatty acid concentration and manure ammonia emissions from finisher pigs.** *Anim Sci* 2005, **81:**357–364.
26. Ngoc TTB, Len NT, Ogle B, Lindberg JE: **Influence of particle size and multi-enzyme supplementation of fibrous diets on total tract digestibility and performance of weaning and growing pigs.** *Anim Feed Sci Technol* 2011, **169:**86–95.
27. Shi XS, Noblet J: **Digestible and metabolizable energy values of 10 feed ingredients in growing pigs fed ad-libitum and sows fed at maintenance level - comparative contribution of the hindgut.** *Anim Feed Sci Technol* 1993, **42:**223–236.
28. Jørgensen H, Serena A, Hedemann MS, Bach Knudsen KE: **The fermentative capacity of growing pigs and adult sows fed diets with contrasting type and level of dietary fibre.** *Livest Sci* 2007, **109:**111–114.
29. Ivarsson E, Frankow-Lindberg BE, Andersson K, Lindberg JE: **Growth performance, digestibility and faecal coliform bacteria in weaned piglets fed a cereal-based diet including either chicory (*Cichorium intybus* L) or ribwort (*Plantago lanceolata* L) forage.** *Animal* 2011, **5**(4):558–564.
30. Schinckel AP, Mahan DC, Wiseman TG, Einstein ME: **Growth of protein, moisture, lipid, and ash of two genetic lines of barrows and gilts from twenty to one hundred twenty-five kilograms of body weight.** *J Anim Sci* 2008, **86:**460–471.
31. Shan T, Wu T, Reng Y, Wang Y: **Breed differences and regulation of the porcine adipose triglyceride lipase and hormone sensitive lipase by TNFα.** *Anim Genet* 2009, **40:**863–870.
32. Guixin Q, Verstegen MWA, Bosch MW: **Variation of digestive capacity between genetically different pig populations: a review.** *J Anim Physiol Anim Nutr* 1995, **73:**233–242.
33. Schinckel AP, de Lange CFM: **Characterization of growth parameters needed as inputs for pig growth models.** *J Anim Sci* 2021–2036, **1996:**74.
34. Borin K, Lindberg JE, Ogle B: **Effect of variety and preservation method of cassava leaves on diet digestibility by indigenous and improved pigs.** *Anim Sci* 2005, **80:**319–324.
35. Len NT, Lindberg JE, Ogle B: **Digestibility and nitrogen retention of diets containing different levels of fibre in local (Mong Cai), F1 (Mong Cai x Yorkshire) and exotic (Landrace x Yorkshire) growing pigs in Vietnam.** *J Anim Physiol Anim Nutr* 2007, **91:**297–303.
36. Len NT, Hong TTT, Lindberg JE, Ogle B: **Comparison of total tract digestibility, development of visceral organs and digestive tract of Mong Cai and Yorkshire x Landrace piglets fed diets with different fibre sources.** *J Anim Physiol Anim Nutr* 2009, **93:**181–191.
37. Ngoc TTB, Len NT, Lindberg JE: **Impact of fibre intake and fibre source on digestibility, gastro-intestinal tract development, mean retention time and growth performance in indigenous (Mong Cai) and exotic (Landrace x Yorkshire) pigs.** *Animal* 2013, **7**(5):736–745.
38. Freire JPB, Guerreiroa AJG, Cunha LF, Aumaitre A: **Effect of dietary fibre source on total tract digestibility, caecum volatile fatty acids and digestive transit time in the weaned piglet.** *Anim Feed Sci Technol* 2000, **87:**71–83.
39. Morales J, Perez JF, Martin-Orue SM, Fondevila M, Gasa J: **Large bowel fermentation of maize or sorghum-acorn diets fed as a different source of carbohydrates to Landrace and Iberian pigs.** *Br J Nutr* 2002, **88:**489–498.
40. Freire JPB, Peiniau J, Cunha LF, Almeida JAA, Aumaitre A: **Comparative effects of dietary fat and fibre in Alentejano and Landrace piglets: digestibility, digestive enzymes and metabolic data.** *Livest Prod Sci* 1998, **53:**37–47.
41. Ngoc TTB, Hong TTT, Len NT, Lindberg JE: **Effect of fibre level and fibre source on gut morphology and micro-environment in local (Mong Cai) and exotic (Landrace x Yorkshire) pigs.** *Asian-Austr J Animal Sci* 2012, **25**(12):1726–1733.
42. Freire JPB, Dias RIM, Cunha LF, Aumaitre A: **The effect of genotype and dietary fibre level on the caecal bacterial enzyme activity of young piglets: digestive consequences.** *Anim Feed Sci Technol* 2003, **106:**119–130.
43. Montagne L, Pluske JR, Hampson DJ: **A review of interactions between dietary fibre and the intestinal mucosa, and their consequences on digestive health in young non-ruminant animals.** *Anim Feed Sci Technol* 2003, **108:**95–117.
44. Bikker P, Dirkzwager A, Fledderus J, Trevisi P, le Huerou-Luron I, Lalles JP, Awati A: **The effect of dietary protein and fermentable carbohydrates levels on growth performance and intestinal characteristics in newly weaned piglets.** *J Anim Sci* 2006, **84:**3337–3345.
45. Zoetendal EG, Collier CT, Koike S, Mackie RI, Gaskins HR: **Molecular ecological analysis of the gastrointestinal microbiota: a review.** *J Nutr* 2004, **134:**465–472.
46. Vaughan EE, Schut F, Heilig HGHJ, Zoetendal EG, de Vos WM, Akkermans ADL: **A molecular view of the intestinal ecosystem.** *Curr Issues Intest Microbiol* 2000, **1:**1–12.
47. Gibson GR, Roberfroid MB: **Dietary modulation of the human colonic microbiota: introducing the concept of prebiotics.** *J Nutr* 1995, **125:**1401–1412.
48. Gibson GR, Probert HM, van Loo J, Rastall RA, Roberfroid MB: **Dietary modulation of the human colonic microbiota: updating the concept of prebiotics.** *Nutr Res Rev* 2004, **17:**259–275.
49. Hentges DJ: **Gut Flora and Disease Resistance.** In *Probiotics: The Scientific Basis.* Edited by Fuller R. London: Chapman and Hall; 1992:87–110.
50. Wang X, Gibson GR: **Effects of the *in vitro* fermentation of oligofructose and inulin by bacteria growing in the human large intestine.** *J Appl Microbiol* 1993, **75:**373–380.
51. May T, Mackie RI, Fahey GC, Cremin JC, Garleb KA: **Effect of fibre source on short-chain fatty acid production and on the growth and toxin production by Clostridium difficile.** *Scand J Gastroenterol* 1994, **29:**916–922.
52. Hedemann MS, Eskildsen M, Lærke HN, Pedersen C, Lindberg JE, Laurinen P, Bach Knudsen KE: **Intestinal morphology and enzymatic activity in newly weaned pigs fed contrasting fibre concentrations and fibre properties.** *J Anim Sci* 2006, **84:**1375–1386.
53. McDonald DE, Pethick DW, Pluske JR, Hampson DJ: **Adverse effects of soluble non-starch polysaccharide (guar gum) on piglet growth and experimental colibacillosis immediately after weaning.** *Res Vet Sci* 1999, **67:**245–250.
54. Bertschinger HU, Eggenberger E, Jucker H, Pfirter HP: **Evaluation of low nutrient, high-fiber diets for the prevention of porcine *Escherichia coli* Enterotoxaemia.** *Vet Microbiol* 1979, **3:**281–290.
55. Wellock IJ, Fortomaris PD, Houdijk JGM, Wiseman J, Kyriazakis I: **The consequences of non-starch polysaccharide solubility and inclusion level on the health and performance of weaned pigs challenged with enterotoxigenic *Escherichia coli.*** *Br J Nutr* 2008, **99:**520–530.
56. Liu HY, Ivarsson E, Dicksved J, Lundh T, Lindberg JE: **Inclusion of chicory (Cichorium intybus L.) in pigs' diets affects the intestinal microenvironment and the gut microbiota.** *Appl Environ Microbiol* 2012, **78:**4102–4109.
57. Ivarsson E, Liu HY, Dicksved J, Roos S, Lindberg JE: **Impact of chicory inclusion in a cereal-based diet on digestibility, organ size and faecal microbiota in growing pigs.** *Animal* 2012, **6**(7):1077–1085.
58. Liu HY, Lundh T, Dicksved J, Lindberg JE: **Expression of heat shock protein 27 in gut tissue of growing pigs fed diets without and with inclusion of chicory fiber.** *J Anim Sci* 2012, **90**(Suppl. 4):25–27.

10

Treatment of wheat straw using tannase and white-rot fungus to improve feed utilization by ruminants

Shailendra Raghuwanshi, Swati Misra and Rajendra Kumar Saxena*

Abstract

Background: Current research to enrich cattle feed has primarily focused on treatment using white rot fungi, while there are scarce reports using the enzyme tannase, which is discussed only in reviews or in the form of a hypothesis. In this context, the aim of the present study was to evaluate the effect of tannase on wheat straw (WS) and also the effect of lyophilized tannase at concentrations of 0.1%, 0.2%, and 0.3% (w/w) on WS followed by fermentation with *Ganoderma* sp. for 10 d and compared in relation to biochemical parameters, crude protein (CP) content, and nutritional value by calculating the C/N ratio in order to improve the nutritional value of cattle feed.

Results: *Penicillium charlesii,* a tannase-producing microorganism, produced 61.4 IU/mL of tannase in 54 h when 2% (w/v) tannic acid (TA) was initially used as a substrate in medium containing (% w/v) sucrose (1.0), $NaNO_3$ (1.0), and $MgSO_4$ (0.08 pH, 5.0) in a 300-L fermentor (working volume 220 L), and concomitantly fed with 1.0% (w/v) TA after 24 h. The yield of partially purified and lyophilized tannase was 5.8 IU/mg. The tannin-free myco-straw at 0.1% (w/w) tannase showed 37.8% (w/w) lignin degradation with only a 20.4% (w/w) decrease in cellulose content and the *in vitro* feed digestibility was 32.2%. An increase in CP content (up to 1.28-fold) along with a lower C/N ratio of 25.0%, as compared to myco-straw, was obtained.

Conclusions: The use of tannin-free myco-straw has potential to improve the nutritional content of cattle feed. This biological treatment process was safe, eco-friendly, easy to perform, and was less expensive as compared to other treatment methods.

Keywords: *Ganoderma* sp, High energy cattle feed, Tannase, Tannin free myco-straw

Background

The drastic increase in human population has caused intensified pressure on agricultural land use, which has led to the conversion of farmland into commercial landscapes. Therefore, countries with agriculture-based economies have endured economic constraints that have negatively impacted the health of cattle due to nutritional deficiencies of feed. Therefore, it is of the utmost importance to improve the quality of cattle feed so that it can be easily be digested by ruminants; therefore, producing higher milk yields to benefit the ever-increasing demand for milk and milk products. Therefore, to overcome problems associated with the decrease in available farmland and the increased demand for milk and milk products, efforts are being made to reduce the wide gap between the requirement and availability of feed and fodder, including technological interventions to increase yield, expand areas to raise fodder crops, conserve land for growing feed and fodder, improve the nutritional value of poor quality roughages, and develop unconventional feeds at competitive prices [1]. The main focus of cattle feed research has been on the enhancement of the nutritional value of dry roughages using physical, chemical, and biological treatments [2].

Globally, India is the second largest producer of wheat as a cereal crop, which, incidentally, produces a large amount of crop residue. Apart from feed and fodder production, a very limited amount of wheat straw (WS) is used for various other purposes such as the manufacture of ethanol, paper, and fertilizer, while the majority is

* Correspondence: rksmicro@yahoo.co.in
Department of Microbiology, University of Delhi South Campus, Benito Juarez Road, New Delhi 110 021, India

either removed from the field, burned on site, piled, spread over fields, or used as mulch [3]. However, the burning of straw causes atmospheric pollution, although it is a cost-effective disposal method [4]. Although the use of WS for ruminant feeding is constrained by its low digestibility of the energy-rich cell walls because of the high silica content, the tannins, lignin, and the relatively low protein and energy content, when fed to cattle as the only feed source it could significantly influence livestock production [5,6].

Several physical and chemical treatments have been proposed to improve the degradability of WS and subsequent feed intake by cattle. Still, the practical use of these treatments is restricted by safety concerns, cost, and potentially negative environmental impacts [7]. In this respect, an alternative approach is the use of ligninolytic fungi or their enzymes to shorten the incubation period and reduce the use of potentially toxic chemicals [3]. Enzymatic and biological delignification has an edge over other treatment methods to improve the nutritional value of WS, as it includes mild reaction conditions, avoids the use of toxic and corrosive chemicals, produces higher product yield, and has fewer side reactions, lower energy demand, and less reactor resistance [8-10]. For biological delignifiaction, white rot fungi (WRF), a type wood-decaying basidiomycetes, act as lignocellulolytic microorganisms, which contains a group of enzymes of varied functions, including the hydrolytic enzymes cellulase and xylanases that degrade polysaccharides and oxidative ligninolytic enzymes such as lignin peroxidase, manganese peroxidase (MnP), and laccase that degrade lignin. Several WRF species are effective lignin degraders that can be exploited to improve the nutritional value of fodder for ruminant nutrition without compromising the cellulose and hemicellulose constituents [11-13].

Tannin acyl hydrolase, commonly referred to as tannase (E.C. 3.1.1.20), is produced by various micro-organisms, and has been reported to improve the nutritional value of feed by the degradation of tannins. It is also able to hydrolyze ester bonds (i.e., the galloyl ester of an alcohol moiety), as well as depside bonds (the galloyl ester of gallic acid, GA) of substrates (or complex polyphenols) such as tannic acid (TA), epicatechin gallate, epigallocatechin gallate, and chlorogenic acid [14]. The breakdown of complex polyphenols could lead to production of GA, epicatechin, epicatechin gallate, and glucose [15]. To date, studies on feed enrichment have mainly focused on treatment using WRF, whereas reports using tannases remain scarce and are presented only in reviews or in the form of hypotheses.

In the present study, the effect of tannase-treatment on WS subjected to fungal pellet inoculum of *Ganoderma* spp. was examined and compared in relation to biochemical parameters and crude protein (CP) production to improve the nutritional value of cattle feed, as evaluated by the carbon-to-nitrogen (C/N) ratio.

Materials and methods

Chemicals and reagents

All chemicals used in this study were of commercial/analytical reagent grade and purchased from Sisco Research Laboratories Pvt., Ltd. (Mumbai, India), Central Drug House Pvt., Ltd. (Mumbai, India) or SS Fermozyme Pvt. Ltd. (New Delhi, India).

Estimation of tannase content

The spectrophotometric method described by Deschamps et al. [16] was used to estimate tannase activity. One IU of tannase was defined as the amount of enzyme required to release 1.0 μmol of GA/mL of the substrate in 1 min under the specific assay conditions.

Estimation of residual TA content

The remaining TA content in the fermented broth was estimated using the methods of Hagerman and Butler [17] and expressed as a percentage of the initial concentration. The effect of partially purified and lyophilized tannase on the nutritional value of cattle feed (WS) was evaluated.

Application of tannase to improve the nutritional value of cattle feed (WS)

Preparation of tannin-free WS

Pre-dried WS was treated with different concentration of tannase (0.1%–0.3% (w/w) with an activity of 5.8 IU/mg) to remove tannins. Briefly, lyophilized tannase was dissolved in citrate phosphate buffer and sprayed on WS to obtain a moisture ratio of 1:2 (WS: moisture) and then mixed, incubated at 40°C under static conditions for 6.0 h, and analyzed for remaining tannin content.

The untreated and enzyme-treated WS samples were powdered using a laboratory mill (Remi Motors, Delhi, India) and sieved (30 mesh, 0.5 mm) for analytical purposes. After enzyme treatment, the WS was considered tannin-free.

Experimental organisms and inoculum preparation

Ganoderma spp. obtained from a laboratory stock culture collection (isolated from the bark of *Eucalyptus lanceolate* from the forested area near Delhi, India), were maintained on malt extract agar (MEA), containing (g/L) malt extract, 20.0; KH_2PO_4, 0.5; $MgSO_4{\cdot}7H_2O$, 0.5; $Ca(NO_3)_2{\cdot}4H_2O$, 0.5; and agar, 20.0 (pH 5.5), at 30°C. The obtained culture was stored at 4°C and subcultured every 2 wk. A 250-mL Erlenmeyer flask containing 50 mL of sterile malt extract broth (MEB) was inoculated with two mycelial discs (8 mm in diameter) from the periphery of 7 d-old fungal cultures grown on MEA. The inoculum was incubated at 30°C for 5 d under static conditions. The obtained mycelial mat was further homogenized and then inoculated (4% v/v) in 2-L Erlenmeyer flasks containing 500 mL MEB and incubated at 30°C for 5 d in a rotary incubator at 200 rpm. The fungal mass, obtained in the form of pellets, was separated from the culture broth through filtration using pre-weighed Whatman grade no. 1 filter paper. The formed pellets were used as inocula for further experiments. After being thoroughly washed with deionized water, dried at 55°C until a constant mass was achieved, and weighed.

Solid state fermentation

Enamel trays (40 cm × 32 cm × 7.5 cm) were loaded with 500 g of tannin-free or untreated WS (as a control) and moistened with mineral salt solution and sterilized. Each sterilized tray was inoculated with a dry fungal mass of 3.75–4.00 g (0.75% w/w). Sterile mineral salt solution was added to the trays to obtain a final substrate-to-moisture ratio of 1:3 and incubated at 30°C under 60% relative humidity for 10 d. Following incubation, the trays were harvested to determine production of various enzymes and the proximate and chemical composition of the WS. All experiments were performed in three times.

Enzyme assays

The myco-straw (WS, both enzymatically treated and non-treated, along with fungal mycelium; 5 g of wet weight) was removed aseptically from the flasks after incubation for 10 d and then suspended in 20 mL of acetate buffer (20 mmol/L, pH 5.0) and agitated at 200 rpm for 1 h. The extrudates were pressed through a muslin cloth to maximize enzyme extraction and centrifuged at 10,000 rpm for 10 min at 4°C. The obtained enzyme solution was assayed to determine the activities of different enzymes.

Laccase (E.C. 1.10.3.2) activity was determined using the method described by Paavola et al. [18]. MnP (E.C. 1.11.1.13) activity was measured by oxidation of DMP (2, 6-dimethoxy phenol) in sodium tartrate buffer 0.1 M (pH 4.5) in the presence of H_2O_2 and $MnSO_4$ to coerulignone. The increase in absorbance at 469 nm was determined at 30°C and the activity was calculated using an ε469 of 27,500 /mol/cm for DMP [19]. Xylanase activity was determined by measuring the release of reducing sugars from birchwood xylan (1.0% w/v) using the dinitrosalicylic acid method [20].

Proximate and chemical composition of WS

Dried WS samples were analyzed in triplicate for moisture, dry matter, ether extracts, crude fiber, total carbohydrates, CP, and total ash content [21]. The acid detergent fiber (ADF), neutral detergent fiber (NDF), cellulose, hemicellulose, and lignin contents were determined using the methods described by Van Soest et al. [22,23]. The C/N ratios in the fermented samples were estimated using a CHNS analyzer (Elementar Vario EL III; Analysensysteme GmbH, Hanau, Germany). Percent efficiency of the solid state fermentation (SSF) process was expressed as the loss of lignin content compared to carbohydrate breakdown and calculated using the method described by Moyson and Verachtert [24]. Loss of WS components after treatment was calculated using the following formula:

$$U1 = [(G_0C_{0}\text{-}G_1C_1/G_0C_0] \times 100$$

Where, G0 and G1 are the masses of dry substrates at the beginning and end of fermentation, respectively, and C0 and C1 are the dry weights of the components at the beginning and end of fermentation, respectively.

In vitro digestion

In vitro digestion of non-inoculated and fungal-treated WS was estimated according to the methods described by Akhter et al. [25] with some modifications. Briefly, the inoculum was prepared by mixing fresh buffalo fecal matter (100 g/L) in pre-warmed (39°C) artificial saliva containing $NaHCO_3$ (9.80 g), $Na_2HPO_4{\cdot}7H_2O$ (7.0 g); KCl (0.57 g), NaCl (0.47 g), $MgSO_4{\cdot}7H_2O$ (0.12 g), and 1.0 mL of $CaCl_2$ (4% w/v). The solution was filtered through a muslin cloth and a 500-mg aliquot of the wheat straw was placed in a 50-mL centrifuge tube and the prepared fecal suspension was added. Then, CO_2 gas was flushed into the centrifuge tubes for 48 h in a 39°C water bath. The mixture contained in the centrifuge tubes was centrifuged and the supernatant was discarded. Next, 35 mL of acidified pepsin (6.6 g in 1 L of 0.1 mol/L HCl) was added to the pellet and the tubes were again incubated under similar conditions for 48 h and then centrifuged. The obtained pellet was filtered through tared filter paper and dried. *In vitro* digestion was expressed as the loss of dry matter mass during incubation under standard assay conditions.

Results

A previous study reported that the selected soil isolate, identified as *Penicillium charlesii*, produced 61.4 IU/mL in 54 h in 220 L of statistically optimized medium composed of (% w/v) TA (2.0), sucrose (1.0), $NaNO_3$ (1.0), and $MgSO_4$ (0.08; pH 5.0) contained in 300-L vessel and then fed with 1.0% (w/v) TA after 24 h of incubation. The tannase obtained from the fermentation broth was partially purified by ultrafiltration (30 kDa), ammonium sulfate precipitation, and then lyophilized with a final yield of 5.8 IU/mg [26].

Analysis of the raw WS (control) and tannase-treated WS samples (at 0.1% w/w of tannase concentration) revealed tannin contents of 3.6% and 1.8% (w/w), respectively. On further increase in tannase concentration up to 0.3% (w/w), a 1.8-fold higher reduction in tannin content was noted compared to the results obtained at 0.1% tannase. These results clearly indicated that an increased tannase concentration resulted in a decrease in tannin content (Table 1). There was a very strong positive correlation ($R^2 = 0.999$) between enhanced GA production and TA degradation (Figure 1).

After tannin removal, tannin-free WS was then subjected to SSF using *Ganoderma* spp. to further improve the WS composition and to analyze the various enzymes produced and substrates transformed during fungal fermentation (Figure 2).

As in the previous experiment (Table 1), an increase in tannase concentration resulted in incremental tannin degradation. But *Ganoderma* spp. grown on tannin-free WS showed that 0.1% (w/w) tannase was the optimal concentration for a remarkable increase in fungus-produced lignocellulolytic enzymes. Our results also indicated that at the optimal concentration of 0.1% tannase (w/w) resulted in a 1.12-fold increase in laccase activity, a 1.06-fold increase in MnP activity, and a 1.17-fold increase in xylanase activity after incubation for 10 d compared to raw untreated WS (Table 2). Along with the production of various lignocellulolytic enzymes, the *Ganoderma* spp. could also colonize and rapidly degrade WS, thereby showing a variation in dry matter loss (DML). A higher DML of 14.6% (w/w) was obtained in *Ganoderma* grown on WS treated with 0.1% (w/w) tannase. By further increasing the tannase concentration beyond 0.1%, a decline in the lignocellulolytic activities of MnP, xylanase, and laccase along with lower DML was observed. Based on these results, we hypothesized that due to the relatively poor growth of *Ganoderma*, as observed at higher tannin degradation or GA concentrations; lignocellulolytic activity was decreased, indicating a reduction in lignin depolymerization.

The results of WS compositional analysis presented in Table 3 clearly showed a higher percentage of degraded lignin (37.8% w/w) with a lower L/C ratio of 0.18 was obtained in the tannin-free *Ganoderma*-treated WS with 0.1% (w/w) tannase. In addition, the percentage of

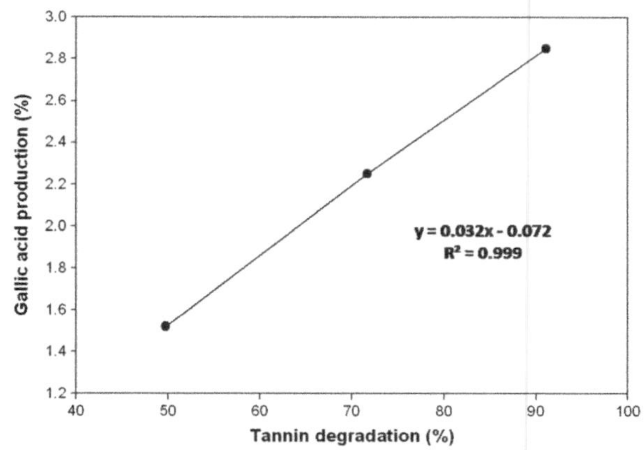

Figure 1 Positive correlation between gallic acid production and tannic acid degradation ($R^2 = 0.999$).

degraded hemicellulose decreased to 17.9% (w/w), whereas a further increase in tannase concentration (up to 0.3%, w/w) resulted in a lower concentration of degraded lignin (13.0%, w/w) with increased cellulose degradation (up to 24.5%, w/w) and a higher L/C ratio of 0.27. Also, 32.2% of the *in vitro* digestibility using 0.1% tannase-treated myco-straw; therefore, 0.1% (w/w) tannase-treated myco-straw was selected, as it showed a lower L/C ratio along with a decent percentage of *in vitro* digestion, which indicated improved quality of cattle feed.

Ganoderma-treated WS (myco-straw) exhibited maximum SSF efficiency (approximately 20%) on incubation day 10. A higher SSF efficiency leads to better nutrient assimilation by fungi and a consistent increase in CP content. It was observed that a higher efficiency of 24.4% was obtained in tannin-free *Ganoderma*-treated WS (0.1% tannase, w/w) on the 10th day of incubation compared to the untreated control WS and 1.22-fold higher compared to myco-straw. Furthermore, the tannin-free *Ganoderma*-treated WS at a tannase concentration of 0.1% (w/w) showed a 1.28-fold higher CP content compared to myco-straw and a 25.0% lower C/N ratio in the fermented substrate (Table 4).

Discussion

Worldwide energy and environmental crises have forced the re-evaluation of efficient utilization of existing resources or to find alternative natural and renewable

Table 1 Enzymatic tannin removal from wheat straw at different concentration of tannase

Tannase concentration	Tannase activity, IU	Remaining tannin, g/100 g WS	Tannin degradation, %	Gallic acid produced, g/100 g WS
Raw wheat straw (control)	0.0 ± 0.00	3.60 ± 0.05 (100%)	0.0 ± 0.00	0.0 ± 0.00
WS + tannase (0.1%)	423 ± 11.6	1.80 ± 0.04 (50.3%)	49.7 ± 0.99	1.5 ± 0.03
WS + tannase (0.2%)	846 ± 27.5	1.02 ± 0.01 (28.3%)	71.7 ± 1.79	2.2 ± 0.06
WS + tannase (0.3%)	1269 ± 47.6	0.33 ± 0.002 (8.9%)	91.1 ± 2.7	2.8 ± 0.09

WS = wheat straw.

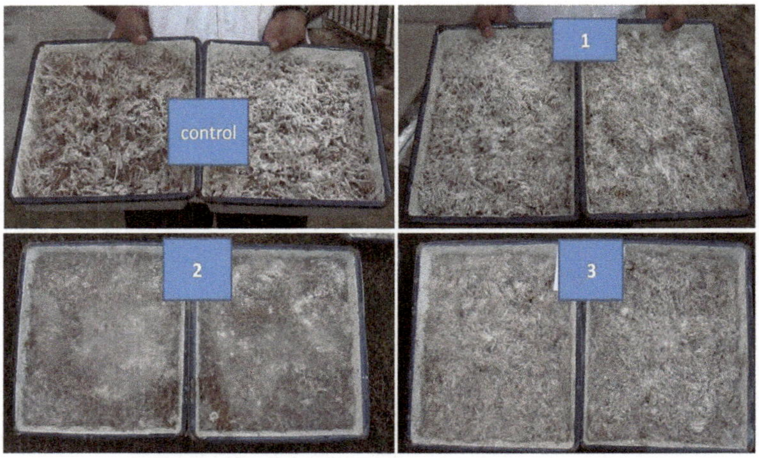

Figure 2 Growth pattern of *Ganoderma* sp. on raw wheat straw and tannin free ssf treated wheat straw (TFWS) Control: raw wheat straw; 1: TFWS treated with 0.1% tannase; 2: TFWS treated with 0.2% tannase; 3: TFWS treated with 0.3% tannase.

resources in nature, particularly to dispose of organic waste, using clean technologies. Every year, large quantities of cellulosic agriculture byproducts are being produced. In the face of global industrialization and the subsequent environmental consequences, as exemplified by the decline in forest belts and global climate changes, forces us to consider the potential benefits of finding alternative sources of cellulose. To date, WS is rarely used for cattle feed because of its relatively low nutritional value, particularly the low protein content, high fiber content, and low digestibility by ruminants. Thus, improvements in the nutrient availability in cattle feed presents a significant challenge to animal nutritionists.

The use of fungi or fungal enzymes that can metabolize lignocellulose presents a potential biological method to improve the nutritional value of WS by selective delignification [11]. The use of tannase or microbes that produce this enzyme in feed preparations increases the bioavailability of nutrients by hydrolyzing phenols, which acts as anti-nutritional factors [27]. However, due to high production costs and our limited knowledge of the catalytic activities of tannase, this enzyme is currently used under very few circumstances.

The concentration of tannins in feed plays a vital role on the digestibility by ruminants [28]. The highest tannin content showed the lowest digestibility, thus showing a strong negative correlation (r = –0.993) [12].

There are several reports in the literature confirming that WS contains appreciable amounts of tannins [29,30]. Enhancement in the *in vitro* digestibility of WS up to 2.2-fold was observed by treatment with 0.1% tannase (w/w), as well as WRF (*Ganoderma* sp.) under SSF, as compared to the digestibility of untreated WS. Similar to our results, Arora and Sharma [12] reported enhanced *in vitro* digestibility up to 1.66-fold by WS treatment with the fungus *Phlebia brevispora*, obtained from northwestern and northeastern regions of India, while a 1.78-fold increase was observed by the use of fungi obtained in central India. In the present investigation, we established that the presence of 0.1% (w/w) tannase enhanced lignin degradation as well as *in vitro* digestibility.

Regarding fibrolytic enzymes, several researchers have reported that the presence of aromatic compounds was required, to a certain extent, to enhance ligninase production in several WRF for lignin degradation. In the present study, it was noted that at higher tannase concentrations beyond optimal, more toxic phenolic compounds with anti-nutritional properties (tannin) were degraded up to approximately 90% with 0.3% (w/w) tannase along with the synthesis of higher amounts of GA, which is comparably

Table 2 Enzyme production and substrate transformation during fungal fermentation of wheat straw

Tannase concentration	Treated with *Ganoderma* sp.		
	Laccase, U/g	MnP, U/g	Xylanase, U/g
Raw wheat straw	4012 ± 160.4	126 ± 2.8	221 ± 6.0
WS + tannase (0.1%)	4518 ± 203.3	133.2 ± 3.6	260 ± 7.8
WS + tannase (0.2%)	3216 ± 120.6	116 ± 2.0	210 ± 5.2
WS + tannase (0.3%)	2213 ± 60.8	56.21 ± 0.84	139 ± 2.4

WS = wheat straw.

Table 3 Disappearance (%) of cell wall components in raw wheat straw and enzyme treated wheat straw subjected to treatment with *Ganoderma* sp.

Substrate	*In vitro* digestibility,%	% Degradation							L/C ratio*
		Wt	ADF	Lignin	Cellulose	NDF	HC	CP	
SSF treated WS (myco- straw)	26.8 ± 0.8	11.6 ± 0.2	17.3 ± 0.4	24.0 ± 0.5	22.6 ± 0.6	18.9 ± 0.4	21.7 ± 0.5	2.1 ± 0.02	0.23 ± 0.003
Tannin free ssf treated WS (0.1%)	32.2 ± 1.1	14.6 ± 0.3	20.3 ± 0.5	37.8 ± 1.3	20.4 ± 0.4	19.5 ± 0.4	17.9 ± 0.3	2.7 ± 0.04	0.18 ± 0.001
Tannin free ssf treated WS (0.2%)	27.9 ± 0.7	12.4 ± 0.2	14.2 ± 0.3	16.6 ± 0.3	21.9 ± 0.5	19.3 ± 0.2	29.6 ± 0.8	2.14 ± 0.03	0.25 ± 0.004
Tannin free ssf treated WS (0.3%)	24.2 ± 0.4	10.9 ± 0.1	10.1 ± 0.1	13.0 ± 0.2	24.5 ± 0.7	19.6 ± 0.3	39.1 ± 1.4	2.13 ± 0.03	0.27 ± 0.005

ADF Acid detergent fiber, NDF Neutral detergent fiber, HC Hemicellulose, CP Crude protein; * data calculated on as such basis (w/w); Uninoculated wheat straw contains *in vitro* digestibility of 146 g/kg (14.6%).

less toxic than NA. However, this increased GA concentration could lead to lower activity of different lignocellulolytic enzymes produced from *Ganoderma* spp. Therefore, an optimal GA concentration is required, as GA acts as an inducer, not an inhibitor, of the lignocellulytic enzymes produced from WRF. In the present report, the amount of GA produced (1.5 g/100 g) by tannin degradation with 0.1% (w/w) tannase was sufficient to promote growth of *Ganoderma* spp. and to induce production of lignocellulytic enzymes. Above this concentration, GA acts as an inhibitor of the lignocellulytic enzymes laccase, and MnP produced by WRF, which are primarily responsible for lignin degradation through inhibition of *Ganoderma* growth. Our results are in accordance with those reported by Gnanamani et al. [31], in which the presence of GA at an optimal concentration of 300 mmol/L could induce activity of the lignocellulytic enzyme laccase in *Phanerochaete chrysosporium* (NCIM 1197). Also, Patel et al. [32] reported an increase in laccase activity of up to 1.3-fold in the presence of an optimal GA concentration. Contrary to our results, Cavallazzi et al. [33] observed no laccase activity in the presence of catechol and GA, even after 18 d of cultivation. The optimal GA concentration for luxurious growth and enhanced enzyme levels in WRF might produce toxic effects to ruminants by feed digestion, although the effect of toxicity will definitely be less than that induced by tannins. Lin et al. [34] postulated that produced GA can easily be assimilated by ruminants due to the sequential actions of different enzymes present in the rumen microflora to pyrogallol, phloroglucinol, and to the less toxic compounds acetate and butyrate.

In the present study, we found that by increasing the tannase concentration beyond optimal (0.1% w/w), a greater amount of tannin was degraded and converted into GA and glucose in WS. Therefore, tannin-free *Ganoderma*-treated WS (beyond 0.1% w/w tannase) contains a lower tannin concentration, which is generally associated with or forms complexes with the hemicellulases of WRF, thereby leading to more free sites available for hemicellulases to degrade greater amounts of hemicellulose in WS by the actions of WRF. On compositional analysis of WS, it was observed that greater amounts of hemicellulose were degraded as compared to cellulose at higher tannase concentrations (beyond 0.1% w/w tannase). Similar results have been reported by several researchers who found that tannin-rich agro-material can easily inhibit hemicellulases over cellulases through association, as cellulases are generally associated with the cell walls of WRF, while hemicellulases are extracellular and are, therefore, more sensitive to tannins present in lignocellulosic materials. In this respect, our results are in accordance with those reported by Hervás et al. [35], who observed increased degradation of cellulose as compared to hemicelluloses in the presence of higher residual tannins (i.e., at lower tannase concentrations).

The lignin to cellulose ratio (L/C ratio) is considered important for cattle feed. This ratio clearly indicates that during treatment greater quantities of lignin are degraded, thereby leaving more cellulose in the residual substrate. Therefore, enzyme-treated feed may contain more cellulose as compared to lignin in residual substrates that can be easily digested and also provides a rich source of energy to ruminants. The L/C ratio of the tannin-free *Ganoderma*-treated WS was 0.18 at a tannase concentration of 0.1% (w/w). Calzada et al. [36] observed lignin degradation in the range of 24% to 55%

Table 4 CP of wheat straw

Substrate	N	C	S	H	CP	C/N ratio
SSF treated WS	0.33 ± 0.007	36.5 ± 1.00	0.24 ± 0.003	3.83 ± 0.07	2.09 ± 0.02	109.16 ± 3.54
Tannin free ssf treated WS (0.1%)	0.43 ± 0.01	34.9 ± 0.9	0.48 ± 0.007	4.89 ± 0.11	2.66 ± 0.4	81.88 ± 2.04
Tannin free ssf treated WS (0.2%)	0.342 ± 0.008	28.7 ± 0.5	0.24 ± 0.004	3.22 ± 0.05	2.14 ± 0.03	83.80 ± 2.30
Tannin free ssf treated WS (0.3%)	0.341 ± 0.008	29.3 ± 0.6	0.29 ± 0.005	4.09 ± 0.09	2.13 ± 0.03	86.04 ± 2.58

Crude Protein (CP) content in raw wheat straw: 0.56; N: nitrogen, C: carbon; S: sulphur; H: hydrogen.

(w/w), whereas cellulose degradation ranged from 34% to 42% (w/w) for WS inoculated with the Pleurotus spp. of WRF. Basu et al. [6] reported that the maximum desirability coefficient after incubation for 6 d was 19.5% (w/w) lignin and 17.8% (w/w) cellulose degradation, although the fungus continued to degrade the substrate thereafter. Fungal fermentation using *Ganoderma* sp. rckk02 decreased the lignin content of WS by 34.95% (w/w) with a subsequent reduction in cellulose content by 34.33% (w/w) after incubation for 15 d [37].

A 78.6% (w/w) increase in CP content was observed in tannin-free *Ganoderma*-treated WS with 0.1% (w/w) tannase as compared to raw WS. While, a 27.3% (w/w) higher CP content was obtained in *Ganoderma*-treated WS (myco-straw). Various studies on fungal fermentation have reported the enrichment of WS with fungal biomass as the protein source under SSF [14,38]. While, Srivastava et al. [13] reported a 57.11% (w/w) increase in CP content by fermenting WS with *Ganoderma* sp. rckk02 compared to raw WS (control). The increase in CP content is partially attributable to fungal proteins, particularly chitin [38], although the observed decrease in certain cases cannot be fully explained. The C/N ratio is an important marker of biomass pretreatment because degradation of lignocellulosic material is dependent on the C/N ratio of the material. Deshpande et al. [39] reported that an optimal C/N ratio promotes good fungal growth and, consequently, higher enzyme production.

It could be concluded that the use of a combination of various lignocellulytic enzymes present in WRF and tannase obtained from *Penicillium charlesii* resulted in complete fermentation over a short period, which indicates the scope of enriched feed production at an industrial scale. The biological treatment processes are safe, eco-friendly, easy to perform, and involves lower operational costs as compared to other treatment methods. Besides, in this process, the inoculum could be raised from existing growing cultures and reused several times to inoculate bulk amounts of residues. While, the use of chemicals is limited because once used, these substances leach into water sources, thereby requiring further treatments involving large amounts of other chemicals. If this technology could be commercialized in the future, then the enzyme tannase (lyophilized or liquid form) along with the *Ganoderma* sp. (dried or lyophilized form) can be made available to cattle ranchers as enriched feed at costs subsidized by the government.

Competing interests

All authors declare that they have no competing interests.

Authors' contribution

SR and SM designed and performed the experiments, including chemical analysis, collected and analyzed the data, and wrote the manuscript. RKS verified the validity of the experimentation and checked the results. All of the authors read and approved the final version of this manuscript.

Acknowledgments

This study was supported by a grant from the Council of Scientific and Industrial Research of India to S.R. (grant no.: 9/45(1190)/2012-EMR-1).

References

1. Rao SVN, Ramkumar S, Natchimuthu K, Bezkorowajnyj P: **Dairy cattle feeding- evidence based pro-poor institutional approach.** In *Proceedings of Animal Nutrition World Conference, 14th to 17th February, New Delhi, India.* 2009.
2. Nisa MU, Sarwar M, Khan MA: **Influence of *ad-libitum* feeding of urea treated wheat straw with or without corn steep liquor on intake, in situ digestion kinetics, nitrogen metabolism and nutrient digestion in Nilli-Ravi buffalo bulls.** *Aust J Agric Res* 2004, **55:**1–8.
3. Hanafi EM, El Khadrawy HH, Ahmed WM, Zaabal MM: **Some observations on rice straw with emphasis on updates of its management.** *World Appl Sci J* 2012, **16**(3):354–361.
4. Haysa MD, Finebm PM, Eronam GCD, Kleemanc MJ, Brian K, Gulletta BK: **Open burning of agricultural biomass: Physical and chemical properties of particle-phase emissions.** *Atmos Environ* 2005, **39**(36):6747–6764.
5. Castrillo C, Fondevila M, Alibes X, Joy M: **Chemical treatments for upgrading lignocellulosic resources and strategies for their utilization in ruminant feeding.** In *Production and Utilization of lignocellulosics.* Edited by Galletti GC. London: Elsevier Applied Sciences; 1991:339–313.
6. Basu S, Gaur R, Gomes J, Sreekrishnan TR, Bisaria VS: **Effect of seed culture on solid state bioconversion of wheat straw by *Phanerochaete chrysosporium* for animal feed production.** *J Biosci Bioeng* 2002, **1:**25–30.
7. Liu JX, Orskov ER, Chen XB: **Optimization of steam treatment as a method for upgrading rice straw as feeds.** *Anim Feed Sci Technol* 1999, **76:**345–357.
8. Kuhar S, Nair LM, Kuhad RC: **Pretreatment of lignocellulosic material with fungi capable of higher lignin degradation and lower carbohydrate degradation improves substrate acid hydrolysis and the eventual conversion to ethanol.** *Can J Microbiol* 2008, **54**(4):305–313.
9. Okano K, Ohkoshi N, Nishiyama A, Usagawa T, Kitagawa M: **Improving the nutritive value of madake bamboo, *Phyllostachys bambusoides*, for ruminants by culturing with the white rot fungus *Ceriporiopsis subvermispora*.** *Anim Feed Sci Technol* 2009, **152:**278–285.
10. Jalc D: **Straw enrichment for fodder protection by fungi.** In *The Mycota XI Agricultural Applications.* Edited by Kempken F. Berlin, Heidelberg: [C] Springer-Verlag; 2002:19–38.
11. Eriksson KEL, Blanchette RA, Ander P: *Microbial and enzymatic degradation of wood and wood components.* Berlin, Heidelberg, New York: Springer; 1990.
12. Arora DS, Sharma RK: **Enhancement in *in vitro* digestibility of wheat straw obtained from different geographical regions during solid state fermentation by white rot fungi.** *BioResources* 2009, **4:**909–920.
13. Srivastava BS, Thakur S, Khasa YP, Gupte A, Puniya AK, Kuhad RC: **White rot fungal conversion of wheat straw to energy rich cattle feed.** *Biogeosciences* 2011, **22:**823–831.
14. Garcý'a-Conesa MT, Ostergaard P, Kauppinen S, Williamson G: **Hydrolysis of diethyl diferulates by a tannase from *Aspergillus oryzae*.** *Carbohyd Polym* 2001, **44**(2):319–324.
15. Raghuwanshi S, Dutt K, Gupta P, Misra S, Saxena RK: ***Bacillus sphaericus*: The highest bacterial tannase producer with potential for gallic acid synthesis.** *J Biosci Bioeng* 2011, **111**(6):635–640.
16. Deschamps AM, Otuk G, Lebeault JM: **Production of tannase and degradation of chestnut tannins by bacteria.** *J Ferment Bioeng* 1983, **61:**55–59.
17. Hagerman AE, Butler LG: **Protein precipitation method for quantitative determination of tannin.** *J Agric Food Chem* 1978, **26:**809–812.
18. Niku-Paavola ML, Raaska M, Itävara M: **Detection of white-rot fungi by a nontoxic stain.** *Mycol Res* 1990, **94:**27–31.
19. Martinez MJ, Ruiz-Duenas FJ, Guillen F, Martinez AT: **Purification and catalytic properties of two manganese peroxidase isoenzymes from *Pleurotus eryngii*.** *Eur J Biochem* 1996, **237:**424–432.
20. Miller GL: **Use of dinitrosalicyclic acid reagent for determination of reducing sugar.** *Anal Chem* 1959, **31:**426–428.
21. AOAC: *Official methods of analysis.* 17th edition. Washington D.C: Association of official analytical chemists; 1995.

22. Goering HK, Van Soest PJ: **Forage fiber analysis (Apparatus, reagents, procedures and some applications).** In *Agricultural Research Service, US Dep. Agric.* Washington, DC: Agriculture Handbook; 1970:19. No. 379.
23. Van Soest PJ, Robertson JB, Lewis BA: **Methods for dietary fiber, neutral detergent fiber, and non starch polysaccharides in relation to animal nutrition.** *J Dairy Sci* 1991, **74:**3583–3597.
24. Moyson E, Verachtert H: **Growth of higher fungi on wheat straw and their impact on the digestibility of the substrate.** *Appl Microbiol Biotechnol* 1991, **36:**42–424.
25. Akhter S, Owen E, Theodorou MK, Butler EA, Minson DJ: **Bovine faeces as a source of microorganisms for the invitro digestibility assay of forages.** *Grass Forage Sci* 1999, **54:**219–226.
26. Raghuwanshi S, Misra S, Saxena RK: **Enzymatic treatment of black tea (CTC and Kangra orthodox) using *Penicillium charlesii* tannase to improve the quality of tea.** *J Food Process Preserv* 2012, doi:10.1111/j.1745-4549.2012.00721.x.
27. Graminha EBN, Goncalves AZL, Pirota RDPB, Balsalobre MAA, Silva R, Gomes E: **Enzyme production by solid-state fermentation: application to animal nutrition.** *Anim Feed Sci Technol* 2008, **144**(1–2):1–22.
28. McLeod MN: **Plant tannins-their role in forage quality.** *Nutr Abstr Rev* 1974, **44:**803–815.
29. Patil SS, Ahmed SA, Telang SM, BAIG MMV: **The nutritional value of *Pleurotus ostreatus* Kumm cultivated on different lignocellulosic agro wastes.** *Innovat Rom Food Biotechnol* 2010, **7:**66–76.
30. Hussain I, Uddin MB, Aziz MG: **Optimization of antinutritional factors from germinated wheat and mungbean by Response Surface Methodology.** *Int Food Res J* 2011, **18:**957–963.
31. Gnanamani A, Jayaprakashvel M, Arulmani M, Sadulla S: **Effect of inducers and culturing processes on laccase synthesis in *Phanerochaete chrysosporium* NCIM 1197 and the constitutive expression of laccase isozymes.** *Enzyme Microb Technol* 2006, **38:**1017–1021.
32. Patel H, Gupte A, Gupte S: **Effect of different culture conditions and inducers on production of laccase by a basidiomycete fungal isolate *Pleurotus ostreatus* HP-1 under solid state fermentation.** *BioResources* 2009, **4**(1):268–284.
33. Cavallazzi JRP, Kasuya CM, Soares MA: **Screening of inducers for laccase production by *Lentinula edodes* in liquid medium.** *Braz J Microbiol* 2005, **36:**383–387.
34. Lin HV, Frassetto A, Kowalik EJ Jr, Nawrocki AR, Lu MM, Kosinski JR, Hubert JA, Szeto D, Yao X, Forrest G, Marsh DJ: **Butyrate and Propionate Protect against Diet-Induced Obesity and Regulate Gut Hormones via Free Fatty Acid Receptor 3-Independent Mechanisms.** *PLoS One* 2012, **7**(4):e35240.
35. Hervás G, Frutos P, Giraldez FJ, Mantecón AR, Álvarez Del Pino MC: **Effect of different doses of quebracho tannins extract on rumen fermentation in ewes.** *Anim Feed Sci Technol* 2003, **109:**65–78.
36. Calzada JF, de Leon R, de Arriola MC, Rolz C: **Growth of mushrooms on wheat straw and coffee pulp: strain selection.** *Biol Wastes* 1987, **20:**217–226.
37. Srivastava B, Nandal P, Sharma A, Jain KK, Khasa YP, Das TK, Mani V, Kewalramani NJ, Kundu SS, Kuhad RC: **Solid state bioconversion of wheat straw into digestible and nutritive ruminant feed by *Ganoderma* sp. rckk02.** *Bioresource Technol* 2012, **107:**347–351.
38. Hadar Y, Karem Z, Gorodecki B, Ardon O: **Utilization of lignocellulosic waste by the edible mushroom.** *Pleurotus Biodegr* 1992, **3:**189–205.
39. Deshpande SK, Bhotmange MG, Chakrabarti T, Shastri PN: **Production of cellulase and xylanase by *Trichoderma reesei* (QM 9414 mutant), *Aspergillus niger* and mixed culture by solid state fermentation (SSF) of water hyacinth (*Eichhornia crassipes*).** *Indian J Chem Technol* 2008, **15:**449–456.

11

Chicory (*Cichorium intybus* L.) and cereals differently affect gut development in broiler chickens and young pigs

Haoyu Liu*, Emma Ivarsson, Torbjörn Lundh and Jan Erik Lindberg

Abstract

Dietary fiber, resistant to host-mediated digestion in the small intestine due to lack of endogenous enzymes, impacts many facets of animal health and is associated with gut development especially in young monogastrics. Furthermore, it can be used as in-feed antibiotic alternative. Chicory (*Cichorium intybus* L.) forage with high content of pectin (uronic acids as building blocks) is a novel class of dietary fiber that is chemically different from cereal grains (with high content of arabinoxylans). In the present study, we investigated effects of dietary inclusion of chicory forage on digestibility, gut morphology and microbiota in broilers and young pigs. In the chicken experiment, 160 1-d old broiler chicks were fed 3 nutritionally balanced diets for 30 d including a cereal-based diet and 2 diets with part of the cereals substituted with 60 and 120 g/kg chicory forage (CF60 and CF120), whereas in the pig experiment, 18 seven-wk old Yorkshire pigs were fed 3 diets for 18 d including a cereal-based diet and 2 diets with 80 and 160 g/kg chicory forage inclusion (CF80 and CF160). Our results showed that young pigs were capable to utilize chicory forage well with higher total tract apparent digestibility (TTAD) of all fiber fractions, particularly uronic acid, compared with the control ($P < 0.01$). In contrast, a decreased TTAD of all fiber fractions was observed in chickens fed on diet CF120 ($P < 0.05$). Moreover, diet induced changes in gut morphology were observed in the large intestine of chickens. The alteration of cecal mucosal thickness was further positively correlated with TTAD of non-starch polysaccharides (NSP) and its constituent sugars ($P < 0.05$). In addition, in pigs, terminal restriction fragment length polymorphism (T-RFLP) analysis of intestinal microbiota revealed substantial dietary effects (cereal control diet vs. chicory forage inclusion) on the relative abundance of 2 dominant bacterial phylotypes (*Prevotella* sp. vs. *Roseburia* sp.) respectively ($P < 0.05$). In conclusion, our data showed that chicory forage (*Cichorium intybus* L.), a novel dietary fiber source in animal nutrition, have potential beneficial properties as fiber ingredient in diets for both pigs and chickens.

Keywords: Broiler, Cereal arabinoxylans, Chicory uronic acids, Digestibility, Gut development, Microbiota, Pectin, Young pigs

Background

Dietary fiber is defined by CODEX Alimentarius as 'carbohydrate polymers with 10 or more monomeric units, which are not hydrolyzed by the endogenous enzymes in the small intestine of humans' [1]. This all-encompassing definition includes fiber naturally occurring in foods, as well as processed (physically, chemically or enzymatically) fiber from raw materials and synthetic fractions, in order to meet the needs of regulation and the associated labeling market. A substantial body of evidence demonstrates that fiber ingredients (mainly non-starch polysaccharides, NSP) constitutes an important component of a balanced diet and may affect many facets of animal nutrition and gut health, especially in young monogastrics [2-8]. The impact of dietary fiber on gut health opens a window in search for in-feed antibiotic alternatives. A prebiotic effect of dietary fiber will help to reduce the antibiotic usage in livestock and this will reduce the risk of transferring the antibiotic resistance gene to human pathogens [9]. Furthermore, dietary fiber has been associated with gut disorder management

* Correspondence: Liu.Haoyu@slu.se
Department of Animal Nutrition and Management, Swedish University of Agricultural Sciences, P.O. Box 7024, Uppsala SE-750 07, Sweden

such as *Salmonella* infection in chickens and post-weaning diarrhea in pigs [10].

Arabinoxylan, composed of xylose as backbone and arabinose as side chains, is one major NSP fraction of the dietary fiber in cereal grains [2]. The arabinoxylans are present in both soluble and insoluble form, although a major part is insoluble. Cereals (*e.g.* wheat and barley) high in arabinoxylan content have been widely used in chicken and pig commercial feeds [6,11]. A study on humanized rats suggests that dietary arabinoxylan confers beneficial effects on gut health and may be a good candidate for prebiotics [2].

Chicory (*Cichorium intybus* L.) is a perennial herb that can produce nutritious and high quality forage [12]. The dietary fiber in chicory forage has high content of pectin (80–90 g/kg dry matter, DM), another type of NSP with uronic acid as building blocks, which is highly soluble in comparison with other pectin sources [8]. Chicory forage is a novel source of dietary fiber that has recently been shown to be well accepted and utilized by monogastric animals [8,13]. It is a potential feed resource that could partly replace cereal grain fiber in livestock feed and can be used as roughage source in organic pig farming [14]. Replacing cereal grain fiber may reduce feed cost and the conflict with human needs as there is limited supply of cereals and cereal by-products [15,16].

The present study investigated the impact of dietary inclusion of chicory forage on digestibility, gut morphology and gut microbiota in broiler chickens and young pigs. Our hypothesis was that chicory forage would affect the animals and their gut development differently from cereals.

Methods

Experimental design and animals

The animal experiment was performed at the Swedish University of Agricultural Sciences (SLU) in Uppsala and was approved by the ethical committee for the Uppsala region. The chicken trial was structured as a randomized block design with 3 treatments, 4 blocks for control group and 2 × 8 blocks for experimental groups, respectively. Each block contained 8 birds. In the experiment using young pigs, a split-litter design was used with 18 pigs from 6 different litters (3 pigs per litter). Animals were randomly distributed to one of 3 treatments (6 replicates for each).

In the chicken trial, 160 female and male 1-d-old broiler chicks (Ross 308) with an initial body weight (BW) of 44.0 ± 15.1 g were studied over a 30-d period. The birds were kept on wood shaving floors from d 1 to d 27. Thereafter, a net floor was used for 3 d for excreta sampling. Chickens had free access to feed and water throughout the experiment. The environment temperature and light were strictly controlled. Body weight and feed intake were recorded for each pen on the first d of the experiment and every wk thereafter.

In the pig trial, 18 seven-wk old Yorkshire pigs (castrated male and female) with an initial BW of 11.7 ± 0.16 kg were studied over an 18-d period. The pigs were housed in individual pens without straw bedding and supplied with feed and water *ad libitum*. Feed intake was recorded daily. The BW was registered weekly and on the last d of the experiment.

Diets

Chicory forage (*Cichorium intybus* L.) was used to compose the experimental diets and replaced the cereal fraction (wheat and barley) in the cereal control diet. All diets were supplemented with protein, amino acids, minerals, and vitamins to meet the nutritional requirements of the broilers and the growing pigs, respectively [17,18]. However, in order to keep diet composition constant throughout the experimental period, the crude protein (CP) content for chickens was lower than recommended during the first wk. Prior to mixing with other feed ingredients, chicory forage was dried at low temperature (30°C) with forced air for a wk. All ingredients were milled through a 3-mm screen for chicken feed and a 3.5-mm screen for pig feed and were fed as pellets to the animals in all cases except for the first wk for chicks, when pellets were ground. Titanium oxide (TiO_2) was included in the diets as an internal digesta marker. No antibiotics were administered.

The detailed diet ingredient composition is shown in Table 1. In brief, for the chicken the experimental diets were comprised of the cereal-based basal diet (C^I) and diets with inclusion of 60 and 120 g/kg chicory forage (CF60 and CF120). For the young pigs, the experimental diets were comprised of the cereal-based basal diet (C^{II}) and diets with inclusion of 80 and 160 g/kg chicory forage (CF80 and CF160).

Sampling and analysis

Feed and feces were collected for chemical analysis. At the end of the animal experiment, chickens were killed by an intravenous injection of sodium pentobarbital through the wing vein, whereas pigs were sedated first and killed by a lethal dose of pentobarbital sodium (60 mg/mL; Apoteket, Umeå, Sweden) at 100 mg/kg BW. For histological analysis, intestinal tissues were sampled as previously [13]. In chickens, jejunum and cecum were sampled, whereas in pigs were distal ileum and proximal colon taken, representing the small and large intestine respectively. In addition, in the pig experiment, intestinal digesta were collected correspondingly.

The chemical composition and gross energy of diets were analyzed as previously described [19] and are shown in Table 1. Gross energy was measured with a bomb calorimeter (Parr 6300 Oxygen Bomb Calorimeter, Illinois,

Table 1 Diet ingredients (g/kg), analyzed chemical composition (g/kg DM) and gross energy content (MJ/kg DM) of control and experimental diets

	Chicken diets			Young pig diets		
Items	C[I]	CF60	CF120	C[II]	CF80	CF160
Wheat	550	507.5	460	410	370	320
Barley	187.5	170	160	400	360	330
Protein sources	160	160	160	160	160	160
Vegetable fat	30	30	30	10	10	10
Chicory	0	60	120	0	80	160
Premix	2	2	2	4	4	4
Titanium oxide	5	5	5	2.5	2.5	2.5
Others	65.5	65.5	63	13.5	13.5	13.5
Dietary fiber	171	183	196	137	175	196
Klason lignin	19	23	21	27	38	43
NSP	152	160	175	110	137	153
Arabinose	24	26	22	18	19	19
Xylose	40	38	34	37	36	36
Uronic acid	18	27	37	4	17	24
Fructan	20	21	18	8	10	6
Gross energy	18.0	18.0	18.0	18.7	18.4	18.2

C[I]: control diet in chicken experiment; CF60: 60 g/kg chicory forage; CF120: 120 g/kg chicory forage; C[II]: control diet in pig experiment; CF80: 80 g/kg chicory forage; CF160: 160 g/kg chicory forage.
For broiler diets, soybean meal was used as protein sources, whereas for young pig diets, a mix of fish meal and potato protein was used.
Premix in chicken experiment: vitamin A, (all-trans retinol) 12000 IU; vitamin D_3, 5000 IU; vitamin E, (DL-α-tocopheryl acetate 70 IU); vitamin K_3, 4 mg; vitamin B_1, 3 mg; vitamin B_2,. 8 mg; vitamin B_6, 5 mg; vitamin B_{12}, 0.02 mg; Pantothenic acid, 20 mg; Folic acid, 2 mg; Niacin, 60 mg; Biotin, 0.175 mg; Iron, 20 mg; Copper, 15 mg; Cobalt, 0.25 mg; Manganese, 70 mg; Zink, 70 mg; I, 1 mg; Selenium, 0.035 mg; Molybdenum, 0.50 mg. Premix in pig experiment: Ca 3.77 g, P 1.02 g, Mg 3.10 g, K 1.50 g, Na 0.37 g, Cl 0.11 g, S 84.00 g, Fe 120493 mg, Cu 29791 mg, Co 52.62 mg, Mn 24076 mg, Zn (as ZnO4) 134 mg, I 238 mg, Se 476 mg. Vitamins: retinol 5952000 IE, cholecalciferol 595200 IE, α-tocopherol 101190 mg, thiamin 2381 mg, riboflavin 5952 mg, pyroxidine 3750 mg, B_{12} 24 mg, pantothenic acid 11914 mg, nicotinic acid 23810 mg, biotin 2238 mg.
Others in chicken experiment(g): $NaCl_3$, L-lysine 3.5-4, DL-methionine, 4–4.5, monocalciumphosphate, 17.5-18, calcium carbonate, 19.5-20; Others in pig experiment (g): NaCl 2.5, L-lysine-HCL 0.5, DL-methionine 0.5, monocalciumphosphate 5, limestone 5.

USA). TiO_2 was used for calculation of total tract apparent digestibility (TTAD) of dietary components [20].

Intestinal histological parameters, including villus height and crypt depth in the small intestine and mucosal thickness in the large intestine were determined according to standard procedures: villus height was depicted from its apex to the transition into the crypt zone, whereas crypt depth was measured as the difference between mucosal thickness (the distance from villus top to crypt end) and villus height.

In order to explore dietary fiber impact on the animal gut and associated changes of microbiota composition, terminal restriction fragment length polymorphism (T-RFLP) analysis was carried out on intestinal digesta samples from pigs as previously described [21]. Putative identifications of TRFs were obtained by ribosomal database mining (http://mica.ibest.uidaho.edu/) and by comparisons with our internal database (based on T-RFLP analysis following 10 clone libraries constructed from our previous pig studies). Moreover, the relative abundance of TRFs within a microbial community profile was used to determine the bacterial diversity [21].

Statistical analysis

Statistical analyses were performed with various procedures in SAS (SAS Institute, Cary, NC, USA, version 9.2). Dietary effects were analyzed using PROC GLM in chicken experiment and PROC MIXED in pig experiment. Furthermore, PROC CORR was carried out to identify relationships between variables. Data were presented as least square means ± SEM. Significance was set at $P < 0.05$.

Results and discussion

Cereal arabinoxylan and plant-origin pectin are NSP abundant in animal diets. Although resistant to host-mediated digestion in the small intestine, these substrates can serve as energy source and physiological stimuli for gut development and microbiota modulation in the large intestine [6,11,22]. In the present study, animals maintained feed intake and growth rate irrespective of dietary treatments

Table 2 Effects of diets on total tract apparent digestibility (TTAD) of fiber in chickens and pigs

	In chickens					In pigs				
	Diet					Diet				
Items	C[I]	CF60	CF120	SEM	*P*-value	C[II]	CF80	CF160	SEM	*P*-value
Total NSP*	31.2[a]	30.1[a]	23.8[b]	1.65	0.008	42.0[a]	56.1[b]	62.4[c]	1.42	<0.0001
Uronic acid	43.3[a]	48.0[a]	33.1[b]	3.34	0.02	36.2[a]	85.1[b]	88.8[b]	1.47	<0.0001
Arabinose	31.0[a]	34.3[a]	18.4[b]	2.40	0.0002	46.8[a]	58.0[b]	62.8[b]	1.71	<0.0001
Xylose	28.5[a]	23.8[a]	16.5[b]	1.77	0.0007	39.8[a]	45.7[b]	49.4[b]	1.70	0.004

C[I]: control diet in chicken experiment; CF60: 60 g/kg chicory forage; CF120: 120 g/kg chicory forage; C[II]: control diet in pig experiment; CF80: 80 g/kg chicory forage; CF160: 160 g/kg chicory forage. *NSP, non-starch polysaccharides.
Data are presented as least square means (n = 4 for diet C[I]and 8 for diet CF60 and CF120 in chicken experiment; n = 6 in pig experiment) and standard error of the mean (SEM).
[abc]Different letters within rows, indicate significant difference ($P < 0.05$).

Table 3 Dietary induced changes in gut morphology of broilers and pigs

Item	In chickens					In pigs				
	Diet					Diet				
	C[I]	CF60	CF120	SEM	*P*-value	C[II]	CF80	CF160	SEM	*P*-value
Small intestine VH	1380	1408	1379	55.8	NS	814	833	806	43.4	NS
Small intestine CD	171	170	161	16.4	NS	265	265	257	18.5	NS
Large intestine MT	376[a]	310[b]	285[b]	13.0	0.001	538	551	524	20.2	NS

C[I]: control diet in chicken experiment; CF60: 60 g/kg chicory forage; CF120: 120 g/kg chicory forage; C[II]: control diet in pig experiment; CF80: 80 g/kg chicory forage; CF160: 160 g/kg chicory forage. VH, villus height; CD, crypt depth; MT, mucosal thickness; NS, non-significant.
Data are presented as least square means (n = 4 for diet C[I] and 8 for diet CF60 and CF120 in chicken experiment; n = 6 in pig experiment) and standard error of the mean (SEM).
[ab]Different letters within rows, indicate significant difference ($P < 0.05$).

[6,19], except that during the first 2 wk of rearing period for chickens, the growth rate was decreased on diet CF120 as compared with the breed standard [23]. This may be attributed to the single feed usage throughout the experiment (lower CP content for younger birds). Overall, the present data on chickens and pigs performance are in the same range as in previous studies using various dietary fiber sources [5,7,11,24].

Dietary fiber utilization in chickens and young pigs

The digestibility of dietary fiber is highly variable and is related to its origin [15]. The present study focused on different chemical composition of fiber fractions in the diets, in which pectin (uronic acid as building blocks) was a large NSP component in the chicory diets differently from the control diet (Table 1). In chickens, the average TTAD of total NSP was 28.4%, which decreased with increasing fiber inclusion level, ranging from 23.8% on diet CF120 to 31.2% on diet C[I] (Table 2; $P = 0.008$). A similar response was also detected for the TTAD of uronic acid ($P = 0.02$), arabinose ($P = 0.0002$) and xylose ($P = 0.0007$), which agreed with results from studies on chickens fed with pea fiber [25]. In contrast, the TTAD of all fiber components increased in pigs with dietary chicory inclusion ($P < 0.05$) (Table 2). Apparently, growing pigs were able to digest chicory fiber to a greater extent than chickens. Ivarsson and co-workers (2011) have reported a TTAD of total NSP of 67% in pigs fed diets with chicory forage inclusion [8]. Intriguingly, the most completely digested NSP fraction in the present study was uronic acid (building blocks of pectin). The TTAD of uronic acid was on average 41.5% in chickens and 70.0% in pigs. Thus, this suggests that chicory forage should be classified as a highly digestible and palatable fibrous feed ingredient for chicken and pigs. However, high inclusion levels of chicory forage (120 g/kg) for broilers may be a challenge [6].

Dietary fiber–induced changes in gut morphology

Animal performance and digestive capacity are suggested to be interrelated with gut development [22,26]. Indeed, fiber utilization is more limited in chickens than in pigs, largely due to its shorter foregut and a higher digesta passage rate than in pigs [10]. We therefore speculated that utilization of the fiber fraction in response to dietary chicory inclusion would also be manifested at gut level in chickens and pigs. We assessed gut morphometry in the small and large intestine and its relationship with TTAD of dietary fiber components. The small intestinal morphology was not altered by diets (Table 3; $P > 0.05$), whereas the large intestinal morphology in chicken's was ($P = 0.001$). This may be attributed to the fact that fiber degradation mainly took place in the large intestine [10]. Furthermore, positive correlations were identified between mucosal thickness in the large intestine and TTAD of NSP (Table 4; $r = 0.44$; $P = 0.05$) and its constituent sugars in chickens ($P < 0.05$). The present results in chickens suggest that it is the extent to which the dietary fiber is digested in the gut that will affect the intestinal morphology rather than the fiber inclusion level in diet *per se*. In contrast, no diet responses or association with diet digestibility was identified in young pigs. Possibly, the porcine gut is less sensitive to dietary fiber ingestion than the gut of chickens [10]. Moreover, the experimental period was only 18 d in pigs compared with 30 d in broiler chickens. Thus, the time of gut exposure to the fiber sources could be the major reason for the species differences in intestinal morphology changes.

Dietary fiber-induced changes in gut microbiota in young pigs

We assumed that the microbiota in the large intestine of our pigs would be shifted by diet intervention, especially

Table 4 Correlations between mucosal thickness and total tract apparent digestibility (TTAD) of fiber components in chickens

Variable	With variable	r	*P*-value
Mucosal thickness	NSP TTAD	0.440	0.05
Mucosal thickness	Xylose TTAD	0.538	0.01
Mucosal thickness	Uronic acid TTAD	0.519	0.02

Data are presented as least square means and standard error of the mean (SEM). Significance was set at $P < 0.05$.

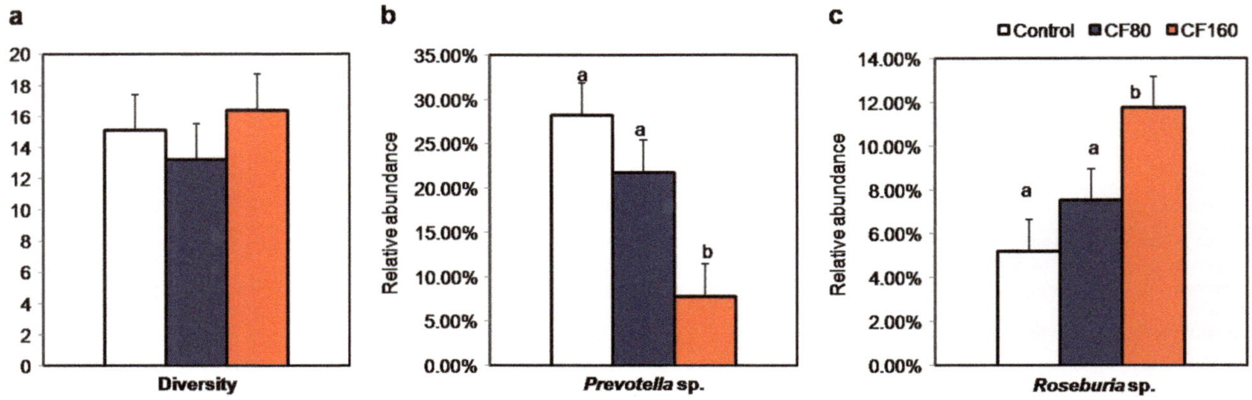

Figure 1 Effects of diet on bacterial diversity (a) and relative abundances of phylotypes corresponding to *Prevotella* sp. (b) and *Roseburia* sp. (c) in the large intestine of pigs. CF80: 80 g/kg chicory forage; CF160: 160 g/kg chicory forage. Data are presented as least square means (n = 6) and standard error of the mean (SEM). [ab] Different letters within rows, indicate significant difference ($P < 0.05$).

by fiber content variation, in a short period of time. This was based on observations in humans, showing that changing the diet from high-fat/low-fiber to low-fat/high-fiber alters the gut microbiota composition within 24 h [27]. Thus, we hypothesized that diets rich in chicory or cereal fiber would result in microbiota of very different species composition.

T-RFLP analysis was done to compare bacterial profile in pigs fed on different diets. In addition, Simpson's index of diversity was calculated based on T-RFLP data to address the bacterial species richness and evenness in the gut ecosystem [21]. However, no significant difference between diets was found on bacterial diversity in the present study (Figure 1a, $P > 0.05$). This was at least partly due to the large individual variation of microbiota composition in our pigs (data not shown), which is in agreement with previous mapping of the porcine gut microbial ecology [28]. In this study, Leser and co-workers (2002) demonstrate that with 400 intestinal bacterial phylotypes identified in pigs, more than 90% have fallen into two phyla, i.e., *Firmicutes* and *Bacteroidetes*. Moreover, members of the *Firmicutes* are most represented by clostridia species, whereas *Prevotella* spp. is the most abundant *Bacteroidetes* bacteria group in the porcine large intestine.

Interestingly, we found that the most dominant TRFs in our pig gut was TRF264, followed by TRF274, which corresponded to one species of *Prevotella* sp. (Figure 1b) and *Roseburia* sp. (bacteria belong to clostridial cluster XIVa; Figure 1c), respectively. Substantial dietary effects on these 2 dominant bacterial phylotypes were revealed. The relative abundance of *Prevotella* sp. was highest on the cereal-based control diet, and decreased with more cereals substituted by chicory forage ($P = 0.004$). This may be because *Prevotella* species are capable of producing enzymes such as xylanases, mannanases, β-glucanase, *etc.* that can degrade dietary xylans in the large intestine [29]. In contrast, changes of the second dominant bacterial phylotype *Roseburia* sp. in our pigs was driven by chicory fiber (with high content of uronic acids), as a marked increase in the relative abundance was observed with increasing inclusion of chicory forage in the diet ($P = 0.004$). Clostridial species have been suggested to play a central role in bacterial cross feeding in the gut microenvironment, converting acetate and lactate to butyrate [30]. Identification of interactions between certain NSP fractions and such bacteria in the gut is of great importance, yet very limited information is available [31]. For pectin in particular, there is a lack of data on its utilization by different microbes. One recent study showed that a member of clostridial bacteria is able to outgrow other microbes on apple pectin *in vitro*, indicating the essential role pectin may play as a substrate for this bacterial group [32]. However, questions remain whether all intestinal *clostridium* species can utilize dietary pectin and whether a common metabolic pathway is shared.

Conclusions

In conclusion, we have demonstrated that chicory (*Cichorium intybus* L.) forage can be used as a highly digestible and palatable fibrous feed ingredient in chicken and pig nutrition. Inclusion of chicory, high in pectin, affects gut morphology and gut microbiota community composition differently from cereal fiber. Furthermore, we found that the extent to which the NSP fractions were digested played a major role for the gut morphology in chickens and not the fiber inclusion level *per se*.

Abbreviations

NSP: Non-starch polysaccharides; DM: Dry matter; BW: Body weight; CP: Crude protein; TiO_2: Titanium oxide; TTAD: Total tract apparent digestibility; T-RFLP: Terminal restriction fragment length polymorphism.

Competing interests
The authors declare that they have no competing interests.

Authors' contributions
JEL and TL conceived the study and experimental design. HYL and EI carried out the experiment trials. HYL and EI performed lab analysis. HYL performed the statistics and drafted the manuscript. JEL revised initial draft manuscript. All authors read and approved the final manuscript.

Acknowledgements
This work was supported by Formas (Swedish Research Council for Environment, Agricultural Sciences and Spatial Planning) and by funding from the Swedish Foundation Cerealia FoU.

References
1. McCleary BV, DeVries JW, Rader JI, Cohen G, Prosky L, Mugford DC, Champ M, Okuma K: **Determination of insoluble, soluble, and total dietary fiber (CODEX definition) by enzymatic-gravimetric method and liquid chromatography: collaborative study.** *J AOAC Int* 2012, **95:**824–844.
2. Van den Abbeele P, Gérard P, Rabot S, Bruneau A, El Aidy S, Derrien M, Kleerebezem M, Zoetendal EG, Smidt H, Verstraete W, *et al*: **Arabinoxylans and inulin differentially modulate the mucosal and luminal gut microbiota and mucin-degradation in humanized rats.** *Environ Microbiol* 2011, **13:**2667–2680.
3. Bach Knudsen KE, Hedemann MS, Lærke HN: **The role of carbohydrates in intestinal health of pigs.** *Anim Feed Sci Technol* 2012, **173:**41–53.
4. Brownawell AM, Caers W, Gibson GR, Kendall CWC, Lewis KD, Ringel Y, Slavin JL: **Prebiotics and the health benefits of fiber: current regulatory status, future research, and goals.** *J Nutr* 2012, **142:**962–974.
5. González-Alvarado JM, Jiménez-Moreno E, Lázaro R, Mateos GG: **Effect of type of cereal, heat processing of the cereal, and inclusion of fiber in the diet on productive performance and digestive traits of broilers.** *Poult Sci* 2007, **86:**1705–1715.
6. Liu HY, Ivarsson E, Jönsson L, Holm L, Lundh T, Lindberg JE: **Growth performance, digestibility, and gut development of broiler chickens on diets with inclusion of chicory (*Cichorium intybus* L.).** *Poult Sci* 2011, **90:**815–823.
7. Hetland H, Svihus B: **Effect of oat hulls on performance, gut capacity and feed passage time in broiler chickens.** *Br Poult Sci* 2001, **42:**354–361.
8. Ivarsson E, Frankow-Lindberg BE, Andersson HK, Lindberg JE: **Growth performance, digestibility and faecal coliform bacteria in weaned piglets fed a cereal-based diet including either chicory (*Cichorium intybus* L) or ribwort (*Plantago lanceolata* L) forage.** *Animal* 2011, **5:**558–564.
9. Looft T, Johnson TA, Allen HK, Bayles DO, Alt DP, Stedtfeld RD, Sul WJ, Stedtfeld TM, Chai B, Cole JR, *et al*: **In-feed antibiotic effects on the swine intestinal microbiome.** *Proc Natl Acad Sci U S A* 2012, **109:**1691–1696.
10. Montagne L, Pluske JR, Hampson DJ: **A review of interactions between dietary fibre and the intestinal mucosa, and their consequences on digestive health in young non-ruminant animals.** *Anim Feed Sci Technol* 2003, **108:**95–117.
11. Högberg A, Lindberg JE: **The effect of level and type of cereal non-starch polysaccharides on the performance, nutrient utilization and gut environment of pigs around weaning.** *Anim Feed Sci Technol* 2006, **127:**200–219.
12. Li G, Kemp PD: **Forage chicory (*Cichorium intybus* L.): a review of its agronomy and animal production.** In *Adv Agron 88.* Edited by Donald LS. Palmerston North, New Zealand: Academic Press; 2005:187–222.
13. Liu HY: *Interactions between dietary chicory, gut microbiota and immune responses*, PhD thesis. Uppsala, Sweden: Swedish University of Agricultural Sciences, Animal Nutrition and Management Department; 2013.
14. European Union: *Commissison Regulation No 889/2008 laying down detailed rules for the implementation of Council Regulation (EC) No 834/2007 on organic production and labelling of organic products with regard to organic.* 2008. http://faolex.fao.org/cgi-bin/faolex.exe?rec_id=066120&database=FAOLEX&search_type=link&table=result&lang=eng&format_name=@ERALL.
15. Noblet J, Le Goff G: **Effect of dietary fibre on the energy value of feeds for pigs.** *Anim Feed Sci Technol* 2001, **90:**35–52.
16. Varel VH, Yen JT: **Microbial perspective on fiber utilization by swine.** *J Anim Sci* 1997, **75:**2715–2722.
17. NRC: *Nutrient requirements of domestic animals. Nutrient requirements of poultry.* Washington, DC: National Academic Press; 1994.
18. NRC: *Nutrient requirements of domestic animals. Nutrient requirements of swine.* Washington, D.C., USA: National Academic Press; 1998.
19. Ivarsson E, Liu HY, Dicksved J, Roos S, Lindberg JE: **Impact of chicory inclusion in a cereal-based diet on digestibility, organ size and faecal microbiota in growing pigs.** *Animal* 2012, **6:**1077–1085.
20. Short FJ, Gorton P, Wiseman J, Boorman KN: **Determination of titanium dioxide added as an inert marker in chicken digestibility studies.** *Anim Feed Sci Technol* 1996, **59:**215–221.
21. Dicksved J, Halfvarson J, Rosenquist M, Jarnerot G, Tysk C, Apajalahti J, Engstrand L, Jansson JK: **Molecular analysis of the gut microbiota of identical twins with Crohn's disease.** *ISME J* 2008, **2:**716–727.
22. Drochner W, Kerler A, Zacharias B: **Pectin in pig nutrition, a comparative review.** *J Anim Physiol Anim Nutr* 2004, **88:**367–380.
23. ROSS: *ROSS: 308 Broiler: performance objectives.* Scotland, UK: Aviagen; 2007.
24. Wellock IJ, Fortomaris PD, Houdijk JGM, Wiseman J, Kyriazakis I: **The consequences of non-starch polysaccharide solubility and inclusion level on the health and performance of weaned pigs challenged with enterotoxigenic *Escherichia coli*.** *Br J Nutr* 2008, **99:**520–530.
25. Jorgensen H, Zhao XQ, Knudsen KEB, Eggum BO: **The influence of dietary fibre source and level on the development of the gastrointestinal tract, digestibility and energy metabolism in broiler chickens.** *Br J Nutr* 1996, **75:**379–395.
26. Leser TD, Mølbak L: **Better living through microbial action: the benefits of the mammalian gastrointestinal microbiota on the host.** *Environ Microbiol* 2009, **11:**2194–2206.
27. Wu GD, Chen J, Hoffmann C, Bittinger K, Chen Y-Y, Keilbaugh SA, Bewtra M, Knights D, Walters WA, Knight R, *et al*: **Linking long-term dietary patterns with gut microbial enterotypes.** *Science* 2011, **334:**105–108.
28. Leser TD, Amenuvor JZ, Jensen TK, Lindecrona RH, Boye M, Moller K: **Culture-independent analysis of gut bacteria: the pig gastrointestinal tract microbiota revisited.** *Appl Environ Microbiol* 2002, **68:**673–690.
29. Flint HJ, Bayer EA: **Plant cell wall breakdown by anaerobic microorganisms from the mammalian digestive tract.** *Ann N Y Acad Sci* 2008, **1125:**280–288.
30. Barcenilla A, Pryde SE, Martin JC, Duncan SH, Stewart CS, Henderson C, Flint HJ: **Phylogenetic relationships of butyrate-producing bacteria from the human gut.** *Appl Environ Microbiol* 2000, **66:**1654–1661.
31. Walker AW, Ince J, Duncan SH, Webster LM, Holtrop G, Ze X, Brown D, Stares MD, Scott P, Bergerat A, *et al*: **Dominant and diet-responsive groups of bacteria within the human colonic microbiota.** *ISME J* 2011, **5:**220–230.
32. Lopez-Siles M, Khan TM, Duncan SH, Harmsen HJM, Garcia-Gil LJ, Flint HJ: **Cultured representatives of two major phylogroups of human colonic *Faecalibacterium prausnitzii* can utilize pectin, uronic acids, and host-derived substrates for growth.** *Appl Environ Microbiol* 2012, **78:**420–428.

12

Dietary requirements of synthesizable amino acids by animals: a paradigm shift in protein nutrition

Guoyao Wu

Abstract

Amino acids are building blocks for proteins in all animals. Based on growth or nitrogen balance, amino acids were traditionally classified as nutritionally essential or nonessential for mammals, birds and fish. It was assumed that all the "nutritionally nonessential amino acids (NEAA)" were synthesized sufficiently in the body to meet the needs for maximal growth and optimal health. However, careful analysis of the scientific literature reveals that over the past century there has not been compelling experimental evidence to support this assumption. NEAA (e.g., glutamine, glutamate, proline, glycine and arginine) play important roles in regulating gene expression, cell signaling, antioxidative responses, fertility, neurotransmission, and immunity. Additionally, glutamate, glutamine and aspartate are major metabolic fuels for the small intestine to maintain its digestive function and to protect the integrity of the intestinal mucosa. Thus, diets for animals must contain all NEAA to optimize their survival, growth, development, reproduction, and health. Furthermore, NEAA should be taken into consideration in revising the "ideal protein" concept that is currently used to formulate swine and poultry diets. Adequate provision of all amino acids (including NEAA) in diets enhances the efficiency of animal production. In this regard, amino acids should not be classified as nutritionally essential or nonessential in animal or human nutrition. The new Texas A&M University's optimal ratios of dietary amino acids for swine and chickens are expected to beneficially reduce dietary protein content and improve the efficiency of their nutrient utilization, growth, and production performance.

Keywords: Diet, Metabolism, Nutrition, Protein, Requirements

Introduction

Amino acids (AA) are building blocks for proteins and must be present in cells for synthesis of polypeptides [1]. The carbon skeletons of eleven of these AA (namely cysteine, histidine, isoleucine, leucine, lysine, methionine, phenylalanine, threonine, tryptophan, tyrosine, and valine) are not synthesized from non-AA molecules in cells of any animals [2]. Therefore, they are classified as nutritionally essential AA (EAA) and must be included in diets for nonruminants to maintain physiological functions of cells, tissues, and the whole body [3,4]. This assumes particular importance for the small intestine because its basal membrane lacks an ability to take up a nutritionally significant quantity of all AA, except for glutamine, from the arterial circulation [5,6].

Classical animal nutrition textbooks do not consider cysteine or tyrosine as an EAA [7-10], because they can be synthesized from methionine and phenylalanine in the liver, respectively. However, the inability of all animals to form the carbon skeletons for methionine and phenylalanine means that there is no *de novo* synthesis of cysteine or tyrosine [2]. Also, intestinal mucosal cells must depend on cysteine and tyrosine as essential precursors to synthesize polypeptides [6]. Moreover, sulfur-containing or aromatic AA in arterial blood are largely not available to enterocytes (absorptive columnar cells in the small intestine). Thus, the presence of cysteine and tyrosine in diets, which can reduce the dietary needs of their precursor AA, are *necessary* to maintain the normal structure and function of the intestine [5,6].

Rose did consider dietary needs of some of the traditionally classified NEAA in his human studies in the 1940s and 1950s, and reported that the omission of NEAA from the diet did not affect nitrogen balance in healthy adults during an eight-day experimental period [8]. Thus, it has

Correspondence: g-wu@tamu.edu
Department of Animal Science, Texas A&M University, College Station, Texas 77843, USA

long been assumed that all of the "nutritionally nonessential amino acids (NEAA)" are synthesized sufficiently in the body to meet the needs for maximal growth and optimal health [7-10]. However, careful analysis of the scientific literature reveals that over the past century there has not been compelling experimental evidence to support this assumption [2]. Indeed, in the 1960s and 1970s, A.E. Harper and other investigators found that the absence of NEAA from chicken and rat diets could not support maximal growth of these animals [11-15]. Growing evidence shows that nearly all of these synthesizable AA are inadequately present in typical plant protein (e.g., corn- and soybean meal)-based diets for growing swine relative to optimal whole-body protein synthesis [16]. Results of recent research revealed that the NEAA have important regulatory roles in nutrient metabolism to favor lean tissue growth and reduction of white adipose tissue [17-20]. Clearly, animals have dietary requirements for not only EAA, but also NEAA to achieve maximum growth and production performance [21-23]. This new concept has resulted in a paradigm shift in our understanding of protein nutrition and is highlighted in the present review article.

Definitions of requirements of dietary AA

Requirements of dietary AA can be classified as qualitative and quantitative [2]. Qualitative requirements are related to the question of "*what* AA are required for maintenance, optimum performance (e.g., growth, lactation, reproduction, and sports competition), and optimum health (e.g., prevention of chronic metabolic disorders, resistance to infectious disease, and recovery from illness)?" Quantitative requirements refer to the question of "*how much* of an AA is required for maintenance, optimum growth, and optimum health?" Feeding experiments have traditionally been employed to determine both qualitative and quantitative requirements of dietary AA by animals [24]. Minimal requirements of AA can also be estimated using so-called factorial analysis, that is, measurements of the loss of N by animals fed a nitrogen- or AA-free diet via urine, feces, gas, and other routes (maintenance) + AA deposited in animals + AA excreted as animal products (e.g., milk, egg, wool, and fetus) [2]. Over the past three decades, studies involving radioactive and stable AA tracers have been used along with the N balance technique to determine dietary requirements for EAA by humans and farm animals [25,26]. The more modern methods involve the use of direct and indirect indicators of AA oxidation during a period of several hours [26]. For yet unknown reasons, the AA oxidation methods generally yielded much higher values of dietary EAA requirements by humans than the nitrogen-balance studies. Readers are referred to recent articles [2,3,24] for insight into historical developments of dietary AA requirements. At present, little is known about dietary requirements for NEAA by mammals, birds, or fishes.

Development of the ideal protein concept in animal nutrition

Chickens

Beginning in the late 1950s, Mitchell and Scott at the University of Illinois conceptualized an ideal protein (optimal proportions and amounts of EAA) for diets of chickens [27,28]. NEAA were not considered by these authors. Early attempts to define an ideal protein were based on the EAA composition of eggs and casein, but were largely unsuccessful because of the excess of many EAA. In 1960, Scott's group simulated the profile of EAA in the chick carcass [29] to design a revised pattern of dietary EAA in an ideal protein for improving growth performance of chicks [30]. An improvement in the ideal protein was indeed achieved using this approach, but remained unsatisfactory due to the lack of NEAA in the diet. However, data on the composition of all EAA or NEAA in chicks were not available [29]. Subsequently, a mixture of several AA (cystine, glycine, proline and glutamate), which are synthesized from pre-existing AA (including EAA) by birds and had previously been thought to be NEAA in chicken nutrition, was used in dietary formulations to yield better results on growth performance [31,32]. This extensive research during the 1960s and the 1970s culminated in several versions of the "chick AA requirement standard" for the first three weeks post-hatching [33-36]. Reference values were given in the Dean and Scott Standard [33], the Huston and Scott Reference Standard [34], the modified Sasse and Baker Reference Standard [35], and the Baker and Han's Ideal Chick Protein [36] (Table 1). The common features shared by these different recommended standards of dietary AA requirements by chickens are that the diets included: (a) all EAA that are not synthesized by chickens; (b) several AA (cystine, glutamate, glycine, proline, and tyrosine) that are synthesized from either EAA or α-ketoglutarate plus ammonia by animals to various extents; and (c) no data on alanine, aspartate, asparagine, glutamine, or serine. Note that the patterns of AA composition in the ideal protein for chicks, as proposed by the Scott [33,34] and Baker [25,36], differ substantially for glycine and proline, and, to a lesser extent, for branched-chain AA, histidine, and sulfur-containing AA. These differences may reflect variations in AA composition of chickens reported in the literature. Because the content of proline plus hydroxyproline in the body of chickens was not known at that time, the relatively small amount of proline in the recommended ideal protein was only arbitrarily set and could limit responses of the animals to dietary EAA in their maximal growth and production performance. In contrast, very large amounts of glutamate (e.g., 13 times the lysine value in the modified Sasse and Baker Reference Standard) [35] were used to presumably provide for the entire need for "nonspecific AA N". However, key questions regarding whether glutamate fulfilled this role and whether

Table 1 The University of Illinois patterns of amino acid compositions in ideal proteins for chicks during the first three wk post-hatching[1]

Amino acid	Amino acid content in the carcass Ref. [29] (% of crude protein)[2]	Dean & Scott standard (1965) Ref. [33]		Huston & Scott standard (1968) Ref. [34]		Sasse & Baker standard (1973) Ref. [35]		Baker & Han standard (1994) Ref. [36]	
		Amount in diet[3]	% of lysine	Amount in diet[3]	% of lysine	Amount in diet[3]	% of lysine	Amount in diet[3]	% of lysine
Arginine	6.65	1.10	98.2	1.00	105	0.95	104	0.95	106
Cystine	—	0.35	31.3	0.35	36.8	0.35	38.5	0.325	36.1
Glycine	—	1.60	143	1.20	126	0.60	65.9	0.60	66.7
Isoleucine	4.35	0.80	71.4	0.60	63.2	0.60	65.9	0.60	66.7
Histidine	1.80	0.30	26.8	0.30	31.6	0.33	36.3	0.32	35.6
Leucine	7.2	1.20	107	1.20	126	1.00	110	0.98	109
Lysine	6.6	1.12	100	0.95	100	0.91	100	0.90	100
Methionine	1.98[†]	0.55	49.1	0.35	36.8	0.35	38.5	0.325	36.1
Phenylalanine	4.25	0.68	60.7	0.50	52.6	0.50	55.0	0.50	55.6
Proline	—	1.00	89.3	0.20	21.1	0.40	44.0	0.40	44.4
Threonine	4.4	0.65	58.0	0.65	68.4	0.65	71.4	0.60	66.7
Tryptophan	0.98	0.23	20.5	0.15	15.8	0.15	16.5	0.145	16.1
Tyrosine	—	0.63	56.3	0.45	47.4	0.45	49.5	0.45	50.0
Valine	5.0	0.82	73.2	0.82	86.3	0.69	75.8	0.69	76.7
Glutamic acid[4]	—	12.0	1071	10.0	1053	12.0	1319	12.0	1333
Total amino acids		23.0		18.7		19.9		19.8	
Total nitrogen		2.83		2.33		2.37		2.35	

[1]These ideal protein models were developed for 0- to 21-d-old broilers using crystalline amino acids. It was assumed that all of these amino acids were 100% available for absorption into enterocytes in chicks. Except for glycine and methionine, all amino acids are L-isomers. DL-methionine is used herein.
[2]Average values for 1-wk-old and 4- to 5-wk-old chicks.
[3]% of diet (as-fed basis; 90% dry matter).
[4]Provided as the nitrogenous source for synthesis of NEAA in chicks.
[†]This value refers to L-methionine.

excess glutamate might interfere with the transport, metabolism and utilization of other AA in chickens were not addressed by the Illinois investigators [33-36]. Possibly due to these concerns and the publication of the NRC nutrient requirements for poultry in 1994 [37], Baker [38] did not include glutamate, glycine or proline in an ideal protein for diets of 0- to 56-d-old chickens in his modified University of Illinois Ideal Ratios of Amino Acids for broiler chickens in 1997 (Table 2).

Swine

Work on the ideal protein for poultry diets laid a foundation for subsequent studies with growing pigs. Thus, the British nutritionist Cole suggested in 1980 that swine diets could be formulated to contain ideal ratios of EAA (with lysine as the reference AA) based on their concentrations in the pig carcass (almost exclusively tissue proteins) [39]. This idea was adopted first by the British Agricultural Research Council (ARC) in 1981 [40] and then by the U.S. National Research Council (NRC) in 1988 [41]. Unfortunately, histidine, arginine, and all synthesizable AA were not included in the ARC's concept of ideal protein (Table 3). Also, its conceptual foundation based solely on the EAA composition of the body was flawed, because the pattern of AA in the diet does not reflect the composition of AA in the animal [16,42]. This mismatch can be explained as follows: (a) individual AA in the diet undergo extensive catabolism and transformations at different rates in the small intestine; (b) the concentrations of AA in the circulation differ markedly from the relative abundance of AA in the diet; (c) individual AA in plasma have different metabolic fates in different animal tissues; and (d) the abundance of AA in tissue proteins differs greatly from that in the diet [2,16,43]. These major shortcomings limit the usefulness of the early versions of the ideal protein in formulating swine diets for maximal growth or production performance of pigs.

Dietary AA are required by animals primarily for maintenance (including the synthesis of nonprotein metabolites) and protein accretion [2]. However, the ARC's ideal protein concept did not take into consideration the relative contribution of maintenance to the total AA needs of the pig [40]. This was due, in part, to technical challenges to accurately determine maintenance requirements of AA,

Table 2 The modified Baker and NRC patterns of changes in amino acid compositions in ideal proteins for 0- to 56-d-old chicks (% of lysine in diet)

Amino acid	Baker's modified models (1997) for chicks [38][1]			NRC (1994) [37] (0- to 21-d-old chicks)[5]
	0 to 21 d[2]	21 to 42 d[3]	42 to 56 d[4]	
Lysine	100	100	100	100
Methionine	36	37	37	42
Cystine	36	38	38	33
Threonine	67	70	70	67
Valine	77	80	80	75
Arginine	105	108	108	104
Tryptophan	16	17	17	17
Isoleucine	67	69	69	67
Leucine	109	109	109	100
Histidine	35	35	35	29
Phe + Tyr	105	105	105	112

[1]These ratios are based on true digestible levels of amino acids in diet (as-fed basis; 90% dry matter). Adapted from Baker [38]. Except for glycine, all amino acids are L-isomers.
[2]Patterns of amino acid composition in the ideal protein are the same for male and female chickens. The amounts of digestible lysine in diet (as-fed basis; 90% dry matter) are 1.12% and 1.02% for male and female chickens, respectively.
[3]Patterns of amino acid composition in the ideal protein are the same for male and female chickens. The amounts of digestible lysine in diet (as-fed basis; 90% dry matter) are 0.89% and 0.84% for male and female chickens, respectively.
[4]Patterns of amino acid composition in the ideal protein are the same for male and female chickens. The amounts of digestible lysine in diet (as-fed basis; 90% dry matter) are 0.76% and 0.73% for male and female chickens, respectively.
[5]These ratios are based on total amino acids in a typical corn- and soybean meal-based diet. The amount of digestible lysine in diet (as-fed basis; 90% dry matter) is 1.2% for 0- to 21-d-old chicks.

Table 3 Previously proposed amino acid compositions for ideal proteins for 10–20 kg growing pigs[1] (% of lysine)

Amino acid	Amino acid content in the carcass[2]	ARC (1981) Ref. [39][3]	Wang-Fuller (1989) Ref. [44][4,a]	Chung-Baker (1992) Ref. [45][4,b]	NRC (1998) Ref. [40][5,b]	Baker (2000) Ref. [46][4,b]
Arginine	91	—	—	42	42	42
Glycine	—	—	—	100	—	—
Histidine	47	33	—	32	32	32
Isoleucine	53	55	60	60	54	60
Leucine	111	100	110	100	102	100
Lysine	100	100	100	100	100	100
Met + Cys	49	50	63	60[c]	57[e]	60[c]
Phe + Tyr	100	96	120	95[d]	94[f]	95[d]
Proline	—	—	—	33	—	—
Tryptophan	12	15	18	18	19	17
Threonine	61	60	72	65	62	65
Valine	72	70	75	68	68	68
Glutamate[6]	—	—	826	878	—	—

[1]These ratios are based on true digestible levels of amino acids in diets [40,44-46], except for ARC (1981) [36]. Except for glycine, all amino acids are L-isomers.
[2]Taken from Baker (1997) [38]. The body proteins in 20–45 kg pigs contain 63 g lysine/16 g nitrogen [38].
[3]These ratios are based on total amino acids in the diet. The total level of lysine in the diet is 1.10% (as-fed basis; 90% dry matter).
[4]The diet contains 1.20% true digestible lysine (as-fed basis; 90% dry matter).
[5]The diet contains 1.01% true digestible lysine (as-fed basis; 90% dry matter).
[6]Provided as the nitrogenous source for synthesis of other NEAA in animals.
[a]Dietary requirements are for 25–50 kg gilts.
[b]Dietary requirements are for 10–20 kg pigs.
[c]The ratio of L-methionine to L-cystine is 1:1.
[d]The ratio of L-phenylalanine to L-tyrosine is 53:47.
[e]The ratio of L-methionine to L-cystine is 47:53.
[f]The ratio of L-phenylalanine to L-tyrosine is 64:36.

which include replacement of degraded proteins, as well as the use of AA for synthesis of low-molecular-weight substances and ATP production [1]. Between 1989 and 1990, in attempts to improve the original ideal protein concept [39,40], T.C. Wang and M.F. Fuller [45] used gilts in the weight range of 25 to 50 kg to estimate an ideal pattern of dietary AA that included requirements for both maintenance and tissue protein accretion. However, these two authors also failed to consider arginine, histidine or all so-called NEAA in the ideal protein although they used glutamate at 826% of the lysine value to provide nonspecific AA-nitrogen [45]. As for the studies with chickens in the 1960s and 1970s, there were also concerns over the assumptions for inclusion of this high level of glutamate in the swine diet that lacks all other NEAA. While glutamate was used to prepare isonitrogenous diets in the previous studies, none of these investigators considered that animals have a dietary requirement of glutamate for optimal growth and production performance.

Having recognized the need to modify the ideal protein concept for formulating swine diets, D.H. Baker took great efforts between 1990 and 2000 to evaluate dietary requirements of EAA by 10–20 kg swine. In their original study, D.H. Baker and his student T.K. Chung added arginine (42% of lysine), glycine (100% of lysine), histidine (32% of lysine), and proline (33% of lysine) to the basal diet containing 1.2% true digestible lysine and using glutamate at 878% of the lysine value to provide nonspecific AA-nitrogen [46]. However, other synthesizable AA (including alanine, aspartate, asparagine, cysteine, glutamine, serine, and tyrosine) were not considered in the revised version of the ideal protein and the rationale for the use of arginine, glycine, histidine, and proline at different proportions to lysine was not explained [46]. Furthermore, the bases for other assumptions were unknown, including: (a) whether glutamate is an effective precursor for sufficient synthesis of all other AA (including aspartate, glutamine, and serine) in specific tissues (e.g., the small intestine, spleen, and lymph nodes) and

Table 4 Compositions of amino acids in the whole-body proteins of chicks and pigs

Amino acid	Chicks[1]		Pigs[2]		Corn grain[3]	Soybean meal[4]	Sorghum grain[5]	Meat & bone meal[6]
	mg AA/g protein	% of lysine	mg AA/g protein	% of lysine	g AA/100 g foodstuff			
Alanine	66.3	108	65.7	109	0.71	1.95	0.96	4.78
Arginine	68.5	111	67.7	112	0.38	3.18	0.41	3.67
Asparagine	36.5	59.3	36.0	59.7	0.35	2.10	0.31	2.21
Aspartate	43.1	70.1	42.8	71.0	0.43	3.14	0.36	3.08
Cysteine	15.0	24.4	13.2	21.9	0.20	0.70	0.19	0.49
Glutamine	50.5	82.1	51.2	84.9	1.02	3.80	0.85	2.81
Glutamate	82.9	135	84.6	140	0.64	4.17	1.18	4.05
Glycine	115	187	117	19.4	0.40	2.30	0.39	8.67
Histidine	21.1	34.3	20.8	34.5	0.23	1.13	0.23	1.19
Isoleucine	35.9	58.4	35.3	58.5	0.34	2.03	0.38	1.92
Leucine	69.2	113	68.3	113	1.13	3.44	1.21	3.56
Lysine	61.5	100	60.3	100	0.25	2.80	0.21	3.13
Methionine	18.9	30.7	18.7	31.0	0.21	0.60	0.20	1.10
Phenylalanine	34.8	56.6	34.3	56.9	0.46	2.21	0.51	1.85
Proline	85.3	139	86.1	143	1.06	2.40	0.96	5.86
OH-Proline	34.8	56.6	37.9	62.9	0.00	0.09	0.00	2.88
Serine	45.0	73.2	44.3	73.5	0.45	2.12	0.46	2.08
Threonine	36.3	59.0	35.1	58.2	0.31	1.76	0.32	2.42
Tryptophan	11.6	18.9	11.1	18.4	0.07	0.62	0.10	0.39
Tyrosine	26.6	43.3	27.2	45.1	0.43	1.66	0.45	1.45
Valine	41.8	68.0	42.2	70.0	0.44	2.09	0.50	2.23

Except for glycine, all amino acids are L-isomers. Adapted from Wu et al. [21]. Calculations were based on the molecular weights of intact amino acids. OH-Pro = hydroxyproline.

[1]Chickens (10-d-old). The content of protein in the body is 14.3 g/100 g wet tissue.

[2]Pigs (30-d-old). The content of protein in the body is 14.1 g/100 g wet tissue.

[3]As-fed basis (89.0% dry matter). Corn grain contains 9.3% crude protein (as-fed basis) [91].

[4]As-fed basis (89.0% dry matter). Soybean meal contains 43.6% crude protein (as-fed basis) [91].

[5]As-fed basis (89.1% dry matter). Sorghum grain contains 10.1% crude protein (as-fed basis) [91].

[6]As-fed basis (96.1% dry matter). Meat & Bone meal contains 52.0% crude protein (as-fed basis) [91].

the whole body; or (b) whether the high content of glutamate in the diet may affect the transport, metabolism and utilization of other AA in the diet. Furthermore, little attention was paid to inter-organ fluxes of amino acids relative to their intracellular metabolism. For example, only ~5% of dietary glutamate enters the portal circulation in growing pigs [5,6]. In addition, although intracellular glutamate is used to synthesize aspartate, many extra-intestinal tissues and cells (e.g., liver and red blood cells) have a limited ability to take up glutamate from the blood circulation [2]. The 10th edition of the NRC Swine Nutrient Requirements published in 1998 [41] did not recognize the needs of pigs for dietary proline or glycine; therefore, Baker [47] omitted glycine and proline from the last version of his "ideal protein" for swine diet formulations in 2000, as he did in 1997 [38]. Over the past two decades, there have been successful attempts to refine the patterns of some AA in diets for lactating, suckling, weanling, finishing, and gestating pigs by addition of arginine [48-53], glutamine [54-59], glutamate [60-64], proline [65-67], or glycine [68,69], or by determining mammary gland growth, changes of whole-body AA composition, and milk yields in lactating sows [70,71]. The outcomes are increases in neonatal and postweaning growth, lactation performance, and litter size in pigs.

Growing evidence shows that both EAA and NEAA (e.g. arginine, glutamine, glutamate, glycine, and proline) play important roles in regulating gene expression, cell

Table 5 True ideal digestibilities of amino acids in corn and soybean meals for poultry and swine diets (%)

Amino acid	Chicks[1]				Pigs[2]			
	Corn grain	Soybean meal	Sorghum grain	Meat & Bone meal	Corn grain	Soybean meal	Sorghum grain	Meat & bone meal
Alanine	87.6	88.9	85.4	90.2	88.5	89.0	87.2	90.5
Arginine	88.4	90.6	87.2	91.4	89.3	90.2	88.4	91.3
Asparagine	86.5	88.3	85.9	90.6	86.8	88.5	86.0	90.2
Aspartate	87.2	89.5	86.1	90.2	86.3	88.2	85.8	89.7
Cysteine	85.1	86.4	84.8	89.4	86.0	87.1	85.1	89.0
Glutamine	88.6	89.5	87.6	90.8	87.7	89.2	86.8	90.8
Glutamate	89.2	90.2	88.4	91.2	88.1	89.6	87.4	91.0
Glycine	86.4	88.3	85.7	90.5	86.6	88.0	85.7	89.5
Histidine	85.5	87.4	84.9	89.6	87.0	88.5	86.2	90.2
Isoleucine	88.7	89.3	88.0	90.8	88.2	88.9	87.6	90.5
Leucine	88.2	89.0	87.6	90.3	87.8	89.6	86.4	90.6
Lysine	85.0	88.4	84.3	90.0	84.5	89.8	83.7	90.4
Methionine	87.5	90.1	86.8	90.6	88.6	89.1	87.4	90.5
Phenylalanine	89.1	90.3	88.5	90.9	89.5	90.0	88.9	91.0
Proline	86.8	88.0	85.9	89.4	86.4	87.2	86.0	89.2
Hydroxyproline	—	—	—	88.7	—	—	—	88.4
Serine	88.4	90.2	87.5	91.1	88.6	89.0	87.9	90.6
Threonine	85.2	86.5	84.8	89.3	84.9	86.8	84.3	88.5
Tryptophan	86.0	87.2	85.3	89.0	85.2	88.1	84.6	89.7
Tyrosine	88.5	89.6	88.0	91.4	89.0	90.2	88.2	91.0
Valine	88.2	89.8	87.6	90.7	87.1	88.7	86.1	90.3

[1]Except for glycine, all amino acids are L-isomers. Broiler chickens (21-wk-old). Ileal digestae were obtained from 21-d-old broiler chicks [92] at 6 h after they were fed either a nitrogen-free purified diet (consisting of 94.9% cornstarch, 2% soybean oil, 1.65% dicalcium phosphate, 0.5% vitamin premix, 0.5% mineral premix, 0.25% NaCl, and 0.2% chromic oxide) or a diet containing the test feed ingredient (consisting of 79.9% cornstarch, 15% test ingredient, 2% soybean oil, 1.65% dicalcium phosphate, 0.5% vitamin premix, 0.5% mineral premix, 0.25% NaCl, and 0.2% chromic oxide) and then euthanized by cervical dislocation. The amount of the diet was provided to overnight (16 h)-fasted chicks at 25 g/kg body weight. Amino acids in the diet, as well as the digestae from chicks fed either the nitrogen-free purified diet or the diet containing the test feed ingredient were analyzed as described by Li et al. [91,94] to calculate the true ideal digestibilities of AA [2]. Data are means for 8 chicks per ingredient, with pooled SEM values being less than 0.8%.

[2]Pigs (50- to 65-d-old). Ileal digestae were obtained from 50- to 65-d-old cannulated pigs [93] at 6 h after they were fed either a nitrogen-free purified diet (consisting of 94.9% cornstarch, 2% soybean oil, 1.65% dicalcium phosphate, 0.5% vitamin premix, 0.5% mineral premix, 0.25% NaCl, and 0.2% chromic oxide) or a diet containing the test feed ingredient (consisting of 79.9% cornstarch, 15% test ingredient, 2% soybean oil, 1.65% dicalcium phosphate, 0.5% vitamin premix, 0.5% mineral premix, 0.25% NaCl, and 0.2% chromic oxide). The amount of the diet was provided to overnight (16 h)-fasted pigs at 12 g/kg body weight. Amino acids in the diet, as well as the digestae from pigs fed either the nitrogen-free purified diet or the diet containing the test feed ingredient were analyzed as described by Li et al. [91,94] to calculate the true ideal digestibilities of AA [2]. Data are means for 6 pigs per ingredient, with pooled SEM values being less than 1.2%.

signaling, nutrient transport and metabolism, intestinal microbiota, anti-oxidative responses, and immune responses [1,2]. Based on these lines of compelling evidence from animal studies, Wu and colleagues proposed the new concept of functional AA, which are defined as those AA that participate in and regulate key metabolic pathways to improve health, survival, growth, development, lactation, and reproduction of the organisms [1,2,16]. Metabolic pathways include: (a) intracellular protein turnover (synthesis and degradation) and associated events; (b) AA synthesis and catabolism; (c) generation of small peptides, nitrogenous metabolites, and sulfur-containing substances (e.g., H_2S); (d) urea cycle and uric acid synthesis; (e) lipid and glucose metabolism; (f) one-carbon-unit metabolism and DNA synthesis; and (g) cellular redox signaling. Functional AA can be nutritionally "essential", "nonessential", or conditionally essential AA. Notably, the concept of functional AA in nutrition has also been adopted for fish [72-74], poultry [75-79], and small laboratory animals (e. g., mice and rats) [80-83]. Readers are referred to recent reviews and original research article on these new developments [24,84-101].

Texas A&M University's optimal ratios of amino acids in diets for swine and chickens

The carbon skeletons of EAA (including tyrosine and cysteine) are not synthesized from non-AA substances in animals [2]. As noted previously, synthesis of NEAA from EAA in animals is inadequate for their maximal growth, milk production, and reproduction performance or for optimal development and health. Thus, the traditional classification of AA as EAA or NEAA is purely a matter of definition. For example, emerging evidence shows that arginine, glutamine, glutamate, and glycine play important roles in regulating gene expression, cell signaling, antioxidative responses, and immunity [51-56]. Additionally, glutamate, glutamine, and aspartate are major metabolic fuels for enterocytes [6] and also regulate intestinal and neurological development and function [2]. In addition, glutamine is essential for ATP production, synthesis of nucleotides,

Table 6 Texas A&M University's optimal ratios of true digestible amino acids in diets for swine[1] (% of diet (as-fed basis))

Amino acid	Growing pigs, kg[2]				Gestating pigs[3]		Lactating sows[2]
	5-10	10-20	20-50	50-110	d 0-90	d 90-114	
Alanine	1.14	0.97	0.80	0.64	0.69	0.69	0.83
Arginine	1.19	1.01	0.83	0.66	1.03	1.03	1.37
Asparagine	0.80	0.68	0.56	0.45	0.50	0.50	0.66
Aspartate	1.14	0.97	0.80	0.64	0.61	0.61	0.94
Cysteine	0.32	0.28	0.24	0.20	0.19	0.19	0.26
Glutamate	2.00	1.70	1.39	1.12	0.89	0.89	1.81
Glutamine	1.80	1.53	1.25	1.00	1.00	1.60	1.38
Glycine	1.27	1.08	0.89	0.71	0.48	0.48	0.75
Histidine	0.46	0.39	0.32	0.26	0.29	0.29	0.39
Isoleucine	0.78	0.66	0.54	0.43	0.45	0.45	0.66
Leucine	1.57	1.33	1.09	0.87	1.03	1.03	1.41
Lysine	1.19	1.01	0.83	0.66	0.51	0.51	0.80
Methionine	0.32	0.28	0.24	0.20	0.16	0.16	0.25
Phenylalanine	0.86	0.73	0.60	0.48	0.54	0.54	0.77
Proline	1.36	1.16	0.95	0.76	0.89	0.89	1.24
Serine	0.70	0.60	0.49	0.39	0.45	0.45	0.74
Threonine	0.74	0.65	0.55	0.46	0.41	0.41	0.56
Tryptophan	0.22	0.19	0.17	0.14	0.11	0.11	0.18
Tyrosine	0.67	0.57	0.46	0.37	0.40	0.40	0.62
Valine	0.85	0.72	0.59	0.47	0.55	0.55	0.72

[1]Data are taken from Wu [95]. Except for glycine, all amino acids are L-isomers. Values are based on true ileal digestible amino acids. Crystalline amino acids (e.g., feed-grade arginine, glutamate, glutamine and glycine), whose true ileal digestibility is 100%, can be added to a diet to obtain their optimal ratios. The molecular weights of intact amino acids were used for all the calculations. The content of dry matter in all the diets was 90%. The content of metabolizable energy in the diets of growing pigs, gestating pigs, and lactating pigs is 3,330, 3,122, and 3,310 Kcal/kg diet, respectively.
[2]Fed ad libitum (90% dry matter).
[3]Fed 2 kg/d on d 0–90, and 2.3 kg/d on d 90–114 (90% dry matter).

expression of anti-oxidative genes, and redox signaling in enterocytes [57]. Furthermore, glutamate activates chemical sensing in the gastrointestinal tract and may inhibit degradation of both EAA and NEAA by intestinal microbes [2,60]. Finally, proline and arginine, which are major sources of ornithine for intestinal and placental synthesis of polyamines [42], are essential for DNA and protein synthesis and also participate in protein and DNA methylation, and, thus genetic and epigenetic regulation of cell growth and development [2]. Collectively, animals have dietary requirements for all synthesizable AA to achieve their full genetic potential for growth, development, reproduction, lactation, and resistance to infectious disease [21].

Composition of EAA in feed ingredients and true ileal digestibilities of EAA in swine [41] and poultry [37,90] have been published. As an initial step to define NEAA requirements by animals, we recently determined the composition of all protein-AA in major feedstuffs [86] and in animal tissues [21]. Examples are given in Table 4 for corn grain, soybean meal, sorghum grain, and meat & bone meal. Based on the previous studies of AA biochemistry and nutrition (including AA metabolism and tissue protein gains) in poultry e.g., 3, [37] and swine e.g., [48-69], the author of the present work would like to propose Texas A&M University's optimal ratios of true digestible AA in diets for swine (Table 5) and chickens (Table 6) during different phases of growth and production. The values for 5- to 10-kg young pigs are based primarily on consideration of: (a) the entry of dietary AA into the portal vein for 30-day-old postweaning pigs, as compared to the accretion of AA in the body [16]; (b) the published data of Baker [47] and NRC [41] on dietary EAA requirements; and (c) the estimated rates of AA synthesis, catabolism and accretion in the body [2,16,88]. Specifically, these estimations are that: (a) rates of *net* synthesis of aspartate, arginine, glutamate, glutamine, glycine, and proline in extra-intestinal tissues of 5–10 kg postweaning pigs are 195, 361, 415, 1149, 331, and 276 mg/kg body weight per day, respectively; (b) rates of catabolism (including oxidation and synthesis of low-molecular-weight substances) of alanine and tyrosine in extra-intestinal tissues are 30% of their rates of accretion in body proteins; (c) rates of catabolism (including oxidation) of leucine and isoleucine in extra-intestinal tissues are 30% and 25% of their rates of accretion in body proteins, respectively; (d) the rate of catabolism (including oxidation) of asparagine in extra-intestinal tissues is 124 mg/kg body weight per day; (e) rates of catabolism (including oxidation) of valine and serine in extra-intestinal tissues are 15% of their rates of accretion in body proteins; (f) the rate of catabolism (including oxidation) of phenylalanine in extra-intestinal tissues is 124 mg/kg body weight per day; and (g) the ratio of phenylalanine to tyrosine is 60:40, whereas the ratio of methionine to cysteine is 1:1. Additionally, the Texas A&M University's optimal ratios of true digestible AA in diets for gestating and lactating sows are based on previous studies of dietary AA composition [89], ileal true digestibility of AA in feed ingredients (Table 4 and Ref. [41]), composition of EAA and NEAA in the body (Table 4), embryonic/fetal survival and growth [42,49,57], as well as milk production and piglet growth [42,50,57].

Several additional comments on the Texas A&M University's optimal ratios of dietary AA are warranted. First, the author adopts the NRC [41] values for lysine, methionine, threonine, and tryptophan in the Texas A&M University's model for 5- to 10-kg pigs. Second, optimal ratios of EAA in diets of older pigs are based on the suggestions of the NRC [41] and Baker [47] in that the ratios of

Table 7 Texas A&M University's optimal ratios of true digestible amino acids in diets for chickens[1] (% of lysine in diet)

Amino acid	Age of chickens		
	0 to 21 d[2]	21 to 42 d[3]	42 to 56 d[4]
Alanine	102	102	102
Arginine	105	108	108
Asparagine	56	56	56
Aspartate	66	66	66
Cysteine	32	33	33
Glutamate	178	178	178
Glutamine	128	128	128
Glycine	176	176	176
Histidine	35	35	35
Isoleucine	67	69	69
Leucine	109	109	109
Lysine	100	100	100
Methionine	40	42	42
Phenylalanine	60	60	60
Proline	184	184	184
Serine	69	69	69
Threonine	67	70	70
Tryptophan	16	17	17
Tyrosine	45	45	45
Valine	77	80	80

[1]Except for glycine, all amino acids are L-isomers. Values are based on true ileal digestible amino acids.
[2]Patterns of amino acid composition in the ideal protein are the same for male and female chickens. The amounts of digestible lysine in diet (as-fed basis; 90% dry matter) are 1.12% and 1.02% for male and female chickens, respectively.
[3]Patterns of amino acid composition in the ideal protein are the same for male and female chickens. The amounts of digestible lysine in diets (as-fed basis; 90% dry matter) are 0.89% and 0.84% for male and female chickens, respectively.
[4]Patterns of amino acid composition in the ideal protein are the same for male and female chickens. The amounts of digestible lysine in diets (as-fed basis; 90% dry matter) are 0.76% and 0.73% for male and female chickens, respectively.

tryptophan, sulfur-AA, and threonine to lysine (all based on true digestibility of EAA) increase slightly with age, whereas the ratios of other EAA to lysine are not altered substantially during postnatal development. Third, this is the first time that NEAA are included in optimal ratios of dietary AA for pigs and poultry at various physiological stages. Fourth, for EAA, the ratios of BCAA, histidine, phenylalanine or tyrosine to lysine in the Texas A&M University's optimal ratios of dietary are higher than those proposed by the NRC [41] and Baker [47] for swine. This is based on the following considerations [2]: (a) BCAA are actively degraded in extra-hepatic and extra-intestinal tissues; (b) leucine can stimulate muscle protein synthesis in young pigs; (c) leucine, isoleucine and valine should be in an appropriate ratio to prevent AA imbalance; (d) large amounts of histidine-containing dipeptides are present in skeletal muscle; and (e) tyrosine is actively utilized in multiple metabolic pathways and its carbon skeleton is formed only from phenylalanine in animals. Finally, the data on dietary EAA requirements by chickens [38], along with composition of EAA and NEAA as well as AA accretion in the body (85 and Table 4) and new knowledge of AA metabolism in birds [2], provided the basis for the proposed Texas A&M University's optimal ratios of amino acids in chicken diets (Table 7). The recommended values for EAA and NEAA requirements must be revised as new and compelling experimental data become available.

Conclusion and perspectives

Amino acids have versatile and important physiological functions beyond their roles as the building blocks of protein [101]. Thus, dietary NEAA and EAA are necessary for the survival, growth, development, reproduction and health of animals. Growing evidence shows that pigs and poultry cannot synthesize sufficient amounts of all NEAA to achieve their maximum genetic potential [95-100]. NEAA (e.g., glutamine, glutamate, proline, glycine and arginine) play important roles in regulating gene expression, cell signaling, antioxidative responses, neurotransmission, and immunity. Additionally, glutamate, glutamine and aspartate are major metabolic fuels for the small intestine to maintain its digestive function and to protect its mucosal integrity. While metabolic needs for an AA by animals do not necessarily translate into its dietary needs, results of recent studies indicate that animals have both metabolic and dietary needs for AA that are synthesized in the body [101-115]. Thus, "synthesizable amino acids" should not be considered as "nutritionally nonessential amino acids" in animal feeding. Besides EAA, NEAA and conditionally essential AA should all be taken into consideration in: (a) revising the classical "ideal protein" concept; and (b) formulating balanced diets to improve protein accretion, feed efficiency, and health in animals. The Texas A&M University's optimal ratios of AA in diets for swine or poultry, as proposed herein, are based on experimental data from published biochemical and nutritional studies. This new initiative will provide a much-needed framework for both qualitative and quantitative analysis of dietary requirements for all AA by livestock, poultry and fish through conduct of additional research. Adoption of these recommended values in animal feeding are expected to beneficially reduce dietary protein content and improve the efficiency of nutrient utilization, growth, and production performance of farm animals. The concept of dietary requirements for NEAA also has important implications in human nutrition and health.

Competing interests

The author declares no competing interests.

Authors' contributions

The author wrote this paper and approved the final version of the manuscript.

Acknowledgments

The author thanks Dr. Fuller W. Bazer for helpful discussions and review of the manuscript, colleagues (particularly Drs. Fuller W. Bazer, Robert C. Burghardt, Yongqing Hou, Gregory A. Johnson, Sung Woo Kim, Darrell A. Knabe, Defa Li, Wilson G. Pond, Stephen B. Smith, Junjun Wang, Zhenlong Wu, and Yulong Yin) for enjoyable collaborations on swine protein digestion and nutrition research, as well as graduate students, postdoctoral fellows, technicians, and visiting scholars for their important contributions to our research program. The author's work at Texas A&M University was supported by National Research Initiative Competitive Grants from the Animal Reproduction Program (2008-35203-19120 and 2011-67015-20028) and Animal Growth & Nutrient Utilization Program (2008-35206-18764, 2008-35206-18762, 2009-35206-05211, and 2014-67015-21770) of the USDA National Institute of Food and Agriculture, Texas A&M AgriLife Research (H-8200), Ajinomoto Inc. (Tokyo, Japan), and Gentech Inc. (Shanghai, China). This paper is dedicated to the memory of Dr. Darrell A. Knabe, a dear friend of the author, a wonderful human being, and a well-respected scientist.

References

1. Wu G: **Amino acids: metabolism, functions, and nutrition.** *Amino Acids* 2009, **37**:1–17.
2. Wu G: *Amino Acids: Biochemistry and Nutrition.* Boca Raton, Florida: CRC Press; 2013.
3. Baker DH: **Advances in protein-amino acid nutrition of poultry.** *Amino Acids* 2009, **37**:29–41.
4. Wu G: **Recent advances in swine amino acid nutrition.** *J Anim Sci Biotechnol* 2010, **1**:49–61.
5. Stoll B, Burrin DG: **Measuring splanchnic amino acid metabolism in vivo using stable isotopic tracers.** *J Anim Sci* 2006, **84**(Suppl):E60–E72.
6. Wu G: **Intestinal mucosal amino acid catabolism.** *J Nutr* 1998, **128**:1249–1252.
7. Rose WC: **The sequence of events leading to the establishment of the amino acid needs of man.** *Am J Public Health* 1968, **58**:2020–2027.
8. Morrison FB: *Feeds and Feeding.* 22nd edition. Ithaca, New York: The Morrison Publishing Company; 1956.
9. Maynard LA, Loosli JK, Hintz HF, Warner RG: *Animal Nutrition.* 7th edition. New York: McGraw-Hill Book Company; 1979.
10. Crampton EW, Harris LE: *Applied Animal Nutrition.* 2nd edition. San Francisco: W.H. Freeman and Company; 1969.
11. Rogers QR, Chen DM, Harper AE: **The importance of dispensable amino acids for maximal growth in the rat.** *Proc Soc Exp Biol Med* 1970, **134**:517–522.
12. Breuer LH Jr, Pond WG, Warner RG, Loosli JK: **The role of dispensable amino acids in the nutrition of the rat.** *J Nutr* 1964, **82**:499–506.
13. Womack M: **Amino acid feeding studies: effects of various nonessential nitrogen sources and of added water.** *Proc Soc Exp Biol Med* 1969, **131**:977–979.

14. Stucki WP, Harper AE: **Importance of dispensable amino acids for normal growth of chicks.** *J Nutr* 1961, **74:**377–383.
15. Maruyama K, Sunde ML, Harper AE: **Is L-glutamic acid nutritionally a dispensable amino acid for the young chick?** *Poultry Sci* 1976, **55:**45–60.
16. Wu G, Bazer FW, Burghardt RC, Johnson GA, Kim SW, Knabe DA, Li XL, Satterfield MC, Smith SB, Spencer TE: **Functional amino acids in swine nutrition and production.** In *Dynamics in Animal Nutrition.* Edited by Doppenberg J, van der Aar P. The Netherlands: Wageningen Academic Publishers; 2010:69–98.
17. Dai ZL, Li XL, Xi PB, Zhang J, Wu G, Zhu WY: **Regulatory role for L-arginine in the utilization of amino acids by pig small-intestinal bacteria.** *Amino Acids* 2012, **43:**233–244.
18. San Gabriel A, Uneyama H: **Amino acid sensing in the gastrointestinal tract.** *Amino Acids* 2013, **45:**451–461.
19. Bauchart-Thevret C, Cui LW, Wu G, Burrin DG: **Arginine-induced stimulation of protein synthesis and survival in IPEC-J2 cells is mediated by mTOR but not nitric oxide.** *Am J Physiol* 2010, **299:**E899–E909.
20. Wu ZL, Satterfield MC, Bazer FW, Wu G: **Regulation of brown adipose tissue development and white fat reduction by L-arginine.** *Curr Opin Clin Nutr Metab Care* 2012, **15:**529–538.
21. Wu G, Wu ZL, Dai ZL, Yang Y, Wang WW, Liu C, Wang B, Wang JJ, Yin YL: **Dietary requirements of "nutritionally nonessential amino acids" by animals and humans.** *Amino Acids* 2013, **44:**1107–1113.
22. Phang JM, Liu W, Hancock C: **Bridging epigenetics and metabolism: Role of non-essential amino acids.** *Epigenetics* 2013, **8**(3):231–236.
23. Wang WW, Wu ZL, Dai ZL, Yang Y, Wang JJ, Wu G: **Glycine metabolism in animals and humans: implications for nutrition and health.** *Amino Acids* 2013, **45:**463–477.
24. Kim SW, Wu G, Baker DH: **Ideal protein and amino acid requirements by gestating and lactating sows.** *Pig News Inform* 2005, **26:**89N–99N.
25. Ball RO, Atkinson JL, Bayley HS: **Proline as an essential amino acid for the young pig.** *Br J Nutr* 1986, **55:**659–668.
26. Elango R, Ball RO, Pencharz PB: **Recent advances in determining protein and amino acid requirements in humans.** *Br J Nutr* 2012, **108**(Suppl 2):S22–S30.
27. Glista WA, Mitchell HH, Scott HM: **The amino acid requirements of the chick.** *Poultry Sci* 1951, **30:**915.
28. Fisher H, Scott HM: **The essential amino acid requirements of chicks as related to their proportional occurrence in the fat-free carcass.** *Arch Biochem Biophys* 1954, **51:**517–519.
29. Price WA Jr, Taylor MW, Russell WC: **The retention of essential amino acids by the growing chick.** *J Nutr* 1953, **51:**413–422.
30. Klain GJ, Scott HM, Johnson BC: **The amino acid requirements of the growing chick fed a crystalline amino acid diet.** *Poultry Sci* 1960, **39:**39–44.
31. Baker DH, Sugahara M, Scott HM: **The glycine-serine interrelationship in chick nutrition.** *Poultry Sci* 1968, **47:**1376–1377.
32. Graber G, Baker DH: **The essential nature of glycine and proline for growing chickens.** *Poultry Sci* 1973, **52:**892–896.
33. Dean WF, Scott HM: **The development of an amino acid reference diet for the early growth of chicks.** *Poultry Sci* 1965, **44:**803–808.
34. Huston RL, Scott HM: **Effect of varying the composition of crystalline amino acid mixture on weight gain and pattern of free amino acids in chick tissue.** *Fed Proc* 1968, **27:**1204–1209.
35. Sasse CE, Baker DH: **Modification of the Illinois reference standard amino acid mixture.** *Poultry Sci* 1973, **52:**1970–1972.
36. Baker DH, Han Y: **Ideal amino acid profile for broiler chicks during the first three weeks posthatching.** *Poultry Sci* 1994, **73:**1441–1447.
37. NRC (National Research Council): *Nutrient Requirements of Poultry.* Washington, D.C: National Academy Press; 1994.
38. Baker DH: **Ideal amino acid profiles for swine and poultry and their applications in feed formulation.** *BioKyowa Tech Rev No* 1997, **9:**1–24.
39. Cole DJA: **The amino acid requirements of pigs: the concept of an ideal protein.** *Pig News Info* 1980, **1:**201–205.
40. ARC (Agricultural Research Council): *The Nutrient Requirements of Pigs: Technical review.* Slough, U.K: Commonwealth Agricultural Bureaux; 1981.
41. NRC (National Research Council): *Nutrient Requirements of Swine.* Washington, D.C: National Academy Press; 1988.
42. Wu G, Bazer FW, Burghardt RC, Johnson GA, Kim SW, Li XL, Satterfield MC, Spencer TE: **Impacts of amino acid nutrition on pregnancy outcome in pigs: mechanisms and implications for swine production.** *J Anim Sci* 2010, **88:**E195–E204.
43. Wu G, Ott TL, Knabe DA, Bazer FW: **Amino acid composition of the fetal pig.** *J Nutr* 1999, **129:**1031–1038.
44. Davis TA, Nguyen HV, Garcia-Bravo R, Fiorotto ML, Jackson EM, Lewis DS, Lee DR, Reeds PJ: **Amino acid composition of human milk is not unique.** *J Nutr* 1994, **124:**1126–1132.
45. Wang TC, Fuller MF: **The optimum dietary amino acid patterns for growing pigs. 1. Experiments by amino acid deletion.** *Br J Nutr* 1989, **62:**77–89.
46. Chung TK, Baker DH: **Ideal amino acid pattern for ten kilogram pigs.** *J Anim Sci* 1992, **70:**3102–3111.
47. Baker DH: **Recent advances in use of the ideal protein concept for swine feed formulation.** *Asian-Aus J Anim Sci* 2000, **13:**294–301.
48. Wu G, Knabe DA, Kim SW: **Arginine nutrition in neonatal pigs.** *J Nutr* 2004, **134:**2783S–2790S.
49. Mateo RD, Wu G, Bazer FW, Park JC, Shinzato I, Kim SW: **Dietary L-arginine supplementation enhances the reproductive performance of gilts.** *J Nutr* 2007, **137:**652–656.
50. Mateo RD, Wu G, Moon HK, Carroll JA, Kim SW: **Effects of dietary arginine supplementation during gestation and lactation on the performance of lactating primiparous sows and nursing piglets.** *J Anim Sci* 2008, **86:**827–835.
51. Yao K, Yin YL, Chu WY, Liu ZQ, Deng D, Li TJ, Huang RL, Zhang JS, Tan BE, Wang W, Wu G: **Dietary arginine supplementation increases mTOR signaling activity in skeletal muscle of neonatal pigs.** *J Nutr* 2008, **138:**867–872.
52. Tan BE, Li XG, Kong XF, Huang RL, Ruan Z, Yao K, Deng ZY, Xie MY, Shinzato I, Yin YL, Wu G: **Dietary L-arginine supplementation enhances the immune status in early-weaned piglets.** *Amino Acids* 2009, **37:**323–331.
53. Kim SW, Wu G: **Regulatory role for amino acids in mammary gland growth and milk synthesis.** *Amino Acids* 2009, **37:**89–95.
54. Wu G, Meier SA, Knabe DA: **Dietary glutamine supplementation prevents jejunal atrophy in weaned pigs.** *J Nutr* 1996, **126:**2578–2584.
55. Wang JJ, Chen LX, Li P, Li XL, Zhou HJ, Wang FL, Li DF, Yin YL, Wu G: **Gene expression is altered in piglet small intestine by weaning and dietary glutamine supplementation.** *J Nutr* 2008, **138:**1025–1032.
56. Haynes TE, Li P, Li XL, Shimotori K, Sato H, Flynn NE, Wang JJ, Knabe DA, Wu G: **L-Glutamine or L-alanyl-L-glutamine prevents oxidant- or endotoxin-induced death of neonatal enterocytes.** *Amino Acids* 2009, **37:**131–142.
57. Wu G, Bazer FW, Johnson GA, Knabe DA, Burghardt RC, Spencer TE, Li XL, Wang JJ: **Important roles for L-glutamine in swine nutrition and production.** *J Anim Sci* 2011, **89:**2017–2030.
58. Manso HE, Filho HC, de Carvalho LE, Kutschenko M, Nogueira ET, Watford M: **Glutamine and glutamate supplementation raise milk glutamine concentrations in lactating gilts.** *J Anim Sci Biotechnol* 2012, **3**(1):2.
59. Jiang ZY, Sun LH, Lin YC, Ma XY, Zheng CT, Zhou GL, Chen F, Zou ST: **Effects of dietary glycyl-glutamine on growth performance, small intestinal integrity, and immune responses of weaning piglets challenged with lipopolysaccharide.** *J Anim Sci* 2009, **87:**4050–4056.
60. Rezaei R, Knabe DA, Tekwe CD, Dahanayaka S, Ficken MD, Fielder SE, Eide SJ, Lovering SL, Wu G: **Dietary supplementation with monosodium glutamate is safe and improves growth performance in postweaning pigs.** *Amino Acids* 2013, **44:**911–923.
61. Zhang J, Yin YL, Shu XG, Li TJ, Li FN, Tan BE, Wu ZL, Wu G: **Oral administration of MSG increases expression of glutamate receptors and transporters in the gastrointestinal tract of young piglets.** *Amino Acids* 2013, **45:**1169–1177.
62. Xin W, Xugang S, Xie C, Li J, Hu J, Yin YL, Deng ZY: **The acute and chronic effects of monosodium L-glutamate on serum iron and total iron-binding capacity in the jugular artery and vein of pigs.** *Biol Trace Elem Res* 2013, **153:**191–195.
63. Wu X, Zhang Y, Liu Z, Li TJ, Yin YL: **Effects of oral supplementation with glutamate or combination of glutamate and N-carbamylglutamate on intestinal mucosa morphology and epithelium cell proliferation in weanling piglets.** *J Anim Sci* 2012, **90**(Suppl 4):337–339.
64. Liu T, Peng J, Xiong Y, Zhou S, Cheng X: **Effects of dietary glutamine and glutamate supplementation on small intestinal structure, active absorption and DNA, RNA concentrations in skeletal muscle tissue of weaned piglets during d 28 to 42 of age.** *Asian-Aust J Anim Sci* 2002, **15:**238–242.
65. Wu G, Bazer FW, Burghardt RC, Johnson GA, Kim SW, Knabe DA, Li P, Li XL, McKnight JR, Satterfield MC, Spencer TE: **Proline and hydroxyproline metabolism: implications for animal and human nutrition.** *Amino Acids* 2011, **40:**1053–1063.
66. Brunton JA, Baldwin MP, Hanna RA, Bertolo RF: **Proline supplementation to parenteral nutrition results in greater rates of protein synthesis in the muscle, skin, and small intestine in neonatal Yucatan miniature piglets.** *J Nutr* 2012, **142:**1004–1008.

67. Brunton JA, Bertolo RF, Pencharz PB, Ball RO: **Proline ameliorates arginine deficiency during enteral but not parenteral feeding in neonatal piglets.** *Am J Physiol* 1999, **277:**E223–E231.
68. Powell S, Bidner TD, Payne RL, Southern LL: **Growth performance of 20- to 50- kilogram pigs fed low-crude-protein diets supplemented with histidine, cystine, glycine, glutamic acid, or arginine.** *J Anim Sci* 2011, **89:**3643–3650.
69. Fickler J, Roth FX, Kirchgessner M: **Use of a chemically-defined amino acid diet in a N-balance experiment with piglets. 1. Communication on the importance of nonessential amino acids for protein retention.** *J Anim Physiol Anim Nutr* 1994, **72:**207–214.
70. Kim SW, Baker DH, Easter RA: **Dynamic ideal protein and limiting amino acids for lactating sows: the impact of amino acid mobilization.** *J Anim Sci* 2001, **79:**2356–2366.
71. Kim SW, Osaka I, Hurley WL, Easter RA: **Mammary gland growth as influenced by litter size in lactating sows: impact on lysine requirement.** *J Anim Sci* 1999, **77:**3316–3321.
72. Andersen SM, Holen E, Aksnes A, Rønnestad I, Zerrahn JE, Espe M: **Dietary arginine affects energy metabolism through polyamine turnover in juvenile Atlantic salmon (Salmo salar).** *Br J Nutr* 2013, **110:**1968–1977.
73. Pohlenz C, Buentello A, Criscitiello MF, Mwangi W, Smith R, Gatlin DM: **Synergies between vaccination and dietary arginine and glutamine supplementation improve the immune response of channel catfish against Edwardsiella ictaluri.** *Fish Shellfish Immunol* 2012, **33:**543–551.
74. Zhang KK, Ai QH, Mai KS, Xu W, Liufu ZG, Zhang YJ, Peng M: **Effects of dietary hydroxyproline on growth performance, body composition, hydroxyproline and collagen concentrations in tissues in relation to prolyl 4-hydroxylase alpha(I) gene expression of juvenile turbot, Scophthalmus maximus L. fed high plant protein diets.** *Aquaculture* 2013, **404:**77–84.
75. Fouad AM, El-Senousey HK, Yang XJ, Yao JH: **Dietary L-arginine supplementation reduces abdominal fat content by modulating lipid metabolism in broiler chickens.** *Animal* 2013, **7:**1239–1245.
76. Wu LY, Fang YJ, Guo XY: **Dietary L-arginine supplementation beneficially regulates body fat deposition of meat-type ducks.** *Br Poult Sci* 2011, **52:**221–226.
77. Dai SF, Gao F, Xu XL, Zhang WH, Song SX, Zhou GH: **Effects of dietary glutamine and gamma-aminobutyric acid on meat colour, pH, composition, and water-holding characteristic in broilers under cyclic heat stress.** *Br Poult Sci* 2012, **53:**471–481.
78. Powell S, Bidner TD, Southern LL: **Effects of glycine supplementation at varying levels of methionine and cystine on the growth performance of broilers fed reduced crude protein diets.** *Poult Sci* 2011, **90:**1023–1027.
79. Li P, Kim SW, Nakagawa K, Zhou HJ, Wu G: **Dietary supplementation with L- glutamine and AminoGut™ enhances protein synthesis in skeletal muscle of growing broiler chickens.** *FASEB J* 2010, **24:**740.21.
80. Ren W, Yin YL, Liu G, Yu X, Li Y, Yang G, Li T, Wu G: **Effect of dietary arginine supplementation on reproductive performance of mice with porcine circovirus type 2 infection.** *Amino Acids* 2012, **42:**2089–2094.
81. Ren WK, Zou LX, Li NZ, Wang Y, Peng YY, Ding JN, Cai LC, Yin YL, Wu G: **Dietary arginine supplementation promotes immune responses to inactivated Pasteurella multocida vaccination in mice.** *Br J Nutr* 2013, **109:**867–872.
82. Ren WK, Liu SP, Chen S, Zhang FM, Li NZ, Yin J, Peng YY, Wu L, Liu G, Yin YL, Wu G: **Dietary L-glutamine supplementation increases *Pasteurella multocida* burden and expression of major virulence factors in mice.** *Amino Acids* 2013, **45:**947–955.
83. Ren WK, Zou LX, Ruan Z, Li NZ, Wang Y, Peng Y, Liu G, Yin YL, Huang RL, Hou YQ, Wu G: **Dietary L-proline supplementation confers immuno-stimulatory effects on inactivated *Pasteurella multocida* vaccine immunized mice.** *Amino Acids* 2013, **45:**555–561.
84. Kim SW, Wu G, Zhao Y, Li P: **Applications of ideal protein and selection of protein sources for sows and piglets.** In *Progress in Functional Amino Acids and Carbohydrates for Animal Production. (Proceedings of 4th International Symposium on Animal Nutrition, Health and Feed Additive, July 4–5, 2009, Guangzhou, China.* Edited by Yin YL, Liu G, Spencer TE. Changsha, Hunan, China: Institute of Subtropical Agriculture, Chinese Academy of Sciences; 2009:32–40.
85. Li P, Mai KS, Trushenski J, Wu G: **New developments in fish amino acid nutrition: towards functional and environmentally oriented aquafeeds.** *Amino Acids* 2009, **37:**43–53.
86. Kim SW, Hurley WL, Wu G, Ji F: **Ideal amino acid balance for sows during gestation and lactation.** *J Anim Sci* 2009, **87:**E123–E132.
87. Ruth MR, Field CJ: **The immune modifying effects of amino acids on gut-associated lymphoid tissue.** *J Anim Sci Biotechnol* 2013, **4**(1):27.
88. Wu G: **Functional amino acids in growth, reproduction and health.** *Adv Nutr* 2010, **1:**31–37.
89. Wu G: **Functional amino acids in nutrition and health.** *Amino Acids* 2013, **45:**407–411.
90. Parsons CM: **Amino acid digestibility in feedstuffs for poultry: Feedstuff evaluation and requirements.** *BioKyowa Tech Rev No* 1991, **1:**1–15.
91. Li XL, Rezaei R, Li P, Wu G: **Composition of amino acids in feed ingredients for animal diets.** *Amino Acids* 2011, **40:**1159–1168.
92. Huang RL, Yin YL, Wu G, Zhang YG, Li TJ, Li LL, Li MX, Tang ZR, Zhang J, Wang B, He JH, Nie XZ: **Effect of dietary oligochitosan supplementation on ileal digestibility of nutrients and performance in broilers.** *Poultry Sci* 2005, **84:**1383–1388.
93. Traylor SL, Cromwell GL, Lindemann MD, Knabe DA: **Effects of level of supplemental phytase on ileal digestibility of amino acids, calcium, and phosphorus in dehulled soybean meal for growing pigs.** *J Anim Sci* 2001, **79:**2634–2642.
94. Dai ZL, Wu ZL, Jia SC, Wu G: **Analysis of amino acid composition in proteins of animal tissues and foods as pre-column o-phthaldialdehyde derivatives by HPLC with fluorescence detection.** *J Chromatogr B* 2014, doi:10.1016/j.jchromb.2014.03.025.
95. Wu G: **Dietary requirements of "nutritionally nonessential amino acids" by mammals.** *Amino Acids Res (Japan)* 2014, **7:**67–76.
96. Rezaei R, Wang WW, Wu ZL, Dai ZL, Wang JJ, Wu G: **Biochemical and physiological bases for utilization of dietary amino acids by young pigs.** *J Anim Sci Biotechnol* 2013, **4:**7.
97. Li XL, Bazer FW, Johnson GA, Burghardt RC, Frank JW, Dai ZL, Wang JJ, Wu ZL, Shinzato I, Wu G: **Dietary supplementation with L-arginine between days 14 and 25 of gestation enhances embryonic development and survival in gilts.** *Amino Acids* 2014, **46:**375–384.
98. Lin G, Wang XQ, Wu G, Feng CP, Zhou HJ, Li DF, Wang JJ: **Improving amino acid nutrition to prevent intrauterine growth restriction in mammals.** *Amino Acids* 2014, **46:**1605–1623.
99. Wang XQ, Frank JW, Little DR, Dunlap KA, Satterfiled MC, Burghardt RC, Hansen TR, Wu G, Bazer FW: **Functional role of arginine during the peri-implantation period of pregnancy. I. Consequences of loss of function of arginine transporter *SLC7A1* mRNA in ovine conceptus trophectoderm.** *FASEB J* 2014, doi:fj.13-248757.
100. Wang XQ, Ying W, Dunlap KA, Lin G, Satterfield MC, Burghardt RC, Wu G, Bazer FW: **Arginine decarboxylase and agmatinase: An alternative pathway for de novo biosynthesis of polyamines for development of mammalian conceptuses.** *Biol Reprod* 2014, **90:**84.
101. Wu G, Bazer FW, Dai ZL, Li DF, Wang JJ, Wu ZL: **Amino acid nutrition in animals: protein synthesis and beyond.** *Annu Rev Anim Biosci* 2014, **2:**387–417.
102. Bauchart-Thevret C, Stoll B, Benight NM, Olutoye O, Lazar D, Burrin DG: **Supplementing monosodium glutamate to partial enteral nutrition slows gastric emptying in preterm pigs.** *J Nutr* 2013, **143:**563–570.
103. Wang WW, Dai ZL, Wu ZL, Lin G, Jia SC, Hu SD, Dahanayaka S, Wu G: **Glycine is a nutritionally essential amino acid for maximal growth of milk-fed young pigs.** *Amino Acids* 2014, doi:10.1007/s00726-014-1758-3.
104. Cabrera RA, Usry JL, Arrellano C, Nogueira ET, Kutschenko M, Moeser AJ, Odle J: **Effects of creep feeding and supplemental glutamine or glutamine plus glutamate (Aminogut) on pre- and post-weaning growth performance and intestinal health of piglets.** *J Anim Sci Biotechnol* 2013, **4:**29.
105. Pi D, Liu Y, Shi H, Li S, Odle J, Lin X, Zhu H, Chen F, Hou Y, Leng W: **Dietary supplementation of aspartate enhances intestinal integrity and energy status in weanling piglets after lipopolysaccharide challenge.** *J Nutr Biochem* 2014, **25:**456–462.
106. Bazer FW, Song GW, Kim JY, Dunlap KA, Satterfield MC, Johnson GA, Burghardt RC, Wu G: **Uterine biology in sheep and pigs.** *J Anim Sci Biotechnol* 2012, **3:**23.
107. Brosnan JT, Brosnan ME: **Glutamate: a truly functional amino acid.** *Amino Acids* 2013, **45:**413–418.
108. Dai ZL, Li XL, Xi PB, Zhang J, Wu G, Zhu WY: **L-Glutamine regulates amino acid utilization by intestinal bacteria.** *Amino Acids* 2013, **45:**501–512.
109. Ren WK, Luo W, Wu MM, Liu G, Yu XL, Jun F, Li TJ, Yin YL, Wu G: **Dietary L-glutamine supplementation improves pregnancy outcome in mice infected with type-2 porcine circovirus.** *Amino Acids* 2013, **45:**479–488.

110. Satterfield MC, Dunlap KA, Keisler DH, Bazer FW, Wu G: **Arginine nutrition and fetal brown adipose tissue development in diet-induced obese sheep.** *Amino Acids* 2012, **43:**1593–1603.
111. Kalhan SC, Hanson RW: **Resurgence of serine: an often neglected but indispensable amino Acid.** *J Biol Chem* 2012, **287:**19786–19791.
112. Ko HM, Oh SH, Bang HS, Kang NI, Cho BH, Im SY, Lee HK: **Glutamine protects mice from lethal endotoxic shock via a rapid induction of MAPK phosphatase-1.** *J Immunol* 2009, **182:**7957–7962.
113. Gupta MK, Uhm SJ, Lee SH, Lee HT: **Role of nonessential amino acids on porcine embryos produced by parthenogenesis or somatic cell nuclear transfer.** *Mol Reprod Dev* 2008, **75:**588–597.
114. Satterfield MC, Dunlap KA, Keisler DH, Bazer FW, Wu G: **Arginine nutrition and fetal brown adipose tissue development in nutrient-restricted sheep.** *Amino Acids* 2013, **45:**489–499.
115. Sawant OB, Ramadoss J, Hankins GD, Wu G, Washburn SE: **Effects of L-glutamine supplementation on maternal and fetal hemodynamics in gestating ewes exposed to alcohol.** *Amino Acids* 2014, doi:10.1007/s00726-014-1751-x.

13

Effect of reduced energy density of close-up diets on dry matter intake, lactation performance and energy balance in multiparous Holstein cows

Wenming Huang[†], Yujia Tian[†], Yajing Wang, Aminamu Simayi, Amingguli Yasheng, Zhaohai Wu, Shengli Li[*] and Zhijun Cao[*]

Abstract

Energy intake prepartum is critically important to health, milk performance, and profitability of dairy cows. The objective of this study was to determine the effect of reduced energy density of close-up diets on dry matter intake (DMI), lactation performance and energy balance (EB) in multiparous Holstein cows which were housed in a free-stall barn and fed for ad libitum intake. Thirty-nine dry cows were blocked and assigned randomly to three groups fed a high energy density diet [HD, n = 13; 6.8 MJ of net energy for lactation (NE_L)/kg; 14.0% crude protein (CP)], or a middle energy density diet (MD, n = 13; 6.2 MJ NE_L/kg; 14.0% CP), or a low energy density diet (LD, n = 13; 5.4 MJ NE_L/kg; 14.0% CP) from d 21 before expected day of calving. After parturition, all cows were fed the same lactation diet to d 70 in milk (DIM). The DMI and NE_L intake prepartum were decreased by the reduced energy density diets ($P < 0.05$). The LD group consumed 1.3 kg/d (DM) more diet compared with HD group in the last 24 h before calving. The milk yield and the postpartum DMI were increased by the reduced energy density diet prepartum ($P < 0.05$). The changes in BCS and BW prepartum and postpartum were not affected by prepartum diets. HD group had higher milk fat content and lower lactose content compared with LD group during the first 3 wk of lactation ($P < 0.05$). The energy consumption for HD, MD and LD groups were 149.8%, 126.2% and 101.1% of their calculated energy requirements prepartum ($P < 0.05$), and 72.7%, 73.1% and 75.2% during the first 4 wk postpartum, respectively. In conclusion, the low energy density prepartum diet was effective in controlling NE_L intake prepartum, and was beneficial in increasing DMI and milk yield, and alleviating negative EB postpartum.

Keywords: Dietary energy density, Dry matter intake, Energy balance, Lactation performance, Transition cow

Introduction

The transition period is the most challenging time in the production cycle of dairy cows because of depressed dry matter intake (DMI) and marked changes in metabolism that support late gestation, parturition and the onset of milk synthesis. The abrupt increase in energy demands after parturition results in a negative energy balance (NEB) which can be met by mobilization of body fat reserves and by a decrease in insulin-dependent glucose utilization in non-mammary tissues [1]. Extensive mobilization of body fat reserves is associated with metabolic disorders such as fatty liver and ketosis. DMI rather than milk yield is the major driver of NEB [2]. Therefore, nutritional management during the transition period designed to increase postpartum DMI might be a potential strategy to alleviate the negative effect of NEB on performance as well as related metabolic disorders.

Approximately 3 wk before parturition, DMI starts to decline and further reduces dramatically in the last week before parturition [3-5]. On the other hand, in the last month of pregnancy the energy requirement of the dairy cow increases by about 20% to support gravid uterus growth [6]. As DMI decline in late gestation is practically unavoidable, increasing the energy density of the close-up dry period diet should help to maintain energy intake [7]. NRC (2001) recommended an approximately 5.2 MJ NE_L/kg of DM diet for far-off dry period cows,

* Correspondence: lisheng0677@163.com; caozhijun@cau.edu.cn
[†]Equal contributors
State Key Laboratory of Animal Nutrition, College of Animal Science and Technology, China Agricultural University, Beijing 100193, China

and a 6.4 to 6.8 MJ NE_L/kg of DM diet for close-up cows, so as to allow the rumen and its microbes to adapt to the freshly calved diet [8]. Some researchers have suggested that prepartum DMI was positively correlated with postpartum DMI and prepartum DMI should be maximized to improve postpartum performance and health [4,9]. In contrast to this viewpoint, most of the last decade's research has shown that over consumption of energy prepartum often resulted in a slower increase in DMI postpartum compared with cows on a restricted intake [5,10,11] or fed on a low energy density diet containing wheat straw [11,12] and this was detrimental to cow health and liver function postpartum [11,13,14]. Lower DMI postpartum intensifies NEB and the mobilization of body fat results in more triacylglycerol and ketone deposition in the liver. Fatty liver has been linked to an increased incidence of metabolic disorders and poor milk yield [15]. Most of the above studies used moderate or high energy density diets with restricted DMI to control energy intake or cows were housed in a tie-stall barn, and the cows may be in a hungry condition and eager to consume more. These experimental results do not reflect the true energy intake and metabolism of transition cows housed in a free-stall barn on commercial farms, however, because transition cows normally have bunk space with free access to feed and water. Further study is needed to address this issue which is critical for dairy management.

It is very common to use moderate or even high energy density transition diets in Chinese dairy farms and the incidence of metabolic diseases is sometimes very high. However, China is short of high quality forage and abundant in low quality forage (Leymus chinensis hay, corn stover and wheat straw). Our hypothesis was that: low energy density diets containing more Leymus chinensis hay could control energy intake prepartum, having the same or better effect on EB postpartum compared to high energy density diets with restricted DMI. The objective of this study was to determine a suitable close-up dietary energy density for multiparous Holstein cows with ad libitum access to feed and housed in a free-stall barn.

Materials and methods

Animals and design

Animal care and use were approved and conducted in accordance with the practices outlined in the Guide for the Care and Use of Agriculture Animals in Agriculture Research and Teaching [16]. Thirty-nine New Zealand multiparous Holstein cows (average contents of milk fat, protein, and lactose were 5.49 ± 0.99%, 3.59 ± 0.38% and 4.98 ± 0.29%, respectively) were enrolled in the study. Cows were dried off at 60 d and moved into an experimental barn 35 d before expected parturition which has 48 free stalls with length × width = 230 cm × 110 cm, and the feed bunk space was 110 cm. The stocking density was about 81% (39 cows/48 free stalls). Cows were grouped according to milk production in the first 3 mo of the previous parity, body weight (BW), body condition score (BCS) and expected calving date and assigned randomly into 1 of 3 dietary treatments. All cows were enrolled in experiment at the same time. The average date before expected parturition for the 3 groups were 21.4 ± 2.5 d, 21.2 ± 2.7 d, and 21.1 ± 2.5 d when the experiment was started. The experiment was conducted from November, 2012 to February, 2013, and the average temperature and humidity during the prepartum and postpartum phases were −3.7 ± 6.3°C, −7.4 ± 4.9°C, and 46.3 ± 7.2%, 43.6 ± 6.3%, respectively.

From dry off to d 22 before expected parturition, all cows were fed the same far-off dry period diet (NE_L5.4 MJ/kg; Table 1). From d 21 before expected day of calving until parturition, cows were blocked and assigned randomly to three groups fed a high energy density diet [HD treatment, n = 13; 6.8 MJ of net energy for lactation (NE_L)/kg; 14.0% crude protein (CP)], or a middle energy density diet (MD treatment, n = 13; 6.2 MJ NE_L/kg; 14.0% CP), or a low energy density diet (LD treatment, n = 13; 5.4 MJ NE_L/kg; 14.0% CP; Table 1). The Leymus chinensis hay was the first batch of materials loaded into TMR mixer and being chopped into small particles, and then mixed with corn silage and concentrate. Different amounts of water were added to each of the three diets to adjust DM content within 51%-52%. After parturition, all the cows were provided with the same lactation diet to 70 DIM (Table 1).

Cows were housed in a free-stall barn with delivery room, and were fed the diets as a TMR ad libitum throughout the experiment, which was offered once (at 1600 h) daily prepartum and twice (at 0730 and 1430 h) postpartum. Orts were controlled less than 8%. Cows were milked 3 times daily (at 0070, 1400, and 2030 h) in a double-48 parallel milking parlor.

Data collections, sampling procedures, and data analysis

DMI of individual cow was determined every day from d -21 to d 35 by roughage intake control feeders (Insentec B.V., Marknesse, the Netherlands) with 48 feed bins. All cows were kept in one group and with access to different feed bins. For each visit to the bin, the system recorded the cow number, bin number, initial and final times, weights and calculated the visit duration and intake. The system's minimum range is 0.1 kg, and it showed a high specificity (100%) and sensitivity (100%) for the feed bins and cow identification [17]. Diets were sampled weekly to determine DM contents. The particle size distribution of the TMR was determined weekly using a Penn State Particle Separator to detect sorting. Briefly, the

Table 1 Composition and analysis of diets fed to Holstein cows during the dry and lactating periods

Item	Far-off diet	Close-up diets[1]			Lactation diet[2]
		HD	MD	LD	
Ingredients, % DM					
Leymus chinensis hay	42.8	13.3	21.7	31.3	-
Alfalfa hay	-	8.4	8.4	8.4	17.2
Whole corn silage	24.7	21.0	21.1	21.0	21.0
Corn	12.0	23.2	14.4	2.0	20.0
Extruded soybean	-	-	-	-	2.6
Soybean meal	2.1	4.5	4.0	3.7	5.8
Rapeseed meal	4.5	4.0	4.0	4.0	1.3
Cottonseed meal	5.0	4.0	4.3	4.5	1.3
Apple pomace	9.2	5.7	7.3	11.4	2.5
Whole cottonseed	-	3.3	3.3	3.3	10.4
DDGS	4.1	4.0	4.0	4.0	10.0
Palm kernel meal	5.1	3.3	3.3	3.3	-
Beet pulp meal	-	-	-	-	3.5
Calcium carbonate	1.2	1.9	1.9	1.9	1.0
Close-up premix[3]	0.1	0.4	0.4	0.4	-
Lactation premix[4]	-	-	-	-	7
Bergafat[5]	-	2.2	0.9	-	1.6
Yeast culture[6]	0.2	0.9	0.9	0.9	0.4
Sodium bicarbonate	-	-	-	-	0.9
Analysis					
DM, %	51.9	51.5	51.5	51.6	52.1
CP, % of DM	12.6	14.0	14.0	14.0	18.2
NDF, % of DM	53.3	37.4	42.9	49.8	31.7
ADF, % of DM	32.4	23.2	26.8	31.5	19.8
NFC, % of DM	22.8	35.0	30.4	23.9	36.9
Crude fat, % of DM	3.1	5.6	4.3	3.3	5.7
NE_L, MJ/kg DM	5.4	6.8	6.2	5.4	7.2
ME, MJ/kg DM	8.7	10.7	10.0	8.8	11.1

[1]*HD* high energy density diet, *MD* middle energy density diet, *LD* low energy density diet.
[2]Lactation diet was fed to cows from parturition to 70 d in milk.
[3]Close-up premix contained (per kg of premix; DM basis): 2,200,000 IU of vitamin A, 550,000 IU of vitamin D_3, 20,000 IU of vitamin E, 2,000 mg of vitamin PP, 3,750 mg Cu, 5,720 mg Mn, 14,850 mg Zn, 150 mg I, 180 mg Se, 120 mg Co.
[4]Lactation premix contained (per kg of premix; DM basis): 1,000,000 IU of vitamin A, 280,000 IU of vitamin D_3, 10,000 IU of vitamin E, 1,000 mg of vitamin PP, 3,250 mg Cu, 4,800 mg Mn, 12,850 mg Zn, 140 mg I, 150 mg Se, 110 mg Co.
[5]Fractionated palm fatty acids (Berg + Schmidt, Hamburg, Germany).
[6]Diamond V XP, Diamond V Mills, Inc.

particle size distribution of diets and orts were compared to determine if sorting occurred (≥ 15 percentage unit change on the19-mm screen) [18, 19].

BW was measured at d −21, −14, −7 d relative to expected parturition and at 1, 7, 14, 21, 28, 35 DIM (WOW, XR3000, Tru-Test, New Zealand). Dry cows were weighed at 1430 h, and lactating cows were also weighed at 1430 h after milking and before diets deliver. BCS was assessed independently by two individuals on a 1 to 5 scale [20].

Cows were milked three times daily, and individual milk production except colostrum was recorded at each milking (Bou-Matic, United States). The quantity of milk was adjusted to an equal energy basis using the formula: four-percent FCM yield = (0.4 × milk yield) + [15 × (% fat/100) × milk yield]. Milk samples were obtained using milk meters at the afternoon milking on 7, 14, 21, 28, 35 and 42 DIM. Milk components (protein, fat, lactose) were determined at Beijing Dairy Cattle Center using a near-infrared reflectance spectroscopy analyzer (Seris300 CombiFOSS; Foss Electric, Hillerød, Denmark).

Energy balance (EB) was calculated individually for each cow according to NRC (2001) [8]. All equations used units of megajoules per kilogram. NE_L intake (NEI) was determined by multiplying DMI by the mean NE_L density of the diet. Net energy required for maintenance (NE_M) was calculated as $BW^{0.75} \times 0.08 \times 120\%$ prepartum and $BW^{0.75} \times 0.08 \times 110\%$ postpartum. Net energy required for pregnancy (NE_P) was calculated as $[(0.00318 \times \text{day of gestation} - 0.0352) \times (\text{calf birth weight}/45)]/ 0.218$. Net energy required for lactation (NE_L) was calculated as $(0.0929 \times \text{fat\%} + 0.0563 \times \text{protein\%} + 0.0395 \times \text{lactose\%}) \times \text{milk}$. The equation used to calculate prepartum EB was $EBPRE = NEI - (NE_M + NE_P)$. The equation used to calculate postpartum EB was $EBPOST = NEI - (NE_M + NE_L)$.

Statistical analysis

Five cows did not complete the study: 1 cow had a dead fetus, 3 cows had mastitis and 1 cow had a displaced abomasum. Thus, 34 cows (HD, n = 11; MD, n = 11; LD, n = 12) completed the experiment and were used in the analysis.

Individual daily DMI, NEI, NDF intake and milk yield values were condensed to weekly means before analysis; the yields and contents of milk fat, protein, lactose and EB were calculated using the weekly mean. To avoid problems with fitting covariance structure, data for DMI, NEI, NDF intake, and EB were analyzed separately for the prepartum and postpartum periods. The DMI, NEI, NDF intake, milk yield, milk components and EB were evaluated using the MIXED procedure of SPSS 16.0 (SPSS Inc.) for repeated measures with the following model: $Y_{ikn} = \mu + W_i + T_k + WT_{ik} + C_{(ik)n}$, where Y_{ikn} = an observation from the ith week relative to calving, k^{th} treatment, and n^{th} cow; μ = the grand mean; W_i = effect of the ith week; T_k = effect of the kth treatment; WT_{ik} = effect of the week by treatment interaction; and $C_{(ik)n}$ = random experimental error from the nth cow nested within the ith week and k^{th} treatment. Repeated statement was used for variables measured over time. Covariance structures including autoregressive (1), compound symmetry, and unstructured were tested. Autoregressive (1) yielded the lowest Akaike's criterion and was finally used in our model. DMI, NEI and NDF intake for −48 h to 48 h, BW, BCS, and weekly EB were evaluated by One-Way ANOVA procedure of SPSS 16.0 (SPSS Inc.). Least squares means were computed and are presented throughout. A value of $P < 0.05$ was set as the significance level.

Results

DMI and NEI

The total average DMI and NEI prepartum were decreased by the reduced energy density diets ($P < 0.05$; Table 2). During the last week before parturition, the DMI of the 3 groups was not affected ($P = 0.22$) and all declined markedly, but the NEI for LD group was lower than that of HD and MD groups ($P < 0.05$). Although HD group was fed a high energy density diet, the DMI ($P = 0.70$) and NEI ($P = 0.99$) were not affected in the last 24 h before parturition. The average NDF intake was increased by the reduced energy density diets during the whole close-up period and the last week before parturition ($P < 0.05$). The total average DMI and NEI of LD group were higher than that of HD group ($P < 0.05$) during the first 5 wk of lactation, and higher in LD group compared with HD and MD groups from 4 to 5 wk of lactation ($P < 0.05$).

BCS and BW

Pretrial BCS and BW were not different among dietary treatment groups (Table 3; $P > 0.05$). The change in BCS and BW prepartum showed no significant differences among the 3 treatments ($P > 0.05$). Although all the 3 treatments gained BW prepartum, the MD group had no change in BCS and LD group lost a little BCS. The loss of BCS ($P = 0.71$) and BW ($P = 0.55$) for LD group was numerically lower than HD and MD groups during the first 5 wk of lactation.

Milk yield and composition

The average milk yield for MD and LD groups was higher than HD group during the first 10 wk (Table 4; $P < 0.05$). The 4% FCM yield for LD group was 2.1 kg/d less than HD group during the first 3 wk ($P > 0.05$), and was similar among the 3 groups during the first 6 wk ($P > 0.05$). The milk fat content of HD group was higher than MD and LD groups during the first 3 wk and first 6 wk ($P < 0.05$), and milk lactose content during the first 3 wk and milk lactose yield during the first 6 wk were lower than these 2 treatments ($P < 0.05$). Energy density of the diets demonstrated no effect on milk protein content, milk fat yield and protein yield during the experiment ($P > 0.05$).

Energy balance

In the close-up dry period, there was significant difference among the 3 treatments in EB (Table 5; $P < 0.001$). HD, MD and LD groups consumed 30.1, 15.9 and 1.3 MJ/d more than the prepartum NE_L requirement, respectively; and achieved 149.8%, 126.2% and 101.1% of their calculated energy requirement, respectively. During the last week before calving, LD group consumed 87.4% of their energy requirement and the other 2 treatments both exceeded 100%. LD group had a milder NEB compared to the other 2 treatments during the first week of lactation. During the first 4 wk postpartum, HD, MD and LD groups achieved 72.7%, 73.1% and 75.2% of their calculated energy requirement, respectively. In all treatments, the most marked NEB appeared in approximately the third and fourth week.

Discussion

DMI and NEI

The average DMI for the 3 treatments was 12.6-14.3 kg/d prepartum, which was higher than our previous study with cows in a tie-stall barn (12.3 kg/d in cool autumn and 8.0 kg/d in hot summer) [21]. In Janovick's study [12], the cows housed in tie stalls and fed the same high NE_L diet (6.8 MJ/kg) had less DMI (13.1 v. 14.3 kg/d) compared with our study, but those fed a similar low NE_L diet (5.2 v. 5.4 MJ/kg) had a slightly higher DMI (12.8 v. 12.6 kg/d), and the BW of our experimental cows were much lower (approximately 650 v. 730 kg). Leymus chinensis hay has high NDF (68% ~ 74%, DM) and ADF (35% ~ 40%, DM) contents, and it's effective ruminal NDF degradability is 26% ~ 31%, and is abundant in the Northeast of China. In the current study, cows fed a low energy density diet containing much Leymus chinensis hay had a significantly lower DMI prepartum. In the last 48 h before calving, however, the DMI for the 3 treatments had no significant difference. The results revealed that a high NDF content diet gave a small change in DMI prepartum. Other studies have shown that diets supplemented with much wheat straw had lower DMI and DMI decline rate prepartum [11,22,23]. The average proportion of the19-mm screen for diets of HD, MD, LD groups were 23.2%, 20.4% and 14.7% as fed basis prepartum, respectively; and they were 35.1%, 29.8% and 19.3% for orts of HD, MD, LD groups, respectively. All the changes on the19-mm screen were less than 15 percentage unit, indicating no sorting has occurred.

58.6-60.7 MJ/d is recommended by NRC (2001) for NEI of close-up dairy cows [8]. In our study, the average prepartum NEI was 97.1, 83.3 and 68.6 MJ/d for HD,

Table 2 Effect of energy density of close-up dry period diets on DMI and (NEI) in Holstein cows

Item	Dietary treatment[1]			SEM	P-value		
	HD	MD	LD		Diet	Time	Interaction
DMI, kg/d							
Prepartum							
Wk −3 to −1	14.3^a	13.6^b	12.6^c	0.14	<0.001	<0.001	0.87
Wk−1	12.3	11.8	11.1	0.29	0.22		
H −48 to −25	10.9	11.0	9.6	0.46	0.37		
H −24 to 0	5.0	5.6	6.3	0.61	0.7		
Postpartum							
Wk 1 to 5	18.8^b	19.1^{ab}	19.7^a	0.13	0.04	<0.001	0.37
H 0 to 24	10.5	10.4	11.1	0.82	0.95		
H 25 to 48	11.9	12.1	12.6	0.7	0.93		
Wk 1 to 3	17.9	18.1	18.3	0.18	0.86	<0.001	0.67
Wk 4 to 5	20.1^b	20.7^b	21.7^a	0.17	0.001	0.12	0.66
NEI, MJ/d							
Prepartum							
Wk −3 to −1	97.1^a	83.3^b	68.6^c	0.83	<0.001	<0.001	0.42
Wk −1	83.3^a	72.4^b	60.2^c	2.22	< 0.001		
H −48 to −25	73.9^a	67.3^a	52.0^b	3.31	0.01		
H −24 to 0	34.1	33.9	34.3	3.64	0.99		
Postpartum							
Wk 1 to 5	136.0^b	138.1^{ab}	142.3^a	0.95	0.03	<0.001	0.67
H 0 to 24	76.3	75.3	80.1	6.23	0.87		
H 25 to 48	86.1	97.9	91.0	5.06	0.93		
Wk 1 to 3	129.2	130.5	132.2	1.40	0.72	<0.001	0.9
Wk 4 to 5	145.2^b	149.4^b	156.9^a	1.25	0.001	0.12	0.64
NDF, kg/d							
Prepartum							
Wk −3 to −1	5.3^c	5.8^b	6.3^a	0.07	<0.001	<0.001	0.99
Wk −1	4.6^b	5.1^{ab}	5.5^a	0.14	0.03		
H −48 to −25	4.1	4.7	4.8	0.20	0.32		
H −24 to 0	1.9	2.4	3.1	0.28	0.19		
Postpartum							
Wk 1 to 5	6.0^b	6.1^{ab}	6.2^a	0.04	0.04	<0.001	0.40
H 0 to 24	3.3	3.3	3.5	0.26	0.95		
H 25 to 48	3.8	3.9	4.0	0.22	0.95		

[1] *HD* high energy density diet, *MD* middle energy density diet, *LD* low energy density diet.
[a, b, c] Values within a row with different superscripts differ significantly at $P < 0.05$.

MD and LD, respectively, which were higher than the energy requirements recommended by NRC (2001) [8]. Other studies have also confirmed that close-up cows fed with moderate or low energy density diets would easily consume more energy relative to their requirements if DMI is not limited [11,24]. But the energy consumption of LD group fed a higher level of Leymus chinensis hay was closest to the NRC (2001) recommendation [8]. Diets high in wheat straw in other studies have also shown the effect of controlling energy intake prepartum [11,12]. The LD group had significantly higher NDF intake compared with HD group during the last week before parturition. In the last 24 h before calving, the NDF intake of LD group increased 63% compared with HD group, but the NEI was similar among the 3 treatment groups. Therefore we could conclude that DMI in the last 24 h before calving

Table 3 Effect of energy density of close-up dry period diets on body condition score (BCS) and BW in Holstein cows

Item	Dietary treatment[1]			SEM	*P*-value
	HD	MD	LD		
BCS					
Initial BCS	3.54	3.52	3.54	0.06	0.93
Wk −3 to −1 change	0.02	0.00	−0.04	0.02	0.27
Wk 1 to 5 change	−0.42	−0.34	−0.33	0.04	0.71
BW, kg					
Initial BW	632	647	659	8.11	0.42
Wk −3 to −1 change	33.6	29.1	22.7	2.52	0.20
Wk 1 to 3 change	−32.1	−25.3	−27.4	3.25	0.73
Wk 1 to 5 change	−57.8	−49.6	−47.4	3.88	0.55

[1]*HD* high energy density diet, *MD* middle energy density diet, *LD* low energy density diet.

Table 4 Effect of energy density of close-up dry period diets on milk yield and composition in Holstein cows

Item	Dietary treatment[1]			SEM	*P*-value		
	HD	MD	LD		Diet	Time	Interaction
Milk yield, kg/d							
Wk 1 to 3	32.0	32.1	32.2	0.51	0.53	<0.001	0.78
Wk 1 to 10	36.6^b	38.6^a	38.7^a	0.36	0.03	<0.001	0.99
4% FCM yield, kg/d							
Wk 1 to 3	44.0	42.4	41.9	0.78	0.55	<0.001	0.72
Wk 1 to 6	45.4	45.7	45.4	0.54	0.90	<0.001	0.91
Fat, %							
Wk 1 to 3	6.43^a	5.87^b	5.67^b	0.11	0.02	0.03	0.89
Wk 1 to 6	5.96^a	5.60^b	5.45^b	0.06	0.006	<0.001	0.90
Protein, %							
Wk 1 to 3	3.66	3.65	3.67	0.05	0.97	<0.001	0.54
Wk 1 to 6	3.50	3.45	3.48	0.02	0.69	<0.001	0.90
Lactose, %							
Wk 1 to 3	4.89^b	5.07^a	5.12^a	0.03	0.02	0.02	0.28
Wk 1 to 6	4.99	5.10	5.12	0.04	0.06	0.002	0.12
Fat, kg/d							
Wk 1 to 3	2.05	1.94	1.90	0.04	0.33	<0.001	0.70
Wk 1 to 6	2.07	2.06	2.01	0.03	0.68	<0.001	0.89
Protein, kg/d							
Wk 1 to 3	1.17	1.20	1.21	0.03	0.57	<0.001	0.55
Wk 1 to 6	1.22	1.26	1.28	0.01	0.19	<0.001	0.94
Lactose, kg/d							
Wk 1 to 3	1.58	1.69	1.72	0.03	0.12	<0.001	0.69
Wk 1 to 6	1.77^b	1.90^a	1.91^a	0.02	0.04	<0.001	0.99

[1]*HD* high energy density diet, *MD* middle energy density diet, *LD* low energy density diet.
$^{a,\,b}$Values within a row with different superscripts differ significantly at $P < 0.05$.

was not decided by rumen volume, but rather by energy requirement.

After parturition, LD group had higher DMI compared with HD and MD groups. These results concur with another two studies showing that large changes in DMI prepartum have been related to lower DMI postpartum [25,26]. This effect can be explained as follows, 1) the dietary palatability changed more for the LD group from dry cow diet to lactation diet, and the average DMI of the first and second 24 h periods after calving were 0.6 kg and 0.7 kg higher than HD group in this study. 2) Small changes of DMI prepartum may lead to less metabolic disorders which could decrease DMI. Some studies have demonstrated that overfeeding energy prepartum results in a greater number of health problems postpartum [11,24,27].

Lactation performance

The reduced energy density of close-up diets increased the milk yield, but did not significantly affect 4% FCM yield. During the first 3 wk, average milk yields of the 3 treatments were similar, but the 4% FCM yield was 2.1 kg/d more for HD group than LD group due to a significantly higher milk fat content. Douglas et al. [24] also showed that cows on restricted intake diets prepartum had a numerically higher milk yield (1.5-2.5 kg/d) compared to those on ad libitum intake. On the contrary, cows fed a 150% of requirement energy diet prepartum tended to produce more milk in wk 1 to 8 than cows fed a 100% and 80% requirement energy diets [12]. In the above two studies, milk fat content for cows with more energy consumption prepartum was significantly or numerically higher than lower energy consumption cows which was in agreement with our study. Janovick et al. [12] reported that high milk fat content might be due to a greater mobilization of body stores in the early lactation period. Blood nonesterified fatty acids (NEFA) could account for as much as 40% of milk fat content on 4 DIM [1]. In this study, the plasma NEFA concentration for HD group was significantly higher than that of LD group during the first 2 wk of lactation (605.7 v. 464.8 μEq/L, the data were not published), and the numerically higher BW and BCS loss of HD group could be explained by more mobilization of adipose tissue, which resulted in a high milk fat content. Although milk lactose content is stable, the milk lactose content for HD group was significantly lower than MD and LD groups during the first 3 wk in the current study. A similar result was reported by Janovick et al. [12]. However, this result was not observed by some others [11,28].

Table 5 Effect of energy density of close-up dry period diets on EB in Holstein cows

Item	Dietary treatment[1]			SEM	*P*-value		
	HD	MD	LD		Diet	Time	Interaction
EB, MJ/day							
Prepartum							
Wk −3 to −1	30.1[a]	15.9[b]	1.3[c]	0.94	< 0.001	< 0.001	0.34
Wk −3	42.3[a]	26.8[b]	7.5[c]	3.22	< 0.001		
Wk −2	37.2[a]	20.9[b]	3.3[c]	2.93	< 0.001		
Wk −1	15.9[a]	6.7[b]	−8.4[c]	2.34	< 0.001		
Postpartum							
Wk 1 to 4	−45.2	−46.0	−43.1	2.2	0.66	0.09	0.99
Wk 1	−43.5	−39.7	−31.0	4.27	0.49		
Wk 2	−41.0	−40.2	−38.9	4.06	0.98		
Wk 3	−51.9	−53.1	−52.3	4.48	0.99		
Wk 4	−47.3	−47.7	−43.9	3.05	0.86		
EB, % of requirement							
Prepartum							
Wk −3 to −1	149.8	126.2	101.1				
Wk −1	125.6	109.7	87.4				
Postpartum							
Wk 1 to 4	72.7	73.1	75.2				
Wk 1	72.7	73.5	79.2				

[1]*HD* high energy density diet, *MD* middle energy density diet, *LD*, low energy density diet.
[a, b, c]Values within a row with different superscripts differ significantly at $P < 0.05$.

Energy balance

Douglas et al. [24] reported that cows fed 6.3 MJ/kg NE_L diet in the whole 60 d dry period consumed an average of 159% of the NRC (2001) requirements for NE_L. In the current study, cows on moderate or even low energy diets offered ad libitum intake during the close-up period consumed more energy relative to their requirements. The HD group was almost in positive EB in the whole close-up period, but the NEI was markedly lower than requirements due to a sharp decrease in DMI in the last 3 d before parturation. A 5.4 MJ of NE_L/kg diet could meet the energy requirements of cows before d 12, and the NEB became much more serious with the onset of calving. All the 3 treatment cows gained BW prepartum, but LD group lost BCS. Janovick et al. [12] also reported that dry cows offered a diet meeting 80% of energy requirements lost BCS but gained18 kg over the dry period. We could conclude that the increase of BW was attributable to the rapid growth of the fetus and placenta in late pregnancy, and not to maternal BW increase.

Although the average postpartum EB was not significantly affected by prepartum diets, HD group had 3.8 and 12.5 MJ/d more energy deficiency than MD and LD groups during the first week of lactation, respectively. Other studies have also showed that overfed cows in the dry period were in more NEB postpartum [11,12]. The higher milk fat content for HD group during the first 3 wk of lactation also implied a more marked NEB in the current study.

Conclusions

Feeding a high energy density diet prepartum resulted in large decline in DMI prepartum, higher milk fat content and lower milk lactose content during the first 3 wk of lactation. The low energy density diet was beneficial in controlling NEI prepartum, increasing milk yield and DMI, and alleviating NEB postpartum.

Abbreviations

DMI: Dry matter intake; EB: Energy balance; NE_L: Net energy for lactation; HD: High energy density diet containing 6.8 MJ NE_L /kg; MD: Middle energy density diet containing 6.2 MJ NE_L /kg; LD: Low energy density diet contaning 5.4 MJ NE_L /kg; DIM: Day in milk; DM: Dry matter; NEB: Negative energy balance; BW: Body weight; BCS: Body condition score; CP: Crude protein; NDF: Neutral detergent fibre; ADF: Acid detergent fibre; NEI: NE_L intake; NE_M: Net energy required for maintenance; NE_P: Net energy required for pregnancy; EBPRE: Prepartum EB; EBPOST: Postpartum EB.

Competing interests

The authors declare that they have no competing interests.

Authors' contributions

WH, YT, AS, AY and ZW carried out the experiments. SL, ZC and YW participated in the design of the study, and drafted the manuscript. All authors read and approved the final manuscript.

Acknowledgments

This research was financially supported by National Natural Science Foundation of China (31272469, 31372334) and China Agriculture Research System (CARS-37). The authors greatly appreciate all the people who offered assistance in the experimental procedures and paper writing.

References

1. Bell AW: **Regulation of organic nutrient metabolism during transition from late pregnancy to early lactation.** *J Anim Sci* 1995, **73:**2804–2819.
2. Grummer RR, Wiltbank MC, Fricke PM, Watters RD, Silva-Del-Rio N: **Management of dry and transition cows to improve energy balance and reproduction.** *J Reprod Dev* 2010, **56:**S22–S28.
3. Bertics SJ, Grummer RR, Cadorniga-Valino C, Stoddard EE: **Effect of prepartum dry matter intake on liver triglyceride concentration and early lactation.** *J Dairy Sci* 1992, **75:**1914–1922.
4. Grummer RR: **Impact of changes in organic nutrient metabolism on feeding the transition dairy cow.** *J Anim Sci* 1995, **73:**2820–2833.
5. Guo J, Peters RR, Kohn RA: **Effect of a transition diet on production performance and metabolism in periparturient dairy cows.** *J Dairy Sci* 2007, **90:**5247–5258.
6. Moe PW, Tyrrell HF: **Metabolizable energy requirements of pregnant dairy cows.** *J Dairy Sci* 1972, **55:**480–483.
7. McNamara S, O'Mara FP, Rath M, Murphy JJ: **Effects of different transition diets on dry matter intake, milk production, and milk composition in dairy cows.** *J Dairy Sci* 2003, **86:**2397–2408.
8. NRC: *Nutrient requirements of dairy cattle.* 7th edition. Washington, DC: National Academies Press; 2001.
9. Doepel L, Lapierre H, Kennelly JJ: **Peripartum performance and metabolism of dairy cows in response to prepartum energy and protein intake.** *J Dairy Sci* 2002, **85:**2315–2334.
10. Agenas S, Burstedt E, Holtenius K: **Effects of feeding intensity during the dry period. 1. Feed intake, body weight, and milk production.** *J Dairy Sci* 2003, **86:**870–882.
11. Dann HM, Litherland NB, Underwood JP, Bionaz M, D'Angelo A, McFadden JW, Drackley JK: **Diets during far-off and close-up dry periods affect periparturient metabolism and lactation in multiparous cows.** *J Dairy Sci* 2006, **89:**3563–3577.
12. Janovick NA, Drackley JK: **Prepartum dietary management of energy intake affects postpartum intake and lactation performance by primiparous and multiparous Holstein cows.** *J Dairy Sci* 2010, **93:**3086–3102.
13. Rukkwamsuk T, Wensing T, Geelen MJ: **Effect of overfeeding during the dry period on regulation of adipose tissue metabolism in dairy cows during the periparturient period.** *J Dairy Sci* 1998, **81:**2904–2911.
14. Herdt TH: **Ruminant adaptation to negative energy balance. Influences on the etiology of ketosis and fatty liver.** *Vet Clin North Am Food Anim Pract* 2000, **16:**215–230.
15. Bobe G, Young JW, Beitz DC: **< i > *Invited review:* pathology, etiology, prevention, and treatment of fatty liver in dairy cows.** *J Dairy Sci* 2004, **87:**3105–3124.
16. FASS: *Guide for the care and use of agricultural animals in research and teaching.* 3rd edition. Champaign, IL: Federation of Animal Science Societies; 2010.
17. Chapinal N, Veira DM, Weary DM, von Keyserlingk MAG: **Technical note: validation of a system for monitoring individual feeding and drinking behavior and intake in Group-Housed cattle.** *J Dairy Sci* 2007, **90:**5732–5736.
18. Lammers BP, Buckmaster DR, Heinrichs AJ: **A simple method for the analysis of particle sizes of forage and total mixed rations.** *J Dairy Sci* 1996, **79:**922–928.
19. Dann HM, Carter MP, Cotanch KW, Ballard CS, Takano T, Grant RJ: **Effect of partial replacement of forage neutral detergent fiber with By-Product neutral detergent fiber in Close-Up diets on periparturient performance of dairy cows.** *J Dairy Sci* 2007, **90:**1789–1801.
20. Ferguson JD, Galligan DT, Thomsen N: **Principal descriptors of body condition score in Holstein cows.** *J Dairy Sci* 1994, **77:**2695–2703.
21. Su H, Wang Y, Zhang Q, Wang F, Cao Z, Rahman MA, Cao B, Li S: **Responses of energy balance, physiology, and production for transition dairy cows fed with a low-energy prepartum diet during hot season.** *Trop Anim Health Prod* 2013, **45:**1495–1503.
22. Minor DJ, Trower SL, Strang BD, Shaver RD, Grummer RR: **Effects of nonfiber carbohydrate and niacin on periparturient metabolic status and lactation of dairy cows.** *J Dairy Sci* 1998, **81:**189–200.
23. Beever DE: **The impact of controlled nutrition during the dry period on dairy cow health, fertility and performance.** *Anim Reprod Sci* 2006, **96:**212–226.
24. Douglas GN, Overton TR, Bateman HN, Dann HM, Drackley JK: **Prepartal plane of nutrition, regardless of dietary energy source, affects periparturient metabolism and dry matter intake in Holstein cows.** *J Dairy Sci* 2006, **89:**2141–2157.
25. Grummer RR, Mashek DG, Hayirli A: **Dry matter intake and energy balance in the transition period.** *Vet Clin North Am Food Anim Pract* 2004, **20:**447–470.
26. Drackley JK, Dann HM, Douglas GN, Guretzky NAJ, Litherland NB, Underwood JP, Loor JJ: **Physiological and pathological adaptations in dairy cows that may increase susceptibility to periparturient diseases and disorders.** *Ital J Anim Sci* 2005, **4:**323–344.
27. Janovick NA, Boisclair YR, Drackley JK: **Prepartum dietary energy intake affects metabolism and health during the periparturient period in primiparous and multiparous Holstein cows.** *J Dairy Sci* 2011, **94:**1385–1400.
28. Rabelo E, Rezende RL, Bertics SJ, Grummer RR: **Effects of transition diets varying in dietary energy density on lactation performance and ruminal parameters of dairy cows.** *J Dairy Sci* 2003, **86:**916–925.

14

Amino acid digestibility in low-fat distillers dried grains with solubles fed to growing pigs

Shelby Marie Curry, Diego Mario David Labadan Navarro, Ferdinando Nielsen Almeida, Juliana Abranches Soares Almeida and Hans Henrik Stein*

Abstract

The objective of this experiment was to determine the standardized ileal digestibility (SID) of amino acids (AA) in 3 sources of distillers dried grains with solubles (DDGS) with different concentrations of fat. Twelve growing barrows (initial body weight: 76.1 ± 6.2 kg) were randomly allotted to a replicated 6 × 4 Youden square design with 6 diets and 4 periods. The fat content of the 3 sources of DDGS were 11.5, 7.5, and 6.9% respectively. Diets contained 60% DDGS and fat concentration of the diets were 7.5, 5.2, and 5.2%, respectively. Two additional diets containing the 2 sources of DDGS with 7.5 and 6.9% fat were also formulated, and corn oil was added to these diets to increase the concentration of fat in the diets to levels that were calculated to be similar to the diet containing conventional DDGS with 11.5% fat. A N-free diet was also formulated to calculate endogenous losses of crude protein (CP) and AA from the pigs. Pigs were fed experimental diets during four 7-d periods. The first 5 d of each period were an adaptation period and ileal digesta were collected on d 6 and 7 of each period. The apparent ileal digestibililty (AID) and SID of CP and all indispensable AA, except AID Pro and SID of Trp, were greater ($P < 0.01$) in conventional DDGS than in the 2 sources of DDGS with reduced fat. Adding oil to the diets containing the 2 sources of DDGS with reduced fat did not consistently increase SID of AA. In conclusion, conventional DDGS has greater SID values for most AA compared with DDGS that contains less fat and inclusion of additional oil to diets containing low-fat DDGS does not increase AID or SID of AA. The lower AA digestibility in low-fat DDGS could not be overcome by the inclusion of additional fat to the diets.

Keywords: Amino acid digestibility, Distillers dried grains with solubles, Pigs

Background

Distillers co-products have been used in swine diets for more than 50 years, but the increase in ethanol production in the last few decades has made distillers dried grains an available and attractive ingredient to use in swine diets [1]. Conventional distillers dried grains with solubles (DDGS) contains approximately 27% CP, 10% fat, 9% acid detergent fibre (ADF), and 25% neutral detergent fibre (NDF) [1-3]. However, new technologies have been developed and implemented to remove fat from DDGS to be able to market the oil for biodiesel production or other uses. If oil is removed by centrifugation of the solubles before solubles are added to the distilled grains, then a low-fat DDGS is produced. Low-fat DDGS contains 6 to 9% oil [3]. If oil is extracted from DDGS using a solvent extraction process, de-oiled DDGS is produced [3]. De-oiled DDGS contains less than 4% oil, and therefore, contains less energy than conventional DDGS [4], which may affect its economic value and inclusion level [5]. Oil removal improves handling issues commonly encountered when using DDGS in pig diets such as challenges with flowability [4].

The DE and ME of de-oiled and low-fat DDGS [4,6] are less than the DE and ME of conventional DDGS. There is, however, limited information about the digestibility of AA in low-fat DDGS, but increased concentrations of dietary fat may increase the digestibility of AA in soybean meal and in mixed corn-soybean meal-DDGS diets [7-10]. It is, therefore, possible that the digestibility of AA is also influenced by the concentration of fat in DDGS. The objective of this research was to determine if the concentration of fat in DDGS and diets containing DDGS affects the apparent ileal digestibility (AID) or

* Correspondence: hstein@illinois.edu
Department of Animal Sciences, University of Illinois, Urbana 61801, USA

standardized ileal digestibility (SID) of CP and AA fed to growing pigs.

Materials and methods

The Institutional Animal Care and Use Committee at the University of Illinois reviewed and approved the protocol for this experiment.

Animals, housing, experimental design, and diets

A total of 12 growing barrows (Genetiporc, Alexandria, MN) with an initial body weight (BW) of 76.1 ± 6.2 kg were randomly allotted to a replicated 6 × 4 Youden square design with 6 diets and four 7-d periods. Pigs had a cannula surgically installed in the distal ileum to allow for collection of ileal digesta [11]. Each pig was individually housed in a 1.2 m × 1.5 m that was equipped with a feeder and a nipple drinker. Pens were in a temperature-controlled barn with a propane heater and forced air fans. Each pen has smooth sidings and a fully slatted Tri-bar steel floor.

Three sources of DDGS were produced by Poet Nutrition, Sioux Falls, SD (Table 1). The 3 sources of DDGS were produced at the same facility on 3 consecutive days in the fall of 2011. The corn that was used on these 3 days was from a common source. The DDGS that was produced on the first day consisted of distillers dried grains and all the solubles. However, fat was skimmed off the solubles that were added to the distiller dried grains on d 2 and 3, resulting in production of 2 sources of low-fat DDGS. The conventional DDGS contained 11.5% fat, and the two sources of low-fat DDGS contained 7.5% and 6.9% fat, respectively. Six diets were formulated (Tables 2 and 3). Three diets that contained 60% of each source of DDGS were formulated and these diets contained 7.5, 5.2, and 5.2% fat, respectively. Two additional diets were also formulated and these diets also contained the 2 sources of low-fat DDGS, but corn oil was added to these diets with the intent to bring the concentration of fat in the diets to the same level as in the diet with the conventional DDGS. A N-free diet was used to measure endogenous losses of CP and AA from the pigs. Chromic oxide (0.4%) was included in all the diets as an inert marker. Vitamins and minerals were included in all diets to meet requirements [12].

Table 1 Analyzed nutrient composition of ingredients (as-fed basis)[1]

Item	Ingredients[2]		
	Conventional DDGS	Low- fat DDGS, 7.5% fat	Low-fat DDGS, 6.9% fat
DM, %	90.61	89.04	87.48
CP, %	25.73	28.03	27.93
AEE[3], %	11.47	7.51	6.86
ADF, %	7.78	9.27	8.82
NDF, %	28.64	30.74	29.87
Starch, %	1.50	1.16	1.33
Ca, %	0.07	0.06	0.06
P, %	0.90	0.87	0.92
Indispensable AA, %			
Arg	1.22	1.22	1.24
His	0.69	0.75	0.71
Ile	1.03	1.12	1.06
Leu	2.79	3.17	3.07
Lys	0.91	0.91	0.88
Met	0.52	0.59	0.55
Phe	1.25	1.36	1.35
Thr	0.97	1.02	1.02
Trp	0.21	0.20	0.20
Val	1.30	1.43	1.33
Dispensable AA, %			
Ala	1.72	1.93	1.86
Asp	1.59	1.73	1.68
Cys	0.62	0.62	0.63
Glu	3.28	4.01	3.81
Gly	1.05	1.10	1.06
Pro	1.91	2.08	2.17
Ser	1.07	1.13	1.18
Tyr	1.01	1.09	1.09

[1]Ingredients were obtained from Poet Nutrition, Sioux Falls, SD.
[2]DDGS = distillers dried grains with solubles.
[3]AEE = acid hydrolyzed ether extract.

Feeding and sample collections

Pigs were fed at a level of 3 times the maintenance energy requirement (i.e., 106 kcal ME per kgBW$^{0.75}$; [12]) and the daily feed allotments were divided into 2 equal meals. Amount of feed supplied was recorded daily and pig weights were recorded at the beginning of the experiment and at the end of each 7-d period. The initial 5 d of each period was considered a diet adaptation period. Ileal digesta were collected on d 6 and 7 for 8 h as previously described [13]. All samples were stored at −20°C immediately after collection to avoid bacterial degradation of the AA. At the conclusion of the experiment, samples were thawed and mixed within animal and diet and a sub-sample was collected and analyzed.

Chemical analysis

A sample from each ingredient and each diet was collected at the time of diet mixing. Ileal digesta samples were lyophilized and ground prior to chemical analysis. Samples from diets and digesta were analyzed for dry matter (DM; Method 930.15; [14]), chromium (Method 990.08; [14]), CP (Method 990.03; [14]), and AA (Method

Table 2 Ingredient composition of experimental diets, as-fed basis

Ingredient, %	Diet[1]					
	Conventional DDGS	Low-fat DDGS, 7.5% fat	Low-fat DDGS, 6.9% fat	Low-fat DDGS, 7.5% fat with oil	Low-fat DDGS, 6.9% fat with oil	N-free
Cornstarch	27.60	27.60	27.60	28.20	28.25	69.00
Conventional DDGS	60.00	-	-	-	-	-
Low-fat DDGS-1	-	60.00	-	58.20	-	-
Low-fat DDGS-2	-	-	60.00	-	57.20	-
Corn oil	-	-	-	1.20	2.15	4.00
Sugar	10.00	10.00	10.00	10.00	10.00	20.00
Chromic oxide	0.40	0.40	0.40	0.40	0.40	0.40
Limestone	1.30	1.30	1.30	1.30	1.30	0.60
Salt	0.40	0.40	0.40	0.40	0.40	0.40
Vitamin-mineral premix[2]	0.30	0.30	0.30	0.30	0.30	0.30
Cellulose[3]	-	-	-	-	-	4.00
Dicalcium Phosphate	-	-	-	-	-	1.30

[1]DDGS = distillers dried grains with solubles.
[2]The vitamin-micromineral premix provided the following quantities of vitamins and micro minerals per kilogram of the complete diet: vitamin A as retinyl acetate, 11,128 IU; vitamin D_3 as cholecalciferol, 2,204 IU; vitamin E as DL-alpha tocopheryl acetate, 66 IU; vitamin K as menadione nicotinamide bisulfate, 1.42 mg; thiamin as thiamine mononitrate, 0.24 mg; riboflavin, 6.58 mg; pyridoxine as pyridoxine hydrocloride, 0.24 mg; vitamin B_{12}, 0.03 mg; D-pantothenic acid as D-calcium pantothenate, 23.5 mg; niacin as nicotinamide and nicotinic acid, 44 mg; folic acid, 1.58 mg; biotin, 0.44 mg; Cu, 10 mg as copper sulfate; Fe, 125 mg as iron sulfate; I, 1.26 mg as potassium iodate; Mn, 60 mg as manganese sulfate; Se, 0.3 mg as sodium selenite; and Zn, 100 mg as zinc oxide.
[3]Solka Floc. Fiber Sales and Development Corp., Urbana, OH.

982.30 E (a, b, c); [14]). Prior to AA analysis, samples were hydrolyzed with 6 mol/L HCl for 24 h at 110°C (method 982.30 E(a); [14]). Methionine and Cys were determined as Met sulfone and cysteic acid after cold performic acid oxidation overnight before hydrolysis (method 982.30 E (b); [14]). Tryptophan was determined after NaOH hydrolysis for 22 h at 110°C (method 982.30 E(c); [14]). Each ingredient was also analyzed for DM, CP, AA, ADF (Method 973.18; [14]), NDF [15], for Ca and P by inductively coupled plasma spectroscopy (Method 985.01; [14]), and for starch using the glucoamylase procedure (Method 979.10; [14]). Each ingredient and all diets were also analyzed for total fat by acid hydrolysis using 3 *N* HCl [16] followed by crude fat extraction using petroleum ether (Method 2003.06; [14]) on a Soxtec 2050 automated analyzer (FOSS North America, Eden Prairie, MN).

Calculations and statistical analysis

Basal endogenous losses of CP and AA were determined from pigs after feeding the N-free diet. Values for AID, endogenous losses, and SID of CP and AA were calculated as previously described [17]. The MIXED procedure of SAS was used to analyze the data (SAS Inst. Inc., Cary, NC). The model included diet as a fixed effect whereas pig and period were included as random effects. A pig was used as the experimental unit. The UNIVARIATE procedure of SAS was used to determine if there were any outliers. However, no outliers were identified. An observation was considered an outlier if the value was more than 3 standard deviations away from the grand mean. The LSMeans statement was used to calculate mean values for each diet and the PDIFF option was used to separate means. Moreover, an orthogonal contrast was conducted to verify if the addition of oil to low-fat DDGS diets increases AA digestibility of low-fat DDGS to the same level as conventional DDGS. The contrast was performed by comparing conventional DDGS and the 2 low-fat DDGS sources with addition of supplemental oil. For all analyses, an alpha value of 0.05 was used to determine significance among means.

Results

The AID for CP and all indispensable AA was greater ($P < 0.01$) in conventional DDGS than in the 2 low-fat sources of DDGS, whereas no differences between the 2 low-fat sources of DDGS were observed (Table 4). The mean AID of indispensable AA and the mean of dispensable AA were also greater ($P < 0.01$) in conventional DDGS than in the 2 low-fat sources of DDGS. Addition of oil to the diets containing the low-fat DDGS did not increase AID values for CP or AA.

The SID of CP and all indispensable AA except Trp was greater in conventional DDGS ($P < 0.01$) than in the 2 sources of low-fat DDGS (Table 5). The SID of Lys was greater ($P < 0.01$) in one of the sources of low-fat DDGS than in the other source. However, if fat was added, there was an increase in SID of Lys in the DDGS with 7.5% fat content, but a reduction in SID of Lys in the low-fat DDGS with 6.9% fat content. For Trp, the SID in conventional DDGS did not differ from that of

Table 3 Analyzed nutrient composition of experimental diets, as-fed basis

Item	Diet[1]					
	Conventional DDGS	Low-fat DDGS, 7.5% fat	Low-fat DDGS, 6.9% fat	Low-fat DDGS, 7.5% fat with oil	Low-fat DDGS, 6.9% fat with oil	N-free
CP, %	15.62	16.49	16.59	16.01	15.65	0.28
DM, %	92.26	90.99	89.57	90.20	87.67	91.93
AEE [2], %	7.47	5.20	5.15	6.21	6.95	4.07
Indispensable AA, %						
Arg	0.73	0.72	0.75	0.72	0.70	0.01
His	0.43	0.46	0.46	0.45	0.41	0.00
Ile	0.57	0.62	0.61	0.60	0.55	0.01
Leu	1.69	1.91	1.88	1.88	1.74	0.02
Lys	0.51	0.51	0.51	0.50	0.47	0.01
Met	0.31	0.35	0.33	0.34	0.31	0.00
Phe	0.73	0.81	0.80	0.79	0.74	0.01
Thr	0.58	0.62	0.63	0.63	0.60	0.01
Trp	0.13	0.12	0.13	0.12	0.12	<0.04
Val	0.81	0.88	0.87	0.84	0.77	0.01
Dispensable AA, %						
Ala	1.08	1.19	1.18	1.17	1.09	0.01
Asp	1.01	1.08	1.08	1.06	1.00	0.01
Cys	0.29	0.32	0.31	0.32	0.28	0.00
Glu	2.38	2.62	2.58	2.56	2.40	0.03
Gly	0.64	0.67	0.68	0.66	0.62	0.01
Pro	1.24	1.40	1.38	1.38	1.30	0.06
Ser	0.62	0.67	0.69	0.71	0.68	0.01
Tyr	0.54	0.58	0.58	0.59	0.56	0.01

[1]DDGS = distillers dried grain with solubles.
[2]AEE = acid hydrolyzed ether extract.

the low-fat DDGS (DDGS, 6.9% fat) without added oil and low fat DDGS (DDGS, 7.5%) with added oil, but the SID Trp of conventional DDGS was greater ($P < 0.05$) than in the low-fat DDGS (DDGS, 7.5% fat) without added oil and low-fat DDGS (DDGS, 6.9%) with added oil. For all AA except Ser, the AID and SID of low-fat DDGS with added oil was smaller ($P < 0.05$) than AID and SID of conventional DDGS, indicating that the addition of oil did not increase AID and SID of low-fat DDGS to the same level as AID and SID of conventional DDGS (Table 5). The SID of all dispensable AA was greater ($P < 0.01$) in conventional DDGS than in the 2 sources of low-fat DDGS. Addition of oil to the diets containing low-fat DDGS did not increase SID values.

Discussion

Distillers dried grains with solubles is a co-product from the dry-grind processing of corn and has been used in swine diets for many years. The use of DDGS in swine diets has increased because it is affordable as well as high in energy, AA, and digestible P [1,2,18]. However, the drying process of DDGS may cause heat damage to the ingredient because it involves high temperatures and moisture, and these conditions are favorable for initiating the Maillard reaction, which reduces AA concentration and digestibility [19]. The Lys:CP indicates heat damage, and a ratio greater than 2.8% is desirable in DDGS and implies no heat damage [20]. Samples used in this experiment had Lys:CP of at least 3.15% indicating that the 3 sources of DDGS used in this experiment were not heat damaged.

In recent years, some ethanol plants have centrifuged the solules that are produced to extract oil, which may be sold to the biodiesel industry [5]. The result of the centrifugation is a reduction in fat concentration from 10% in conventional DDGS to 6 to 9% in low-fat DDGS [3]. The 2 sources of DDGS used in this experiment that were produced after the solubles had the oil removed contained 7.5 and 6.9% fat respectively, and these sources of

Table 4 Apparent ileal digestibilitly of CP and AA in distillers dried grains with solubles (DDGS) fed to pigs[1]

Item	Ingredients							Additional oil[2]	
	Conventional DDGS	Low-fat DDGS, 7.5% fat	Low-fat DDGS, 6.9% fat	Low-fat DDGS, 7.5% fat with oil	Low-fat DDGS, 6.9% fat with oil	SEM	*P*-value	SEM	*P*-value
CP, %	71.8[a]	64.6[b]	66.1[b]	66.9[b]	68.0[b]	1.26	<0.01	1.54	<0.01
Indispensable AA, %									
Arg	81.8[a]	75.0[b]	76.9[b]	77.3[b]	75.5[b]	0.93	<0.01	5.88	<0.01
His	78.0[a]	70.8[b]	71.9[b]	72.9[b]	70.3[b]	1.12	<0.01	1.08	<0.01
Ile	75.6[a]	69.1[b]	69.3[b]	71.2[b]	67.8[b]	0.95	<0.01	1.00	<0.01
Leu	85.4[a]	81.4[b]	80.2[b]	82.8[b]	81.2[b]	0.76	<0.01	0.69	<0.01
Lys	62.2[a]	50.8[b]	56.1[b]	56.9[b]	51.4[b]	1.91	<0.01	1.51	<0.01
Met	85.8[a]	82.8[b]	81.5[b]	83.7[b]	82.8[b]	0.68	<0.01	0.63	<0.01
Phe	81.5[a]	77.2[b]	76.8[b]	78.5[b]	76.8[b]	0.75	<0.01	0.71	<0.01
Thr	65.8[a]	59.9[b]	61.4[b]	63.6[b]	61.2[b]	1.30	<0.01	1.18	<0.01
Trp	76.5[a]	70.8[b]	74.7[b]	74.1[b]	71.3[b]	1.23	<0.01	1.36	0.01
Val	75.7[a]	69.8[b]	70.3[b]	71.4[b]	68.2[b]	0.95	<0.01	0.99	<0.01
Mean	78.3[a]	72.8[b]	73.2[b]	74.9[b]	72.5[b]	0.85	<0.01	0.84	<0.01
Dispensable AA, %									
Ala	80.7[a]	76.5[b]	75.8[b]	77.9[b]	76.2[b]	0.89	<0.01	0.84	<0.01
Asp	67.7[a]	62.3[b]	63.3[b]	65.1[b]	61.4[b]	1.13	<0.01	1.16	<0.01
Cys	71.7[a]	63.4[b]	64.8[b]	67.1[b]	63.0[b]	1.60	<0.01	1.41	<0.01
Glu	82.4[a]	77.4[b]	76.8[b]	78.7[b]	77.3[b]	0.96	<0.01	0.85	<0.01
Gly	58.6[a]	48.5[b]	51.0[b]	51.6[b]	47.4[b]	2.13	<0.01	2.24	<0.01
Pro	64.3	57.5	59.7	59.6	57.4	4.69	0.37	3.16	0.08
Ser	73.4[a]	69.4[b]	70.0[b]	72.8[b]	72.0[b]	1.23	<0.01	1.02	0.34
Tyr	82.3[a]	77.7[b]	77.6[b]	79.9[b]	78.5[b]	0.72	<0.01	0.72	<0.01
Mean	74.3[a]	68.7[b]	69.0[b]	70.8[b]	68.7[b]	1.39	<0.01	1.30	<0.01
Total AA	76.1[a]	70.5[b]	70.9[b]	72.7[b]	70.4[b]	1.10	<0.01	1.08	<0.01

[a,b]Means without a common superscript in the same row differ; pairwise comparison.
[1]Data are means of 8 observations.
[2]Orthogonal contrast comparing conventional DDGS and the 2 sources of low-fat DDGS with additional oil.

DDGS were classified as low-fat DDGS, according to the definition by NRC [3].

The starch level in the 3 sources of DDGS used in this experiment was less than previously reported [1], indicating that the fermentation process in the ethanol plant was very efficient. The ADF levels of DDGS used in this experiment were also less than reported by NRC [3]. The CP, fat, and AA levels were, however, in agreement with previous values reported [3]. The values for SID of AA and CP for the low-fat DDGS were within the range of expected values [3]. However, SID of CP, Arg, Met, Trp, Ala, Asp, and Ser of DDGS were greater than previously reported [3]. Moreover, the SID of the other AA are in the upper limit of the range reported [3]. It is possible that the apparent reduction in SID values in low-fat DDGS may be due to the higher coefficients of digestibility of the conventional DDGS used in this experiment compared with previously reported values [3].

There is limited information about the AA digestibility of low-fat DDGS or effects of additional dietary fat on AID and SID of low-fat DDGS. However, data on the effect of low-fat DDGS on digestible energy (DE) and metabolizable energy (ME) indicate that dietary fat is not always a good predictor of ME for swine [5,6,10]. Different estimates for DE and ME have been reported for different sources of DDGS even if fat concentration was similar [4,6].

The AID of AA in nursery and growing pigs is increased by the inclusion of additional fat to the diet [7,21]. The increase in dietary fat delays gastric emptying [22] and although the length of the fatty acid can be different among different sources of vegetable oil, the impact of fatty acids on gastric emptying is similar regardless of chain length [22]. The slower gastric emptying may result in slower rate of passage of the diet, causing an increase in the time of exposure of feed to

Table 5 Standardized ileal digestibility of CP and AA in distillers dried grains with solubles (DDGS) fed to pigs[1]

Item	Ingredients							Additional oil[2]	
	Conventional DDGS	Low-fat DDGS, 7.5% fat	Low-fat DDGS, 6.9% fat	Low-fat DDGS, 7.5% fat with oil	Low-fat DDGS, 6.9% fat with oil	SEM	*P*-value	SEM	*P*-value
CP, %	79.8[a]	72.8[b]	73.6[b]	74.6[b]	75.3[b]	1.22	< 0.01	1.29	<0.01
Indispensable AA, %									
Arg	87.7[a]	81.0[c]	82.5[bc]	83.2[b]	81.4[bc]	0.93	< 0.01	0.92	<0.01
His	80.9[a]	73.5[b]	74.6[b]	75.6[b]	73.3[b]	1.12	< 0.01	1.08	<0.01
Ile	79.8[a]	72.9[bc]	73.1[bc]	75.2[b]	71.9[c]	0.95	< 0.01	1.00	<0.01
Leu	87.7[a]	83.4[bc]	82.2[c]	84.8[b]	83.3[bc]	0.76	< 0.01	0.69	<0.01
Lys	67.9[a]	56.4[c]	61.7[b]	62.6[b]	57.3[c]	1.91	< 0.01	1.51	<0.01
Met	88.1[a]	84.8[bc]	83.6[c]	85.8[b]	85.0[bc]	0.68	< 0.01	0.63	<0.01
Phe	84.9[a]	80.3[bc]	79.8[c]	81.6[b]	80.0[bc]	0.75	< 0.01	0.71	<0.01
Thr	73.4[a]	66.9[c]	68.2[bc]	70.4[b]	68.2[bc]	1.30	< 0.01	1.18	<0.01
Trp	83.1[a]	77.8[c]	81.1[ab]	81.0[abc]	78.0[bc]	1.23	< 0.05	1.36	0.02
Val	80.5[a]	74.2[bc]	74.6[bc]	75.9[b]	73.0[c]	0.95	< 0.01	0.99	<0.01
Mean	82.5[a]	76.7[c]	77.0[bc]	78.8[b]	76.6[c]	0.85	< 0.01	0.85	<0.01
Dispensable AA, %									
Ala	85.4[a]	80.6[bc]	79.9[c]	82.1[b]	80.6[bc]	0.89	< 0.01	0.84	<0.01
Asp	73.6[a]	67.8[c]	68.7[bc]	70.6[b]	67.1[c]	1.13	< 0.01	1.16	<0.01
Cys	76.0[a]	67.3[c]	68.8[bc]	70.9[b]	67.3[c]	1.60	< 0.01	1.41	<0.01
Glu	85.4[a]	80.1[b]	79.5[b]	81.4[b]	80.2[b]	0.96	< 0.01	0.85	<0.01
Gly	75.9[a]	64.8[b]	66.8[b]	68.0[b]	64.3[b]	2.13	< 0.01	2.24	<0.01
Pro	98.5[a]	87.4[b]	89.5[b]	89.7[b]	88.4[b]	4.70	< 0.05	3.16	<0.01
Ser	78.8[a]	74.4[c]	74.7[bc]	77.4[a]	76.7[ab]	1.22	< 0.01	1.02	0.10
Tyr	85.7[a]	80.9[c]	80.7[c]	83.0[b]	81.7[bc]	0.72	< 0.01	0.71	<0.01
Mean	84.3[a]	77.7[b]	78.0[b]	79.8[b]	78.0[b]	1.39	< 0.01	1.30	<0.01
Total AA	83.5[a]	77.2[b]	77.5[b]	79.4[b]	77.4[b]	1.10	< 0.01	1.08	<0.01

[a,b]Means without a common superscript in the same row differ; pairwise comparison.
[1]Data are means of 8 observations. Values for standardized ileal digestibility were calculated by correcting apparent ileal digestibility values for basal endogenous losses (g/kg of DMI), which were determined by feeding pigs a N-free diet; CP, 13.84; Arg, 0.47; His, 0.14; Ile, 0.26; Leu, 0.42; Lys, 0.32; Met, 0.08; Phe, 0.27; Thr, 0.48; Trp, 0.09; Val, 0.39; Ala, 0.55; Asp, 0.65; Cys, 0.14; Glu, 0.78; Gly, 1.20; Pro, 4.60; Ser, 0.37; Tyr, 0.20.
[2]Orthogonal contrast comparing conventional DDGS and the 2 sources of low-fat DDGS with additional oil.

proteolytic enzymes, thus providing longer time for peptides and AA to be digested and absorbed, and increase in AID of AA [7]. The addition of oil to diets fed to growing pigs increase not only the AID but also the SID of AA [9,10].

Results of this experiment indicate that the addition of dietary fat to diets containing low-fat DDGS fed to growing pigs did not improve the AID or SID of AA. These results are not in agreement with previous data [7,9]. However, the difference in fat levels between diets without or with added fat were much greater (4.0 vs. 7.5%, 0.24 vs. 6.7%) in other experiments [7,9] compared with the differences observed in the present experiment (5.2 vs. 6.2%, or 5.2 vs. 7.0%). Moreover, the fat level of the low-fat DDGS diets were relatively high and similar, and that could be the reason additional inclusion of fat did not result in an increase in SID of AA.

Conclusions

Results of this experiment indicate that removal of oil may result in reduced AID and SID of AA in DDGS, and that the 2 sources of DDGS used in this experiment had greater SID of AA compared with previously reported values. The AID and SID of AA in low-fat DDGS were not improved by the inclusion of additional fat in the diet.

Abbreviations

AA: Amino acids; ADF: Acid detergent fibre; AID: Apparent ileal digestibility; BW: Body weight; CP: Crude protein; DDGS: Distillers dried grains with solubles; DE: Digestible energy; DM: Dry matter; ME: Metabolizable energy; NDF: Neutral detergent fibre; SID: Standardized ileal digestibility.

Competing interests

The authors declare that they have no competing interests.

Authors' contributions

All authors read and approved the final manuscript.

Acknowledgements
Financial support for this research from Poet Nutrition, Sioux Falls, SD, is appreciated.

References

1. Stein HH, Shurson GC: **Board-invited review: the use and application of distillers dried grains with solubles in swine diets.** *J Anim Sci* 2009, **87:**1292–1303.
2. Spiehs MJ, Whitney MH, Shurson GC: **Nutrient database for distillers dried grains with solubles produced from new ethanol plants in Minnesota and South Dakota.** *J Anim Sci* 2002, **80:**2639–2645.
3. NRC: *Nutrient requirements of swine.* 11th edition. Washington, DC: Natl Acad Press; 2012.
4. Jacela JY, DeRouchey JM, Dritz SS, Tokach MD, Goodband RD, Nelssen JL, Sulabo RC, Thaler RC, Brandts L, Litle DE, Prusa KJ: **Amino acid digestibility and energy content of deoiled (solvent-extracted) corn distillers dried grains with solubles for swine and effects on growth performance and carcass characteristics.** *J Anim Sci* 2011, **89:**1817–1829.
5. Kerr BJ, Dozier III WA, Shurson GC: **Effects of reduced-oil corn distillers dried grains with solubles composition on digestible and metabolizable energy value and prediction in growing pigs.** *J Anim Sci* 2013, **91:**3231–3243.
6. Anderson PV, Kerr BJ, Weber TE, Ziemer CJ, Shurson GC: **Determination and prediction of digestible and metabolizable energy for chemical analysis of corn coproducts fed to finishing pigs.** *J Anim Sci* 2012, **90:**1242–1254.
7. Li S, Sauer WC: **The effect of dietary fat content on amino acid digestibility in young pigs.** *J Anim Sci* 1994, **72:**1737–1743.
8. Albin DM, Smiricky MR, Wubben JE, Gabert VM: **The effect of dietary level of soybean oil and palm oil on apparent ileal amino acid digestibility and postprandial flow patterns of chromic oxide and amino acids in pigs.** *Can J Anim Sci* 2001, **81:**495–503.
9. Cervantes-Pahm SK, Stein HH: **Effect of dietary soybean oil and soybean protein concentrate on the concentration of digestible amino acids in soybean products fed to growing pigs.** *J Anim Sci* 2008, **86:**1841–1849.
10. Kil DY, Stein HH: **Dietary soybean oil and choice white grease improve apparent ileal digestibility of amino acids in swine diets containing corn, soybean meal, and distillers dried grains with solubles.** *Rev Colomb Cienc Pecu* 2011, **24:**248–253.
11. Stein HH, Shipley CF, Easter RA: **Technical Note: a technique for inserting a T-cannula into the distal ileum of pregnant sows.** *J Anim Sci* 1998, **76:**1433–1436.
12. NRC: *Nutrient requirements of swine.* 10th edition. Washington, DC: Natl Acad Press; 1998.
13. Stein HH, Gibson ML, Pedersen C, Boersma MG: **Amino acid and energy digestibility in ten samples of distillers dried grain with solubles fed to growing pigs.** *J Anim Sci* 2006, **84:**853–860.
14. AOAC: *Official methods of analysis.* 18th edition. Gaithersburg, MD: Association of Official Analytical Chemists; 2007.
15. Holst DO: **Holst filtration apparatus for Van Soest detergent fiber analysis.** *J AOAC* 1973, **56:**1352–1356.
16. Sanderson P: **A new method of analysis of feedstuffs for the determination of crude oils and fats. 1986.** In *Recent Advances in Animal Nutrition.* Edited by Haresign W, Cole DJA. London, UK: Butterworths; 1986:77–81.
17. Stein HH, Seve B, Fuller MF, Moughan PJ, de Lange CFM: **Invited review: Amino acid bioavailability and digestibility in pig feed ingredients: terminology and application.** *J Anim Sci* 2007, **85:**172–180.
18. Almeida FN, Stein HH: **Performance and phosphorus balance of pigs fed diets formulated on the basis of values for standardized total tract digestibility of phosphorus.** *J Anim Sci* 2010, **88:**2968–2977.
19. Almeida FN, Htoo JK, Thomson J, Stein HH: **Amino acid digestibility of heat damaged distillers dried grains with soluble fed to pigs.** *J Anim Sci Biotechnol* 2013, **4:**44–54.
20. Stein HH: *Distillers dried grains with solubles (DDGS) in diets fed to swine.* 2007. http://nutrition.ansci.illinois.edu/sites/default/files/SwineFocus001.pdf.
21. Imbeah M, Sauer WC: **The effect of dietary level of fat on amino acid digestibilities in soybean meal and canola meal and on rate of passage in growing pigs.** *Livest Prod Sci* 1991, **29:**227–239.
22. Hunt JN, Knox MT: **A relation between the chain length of fatty acids and the slowing of gastric emptying.** *J Physiol* 1968, **194:**327–336.

Evaluation of nutritive value and *in vitro* rumen fermentation gas accumulation of de-oiled algal residues

Kun Jun Han[1*] and Michael E McCormick[2]

Abstract

Background: Algae are widely recognized for their high oil content and for exponentially accumulating biomass with particular potential to provide single cell protein for human consumption or animal feed. It is believed that along with biodiesel from algae, the high protein de-oiled algal residue may become an alternative feed supplement option in the future. This study was conducted to investigate de-oiled algal residue obtained from the common *Chlorella* species, *Thalassiosira weissflogii, Selenarstrum capricornutum, Scenedesmus sp.*, and *Scenedesmus dimorphus* for assessment as potential feed supplements for ruminants by comparing with soybean (*Glycine max*) meal and alfalfa (*Medicago sativa*) hay.

Results: With the exception of *T. weissflogii*, algal residue had higher concentrations of Cu, Zn, and Mn and lower concentration of Ca, Mg, and K than soybean meal and alfalfa hay. The algal residue CP (crude protein) concentrations ranged from 140 to 445 g/kg DM and varied among the de-oiled residues. *In vitro* rumen fermentation gas accumulation curves indicated that algal biomass degradation potential was less than that of soybean meal or alfalfa hay by up to 41.7%.
The gas production curve, interpreted with a dual pool logistic model, confirmed that the fraction sizes for fast fermenting and slow fermenting of de-oiled algal residues were smaller than those in soybean meal and alfalfa hay, and the fermenting rate of the fractions was also low.

Conclusions: Inferior *in vitro* rumen gas accumulation from the five de-oiled algal residues suggests that these algal byproducts are less degradable in the rumen.

Keywords: Crude protein, De-oiled algal residue, Feed supplement, *in vitro* rumen fermentation gas, Macro mineral, Micro mineral

Background

An extended grazing period with mature grasses or summer grazing with warm-season grasses usually does not sate the requirement on protein and energy for actively growing heifers and milking cows because of low nutrient concentration and low DM intake [1]. Supplement programs boosting protein, energy, or minerals are commonly practiced in livestock operations when feeding cattle with low quality forage. Well managed supplement programs increase DM intake of forage and ultimately improve livestock performance. Byproducts from algae biomass can be utilized as a protein supplement because of high CP concentration [2]. However, an algae based feed supplement program is not yet fully available for cattle operations due to high production costs [3]. Along with biofuel production from large biomass producing field crops, liquid fuel production from algae has demonstrated great potential due to its high water use efficiency and rapid biomass accumulation ability. Although there are still technical obstacles to overcome to achieve mass production and processing of algae in a sustainable way, theoretically, it has been estimated that in Europe alone, 0.4 billion liters of diesel can be supplied through algae-originated biodiesel in the next ten to fifteen years [4]. Through the production of algae protein, which may

* Correspondence: Khan@agcenter.lsu.edu
[1]Louisiana State University Agricultural Center, School of Plant, Environmental, and Soil Sciences, 104 M.B. Sturgis Hall, Baton Rouge, LA 70803, USA
Full list of author information is available at the end of the article

become one of the most economical protein supplement options for livestock operations. In spite of algae's high protein concentration, low digestibility or lack of sulfur-containing amino acids was reasoned to cause the lower growth rate of monogastric animals fed on algae containing diet when compared with that of those fed on a casein based control diet [5]. Stability of single cell protein protected by a cellulose enriched cell membrane has been implicated as a limiting factor in low nutritional value in monogastric animals [6,2,7,4]. In contrast with testing results with monogastric animals, feeding *Spirulina platensis* as a supplement along with low CP containing guinea grass [*Panicum maximum* (synn. *Urocloa maxima*)] hay improved efficiency of microbial protein production in cattle [8]. The lack of animal performance response to high nutritive value algae byproduct necessitates more research efforts focusing on the factors limiting utilization efficiency of algae byproducts [9]. Unlike the somewhat disappointing utilization results in monogastric animal feeding trials, cell wall protected algae protein may be beneficial to ruminants due to the fiber degradation ability of rumen microbes. The objective of this research was to assess potential nutritional value of de-oiled algal residues as a ruminant feed supplement by comparing CP, minerals, and *in vitro* rumen fermentation gas accumulation characteristics with those of commonly utilized feed supplements such as soybean meal and alfalfa hay.

Methods

Preparation of de-oiled algal residue

The following algae samples were obtained in pure form from the University of Texas Culture Collection of Algae (UTEX) Austin, TX, USA: *S. capricornutum*, *Scenedesmus sp.* and *S. dimorphus*. *Thalassiosira weissflogii*, a diatom, was obtained from the Provasoli-Guillard National Center for Marine Algae and Microbiota, East Boothbay, ME, USA. Common *Chlorella* species collected in Baton Rouge, Louisiana, USA was also cultivated. All species were initially grown indoors in triplicate 1 L flasks in an incubation chamber at 25°C ± 2°C with a 12 h photoperiod illuminated to 500 μmol photons/m^2/s with fluorescent lights. Guillard's (F/2) marine water enrichment solution (Sigma-Aldrich, Cat# G0154) was used as a cultural media. For *T. weissflogii*, a marine species, Instant Ocean™ was used bring the salinity to 35 ppt. After one week, cultures were moved outdoors into 55 L vessels and grown for two additional weeks. Culture density was monitored at 8 h intervals and optimum harvest time was determined after two successive "no growth" biomass measurement recordings. The cultures were centrifuged at 3,000 rpm to concentrate the solids content to approximately 2% ds, and then further concentrated to 10% ds by centrifuging at 7,000 rpm for 10 min. Harvested algal biomass paste was dried on aluminum weigh pans in a forced-air oven at 55°C for 48 h until no further weight loss was measured. Since the study goal is to estimate feed supplement value of algae byproduct, the de-oiling process was applied to dried algae simulating the algal residue utilization after biodiesel extraction. Oil was extracted from the dried algal biomass using a SoxTec 2050 (FOSS, Eden Prairie, MN, USA) with hexane (99.9% HPLC grade) as the solvent for a 12 h reflux time. Residual solvent in biomass was evaporated under a laminar flow hood for 48 h to leave de-oiled algal residue.

Analysis of crude protein (CP), neutral detergent fiber (NDF), and mineral in samples

De-oiled algae samples, commercially available soybean meal, and alfalfa hay were ground through a 1-mm screen and analyzed for Kjeldahl N and P in an automated colorimetric assay adapted for flow-injection analyzer (QuickChem 8000 FIA, Lachat Instruments, Milwaukee, WI, USA) according to [10] procedures. The CP in samples was calculated as total N × 6.25 and NDF was determined with an Ankom Model 200 fiber analyzer (Ankom Technology, Macedon, NY, USA) using a sodium sulphite procedure [11]. Mineral concentration was determined by flame atomic absorption spectrophotometry (Perkin-Elmer Analyst 300, Norwalk, CT, USA) after dry ashing at 500°C overnight in porcelain crucibles.

Determination of coefficient for *in vitro* fermentation gas production

The protocol used in this research was approved by the Louisiana State University Agricultural Center Protocol for Animal Care and Use Committee (IACUC Approval # A2012-14). The rumen fluid was collected from a cannulated Holstein heifer fed mixed alfalfa hay and grass hay. Quadruplicate 0.50-g samples were analyzed for *in vitro* fermentation gas analysis according to [12]. The whole incubation was processed at 39°C, and gas readings for each incubation bottle were recorded at 30 min intervals using Ankom Gas Module (Ankom Technology, Macedon, NY, USA). Rate and extent of fermentation gas production were determined for each sample by fitting the gas production data to the dual pool logistic equation [13]:

$$V = V_{F1}(1 + \exp(2 + 4\ \mu_{m1}/V_{F1} \times (\lambda_1 - t)))^{-1} + V_{F2}(1 + \exp(2 + 4\ \mu_{m2}/V_{F2} \times (\lambda_2 - t)))^{-1}$$

where V is the amount of gas production at time t, and V_{F1} and V_{F2} are the final gas production volumes corresponding to complete substrate digestion for a rapidly fermenting pool and a slowly fermenting pool, respectively. μ_{m1} and μ_{m2} are the points of inflection of the gas curve for the two pools, respectively. λ_1 and λ_2 are the lag times of the two pools.

Statistical analysis

Data for various algal residues were analyzed using the Proc Glimmix of SAS version 9.2 [14]. Differences between least square means were tested using Satterthwaite approximation for the denominator degrees of freedom as an option. Quadruplicate laboratory analyses were conducted for *in vitro* gas production analysis. Four runs of gas measurements within variation less than ± 5% at 72 h incubation were used for the analysis. Data fitting a non-linear model were analyzed using Proc NLIN of SAS [14], which was developed by Dr. P. J. Weimer (personal communication).

Results and discussion

Crude protein in de-oiled algal residue

Since dried algae are bound to contain variable amounts of protein embedded by cellulosic membrane and considerable oil (up to 350 g/kg DM), it follows that the major substrate of de-oiled algal residue would consist of protein (as CP), cell content (mostly non-fibrous carbohydrate), and fibrous carbohydrate (as NDF). The NDF concentration in de-oiled algal residues ranged from 82 to 180 g/kg DM (Figure 1). The NDF in soybean meal used as one of the control feed supplements averaged 106 g/kg DM which did not differ from the NDF concentrations in de-oiled algal residues while the NDF in alfalfa hay was 450 g/kg DM.

There was a substantial variation in CP concentrations among the de-oiled algal residues (Figure 1). The CP concentrations in *S. capricornutum*, *Scenedesmus sp.* and *S. dimorphus* were equivalent to that in soybean meal. The CP concentration in *Chlorella* residue was between that of soybean and alfalfa hay ($P < 0.01$). The CP in *T. weissflogii* was similar to that in alfalfa hay. As the present study and Piorreck et al. [2] pointed out, protein concentrations of most algal residues are above 300 g/kg DM, indicating that tested algal residues may be considered as high protein supplements.

Besides cellulosic membrane embedded protein issue, [7] pointed out that around 10% of *Scenedesmus* and *Spirulina* DM was non-protein nitrogen which is not an available protein form for monogastric animals. Therefore, the determination of CP based on total nitrogen in algae DM resulted in overestimation of available protein for maintenance and growth in monogastric animals, but this form of N is still effective in ruminants as a N source for generation of rumen microbial protein. The benefit of successful byproduct supplement programs can be realized through providing greater RUP (rumen un-degradable protein) in a cost effective manner [15]. Increasing RUP in a diet of early lactation stage cows could produce more milk without additional grain intake [16]. As Becker [7] pointed out, with the cellulosic cell wall component as the most limiting factor for algae protein digestion and utilization, algae protein embedded in the cellulosic membrane may provide more feeding value to ruminant animals than monogastric animals.

Mineral concentration in de-oiled algal residue

Mineral concentration aspects in *T. weissflogii* were unique from other algal residues. De-oiled residue from *T. weissflogii* contained greater concentrations of Ca, Mg, Cu, Zn, and Mn than those in two conventional feed supplements. Comparing with the minerals in tested algal residue, *T. weissflogii* contains much greater concentrations of Ca, Mg, K, and Zn. Generally, the Ca in other de-oiled algal residue was approximately half of that in soybean meal or one fourth of that in alfalfa hay

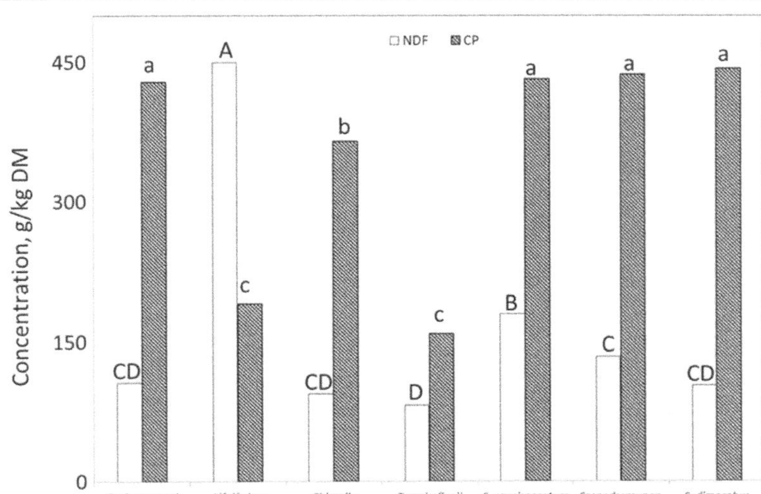

Figure 1 NDF and CP in soybean meal, alfalfa hay, and de-oiled algae residues. Within neutral detergent fiber (NDF) or within crude protein (CP), bars without a common letter differ ($P < 0.05$).

(Table 1). Phosphorus concentrations did not show as much variation as Ca in the five de-oiled algal residues and the two conventional protein supplements. Commonly, in the algal residues and conventional feed supplements, the P concentration converged around 1 g/kg DM. The relatively high P concentration compared with Ca in algal residues may require feeding extra Ca to assure a proper Ca-P ratio (2:1) for high performing ruminants [17]. The Mg concentration in the algal residues was less than that of soybean meal or alfalfa hay except in *T. weissflogii*. Again, except for *T. weissflogii*, the K concentrations in four algal residues were below detectable limits.

Micro mineral concentrations such as Cu, Zn, and Mn in de-oiled algal residues were greater than those of soybean meal or alfalfa hay (Table 1). Except for P, generally low macro mineral concentration but high concentrations of Cu, Zn, and Mn may be of concern for use of tested algal residues in animal nutrition. High Cu in algal residue may be a particular concern with small ruminants and calves because of relatively low tolerant levels appearing in [18]. Also, the high Mn concentration in algal residue may also be a concern for potential Mn toxicity [19].

In vitro rumen fermentation gas accumulation of de-oiled algal residue

Accumulated *in vitro* fermentation gas from algal residues were only 40 to 67% of that from soybean meal at 24 h and the fermentation gas accumulation curves displayed a nonlinear trend over the incubation period (Figure 2). Gas accumulation of soybean meal and alfalfa hay demonstrated distinctively different patterns from de-oiled algal residues. After the first 3 h or so of fermentation, gas production values for two control supplements and de-oiled algal residues diverged. The accumulated gas production from soybean meal and alfalfa hay reached asymptote around 18 to 20 h of incubation.

Based on the assumption of a similar cell wall material and CP concentrations in soybean meal, *S. capricornutum*,

Table 1 Mineral concentration of de-oiled algal residues, soybean meal, and alfalfa hay

Supplement	Ca	P	Mg	K	Cu	Zn	Mn
	g/kg DM				mg/kg DM		
Soybean meal	0.59^{c}[†]	0.70^{c}	0.23^{b}	2.43^{b}	14^{c}	49^{c}	27^{d}
Alfalfa	1.45^{b}	0.19^{d}	0.29^{b}	3.15^{a}	10^{d}	17^{d}	53^{d}
Chlorella	0.21^{e}	1.17^{a}	0.11^{c}	ND[‡]	35^{b}	70^{b}	238^{c}
T. weissflogii	7.41^{a}	0.92^{b}	0.74^{a}	2.38^{b}	44^{b}	422^{a}	430^{b}
S. capricornutum	0.32^{d}	1.00^{ab}	0.14^{c}	ND	90^{a}	79^{b}	497^{a}
Scenedesmus sp.	0.27^{de}	1.06^{ab}	0.17^{c}	ND	93^{a}	78^{b}	466^{ab}
S. dimorphus	0.27^{de}	1.11^{a}	0.15^{c}	ND	77^{a}	75^{b}	431^{b}

†Within a column, means without a common superscript differ ($P < 0.05$).
‡ND, below detectable limit.

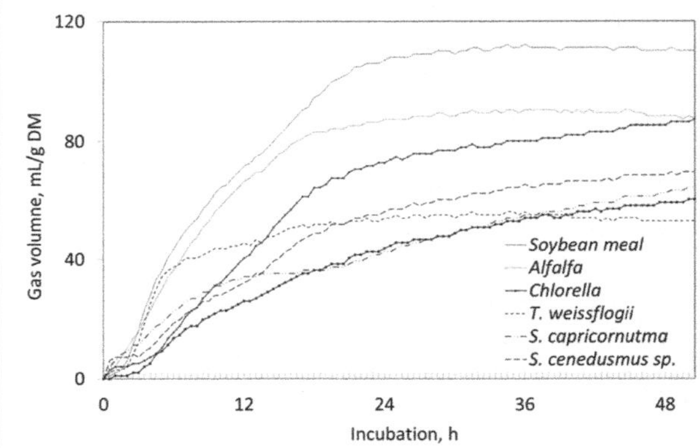

Figure 2 ***In vitro*** **rumen fermentation gas accumulation curves of tested substrates.**

Scendesmus sp., and S. dimorphus, the low gas production from algal residue may indicate less fermentable characteristics of cellulosic structure in the algae cell membrane than in soybean meal. Because gas production maintains a high correlation with volatile fatty acid production from rumen fermentable carbohydrate components in feed samples [20], lower gas production from algal residues than from soybean meal evidenced less fermentable algae cell membrane.

T. weissflogii algal residue demonstrated a distinct pattern from other tested algal residues with an early steep rise of fermentation gas accumulation and approached asymptote after a 5 h incubation. Among the tested algae, *T. weissflogi* grows mainly in a marine environment and in some inland water environments. The different features shown in mineral concentration and gas production in *T. weissflogii* from other tested algae may reflect some unique characteristics of this species related to evolution and growth environment. *Scenedesmus sp.* and *S. dimorphus* accumulated fermenting gas more steadily. The gas curve pattern of *Chlorella* was more similar to soybean meal or alfalfa hay even with less accumulation of fermentation gas. Among the de-oiled algal residues, *Chlorella* produced the most fermentation gas. The de-oiled *S. capricornutum*, *Scenedesmus* sp., and *S. dimorphus* produced almost identical quantities of gas accumulation after a 36 h incubation.

The logistic model consisting of a fast fermenting fraction and a slow fermenting fraction, as determined by the accumulated fermentation gas, could quantify several coefficients of digestion kinetics in Table 2. The size of the fast fermenting fraction of algal residues of *Chlorella* was numerically the greatest but not significantly different from that of *T. weissflogii, Scenedesmus*, and the two control supplements. The fast fermenting fraction of *S. capricornutum* and *S. dimorphus* were smaller than that of *Chlorella* ($P < 0.05$). As anticipated, the fermentation gas accumulation rate from the fast fermenting fraction was greatest in soybean meal, followed by alfalfa hay and *Chlorella* ($P < 0.01$). The gas accumulation rate from the

Table 2 *In vitro* rumen fermentation gas accumulation kinetics of de-oiled algal residues, soybean meal, and alfalfa hay

Item	Control		De-oiled algal residue				
	Soybean meal	Alfalfa hay	Common chlorella	*T. weissflogii*	*S. capricornutum*	*Scenedesmus sp.*	*S. dimorphus*
	Coefficient of *in vitro* ruminal assay						
Fast fermenting fraction, mL/g DM	57.5^{ab†}	53.3^{ab}	70.0^{a}	53.1^{ab}	34.5^{b}	47.6^{ab}	22.7^{b}
Fast rate, mL/h	5.00^{a}	2.81^{b}	2.37^{b}	1.83^{c}	1.55^{c}	1.47^{c}	1.29^{c}
Fast lag, h	1.78^{a}	1.42^{a}	4.47^{a}	0.20^{a}	2.01^{a}	3.34^{a}	3.01^{a}
Slow fermenting fraction, mL/g DM	63.5^{a}	38.0^{b}	22.1^{bc}	17.3^{c}	31.2^{b}	37.4^{b}	37.0^{b}
Slow rate, mL/h	2.15^{a}	1.82^{a}	0.27^{b}	0.33^{b}	0.38^{b}	0.34^{b}	0.60^{b}
Slow lag, h	6.9^{a}	3.3^{a}	8.9^{a}	3.1^{a}	16.0^{a}	14.1^{a}	7.6^{a}

[†]Within a row, means without a common superscript differ ($P < 0.05$).

slow fermenting fraction demonstrated a clearer separation of coefficients between the control supplements and the five algal residues. The fermentation rates of the slow fermenting fractions of soybean meal and alfalfa hay in a rumen were up to eight times greater than those of de-oiled algal residues. The lag times for both fractions were not significantly different even with substantial differences among algal residue and with the two control supplements due to variations in individual batches.

Conclusions

Comparison of fermentation behaviors of de-oiled algal residue with soybean meal and alfalfa hay indicated that carbohydrate in algal residues is less rumen fermentable than those in soybean meal or alfalfa hay. De-oiled algal residue may have potential as a high protein feed supplement which may supply a moderate amount of N for rumen microbial protein production. However, more research efforts are required for partitioning algal proteins based on rumen degradability. Mineral concentrations, particularly several essential trace minerals, are high, and some caution may be required when used for small ruminants and calves.

Abbreviations

CP: Crude protein; NDF: Neutral detergent fiber; DM: Dry matter.

Competing interests

The authors declare that they have no competing interests.

Authors' contributions

KJ conducted the study and data interpretation, drafted the initial version of the manuscript. KJ and MM revised the manuscript. Both authors read and approved the final manuscript.

Acknowledgements

This study was supported by Louisiana Board of Regents Research grant. Authors thanks N. Lemoine and C. Lohrey for sharing dried algae biomass, and Drs. P. Weimer and R. Muck for technical support on digestion kinetics model.

Author details

[1]Louisiana State University Agricultural Center, School of Plant, Environmental, and Soil Sciences, 104 M.B. Sturgis Hall, Baton Rouge, LA 70803, USA. [2]Louisiana State University Agricultural Center, Southeast Region Office, 21549 Old Covington, Hammond, LA 70403, USA.

References

1. Han KJ, McCormick ME, Walz R: **Forage Quality Results from Louisiana and Mississippi producer's Forage Samples, 1999–2006.** In *Southeast Research Station Field Day Summaries, 2007.* Baton Rouge, LA: Louisiana State University Agricultural Center; 2007:35–37.
2. Piorreck M, Baasch K, Pohl P: **Biomass production, total protein, chlorophylls, lipids and fatty acids of freshwater green and blue-green algae under different nitrogen regimes.** *Phytochemistry* 1984, **23:**207–216.
3. Beley A, Kato T, Ota Y, Spirulina (Arthropira): **Potential application as an animal feed supplement.** *J Appl Phycol* 1996, **8:**303–311.
4. Wijffels RH, Barbosa MJ: **An outlook on microalgal biofuels.** *Science* 2010, **329:**796–799.
5. Leveille GA, Sauberich HE, Shockley JW: **Protein value and the amino acid deficiencies of various algae for growth of rats and chicks.** *J Nutr* 1962, **76:**423–428.
6. Domozych DS, Stewart KD, Mattox KR: **The comparative aspects of cell wall chemistry in the green algae (cholorphyta).** *J Mol Evol* 1980, **15:**1–12.
7. Becker EW: **Micro-algae as a source of protein.** *Biotechol Adv* 2007, **25:**207–210.
8. Panjaitan T: **Strategies to Enhance Efficiency of Microbial Protein Production in Cattle Consuming Tropical Forages.** In *Ph.D. Thesis.* Brisbane, Australia: Schools of Animal Studies and Veterinary Science, University of Queensland; 2010.
9. Lum KK, Kim J, Lei XG: **Dual potential of s as a sustainable biofuel feedstock and animal feed.** *J Anim Sci Biotechnol* 2013, **4:**1–7.
10. AOAC: *Official Methods of Analysis.* 15th edition. Arlington, VA: Assoc Offic Anal Chem; 1990.
11. Robertson JB, Van Soest PJ: **The Detergent System of Analysis and its Application to Human Foods.** In *The Analysis of Dietary Fiber.* Edited by James WPT, Theander O. New York: Marcell Dekker; 1981:123–158.
12. Goering HK, Van Soest PJ: *Forage Fiber Analysis (Apparatus, Reagents, Procedures and Some Applications), Agricultural Handbook no. 379.* Washington, D.C: ARS-USDA; 1970.
13. Schofield P, Pitt RE, Pell AN: **Kinetics of fiber digestion from in vitro gas production.** *J Anim Sci* 1994, **72:**2980–2991.
14. SAS Institute: *SAS/STAT 9.2 User's Guide.* 2nd edition. Cary, NC: SAS Inst; 2009.
15. Clark JH, Murphy MR, Crooker BA: **Supplying the protein needs of dairy cattle from by-product feeds.** *J Dairy Sci* 1987, **70:**1092–1109.
16. McCormick ME, French DD, Brown TF, Cuomo GJ, Chapa AM, Fernandez JM, Beatty JF, Blouin DC: **Crude protein and rumen undegradable protein effects on reproduction and lactation performance of Holstein cows.** *J Dairy Sci* 1999, **82:**2697–2708.
17. National Research Council: *Nutrient Requirement of Dairy Cattle.* 7th edition. Washington, D.C: Nat. Acad. Sci; 2001.

18. Gupta UC, Gupta SC: **Trace element toxicity relationships to crop production and livestock and human health.** *Commun Soil Sci Plant Anal* 1998, **29:**1491–1522.
19. National Research Council: *Mineral Tolerance of Domestic Animals.* Washington, D.C: Nat Acad Sci; 1980.
20. Getachew G, Robinson PH, DePeteers EJ, Taylor SJ: **Relationships between chemical composition, dry matter degradation, and in vitro gas production of several ruminant feeds.** *Anim Feed Sci Technol* 2004, **111:**57–71.

16

Effects of creep feeding and supplemental glutamine or glutamine plus glutamate (Aminogut) on pre- and post-weaning growth performance and intestinal health of piglets

Rafael A Cabrera[1,6], James L Usry[2,7], Consuelo Arrellano[3], Eduardo T Nogueira[4], Marianne Kutschenko[4], Adam J Moeser[5] and Jack Odle[1*]

Abstract

Background: Creep feeding is used to stimulate piglet post-weaning feed consumption. L-Glutamine (GLN) is an important source of fuel for intestinal epithelial cells. The objective of this study was to determine the impact of creep feeding and adding GLN or AminoGut (AG; containing glutamine + glutamate) to pre- and post-weaning diets on pig performance and intestinal health. Litters (N = 120) were allotted to four treatments during 14–21 d of lactation: 1) No creep feed (NC, n = 45); 2) creep fed control diet (CFCD, n = 45); 3) creep fed 1% GLN (CFGLN, n = 15); 4) creep fed .88% AG (CFAG, n = 15). After weaning, the NC and CFCD groups were sub-divided into three groups (n = 15 each), receiving either a control nursery diet (NC-CD, CFCD-CD) or a diet supplemented with either GLN (NC-GLN, CFCD-GLN) or with AG (NC-AG, CFCD-AG). Litters that were creep fed with diets containing GLN or AG also were supplemented with those amino acids in the nursery diets (CFGLN-GLN, CFAG-AG). Glutamine was added at 1% in all three post-weaning diet phases and AG was added at .88% in phase 1 and 2 and at .66% in phase 3.

Results: Feed conversion (feed/gain) showed means among treatment means close to significance ($P = 0.056$) and Tukey's test for pairwise mean comparisons showed that Pigs in the CFGLN-GLN group had the best feed conversion (feed/gain) in the first three-week period post-weaning, exceeding ($P = 0.044$) controls (CFCD-CD) by 34%. The NC-AG group had ($P = 0.02$) the greatest feed intake in the last three week of the study, exceeding controls (CFCD-CD) by 12%. CFGLN-GLN, CFCD-GLN and sow reared (SR) pigs had the greatest ($P = 0.049$) villi height exceeding the CFCD-AG group by 18%, 20% and 19% respectively. The CFAG-AG group had the deepest ($P = 0.001$) crypts among all treatments. CFGLN-GLN, CFCD-GLN and SR groups had the greatest ($P = 0.001$) number of cells proliferating (PCNA) exceeding those in the NC-CD group by 43%, 54% and 63% respectively. Sow reared pigs showed the greatest ($P = 0.001$) intestinal absorption capacity for xylose and mannitol.

Conclusion: Supplementation of creep feed and nursery diets with GLN and/or AminoGut in the first three week improved feed conversion possibly due to improved intestinal health.

Keywords: Creep feeding, Glutamine, Growth performance, Intestinal health, SEM, Villi

* Correspondence: Jack_Odle@ncsu.edu
[1]Laboratory of Developmental Nutrition, Department of Animal Science, North Carolina State University, 101 Polk Hall, North Carolina State University, Raleigh, NC 27695, USA
Full list of author information is available at the end of the article

Background

After pigs are weaned from their dams, morphological and functional changes occur in their small intestine. Pluske et al. [1] reported decreased villi height and increased crypt dept. Because newly-weaned pigs are transitioned from milk to dry feed, the pig's intestinal tract is unable to fully digest and absorb the more complex plant-based macronutrients in the feed. Various researchers [2,3] have reported that this accumulation of undigested and unabsorbed feed creates the perfect medium for opportunistic bacteria such as haemolytic *E. coli* to grow. The normal weaning process stimulates pancreatic development and its enzymatic output; however there is a delay until the different enzymes reach sufficient levels [4]. This in turn can cause post-weaning diarrhea. Creep feeding is deemed to be very important during the suckling period for swine practitioners because it (a) increases weaning weight when offered in small and frequent quantities and (b) eases the transition period for the piglets from sow's milk to the dry feed. The latter has a physiological implication in order to avoid digestive upset such as post-weaning diarrhea and poor growth. Some argue [5,6] that the use of creep feed during the suckling period could potentially trigger hypersensitivity to feed antigens that can stimulate post-weaning diarrhea. Barnett et al. [7] observed antibody titers in the blood of weaned piglets and confirmed that feed antigens can induce an immune reaction in creep-fed pigs. The reduction in feed intake associated with weaning has been known to affect intestinal integrity and potentially cause pathological disorders. Klasing [8] argued that dietary supplementation of some nutrients or immune modulators can rectify the intestinal impairment and modulate the immune function of animals contributing to improvements in overall health and performance. Nutrition can regulate the type of immune response by a number of mechanisms [8]. Swine nutritionists have traditionally focused on those amino acids that cannot be synthesized by the animals with little attention given to those that can be synthesize by the animals and yet have a great impact on regulating nutrient metabolism and the immune responses [9,10]. These amino acids include arginine, glutamine, glutamate, proline, leucine, cysteine and tryptophan. Recent studies indicate that these amino acids serve important regulatory functions in nutrient metabolism, protein turnover, and immune function, thereby enhancing growth and feed efficiency in pigs. The underlying mechanisms include activation of nitric oxide, mammalian target of rapamycin (mTOR), gaseous signaling, and AMP-activated protein kinase pathways as well as anti-oxidative [11].

Glutamine is a major metabolic fuel for rapidly dividing cells, including enterocytes and lymphocytes, as well as a key regulator of gene expression and cell signaling pathways [12]. Schrock and Goldstein [13] reported that glutamine serves as precursor for the increased renal ammoniagenesis during chronic metabolic acidosis. The amide nitrogen of glutamine is essential for purine and pyrimidine biosynthesis.

Glutamine has important and unique metabolic functions, and it is considered a conditionally essential amino acid in some species under inflammatory conditions [14] and disease states [15,16]. Souba and others [17] have indicated that the provision of GLN-enriched diets in various stress states associated with bacterial translocation decreases the incidence of translocation of bacteria by decreasing the adherence of bacteria to enterocytes. Reeds and others [18] argued that the high metabolic rate of the intestinal mucosa is very unique when compared to the others organs in the body. First, the enterocytes are specialized in absorbing nutrients from the lumen to the basal lamina. Second, mucosa cells are presented with high quantities of substrates from both the intestinal lumen and the mesenteric arterial circulation. Accordingly, under fed conditions, the quantification of substrate used by the gut can be challenging to quantify given the dual supply from both diet and the arterial circulation. Finally, GLN is the only amino acid in arterial blood that is taken up by the small intestine in swine [19]. The small intestine (even though only represents 3 to 4% of the body weight) utilizes 30% of the arterial GLN and 67% of dietary GLN in swine. For comparison, 95 to 97% of dietary glutamate is extracted by the pig small intestine in first pass [20,21] but only 50% is metabolized to CO_2 [18].

Because the gastrointestinal tract has an obligatory requirement for L-GLN [18] and availability of L-GLN from endogenous tissue production may not be sufficient for the maintenance of the structural and functional integrity of the intestinal mucosa [22,23], We hypothesized that creep feeding of suckling piglets and adding L-glutamine or the combination of L-glutamine and L-glutamate to pre and/or post-weaning diets would alleviate villi atrophy, reduce post-weaning diarrhea and improve post-weaning growth.

The objective of the study was to evaluate the effects of L-GLN and AminoGut (containing L-GLN and L-glutamate) on intestinal histology, intestinal absorptive capacity, enzymatic activity, and growth performance in a commercial swine operation. The effects of these supplements on piglet growth performance have not been evaluated during the whole nursery period following supplementation during the pre-weaning period.

Methods

All protocols were under the supervision of licensed veterinarians. Standard operating procedures for animal care were in accord with published guidelines for animal care [24]. The experimental animals were not subjected to prolonged constraint or surgical procedures and were humanely treated throughout the experiment. The study was conducted during the summer 2010 on a 4800-sow farm in Maple Hill, NC (Murphy-Brown, LLC; Rose Hill, NC). One-hundred and twenty litters were randomly allotted to one of eight dietary treatment scenarios (Figure 1). At one week prior to weaning, four creep-feed treatments were initiated: A) No creep feed; B) Creep feed, control diet; C) Creep feed containing 1% GLN; D) Creep feed containing 0.88% AminoGut. AminoGut is a commercial dietary supplement produced by Ajinomoto do Brazil (São Paulo, Brazil) containing a mixture of L-glutamine (min 10%) and L-glutamate (min 10%). The pelleted creep feed was a phase 1 nursery diet (Table 1), manufactured at the North Carolina State University feed mill. Litters were offered fresh creep feed at 4-h intervals from 8 am to 4 pm each day. Litters weights were recorded at birth (WayPig model 252, Raytec Manufacturing, Ephrata, PA) and weaning (Avery Weight-Tronix model 640, www.agscales.com, Fairmont, MN). Pigs were weaned at an average of 21 d and were transported to the Site 2 nursery (~ 300 meters from farrowing Site 1), and litters from preweaning treatments A and B were each divided among the following nursery diets: 1) Control diet, 2) GLN diet, and 3) AminoGut diet (Figure 1, Table 1). Litters from preweaning treatments C and D were continued on similar diets post-weaning (e.g., GLN and AminoGut respectively). Additional litters were allowed to nurse the sow (without creep feed) until 4 week of age to provide age-matched controls for invasive measures of intestinal health (n = 7). The experimental design is illustrated in Figure 1, showing abbreviations used for each dietary treatment scenario. After weaning, phase 1 diet (Table 1) was budgeted at 2.72 kg /pig, Phase 2 diet was budgeted at 5.45 kg/pig and Phase 3 diets was budgeted at 18.16 kg /pig. At 3 and 6 week postweaning, the pigs and feeders were weighed for growth and feed conversion calculations.

At one week post weaning, one pig per pen was fasted overnight and then intra-gastrically gavaged with a D-xylose/mannitol solution as follows. A solution containing 0.2 g/L of D-xylose (Pfizer, N.Y., NY) and 0.3 g/L of mannitol (Sigma, Saint Louis, MO) was prepared and was given to the pigs on an average of 9.5 h after fasting. The selected dose was 6.5 mL/kg of body weight. Pigs were individually weighed (Berkley FS-50 hanging scale, Somers Point, NJ). At precisely two h post gavage, pigs were bled via jugular venipuncture. The time of bleeding was selected based upon work by Doerfler et al. [25]. After pigs were bled they were humanly euthanized for collection of intestinal tissues. Jejunum samples (25 cm from the stomach) were collected for both light and scanning electron microscopy (SEM). The portion cut specifically for SEM, was cut open and laid flat in a small cartridge in order to obtain a better picture of the intestinal villi. A separate jejunum tissue sample was cut and the intestinal mucosa was scrapped for maltase activity analysis. Blood was centrifuged after 24 h and the serum stored at −20°C for further analysis. Performance data were statistically analyzed using the PROC GLIMMIX of SAS with birth weight and weaning age as covariates. Intestinal and serum metabolite data were analyzed using the Mixed Procedure of SAS with

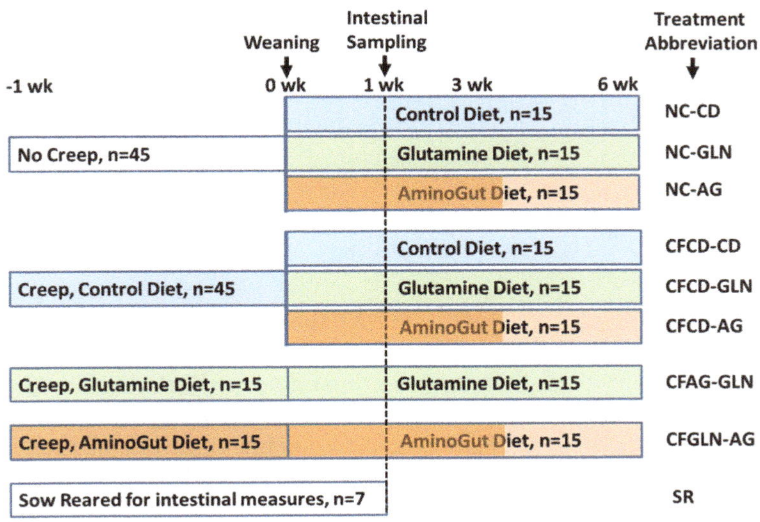

Figure 1 Experimental design showing four pre-weaning creep diet groups and eight post-weaning diet groups together with sow-reared control pigs. Creep feed was initiated 1 week prior to weaning and consisted of pelleted Phase 1 diets. Post-weaning diets consisted of either control basal diet (see Table 1) with additions of either 1% L-GLN (in all feed Phases 1–3) or 0.88% AminoGut in Phases 1 and 2 and 0.66% in Phase 3. Pigs were weighted at birth, weaning, (some at 1 wk post-weaning), 3 wk and 6 wk post-weaning. Selected pigs were euthanized (n = 7/trt) at 1 wk post-weaning for intestinal health measurements. Various treatment abbreviations are also summarized.

Table 1 Ingredients and nutrient composition of the basal diets (Phase 1, 2 and 3)[1]

Ingredients, %	Phase 1	Phase 2	Phase 3
Corn	39.00	51.36	63.73
Soybean meal, 48%	20.00	25.00	30.00
Choice white grease	3.00	3.00	3.00
Dicalcium	0.00	0.60	1.30
Limestone	0.55	0.60	0.50
Salt	0.30	0.30	0.50
l-Lysine	0.25	0.27	0.39
DL-Methionine	0.21	0.18	0.18
L-Threonine	0.16	0.16	0.19
L-Tryptophan	0.02	0.02	0.00
Granular whey	5.00	0.00	0.00
Porcine plasma protein	4.00	2.00	0.00
Dairy Lac 80	20.00	10.00	0.00
Fish meal select	6.00	5.00	0.00
ZnO	0.30	0.30	0.00
Copper sulfate	0.05	0.05	0.05
Vitamins Mix[2]	0.01	0.01	0.01
Trace mineral Mix[3]	0.15	0.15	0.15
Mecadox	1.00	1.00	0.00
Total	**100.00**	**100.00**	**100.00**
Calculated nutrients			
ME (Mcal/kg)	3.52	3.46	3.46
CP, %	21.67	21.77	20.12
TID Lysine, % [4]	1.39	1.35	1.25
TID Met, %	0.52	0.50	0.45
TID M + C, %	0.83	0.81	0.75
TID Threonine, %	0.93	0.88	0.82
TID Tryptophan, %	0.25	0.25	0.21
TID Isoleucine, %	0.77	0.78	0.73
TID Valine, %	0.93	0.91	0.82
Glutamate, %	3.69	3.79	3.62
Total P, %	0.64	0.67	0.63
Available P, %	0.43	0.40	0.30
Ca, %	0.80	0.80	0.60
Na, %	0.32	0.21	0.21
Lactose, %	19.50	8.00	0.00

[1]L-Glutamine was supplemented at 1% of the diet at the expense of corn. Amino-Gut was supplemented 0.8% in phase 1 and 2; and at 0.6% in phase 3.
[2]Vitamin Mix contained per kg of premix: vitamin A Acetate 4,231,292 IU; vitamin D_3 603,174 IU; vitamin E ascorbate 24,172 IU; vitamin B_{12} 14.97 mg; riboflavin spray dried 3,023 mg; niacin 18,140 mg; d-pantothenic 12,093 mg; menadione 1,995 mg; folic acid spray dried 907 mg; biotin spray dried 121 mg.
[3]Trace mineral Mix contained: ferrous sulfate 11.02%; copper sulfate 1.10%; ethylenediamine dihydriodine 198 ppm; manganese oxide 2.64%; sodium selenite 198.4 ppm; zinc sulfate 11.02%.
[4]Total ileal digestible lysine.

the body weight (one week post-weaning in the nursery) used as covariate.

Scanning electron microscopy

Jejunum samples were collected from four-week old piglets (1 wk post-weaning) and immersed in 4 F:1G fixative containing 4% formaldehyde and 1% glutaraldehyde in a phosphate buffer, with an osmolarity of 176 mOsM and a pH of 7.2-7.4 [26]. Samples were cut to be between 2–3 mm in order to minimize chances of bulk charging. Samples were rinsed in 0.1 mol/L phosphate buffer and dehydrated in an ethanolic series to 100% ethanol before subjection to critical point drying after being stored for approximately 7 wk in the 4 F: 1G fixative. Samples were then mounted on SEM stubs with carbon tape and sputter coated with gold-palladium before being viewed with a JEOL JSM-6360LV scanning electron microscope (JEOL, Peabody, MA). This microscope is a fully digital instrument that can view specimens by secondary electron imaging (SEI), backscatter electron imaging (BEI), at high vacuum, or at low vacuum.

Hematoxylin & eosin staining

Jejunum samples were collected (25 cm from the stomach) and preserved in a formalin solution and stored in room temperature for histology analysis. Tissues were trimmed into five millimeter thick sections and placed in processing cassettes. The tissues were processed in a Tissue-Tek VIP5 tissue processor (Sakura Finetek, Torrance, CA) using a standard overnight processing schedule. Tissues were embedded in paraffin and five micron sections were mounted on glass slides. The slides were stained on a DRS-601 slide stainer (Sakura Finetek, Torrance, CA) with hematoxylin and eosin, cleared and mounted with a permanent media. The stained tissues on glass slides were examined using an Olympus AH-2 Vanox-S microscope (Ultrasonic Power Corporation, Freeport, IL) and measured using SPOT™ software (SPOT™ Imaging Solutions, Sterling Heights, MI).

PCNA Staining

Five micron jejunal slices were mounted on glass slides. A primary mouse monoclonal antibody (PC10) was used for the as a proliferation marker. This antibody is specific for proliferating cell nuclear antigen, PCNA, p36 protein expressed at high levels in proliferating cells. It was diluted at 1:1,500 and incubated for 30 min. The remaining steps were completed using the Dako EnVision Mouse kit (Dako, Denmark). Intensively stained and the total number of enterocytes were counted in 8 consecutive well-orientated crypts (those that extended to the muscularis mucosa).

Analysis of mannitol

Samples of serum were frozen, thawed at room temperature and vortexed to mix. Samples were then filtered by centrifugation using Ultrafiltration Spin Columns (0.45 μm, Millipore, Temecula, CA). An aliquot of 200 μL of sample was transferred to HPLC autosampler vials containing 250 μL inserts. An internal standard solution of myo-inositol was added (2 μL). Analysis was done using High Performance Liquid Chromatography (HPLC). The extracts were analyzed using a Dionex BioLC (Dionex Corporation, Sunnyvale, CA) at a controlled temperature of 30°C. The system consisted of a gradient pump, an autosampler, and a pulsed amperometric detector. The mobile phase was 52 mmol/L NaOH (Thermo-Fisher Chemical Corp. Pittsburgh, PA) at an isocratic flow rate of 1.0 mL/min. The column used was a Dionex PA-10, 250 mm length and 4 mm i.d., fitted with Dionex PA-10 and borate guard columns. The detector was programmed to run a quadruple waveform as recommended by the manufacturer. A shift in the detector range was 1 μC. The injection volume was 10 μL. The mannitol was calculated using an authentic standard of d-mannitol and myo-inositol as an internal standard. All the reference standards were purchased from Sigma Chemical Corp (St. Louis, MO).

Analysis of xylose

The collected pig serum (20 μL) was subjected to a modified micro method [27,28] first described by Eberts et al. [29] for determination of plasma D-xylose. To each 20 μL plasma sample, 2 mL of phloroglucinol (Sigma Chemical Co., Saint Louis, MO 63178–9916) color reagent was added and heated for 4 min at 100°C. The samples were allowed to cool at room temperature in a water bath. After cooling, the absorbance of each sample was read on a Gilford UV–vis spectrophotometer (Thermo Fisher Scientific, Inc.; Waltham, MA) set at 554 nm.

Maltase enzyme activity

The maltase assay was performed as described by Dahlqvist [30]. Maltase activity (U/g of protein) was expressed as units, with 1 unit defined as the amount of enzyme transforming 1.0 μmol of substrate per min at 25°C.

Results

Results for pre-weaning performance are summarized in Table 2. We found weaning age to be significant (P = 0.001) among the pre-wean treatments. For subsequent comparisons, this variable was used as covariate. Pigs/litter, sow parity, birth weight, weaning weight, and mortality were not different among the treatments. Creep feed consumption also did not differ for those treatments receiving creep feed. The average creep feed consumptions for the control diet and those supplemented with either glutamine or AminoGut were 49.44, 45.57 and 48.44 g/pig respectively. We did not find an effect of creep feeding on post-weaning performance (Tables 3 & 4). A longer (> 1 wk) creep feed period needs to be examined. Feed conversion (feed/gain) showed means among treatments close to significance ($P = 0.056$) and Tukey's test for pairwise mean comparisons showed that Pigs in the CFGLN-GLN group had the best feed conversion (feed/gain) in the first three-week period post-weaning, exceeding ($P = 0.044$) controls (CFCD-CD) by 34%. All others variables were not significant during this post-weaning period among the treatments (Table 3). The NC-AG group had ($P = 0.02$) the greatest feed intake among all treatments in the last three week of the study (Table 4), exceeding controls (CFCD-CD) by 12%. All others variables were not significant during this post-weaning period among the treatments. CFCD-GLN, Sow-Reared and CFGLN-GLN groups had the greatest ($P = 0.049$) villi height exceeding those which were creep fed with a control diet and later supplemented with AminoGut (CFCD-AG) by 20%, 19% and 18% respectively (Table 5). The Sow-Reared group was added as a point of reference against the other treatments. All tissue samples for all treatments were taken at 28 d of age. We also found that pigs creep fed with a diet supplemented with AminoGut and fed a post-weaning diet supplemented with AminoGut (CFAG-AG) had the deepest ($P = 0.001$) crypts among all the treatments. Sow-Reared, CFCD-

Table 2 Pre-weaning performance of creep-fed piglets

Variables	No-Creep	Creep	Creep + Glutamine	Creep + AminoGut	SEM	P value
No. litters	45	45	15	15	-	-
Wean age, d	21.5	20.5	20.3	18.9	0.13	0.001
Pigs/litter	12.5	12.4	12.5	11.9	0.23	0.286
Sow parity	3.4	3.4	3.4	3.4	0.20	0.998
Birth wt, kg/pig	1.5	1.4	1.4	1.5	0.04	0.250
Wean wt, kg	6.5	6.4	6.4	6.5	0.26	0.980
Creep feed, g/pig[1]	-	49.4	45.6	48.4	2.93	0.617
Mortality, pigs/litter	1.0	1.2	1.5	1.2	0.22	0.488

[1]Creep feed was offered for one week prior to weaning.

Table 3 Pig performance from week 1 to 3 post-weaning

Variables	NC-CD[1]	NC-GLN	NC-AG	CFCD-CD	CFCD-GLN	CFCD-AG	CFGLN-GLN	CFAG-AG	SEM	*P* value
No. pens	6	6	6	6	6	6	6	6	-	-
No. pigs	144	144	144	144	144	144	144	144	-	-
Initial wt, kg	6.50	6.35	6.59	6.14	6.39	6.61	6.42	6.52	0.34	0.984
Final wt, kg	10.4	10.20	10.66	9.51	10.06	10.08	10.65	10.34	0.26	0.128
ADG, kg/pig	0.25	0.24	0.25	0.19	0.23	0.22	0.26	0.24	0.02	0.124
Intake, kg/pig	0.33	0.28	0.30	0.27	0.27	0.28	0.29	0.29	0.02	0.644
Feed/gain	1.35^{bc}	1.17^{ab}	1.17^{ab}	1.48^{c}	1.15^{ab}	1.30^{b}	1.11^{a}	1.21^{ab}	0.09	0.056
Mortality	0.38	1.43	0.94	1.28	0.80	0.63	0.82	0.73	0.37	0.562
Removed	1.23	0.90	2.00	2.49	1.28	1.03	1.81	0.59	0.70	0.584

[1]*NC-CD* No Creep Control Diet, *NC-GLN* No Creep Glutamine, *NC-AG* No Creep AminoGut, *CFCD-CD* Creep Fed Control Diet-Control Diet. *CFCD-GLN* Creep Fed Control Diet-Glutamine, *CFCD-AG* Creep Fed Control Diet-AminoGut, *CFGLN-GLN* Creep Fed Glutamine-Glutamine. *CFAG-AG* Creep Fed AminoGut-AminoGut. See Figure 1.
[abc]Least square means within a row lacking a common superscript are different ($P < 0.05$).

GLN and CFGLN-GLN, and groups had the greatest ($P = 0.001$) number of cells proliferating (PCNA), exceeding those which did not receive creep feed and later receiving a control diet (NC-CD) by 63%, 54% and 43% respectively. We found a correlation between villi height and PNCA: the taller the villi height, the greater the number of proliferating cells. Sow-Reared pigs showed the greatest ($P = 0.001$) intestinal absorption capacity for xylose and mannitol when compared with the others treatments. The levels of xylose and mannitol found in the sow reared pigs blood exceeded the average of the levels found in the other treatments by 3.2 and 7.4 folds respectively. This is consistent with the architecture of the villi of the sow reared pigs when compared to the other treatments (see qualitative SEM images, Figure 2). There was no significance difference among the other treatments on the absorption of these sugars. We found the levels of xylose in the blood to be higher than those of mannitol even though a higher amount of mannitol was diluted in the final solution (0.2 g/L vs. 0.3 g/L). We found no significant differences among the treatments in maltase activity although there was a tendency ($P = 0.18$) for creep fed treatments to be numerically different than those which did not receive creep feed (260 vs. 214 μmoles / min. g of protein respectively).

Discussion

Windmueller and Spaeth [31] determined that in the adult rat small intestine, CO_2, lactate, alanine and glucose account for 56–64, 16–20, 4–8, and 2-10% of the total catabolized carbons of luminal glutamine, glutamate and aspartate, respectively. These results and others showed that amino acids (glutamine, glutamate and aspartate), rather than glucose, are the major fuels for the small intestine mucosa, responsible for providing energy required for intestinal ATP-dependent metabolic processes [32]. Although there seems little doubt that glutamine plays an important, but remarkably poorly characterized role in the metabolism of many proliferating cells, much of the more recent literature on intestinal metabolism has ignored two observations made by Windmueller and Spaeth [33]. Those are, first, that the metabolism of luminal glutamate

Table 4 Pig performance from week 3 to 6 post-weaning

Variables	NC-CD[1]	NC-GLN	NC-AG	CFCD-CD	CFCD-GLN	CFCD-AG	CFGLN-GLN	CFAG-AG	SEM	*P* value
No. pens	6	6	6	6	6	6	6	6	-	-
Initial wt, kg	10.45	10.32	10.72	9.44	10.18	10.10	10.78	10.43	0.43	0.457
Final wt, kg	22.32	21.97	22.74	22.21	22.86	22.50	21.71	22.76	0.41	0.437
ADG, kg/pig	0.52	0.51	0.55	0.50	0.54	0.53	0.50	0.54	0.02	0.444
Intake, kg/pig[2]	0.77^{b}	0.74^{b}	0.86^{a}	0.77^{b}	0.79^{b}	0.79^{b}	0.70^{c}	0.80^{b}	0.03	0.022
Feed/gain[2]	1.39	1.48	1.65	1.49	1.64	1.84	1.66	1.50	0.12	0.271
Mortality,%	0.01	0.00	0.35	0.13	0.16	0.01	0.19	0.01	0.13	0.434
Removed,%	1.01	1.5	1.69	1.62	0.99	1.15	1.69	1.17	0.32	0.497

[1]*NC-CD* No Creep Control Diet, *NC-GLN* No Creep Glutamine, *NC-AG* No Creep AminoGut, *CFCD-CD* Creep Fed Control Diet-Control Diet, *CFCD-GLN* Creep Fed Control Diet-Glutamine, *CFCD-AG* Creep Fed Control Diet-AminoGut, *CFGLN-GLN* Creep Fed Glutamine-Glutamine, *CFAG-AG* Creep Fed AminoGut-AminoGut. See Figure 1.
[2]For the calculation of feed intake and feed/gain, the number of pens was reduced from 6 to 3 pens because two pens shared a common feeder.
[abc]Least square means within a row lacking a common superscript are different ($P < 0.05$).

Table 5 Intestinal morphology, PCNA staining, and maltase activity, and serum xylose and mannitol following oral gavage of pigs 1-wk post-weaning

Variables	NC-CD[1]	NC-GLN	NC-AG	CFCD-CD	CFCD-GLN	CFCD-AG	CFGLN-GLN	CFAG-AG	Sow Reared	SEM	*P* value
No. pigs	6	6	6	6	6	6	6	6	6	-	-
Age pigs, d	28	28	28	28	28	28	28	28	28	-	-
Body wt, kg[2]	6.5	6.7	6.2	5.8	6.4	6.4	6.5	6.5	6.3	0.28	0.824
Villi height, μm	508^{ab}	515^{ab}	510^{ab}	560^{ab}	576^{a}	480^{b}	568^{a}	560^{ab}	570^{a}	23	0.049
Villi width, μm	136	140	120	133	148	158	145	147	132	8.9	0.301
Crypt depth, μm	202^{ab}	188^{b}	177^{b}	192^{ab}	218^{ab}	242^{ab}	225^{ab}	258^{a}	118^{c}	14	0.001
PCNA, cells	56^{c}	81^{a}	67^{b}	71^{ab}	86^{a}	70^{ab}	80^{a}	64^{b}	91^{a}	5.5	0.001
Xylose, mg/dl	14.6^{b}	13.5^{b}	11.1^{b}	14.4^{b}	9.5^{b}	14.1^{b}	15.4^{b}	13.7^{b}	42.8^{a}	1.94	0.001
Mannitol, mg/dl	1.8^{b}	1.5^{b}	2.3^{b}	1.9^{b}	2.8^{b}	3.0^{b}	2.8^{b}	3.0^{b}	17.7^{a}	0.68	0.001
Maltase, μmol/(min. g protein)	256	183	201	299	282	254	244	242	239	32	0.189

[1] *NC-CD* No Creep Control Diet, *NC-GLN* No Creep Glutamine, *NC-AG* No Creep AminoGut, *CFCD-CD* Creep Fed Control Diet-Control Diet. *CFCD-GLN* Creep Fed Control Diet-Glutamine, *CFCD-AG* Creep Fed Control Diet-AminoGut, *CFGLN-GLN* Creep Fed Glutamine-Glutamine. *CFAG-AG* Creep Fed AminoGut-AminoGut. Sow Reared were pigs kept with the sow until they were 28 d of age. See Figure 1.
[2] Body weight was taken one week post-placement in the nursery. Sow reared pigs stayed with their dams for 28 d.
[abc] Least square means within a row lacking a common superscript are different ($P < 0.05$).

was even more extensive than that of arterial glutamine; and second, that the presence of high concentrations of glutamate in the intestinal lumen had only a small (less than 25%) effect on intestinal utilization of glutamine. This perhaps suggests that these two closely related amino acids may have different functional roles in the intestinal mucosa.

It seems that glutamate can partially substitute for Gln in several pathways, including ATP production and synthesis of arginine, alanine, proline and aspartate [34]. Glutamate plays a significant role on avoiding Gln degradation by mitochondrial phosphate-activated glutaminase in extra hepatic tissues and cells yielding a sparing effect on the use of Gln as a metabolic fuel [35] and its availability in cells [36]. Wu [10] rightly notes that that key functions of Gln (syntheses of Gln-tRNA, aminosugars, carbamoylphosphate, NAD, NADP, as well as purines and pyrimidines; renal ammoniagenesis; and regulation of ornithine decarboxylase expression) cannot be supplied by glutamate. Wu and others [37] argued that although both Gln and glutamate provided in the enteral diet are extensively catabolized by the small intestine, this organ takes up Gln, but not glutamate, from the arterial blood. They suggested that due to the complex compartmentalization of cellular metabolism, extracellular glutamate may channel preferentially into the cytoplasm rather than into the mitochondria and, therefore, have different effects than the glutamate generated from Gln in mitochondria.

The vast majority of research showing the benefits of supplementing Gln in the diet can be found in research with swine. Wu and co-workers [38] reported that among all the amino acids, uterine and umbilical uptake of Gln was the greatest in pregnant gilts, implicating an important role for Gln in fetal growth and development. They fed 1% Gln in the diet of gestating gilts between 90 and 114 d of gestation and found that it significantly increased average birth weight. They also found that the number of intrauterine growth retarded piglets, variation in birth weight and pre-weaning mortality were reduced by 39, 33, and 46%, respectively, when compared with the control group.

Kim and Wu [39] reported that lactating sows have a high requirement for Gln and the uptake of Gln by the porcine mammary glands may be inadequate for the synthesis of milk proteins. By d 10 during the lactating period, the mammary glands uptake 16 g Gln/d from the arterial circulation [40], however Haynes and others [41] reported that at that point in time, 36 g Gln/d is being secreted. Wu and colleagues [38] fed 1% Gln from d1 to d 20 to lactating sows and found an increase of Gln concentrations in the plasma, skeletal muscle and whole milk of the sows, as well as piglet growth and survival.

Haynes et al., [41] evaluated the effectiveness of Gln or L-alanyl-L-glutamine (Ala-Gln) *in vivo* with 7-d-old piglets challenged with a single intraperitoneal injection LPS (0.1 mg/kg body weight). Administration of Gln or Ala-Gln to LPS challenged piglets increased Gln concentrations in small intestinal lumen and plasma, reduced intestinal expression of TLR-4, active caspase-3 and NF*k*B, ameliorated intestinal injury, decreased rectal temperature and enhanced growth performance. These results demonstrate a protective effect of Gln or Ala-Gln against LPS-induced enterocytes death. They also reported that the Gln supplementation stimulated the growth of sow-reared piglets by 12%.

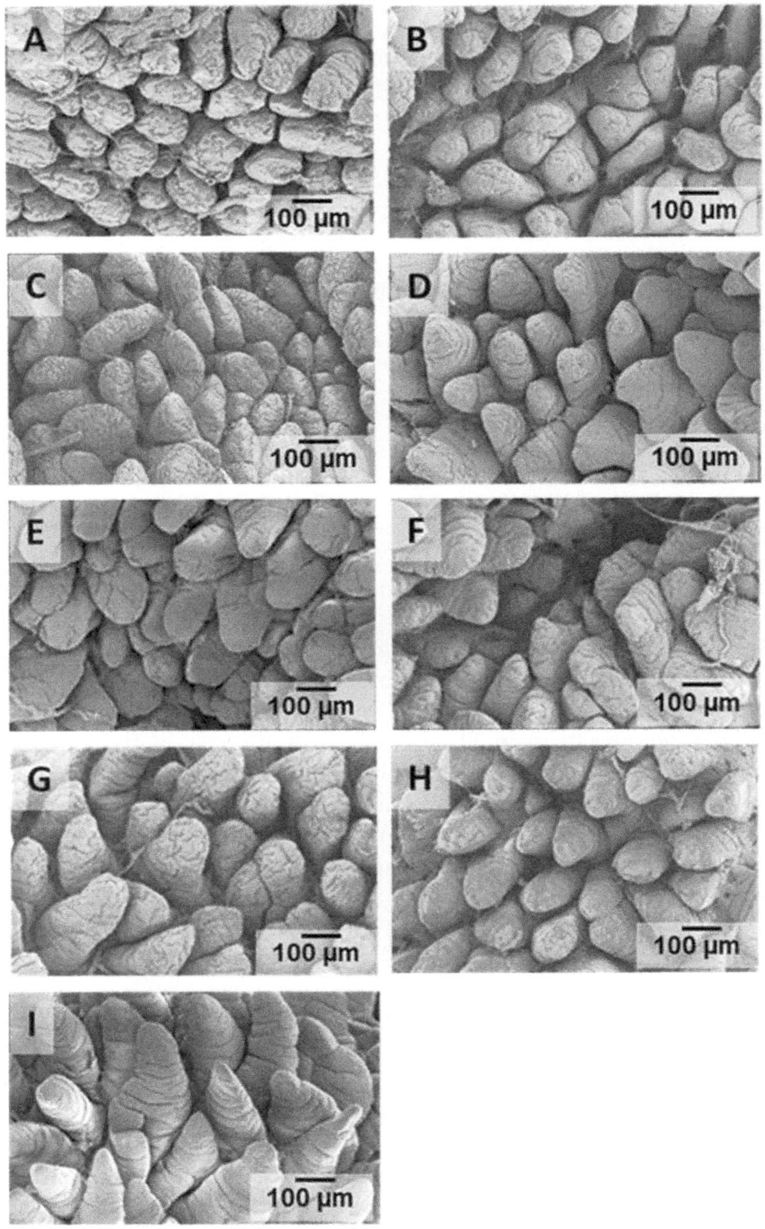

Figure 2 Scanning electron micrographs of jejunal villi of pigs at four wk of age. A. No creep Control Diet (NC-CD). **B**. No creep Glutamine (NC-GLN). **C**. No creep AminoGut (NC-AG). **D**. Creep Fed Control Diet-Control Diet (CFCD-CD). **E**. Creep Fed Control Diet-Glutamine (CFCD-GLN). **F**. Creep Fed Control-AminoGut (CFCD-AG). **G**. Creep Fed Glutamine-Glutamine (CFGLN-GLN). **H**. Creep Fed AminoGut-AminoGut (CFAG-AG). **I**. Sow-Reared Control.

Yi et al. [42] found that feeding glutamine had beneficial effects in alleviating growth depression of *E. coli* K88$^+$-challenged pigs, mainly via maintaining intestinal morphology and function, and/or possible modulating the somatotrophic axis. Jiang and others [43] reported similar results. Wu et al. [44] orally administered Gln (0.5 g/kg BW/d) to low-birth weight piglets from 0 to 21 d of age and found that their growth were improved by 16% and their pre-wean mortality by 48%.

Our results are most consistent with those reported by Wu et al., [45]. They found a 29% improvement in feed conversion (21d post-weaning) when supplementing 1% glutamine. Glutamine (Gln) supplementation (1%) prevented jejunal atrophy (measured as villus height) during the first week post-weaning and increased feed: gain ratio (indicator of growth performance) by 25% during the second week post-weaning. It also increased plasma concentration of aspartate, glutamate and alanine and also reduced the extent to which plasma taurine concentration fell in post-weaning pigs. The prevention of villi atrophy during the first week post-weaning also has been reported by Wang and co-workers [46].

Liu and others [47] reported similar results than those reported by Wu et al. [45]. They fed 1% L-glutamine or 1% L-glutamate to weaned pigs from 28-d to 42-d of age. Jejunal atrophy was prevented during the first week for the groups fed either L-GLN or L-glutamate when compared to the control group. Again these results provide an experimental basis for the use of glutamine and glutamate to improve piglet intestinal health and to support improved growth performance.

D-Xylose absorption test has been used as a tool for the assessment of the effect of anticoccidials on the intestinal absorptive capacity of broilers during experimental coccidiosis [48] and malabsorption in poult enteritis and mortality syndrome [25]. D-xylose, a poorly metabolized pentose sugar, is well absorbed from the small intestine of chickens and readily excreted in the urine [25]. Blood D-xylose concentrations are expected to peak at 30–60 min after intake in poultry [25,48] and 60 min in pigs [49,50].

Mannitol has been clinically applied in diagnostic and therapeutic doses for 1) the determination of extracellular fluid volume and glomerular filtration rate, 2) testing intestinal absorption and mucosal integrity, 3) as a diuretic and 4) as a radical scavenger and osmotically active component of infusions.

There are few studies using these two sugars as markers of gastrointestinal *in vivo* permeability in pigs in a commercial setting. In this study, the uptake of xylose was greater than mannitol regardless of their molecular weight (150 and 182 g/mol respectively) and the amount administered (mannitol higher than xylose). Xylose can be metabolized in the gut by bacteria, and then absorbed whereas mannitol cannot. Therefore we would expect xylose to be absorbed more rapidly than mannitol. Mannitol is partially metabolized, the remainder being excreted in the urine. Nasrallah and Iber [51] administered orally doses of 20 to 100 g of ^{14}C-mannitol to five humans with cirrhosis of the liver and to five subjects who had normal liver function. They found that at least one-sixth of orally ingested mannitol is absorbed and about one-third is metabolized.

The lack of significant differences in performance among the treatments for the entire 6-wk period correlates well with the lack of significant differences among the treatments for the levels xylose and mannitol absorbed and found in their blood. We were not surprised by the high levels of intestinal absorptive capacity shown by the sow reared pigs when compared to the other treatments.

These tests of small intestine permeability to low-molecular-weight carbohydrates have a considerable application in the study of small intestine diseases such as celiac disease in humans [52], diagnosing food allergy and assessing the effectiveness of anti-allergic agents such as sodium cromoglycate [53].

In young animals, lactase activity prevails, however as it gets older then maltase activity (as well amylases, lipases proteases) increases. Low concentration of maltase in the surface of epithelial cells may be an indication of villus atrophy due to disease or malnutrition [54]. We were unable to find any significant differences among the treatments in maltase activity.

Scanning electron microscopy (SEM) allows observation of the surface of the epithelium in three dimensions and gives a fresh dimension in the investigation of gut mucosa [55]. The visual assessment of the SEM graphs showed that pigs which were not creep fed during the suckling period had a rough villi surface with numerous cells shedding (apoptosis) along the entire length of the villi (Figure 2A). They also showed deep transversal furrows in most (if not all) the epithelial cells (Figure 2A, B). Those treatments creep fed either with a control diet or supplemented with glutamine or Aminogut showed longer villus than those treatments which were not creep fed (552 microns vs. 511 microns respectively) (Figure 2 D, E & F). The CFGLN-GLN treatment showed elongated, well defined and high villus (Figure 2G). Increased villus height could increase total luminal villus absorption area and could result in adequate digestive enzyme concentration and/or increased transport of nutrient at the villus surface. Gln has been shown to enhance epithelial repair in several models of intestinal injury and to stimulate epithelial proliferation and protein synthesis or reduce apoptosis in cell culture [56,57]. Increased uptake of Gln in the crypts not only could promote a compensatory increase in Na^+ absorption but also would place this nutrient in the ideal location to promote crypt cell production and restoration of the villus architecture. The CGAG-AG treatment showed deep and wide crypts. This could be explain by the fact that glutamine is donating an amide group for the biosynthesis of purines (adenine and guanine) and pyrimidines (thymine and cytosine) which are the nucleotides bases to support nucleic acid production (DNA) for rapidly dividing cells in the crypts. In RNA, the complement of adenine is uracil instead of thymine. The sow reared pigs showed what may be the perfect villi structure: healthy, well defined villus, no signs of apoptotic cells and sufficient mucin production (Figure 2I). Mucins are a family of high molecular weight, heavily glycosylated proteins produced by epithelial tissues (specifically by the goblet cells) in most metazoans. Two noticeable jejunal villi structure characteristics in all treatments for 28 d pigs were 1) transversal furrows that were present along the entire length of the villi and 2) the shape of the villi were not finger-like but rather wide and tongue-like in shape. It was evident that the small intestinal mucosa undergoes profound structural and developmental changes during the first 4 week of the pig's life and these changes are manifested in shape, size and density of the villi.

Conclusion

The supplementation of glutamine and glutamine plus glutamate (AminoGut) in pre- and post-weaning diets improved feed conversion in the first three week post-weaning when compared to CFCD-CD treatment. These findings are in the agreement with those reporting a reduction in villi atrophy when supplementing glutamine at 1% in diets during the first week post-weaning. Sow reared pigs showed the best intestinal absorptive capacity and villi architecture. More research is needed at the field level to justify the economical feasibility of adding either glutamine or AminoGut in current commercial livestock diets and the European model of weaning pigs at 28 d of age. Consideration should also be given to potential supplementation of the sow to enrich milk concentrations [58].

The existing vast knowledge of the roles of functional AA's such as glutamine and others (arginine, glutamate, proline, leucine, cysteine and tryptophan) provides the scientific basis for nutritionists to revise current nutrient requirements for livestock especially weaned pigs. These findings indicate that strong consideration must be given to GLN and glutamate as nutritionally essential amino acids for post-weaning pigs diets.

Abbreviations

mTOR: Mammalian target of Rapamycin; IgA: Immunoglobulin A; GLN or Gln: Glutamine; AG: AminoGut; NC: Non-Creep fed; CFCD: Creep Fed Control Diet; CFGLN: Creep Fed Glutamine; CFAG: Creep Fed AminoGut; NC-CD: Non-Creep Fed and later receiving a Control Diet; CFCD-CD: Creep Fed Control Diet and later receiving a Control Diet; NC-GLN: Non-Creep Fed and later receiving a diet supplemented with Glutamine.; CFDC-GLN: Creep Fed Control Diet and later receiving a diet supplemented with Glutamine.; NC-AG: Non-Creep Fed and later receiving a diet supplemented with AminoGut; CFCD-AG: Creep Fed Control Diet and later receiving a diet supplemented with AminoGut.; CFGLN-GLN: Creep Fed Glutamine and later receiving a diet supplemented with Glutamine.; CFAG-AG: Creep Fed AminoGut and later receiving a diet supplemented with AminoGut; SR: Sow reared pigs; PCNA: Proliferating cells nuclei antigen; SBM: Soy bean meal; AA: Amino acids; C: Celsius; CO_2: Carbon dioxide; SEM: Scanning electron microscopy; HPLC: High performance liquid Chromatography; PAD: Pulsed amperometric detector; BW: Body weight; ADG: Average daily gain; mM: Millimolar; mL: Milliliters; nm: Nanometer; min: Minutes; µL: Microliters; ATP: Adenosine triphosphate; g: Gram; L: Liter; tRNA: Transfer ribonucleic acid; NAD: Nicotinamide adenine dinucleotide; NADP: Nicotinamide adenine dinucleotide phosphate; d: Day; mg: Milligram; kg: Kilogram; LPS: Lipopolysaccharides; TLR: Toll like receptors; NFκB: Necrotic factor kappa B; CWG: Choice white grease; ZnO: Zinc oxide; NRC: National research council; ME: Metabolizable energy; CP: Crude protein; TID: Total ileal digestible.

Competing interests

Full financial support for this trial was provided by Ajinomoto Heartland Lysine and Ajinomoto do Brazil. Dr. James L. Usry served as Swine Technical Director for Ajinomoto Heartland Lysine (currently serves as Swine Technical Director for Micronutrients), and both Marianne Kutschenko and Dr. Eduardo T. Nogueira serve as Technical Managers for Ajinomoto do Brazil. All three are authors in this manuscript. However their assistance was limited to guiding us on the design of the research protocol and the inclusion rate of both Glutamine and AminoGut. Both products are being evaluated but not currently being used in the US.

Authors' contributions

RAC as the lead author was in charge of designing the research protocol, obtaining all materials needed for it, executing it, carrying out all the lab work, data collection and analysis and finally writing the manuscript. JLU was instrumental on guiding the lead author on the design of the research protocol. CA provided the author with statistical guidance on the data analysis. ETN and MK provided significant literature and previous work references about glutamine and its biological functions as a non-essential amino acid. They were also involved in the design of the research protocol. JO as a co-advisor to the lead author, he was involved in the design, execution and interpretation of the results. He also assisted with blood and tissues collection and with the statistical analysis. AJM as a co-advisor to the lead author, he was involved in the design and interpretation of the results. He helped design our *in vivo* method for determining gut barrier function. All authors read and approved the final version of the manuscript.

Authors' information

RC holds a PhD in Animal Nutrition from North Carolina State University. His area of research is neonatal survival, nutrient digestibility and gastrointestinal health of swine. In 2001, he was awarded the "Innovative Award Applied Research" by National Pork Producer Council at the Midwest Animal Science Meeting in Des Moines, Iowa. He is a member of the North Carolina Pork Council and the American Society of Animal Science. He currently serves as Director of Swine Technical Services for Huvepharma USA, Inc. JLU holds a PhD in Agricultural Engineering from the University of Kentucky in animal growth modeling. He spent 21 years at Ajinomoto Heartland where he became VP of Nutritional Services and currently is employed at Micronutrients as Director of Swine Nutrition. Most of his career has centered on amino acid research and development. CA holds a PhD in Statistics from North Carolina State University. Her research interests include experimental design applied to life sciences, statistical modeling, and discrete data analysis. She is interested in statistical consulting, research methodology and creative learning and teaching. She is a Research Assistant professor in the Department of Statistics at NCSU and a member of American Statistical Society. ETN holds a PhD in Animal Nutrition from Viçosa Federal University (UFV, Brazil)/University of Western Australia (UWA, Australia). His area of research is amino acids nutrition. He currently serves as Latin America Technical General Manager for Ajinomoto do Brazil/Ajinomoto Animal Nutrition. MK holds an MSc in Animal Nutrition from Maringa State University (UEM, Brazil). Her area of research is amino acid nutrition. She currently serves as Latin America Technical Manager for Ajinomoto do Brazil/ Ajinomoto Animal Nutrition.

AJM holds a MS in Swine Nutrition, a PhD in Gastrointestinal Physiology and a Doctor of Veterinary Medicine (DVM) all from NCSU. His main area of research is to study basic mechanisms of stress-induced intestinal dysfunction. Stress is an important contributing factor to enteric disorders of veterinary species and humans however, the mechanisms are poorly understood. His work has focused on the role of mucosal mast cells in psychological stress-induced disturbances in intestinal mucosal barrier function. He believes that this work will have important implications in the understanding of stress-related gut disorders such as infectious diarrhea, Inflammatory Bowel Disease, and Irritable Bowel Syndrome, and will facilitate the design of novel preventative and treatment strategies for veterinary and human patients suffering from these disorders. He is an assistant professor of GI physiology and swine medicine at NC State College of Veterinary Medicine. He is member of several professional societies including the American Physiological Society, American Association of Swine Veterinarians, and American Gastroenterological Association. JO has a PhD in Nutritional Biochemistry from the University of Wisconsin. As a Williams Neal Reynolds Professor in the Department of Animal Science at NCSU, his research interests are molecular and metabolic regulation of lipid digestion and metabolism; neonatal nutrition; intestinal growth and metabolism in normal and pathophysiological states. His program is focused on using the young piglet as a model for the human infant in nutrition and digestive physiology. His most recent awards include "Williams Neal Reynolds Distinguished Professor" and "The Outstanding Graduate Instructor" both given by the College of Agriculture and Life Science at NCSU, the "Animal Growth and

Development Research" given by the American Society of Animal Science. He was a member of the National Research Council (NRC) committee which recently published the new 2012 Nutrient Requirements of Swine. He is an Associate Editor in Advances in Nutrition (American Society for Nutrition) and the Journal of Animal Science and Biotechnology.

Acknowledgements

Research was conducted at the Maple Hill sow farm, NC (Murphy-Brown, LLC; Rose Hill, NC). The authors would like to thank Ajinomoto Heartland Lysine and Ajinomoto do Brazil for the financial support to carry out this trial. The authors would also like to thank Dr. Jeff Hansen, Dr. Ana de Souza and Dr. Dustin Kendall from Murphy-Brown LLC for coordinating this trial with the farm staff. The authors recognize the assistance provided by Indira Gomez, Wilmer Pacheco and Santa Maria Mendoza with the daily feeding, care of the animals, and blood and tissues collection. We are extremely grateful to Lauren Lanier who agreed to make this project her internship assignment with Murphy Brown LLC. We appreciate the technical assistance provided from Dr. Lisa Dean (Food Science department, NCSU) with the xylose assay, Dr. Michael Dykstra (College of Veterinary Medicine, NCSU) with the SEM guidance, Mrs. Debbie Ort (Poultry Science department, NCSU) with the mannitol assay and Dr. Sheila Jacobi (Laboratory of Developmental Nutrition, Department of Animal Science, NCSU) with the maltase assay and the sow reared pigs' blood and tissue collection.

Author details

[1]Laboratory of Developmental Nutrition, Department of Animal Science, North Carolina State University, 101 Polk Hall, North Carolina State University, Raleigh, NC 27695, USA. [2]Ajinomoto Heartland Lysine, Chicago, IL 60631, USA. [3]Department of Statistics, North Carolina State University, Raleigh, NC 27695-8203, USA. [4]Ajinomoto do Brasil. Ajinomoto Animal Nutrition, São Paulo, SP 04015-001, Brazil. [5]Department of Population Health and Pathobiology, College of Veterinary Medicine, North Carolina State University, Raleigh, NC 27695, USA. [6]Author current employment: Huvepharma USA, 525 West Park Drive Suite 230, Peachtree City, GA 30269, USA. [7]Author current employment: Micronutrients, 1550 Research Way, Indianapolis, IN 46231, USA.

References

1. Plunske JR, Hampson DJ, Williams IH: **Factors influencing the structure and function of the small intestine in the weaned pig.** *A review. Livest Prod Sci* 1997, **1997**(51):215–236.
2. Kenworthy R, Crabb WE: **The intestinal flora of young pigs with particular reference to early weaning, E, Coli and scours.** *J Comp Path* 1963, **73:**215–228.
3. Palmer NC, Hulland TK: **Factors predisposing to the development of coliform gastroenteritis in weaned pigs.** *Can Vet J* 1965, **6:**310–316.
4. Cera KR, Mahan DC, Reinhart GA: **Effect of weaning, week postweaning and diet composition on pancreatic and small intestinal luminal lipase response in young swine.** *J Anim Sci* 1990, **68:**384–391.
5. Newby TJ, Miller BG, Stokes CR, Bourne FJ: **Hypersensitivity to dietary antigens as the predisposing factor in post-weaning diarrhea.** *Pig Vet Soc Proc* 1983, **10:**50.
6. Miller BG, Newby TJ, Stokes CR, Bourne FJ: **Influence of diet on post-weaning malabsorption and diarrhea in the pig.** *Res Vet Sci* 1984, **36:**187.
7. Barnett KL, Kornegay ET, Risley CR, Lindemann MD, Schurig GG: **Characterization of creep feed consumption and its subsequent effects on immune response, scouring index and performance of weanling pigs.** *J Anim Sci* 1989, **67:**2698–2708.
8. Klasing KC: **Nutrition and the immune system.** *Br Poul Sci* 2007, **48:**525–537.
9. Li P, Yin YL, Li DF, Kim SW, Wu G: **Amino acids and immune function.** *Br J Nutr* 2007, **98:**237–252.
10. Wu G: **Amino acids: metabolism, functions, and nutrition.** *Amino acids* 2009, **37:**1–17.
11. Wu G: **Recent advances in swine amino acid nutrition.** *J Anim Sci Biotech* 2010, **1:**118–130.
12. Rhoads JM, Wu G: **Glutamine, arginine, and leucine signaling in the intestine.** *Amino acids* 2009, **37:**111–122.
13. Schrock H, Goldstein L: **Inter-organ relationships for glutamine metabolism in normal and acidotic rats.** *Am J Physiol* 1981, **240:**E519–E525.
14. Newsholme P: **Why is L-glutamine metabolism important to cells of the immune system in health, post injury, surgery or infection?** *J Nutr* 2001, **131:**2515S–2522S.
15. Nappert G, Zello GA, Naylor JM: **Intestinal metabolism of glutamine and potential use of glutamine as a therapeutic agent in diarrheic calves.** *J Am Vet Med Assoc* 1997, **211:**547–553.
16. Van de Poll MC, Soeters PB, Deutz NE, Fearon KC, Dejong CH: **Renal metabolism of amino acids: Its role in interogan amino acid exchange.** *Am J Clin Nutr* 2004, **79:**185–197.
17. Souba WW, Herskowitz K, Salloum RM, Chena MK, Augsten TR: **Gut glutamine metabolism.** *J Parenter Enter Nutr* 1990, **14**(Suppl. 4):45S–50S.
18. Reeds PJ, Burrin DG, Stoll B, Jahoor F: **Intestinal glutamate metabolism.** *J Nutr* 2000, **130:**978S–982S.
19. Wu G, Knabe DA, Flynn NE: **Synthesis of citrulline from glutamine in pig enterocytes.** *Biochem J* 1994, **299:**115–121.
20. Stoll B, Burrin DG: **Measuring splanchnic amino acid metabolism in vivo using stable isotopic tracers.** *J Anim Sci* 2006, **84:**E60–E72.
21. Wu G: **Recent advances in swine amino acid nutrition.** *J Anim Sci Biotech* 2010, **1:**49–61.
22. Van der Hulst RRWJ, Van Kreel BK, Von Meyenfeldt MF, Brummer RK, Arends JW, Deutz NE, Soeters PB: **Glutamine and the preservation of gut integrity.** *Lancet* 1993, **431:**1363–1365.
23. James LA, Lunn PG, Elia M: **Glutamine metabolism in the gastrointestinal tract of the rat assessed by the relative activities of glutaminase and glutamine synthetase.** *Br J Nutr* 1998, **79:**365–372.
24. FASS: *Guide for the care and use of agricultural animals in agricultural research and teaching.* 1 revth edition. Savoy, IL: Fed. Anim. Sci. Soc; 1999.
25. Doerfler RE, Cain LD, Edens FW, Parkhurst CR, Qureshi MA, Havenstein GB: **D-Xylose absorption as a measurement of malabsorption in poult enteritis and mortality syndrome.** *Poult Sci* 2000, **79:**656–660.
26. McDowell EM, Trump BF: **Histological fixatives suitable for diagnostic light and electron microscopy.** *Arch Pathol Lab Med* 1976, **100:**405.
27. Goodwin MA, Latimer KS, Fletcher OJ: **Quantification of intestinal D-xylose absorption in normal turkeys.** *Poult Sci* 1984, **63:**1742–1747.
28. Goodwin MA, Latimer LS, Nersessian BN: **Quantification of intestinal D-xylose absorption in normal and reovirus-inoculated turkeys.** *Avian Dis* 1984, **28:**959–967.
29. Eberts TJ, Sample RHB, Glick MR, Ellis GH: **A simplified, colorimetric micro method for xylose in serum or urine, with phloroglucinol.** *Clin Chem* 1979, **25:**1440–1443.
30. Dahlqvist A: **Assay of intestinal disaccharidases.** *Scand J Clin Lab Invest* 1984, **44:**169–172.
31. Windmueller HG, Spaeth AE: **Metabolism of absorbed aspartate, asparagine and arginine by rat small intestine in vivo.** *Arch Biochem Biophys* 1976, **175:**670–676.
32. Burrin DG, Reeds PJ: **Alternative fuels in the gastrointestinal tract.** *Curr Opin Gastroenterol* 1997, **13:**165–170.
33. Windmueller HG, Spaeth AE: **Intestinal metabolism of glutamine and glutamate from the lumen when compared to glutamine from the blood.** *Arch Biochem Biophys* 1975, **171:**662–672.
34. Reeds PJ, Burrin DG, Stoll B, Jahoor F, Wykes L, Henry J, Frazer ME: **Enteral glutamate is the preferential source for mucosal glutathione synthesis in fed piglets.** *Am J Physiol* 1997, **273:**E408–E415.
35. Yin YL, Huang RL, Li TJ, Ruan Z, Xie MY, Deng ZY, Hou YQ, Wu G: **Amino acid metabolism in the portal-drained viscera of young pigs: effects of dietary supplementation with chitosan and pea hull.** *Amino acids* 2010, **39:**1581–1587.
36. Boutry C, Bos C, Matsumoto H, Even P, Azzout-Marniche D, Tome D, Blachier F: **Effects of monosodium glutamate supplementation on glutamine metabolism in rats.** *Front Biosci* 2011, **E3:**279–290.
37. Wu G, Borbolla AG, Knabe DA: **The uptake of glutamine and release of arginine, citrulline and proline by the small intestine of developing pigs.** *J Nutr* 1994, **124:**2437–2444.
38. Wu G, Ott TL, Knabe DA, Bazer FW: **Amino acid composition of the fetal pig.** *J Nutr* 1999, **129:**1031–1038.
39. Kim SW, Wu G: **Regulatory role of amino acids in mammary gland growth and milk synthesis.** *Amino acids* 2009, **37:**89–95.
40. Trottier NL, Shipley CF, Easter RA: **Plasma amino acid uptake by the mammary gland of the lactating sow.** *J Anim Sci* 1997, **75:**1266–1278.

41. Haynes TE, Li P, Li X, Shimotori K, Sato H: **L-Glutamine or L-alanyl-L-glutamine prevents oxidant or endotoxin-induced death of neonatal enterocytes.** *Amino Acids* 2009, **37:**131–142.
42. Yi GF, Carroll JA, Allee GL, Gaines AM, Kendall DC, Usry J, Toride Y, Izuru S: **Effect of glutamine and spray-dried plasma on growth performance, small intestine morphology, and immune responses of Escherichia coli K88$^+$-challenged weaned pigs.** *J Anim Sci* 2005, **83:**634–643.
43. Jiang ZY, Sun LH, Lin YC, Ma XY, Zheng CT, Zhou GL, Chen F, Zou ST: **Effects of dietary glycyl-glutamine on growth performance, small intestinal integrity, and immune responses of weaning piglets challenged with lipopolysaccharide.** *J Anim Sci* 2009, **87:**4050–4056.
44. Wu G: **Functional amino acids in growth, reproduction and health.** *Adv Nutr* 2010, **1:**31–37.
45. Wu G, Meier SA, Knabe DA: **Dietary glutamine supplementation prevents jejunal atrophy in waned pigs.** *J Nutr* 1996, **126:**2578–2584.
46. Wang JJ, Chen X, Li P, Li XL, Zhou HJ, Wang FL, Li DF, Yin YL, Wu G: **Gene expression is altered in piglet small intestine by weaning and dietary glutamine supplementation.** *J Nutr* 2008, **138:**1025–1032.
47. Liu T, Peng J, Xiong Y, Zhou S, Cheng X: **Effects of dietary glutamine and glutamate supplementation on small intestine structure, active absorption and DNA, RNA concentrations in skeletal muscle tissue of weaned piglets during d 28 to 42 d of age.** *Asian-Aust J Anim Sci* 2001, **15:**238–242.
48. Mansoori B, Nodeh H, Modirsanei M, Rahbari S, Aparnak P: **D-xylose absorption test: A tool for the assessment of the effect of anticoccidials on the intestinal absorptive capacity of broilers during experimental coccidiosis.** *Anim Feed Sci Technol* 2009, **148:**301–308.
49. You-Sheng L, Jie-Shou L, Ninj L, Zhi-Wei J, Yun-Zhao Z, Nan-Yun L, Fang-Nan L: **Evaluation of various solutions for small bowel graft preservation.** *World J Gastr* 1998, **4:**140–143.
50. Chang RW, David PJ, Oh J-T, Andreoli S, Kim HB, Fauza D, Jaksic T: **Serial transverse enteroplasty enhances intestinal function in a model of short bowel syndrome.** *Ann Surg* 2006, **243:**223–228.
51. Nasrallah SM, Iber FL: **Mannitol absorption and metabolism in man.** *Am J Med Sci* 1969, **258:**80–88.
52. Cox MA, Lewis KO, Cooper BT: **Measurement of small intestinal permeability markers, lactulose and mannitol in serum.** *Dig Dis Sci* 1999, **44:**402–406.
53. Andre C, Andre F, Colin L, Cavagna S: **Measurement of intestinal permeability to mannitol and lactulose as a means of diagnosing food allergy and evaluating therapeutic effectiveness of disodium cromoglycate.** *Ann Allergy* 1987, **59:**127–130.
54. Kidder DE, Manners MJ: **Carbohydrases in the small intestine mucosa of sow-reared and 3-wk weaned pigs.** *Proc Nutr Soc* 1978, **37:**51A.
55. Skrzypek T, Valverde JL, Skrzypek H, Wolinski J, Kazimierczak W, Szymanczyk S, Pawlowska M, Zabielski R: **Light and scanning electron microscopy evaluation of the postnatal small intestinal mucosa development in pigs.** *J Physiol Pharmacol* 2005, **56:**71–87.
56. Blikslager A, Hunt E, Guerrant R, Rhoads M, Argenzio A: **Glutamine transporter in crypts compensates for loss of villus absorption in bovine cryptosporidiosis.** *Am J Physiol Gastrointest Liver Physiol* 2001, **281:**G645–G653.
57. Xi P, Jiang Z, Dai Z, Li X, Yao K, Zheng C, Lin Y, Wang J, Wu G: **Regulation of protein turnover by L-glutamine in porcine intestinal epithelial cells.** *J Nutr Biochem* 2012, **23:**1012–1017.
58. Manso HE, Filho HC, de Carvalho LE, Kutschenko M, Nogueira ET, Watford M: **Glutamine and glutamate supplementation raise milk glutamine concentrations in lactating gilts.** *J Anim Sci Biotech* 2012, **3:**2.

17

Amino acid digestibility of heat damaged distillers dried grains with solubles fed to pigs

Ferdinando Nielsen Almeida[1], John Kyaw Htoo[2], John Thomson[3] and Hans Henrik Stein[1*]

Abstract

The primary objective of this experiment was to determine the effects of heat treatment on the standardized ileal digestibility (SID) of amino acids (AA) in corn distillers dried grains with solubles (DDGS) fed to growing pigs. The second objective was to develop regression equations that may be used to predict the concentration of SID AA in corn DDGS. A source of corn DDGS was divided into 4 batches that were either not autoclaved or autoclaved at 130°C for 10, 20, or 30 min. Four diets containing DDGS from each of the 4 batches were formulated with DDGS being the only source of AA and CP in the diets. A N-free diet also was formulated and used to determine the basal endogenous losses of CP and AA. Ten growing pigs (initial BW: 53.5 ± 3.9 kg) were surgically equipped with a T-cannula in the distal ileum and allotted to a replicated 5 × 4 Youden square design with 5 diets and 4 periods in each square. The SID of CP decreased linearly ($P < 0.05$) from 77.9% in non-autoclaved DDGS to 72.1, 66.1, and 68.5% in the DDGS samples that were autoclaved for 10, 20, or 30 min, respectively. The SID of lysine was quadratically reduced ($P < 0.05$) from 66.8% in the non-autoclaved DDGS to 54.9, 55.3, and 51.9% in the DDGS autoclaved for 10, 20, or 30 min, respectively. The concentrations of SID Arginine, Histidine, Leucine, Lysine, Methionine, Phenylalanine, or Threonine may be best predicted by equations that include the concentration of acid detergent insoluble N in the model ($r^2 = 0.76$, 0.68, 0.67, 0.84, 0.76, 0.73, or 0.54, respectively). The concentrations of SID Isoleucine and Valine were predicted ($r^2 = 0.58$ and 0.54, respectively) by the Lysine:CP ratio, whereas the concentration of SID Tryptophan was predicted ($r^2 = 0.70$) by the analyzed concentration of Tryptophan in DDGS. In conclusion, the SID of AA is decreased as a result of heat damage and the concentration of SID AA in heat-damaged DDGS may be predicted by regression equations developed in this experiment.

Keywords: Amino acids, Digestibility, Distillers dried grains with solubles, Heat damage

Background

Production of corn distillers dried grains with solubles (DDGS) involves a drying step in which the temperature at the dryer inlet may be above 500°C, while the temperature at the dryer discharge may be above 100°C [1]. Application of heat to feed ingredients may initiate Maillard reactions, which decrease the concentration and digestibility of Lysine and other amino acids (AA) [2-4]. Lysine is particularly susceptible to undergo Maillard reactions because of its free amino group, which easily reacts with reducing sugars. If the amino group of an AA reacts with a reducing sugar to form early or advanced Maillard reaction products, it becomes unavailable to pigs [2,5]. During the acid hydrolysis step of AA analysis, however, Lysine that has reacted with reducing sugars is partially recovered, thus leading to an overestimation of available Lysine. For this reason, determination of reactive Lysine, color, and the Lysine:CP ratio have been suggested as approaches to estimate the availability of Lysine in DDGS [2,6-9]. Heat damage and Maillard reactions have been described in different sources of corn DDGS [2,10,11]. However, if different sources of DDGS are used, it is difficult to distinguish between effects of heat damage and other factors influencing AA digestibility. Gradual increases in heating of a specific source of DDGS will result in reduced concentrations of reactive Lysine [2,6]. There is, however, no information about effects of increasing time of heating of a specific source of DDGS on *in vivo* AA digestibility, and on the changes in reactive Lysine, and the Lysine:CP ratio. Prediction of the concentration of

* Correspondence: hstein@illinois.edu
[1]Department of Animal Sciences, University of Illinois, 1207 W Gregory Dr, Urbana, IL 61801, USA
Full list of author information is available at the end of the article

digestible AA and CP from the concentration of reactive Lysine and the Lysine:CP ratio in corn DDGS that was purposefully heat-damaged have not been reported. Thus, the primary objective of this experiment was to determine effects of heat treatment on the apparent ileal digestibility (AID) and the standardized ileal digestibility (SID) of AA in corn DDGS fed to growing pigs. A second objective was to develop prediction equations that may be used to predict the concentration of SID AA in corn DDGS.

Materials and methods

The Institutional Animal Care and Use Committee at the University of Illinois reviewed and approved the protocol for this experiment. Pigs used in the experiment were the offspring of G-performer boars and F-25 females (Genetiporc, Alexandria, MN).

Animals, housing, and experimental design

Ten growing pigs (initial BW: 53.5 ± 3.9 kg) were surgically equipped with a T-cannula in the distal ileum [12] and allotted to a replicated 5 × 4 Youden square design with 5 diets and 4 periods in each square. Pigs were individually housed in a controlled environment in pens (1.2 m ×1.5 m) equipped with a feeder and a nipple waterer.

Diets and feeding

Distillers dried grains with solubles was obtained from Poet Nutrition (Sioux Falls, SD) and analyzed for CP and AA. The DDGS was divided into 4 batches that were not autoclaved or autoclaved at 130°C for 10, 20, or 30 min (Table 1). Four diets containing DDGS from each of the 4 batches were formulated, with DDGS being the only source of AA and CP in the diets (Tables 2 and 3). A N-free diet also was formulated and used to determine the basal endogenous losses of CP and AA in the pigs. Diets were supplied with vitamins and minerals to meet or exceed the requirement estimates for growing pigs [13]. Chromic oxide also was included (0.4%) in diets and used as an indigestible marker.

The amount of feed provided was calculated as 2.5 times the maintenance requirement of energy (i.e., 106 kcal of ME/kg $BW^{0.75}$; [13]). Pigs were fed once daily at 0800 h. At the beginning of each period, feed allowance was adjusted based on the BW of each pig. Water was available at all times.

Sample collection

Each period consisted of 7 d. The initial 5 d were considered an adaptation period to the diet. On d 6 and d 7, ileal digesta were collected for 8 h using standard operating procedures. A plastic bag was attached to the cannula barrel and digesta flowing into the bag were collected. Bags were replaced whenever they were filled with digesta, or at least once every 30 min and immediately frozen at − 20°C to prevent bacterial degradation of AA in the digesta.

Chemical analyses

At the conclusion of the experiment, ileal digesta samples were thawed, mixed within animal and diet, and a sub-sample was lyophilized, ground to pass a 1 mm screen, and analyzed. A sample of each diet and of each batch of DDGS was collected at the time of diet mixing. Diets, ingredients, and ileal samples were analyzed for AA by ion-exchange chromatography with postcolumn derivatization with ninhydrin. Amino acids were oxidized with performic acid, which was neutralized with Na metabisulfite [14,15]. Amino acids were liberated from the protein by hydrolysis with 6 mol/L HCL for 24 h at 110°C and quantified with the internal standard by measuring the absorption of reaction products with ninhydrin at 570 nm. Tryptophan was determined by HPLC with fluorescence detection (extinction 280 nm, emission 356 nm), after alkaline hydrolysis with barium hydroxide octahydrate for 20 h at 110°C [16]. Diets, ingredients, and ileal samples also were analyzed for DM (method 935.29; [17]), and for CP following the Dumas procedure (method 968.06; [17]). Diets and ileal samples were analyzed for chromium (method 990.08; [17]). Ingredients were analyzed for acid detergent fibre (ADF; Method 973.18; [17]), neutral detergent fibre (NDF) [18], lignin (method 973.18 (A-D); [17]), ash (method 942.05; [17]), for total fat by acid hydrolysis using 3 mol/L HCl [19] followed by crude fat extraction with petroleum ether (method 2003.06; [17]), total reducing sugars [20], and furosine [8]. The concentration of acid detergent insoluble nitrogen (ADIN) in ingredients was determined as the concentration of N in the ADF fraction (method 990.03; [17]).

Calculations and statistical analysis

Values for AID and SID of CP and AA were calculated [21]. The Lysine:CP ratio in each DDGS sample was calculated by expressing the concentration of Lysine in the sample as a percentage of the CP in the sample. The concentration of reactive Lysine (%) was calculated by the following equation:

$$\text{Reactive Lysine} = \text{total Lysine}\ (\%) - [\text{furosine}\ (\%) \div 0.32 \times 0.40];$$

Data were analyzed using the MIXED procedure (SAS Institute Inc., Cary, NC). Normality of the data and the presence of outliers were evaluated using the UNIVARIATE procedure of SAS. One outlier was identified and removed from the data. The model included dietary treatment as fixed effect and pig and period as random effects. Linear and quadratic effects of increasing time of heat treatment

Table 1 Chemical composition of distillers dried grains with solubles

	Distillers dried grains with solubles			
		Autoclaved at 130°C		
Item	Non-autoclaved	10 min	20 min	30 min
Dry matter, %	93.21	92.43	90.91	94.47
Ash, %	5.25	5.08	5.17	5.06
Crude protein (CP), %	27.91	27.44	26.51	27.05
AEE,[1] %	10.20	12.73	12.83	9.25
Lysine: CP ratio[2]	2.94	2.37	2.75	2.51
Furosine, %	0.015	0.009	0.006	0.008
Reactive Lysine[3]	0.80	0.64	0.72	0.67
Acid detergent fibre, %	7.96	11.05	9.85	10.89
Neutral detergent fibre, %	31.29	33.23	33.32	32.40
Acid detergent lignin, %	0.88	2.06	1.73	2.57
Acid detergent insoluble N,[4] %	0.12	0.53	0.42	0.55
Reducing sugars, %	0.78	0.60	0.88	0.65
Indispensable AA, %				
Arginine	1.24	1.10	1.19	1.10
Histidine	0.71	0.67	0.70	0.67
Isoleucine	0.97	0.91	0.96	0.93
Leucine	2.92	2.82	2.89	2.78
Lysine	0.82	0.65	0.73	0.68
Methionine	0.53	0.49	0.52	0.50
Phenylalanine	1.24	1.17	1.21	1.17
Threonine	1.02	0.98	1.01	0.98
Tryptophan	0.22	0.20	0.20	0.20
Valine	1.26	1.19	1.26	1.23
Dispensable AA, %				
Alanine	1.90	1.83	1.87	1.82
Aspartic acid	1.75	1.66	1.73	1.66
Cysteine	0.55	0.50	0.52	0.50
Glutamine	4.43	4.31	4.41	4.26
Glycine	1.08	1.04	1.09	1.04
Proline	2.21	2.06	2.13	2.04
Serine	1.31	1.26	1.29	1.23

[1]AEE = acid hydrolyzed ether extract.
[2]Calculated by expressing the concentration of Lysine in each ingredient as a percentage of the concentration of CP [23].
[3]Reactive Lysine (%) = [Lysine (%) − (Furosine (%) ÷ 0.32 × 0.40)]; [2].
[4]*ADIN*, acid detergent insoluble nitrogen.

on the AID and SID of AA were analyzed by orthogonal polynomial contrasts. Correlations among predictor variables or between predictor variables and dependent variables were determined using the CORR procedure of SAS. The coefficients of correlation (r_p) were divided into 4 groups: no correlation ($P > 0.05$), low correlation ($r_p < 0.30$), moderate correlation ($0.60 > r_p \geq 0.30$), or high correlation ($r_p \geq 0.60$). Regression equations to estimate the relationship between the concentration of SID AA and nutrient concentration were developed using the REG procedure in SAS. The pig was the experimental unit for all analyses and significance among means was assessed with an α level of 0.05.

Results

Heat treatment did not change the concentrations of DM, ash, or CP in DDGS (Table 1). The concentration of total Lysine, however, was 0.82% in non-autoclaved

Table 2 Ingredient composition of experimental diets, as-fed basis

	Distillers dried grains with solubles diets				
		Autoclaved at 130°C			
Ingredient, %	Non-autoclaved	10 min	20 min	30 min	N-free diet
DDGS[1]	60.00	60.00	60.00	60.00	-
Cornstarch	27.00	27.00	27.00	27.00	66.40
Sucrose	10.00	10.00	10.00	10.00	20.00
Solka floc[2]	-	-	-	-	5.00
Soybean oil	-	-	-	-	4.00
Ground limestone	1.20	1.20	1.20	1.20	-
Dicalcium phosphate	0.60	0.60	0.60	0.60	3.00
Sodium chloride	0.40	0.40	0.40	0.40	0.40
Magnesium oxide	-	-	-	-	0.10
Potassium carbonate	-	-	-	-	0.40
Chromic oxide	0.40	0.40	0.40	0.40	0.40
Vitamin-mineral premix[3]	0.40	0.40	0.40	0.40	0.30

[1]*DDGS*, distillers dried grains with solubles.
[2]Fiber Sales and Development Corp., Urbana, OH.
[3]Provided the following per kilogram of complete diet: Vitamin A as retinyl acetate, 11,128 IU; vitamin D_3 as cholecalciferol, 2,204 IU; vitamin E as DL-alphatocopheryl acetate, 66 IU; vitamin K as menadione nicotinamide bisulfite, 1.42 mg; thiamin as thiamine mononitrate, 0.24 mg; riboflavin, 6.58 mg; pyridoxine as pyridoxine hydrochloride, 0.24 mg; vitamin B_{12}, 0.03 mg; D-pantothenic acid as D-calcium pantothenate, 23.5 mg; niacin as nicotinamide, 1.0 mg, and nicotinic acid, 43.0 mg; folic acid, 1.58 mg; biotin, 0.44 mg; Cu, 10 mg as copper sulfate; Fe, 125 mg as iron sulfate; I, 1.26 mg as potassium iodate; Mn, 60 mg as manganese sulfate; Se, 0.3 mg as sodium selenite; and Zn, 100 mg as zinc oxide.

DDGS and 0.65, 0.73, and 0.68% in the DDGS that was autoclaved for 10, 20, and 30 min, respectively. The calculated concentration of reactive Lysine was 0.80, 0.64, 0.72, and 0.67% for non-autoclaved DDGS and DDGS that was autoclaved for 10, 20, and 30 min, respectively. The Lysine:CP ratio was 2.94 in non-autoclaved DDGS, whereas the Lysine:CP ratio was 2.37, 2.75, and 2.51 in the batches that were autoclaved for 10, 20, and 30 min, respectively. The ADF concentration was 7.96% in non-autoclaved DDGS, whereas for autoclaved DDGS, the concentration of ADF was 11.05, 9.85, and 10.89% (10, 20, and 30 min, respectively). The concentration of ADIN in non-autoclaved DDGS was 0.12%, whereas the concentration of ADIN ranged from 0.42 to 0.55% in the autoclaved DDGS. Reducing sugar concentration was 0.78 in non-autoclaved DDGS, but DDGS that was autoclaved for 10, 20, and 30 min contained 0.60, 0.88, and 0.65% reducing sugars, respectively.

Digestibility of CP and AA

The AID of CP decreased (linear, $P < 0.05$) from 64.4% in the non-autoclaved DDGS to 59.0, 52.2, and 55.7% in the DDGS that was autoclaved for 10, 20, or 30 min, respectively (Table 4). The AID of Ile, Lysine, Methionine, Phenylalanine, and Aspartic acid was reduced (quadratic, $P < 0.05$) with increasing time of autoclaving, whereas the AID of all other AA was reduced (linear, $P < 0.05$) with increasing time of autoclaving. Lysine was the AA most affected by increasing time of autoclaving. The AID of Lysine was reduced (quadratic, $P <$ 0.05) from 61.2 to 48.0, 48.7, and 44.9% for the non-autoclaved DDGS and the DDGS that was autoclaved for 10, 20, or 30 min, respectively. The mean AID of indispensable AA was reduced (linear, $P < 0.05$) as time of autoclaving increased.

The SID of CP decreased (linear, $P < 0.05$) from 77.9% in the non-autoclaved DDGS to 72.1, 66.1, and 68.5% in the DDGS that was autoclaved for 10, 20, or 30 min, respectively (Table 5). The SID of Lysine was reduced (quadratic, $P < 0.05$) from 66.8% in the non-autoclaved DDGS to 54.9, 55.3, and 51.9% in the DDGS that was autoclaved for 10, 20, or 30 min, respectively.

The SID of Isoleucine, Methionine, Phenylalanine, and Aspartic acid also was reduced (quadratic, $P < 0.05$) with increasing time of autoclaving, whereas the SID of all other AA was reduced (linear, $P < 0.05$) with increasing time of autoclaving. The mean SID of indispensable AA also was reduced (linear, $P < 0.05$) as time of autoclaving increased.

Coefficients of linear correlation and regression equations

The concentrations of SID AA were correlated ($r_p >$ 0.70; $P < 0.01$) with the Lysine:CP ratio, except for the concentration of SID Threonine, which was only

Table 3 Analyzed composition of experimental diets, as-fed basis

Item	Distillers dried grains with solubles				N-free diet
		Autoclaved at 130°C			
	Non-autoclaved	10 min	20 min	30 min	
Dry matter, %	92.42	93.18	92.59	92.32	92.53
Crude protein (CP), %	14.96	15.59	14.62	15.76	0.28
Indispensable AA, %					
Arginine	0.73	0.68	0.70	0.67	-
Histidine	0.42	0.41	0.42	0.41	-
Isoleucine	0.56	0.57	0.58	0.55	-
Leucine	1.69	1.75	1.76	1.72	-
Lysine	0.51	0.42	0.44	0.41	-
Methionine	0.31	0.29	0.29	0.28	-
Phenylalanine	0.71	0.73	0.74	0.72	-
Threonine	0.61	0.61	0.62	0.60	-
Tryptophan	0.13	0.12	0.12	0.12	-
Valine	0.74	0.75	0.76	0.73	-
Dispensable AA, %					
Alanine	1.11	1.14	1.15	0.12	-
Aspartic acid	1.05	1.04	1.06	1.03	-
Cysteine	0.33	0.32	0.31	0.31	-
Glutamic acid	2.60	2.69	2.70	2.65	-
Glycine	0.64	0.65	0.66	0.64	-
Proline	1.30	1.24	1.27	1.23	-
Serine	0.78	0.79	0.79	0.78	-

moderately correlated with the Lysine:CP ratio (Table 6). The concentration of SID AA was poorly correlated ($r_p <$ 0.50) with the concentration of reducing sugars. The concentrations of SID AA were negatively correlated with the concentrations of ADF, NDF, lignin, and ADIN ($P < 0.01$). The concentration of SID of each AA also was well correlated ($r_p > 0.65$; $P < 0.01$) with its respective concentration of total AA, except for SID Threonine, which was only moderately correlated ($r_p < 0.65$; $P < 0.01$) with the concentration of total Threonine.

The concentrations of digestible AA may be predicted by regression equations presented in Table 7. The concentration of SID Arginine or SID Lysine were best predicted by equations that included the concentration of ADIN in the model ($r^2 = 0.76$ and 0.84, respectively). The concentration of most SID AA was predicted ($r^2 >$ 0.50) by the concentration of each total AA, but the concentration of SID Threonine was poorly ($r^2 = 0.36$) predicted by the concentration of Threonine.

Discussion

Composition of distillers dried grains with solubles

The concentrations of DM, ash, and CP were not affected by autoclaving DDGS and this observation supports results of González-Vega et al. [4] who reported that autoclaving soybean meal did not change concentrations of DM, ash, or CP. Changes in the concentration of Lysine observed for autoclaved DDGS in this experiment also agree with results of previous experiments in which the concentration of Lysine was decreased by heat treatment [2,3,6,22]. Based on the concentration of Lysine, these results indicate that autoclaving corn DDGS for 10 min was sufficient to cause heat damage, but increasing time of autoclaving to 20 and 30 min did not result in further decreases in Lysine concentration. In contrast, in a previous experiment, the concentration of Lysine in corn DDGS autoclaved for 0, 10, 20, or 30 min at 135°C was linearly decreased from 0.82 to 0.59% [6]. These differences may be a result of differences in sample preparation and autoclaving procedures. Although temperature and time of autoclaving were similar between the 2 experiments, in this experiment, DDGS was autoclaved on an as-is basis in quantities of 2.5 kg per tray whereas Fontaine et al. [6] ground DDGS to < 3 mm particle size and then autoclaved DDGS in 250 g quantities. The concentrations of furosine in DDGS used in this experiment were slightly less than the average concentration of furosine (0.02%;

Table 4 Apparent ileal digestibility of crude protein (CP) and AA in distillers dried grains with solubles subjected to increasing levels of heat treatment by weanling pigs[1]

	Distillers dried grains with solubles				SEM	P-value[2]	
		Autoclaved at 130°C					
Item	Non-autoclaved	10 min	20 min	30 min		Linear	Quadratic
CP, %	64.4	59.0	52.2	55.7	2.0	< 0.01	0.12
Indispensable AA, %							
Arginine	79.3	71.9	72.0	69.1	1.7	< 0.01	0.06
Histidine	73.9	66.8	65.9	66.0	1.3	< 0.01	0.23
Isoleucine	72.3	66.2	66.6	63.1	1.4	< 0.01	0.02
Leucine	84.0	80.7	80.4	79.1	0.8	< 0.01	0.12
Lysine	61.2	48.0	48.7	44.9	1.9	< 0.01	0.02
Methionine	83.6	77.7	77.8	75.7	0.9	< 0.01	< 0.01
Phenylalanine	79.8	75.1	75.4	73.6	1.0	< 0.01	0.04
Threonine	60.2	53.9	53.2	51.2	1.6	< 0.01	0.27
Tryptophan	55.1	43.7	42.8	41.5	2.0	< 0.01	0.15
Valine	71.9	65.7	65.5	62.6	1.4	< 0.01	0.08
Mean	75.6	70.0	69.9	67.8	1.1	< 0.01	0.07
Dispensable AA, %							
Alanine	77.5	73.3	71.4	71.0	1.3	< 0.01	0.86
Aspartic acid	65.5	55.2	55.4	53.3	1.5	< 0.01	0.02
Cysteine	71.9	64.7	60.5	62.7	1.5	< 0.01	0.54
Glutamic acid	81.7	76.7	75.6	75.3	0.9	< 0.01	0.34
Glycine	43.4	36.3	29.0	31.7	4.6	< 0.01	0.47
Proline	69.6	65.1	64.1	63.2	1.3	< 0.01	0.47
Mean	68.3	61.8	59.2	59.5	1.5	< 0.01	0.85

[1]Data are means of 8 observations, except for the distillers dried grains with solubles diet that was autoclaved for 10 min (n = 7).
[2]Linear and quadratic effects of time of autoclaving.

CV = 91.4%) measured in 21 sources of DDGS, and these differences may be a result of the high degree of variation when determining the concentration of furosine in DDGS [8]. The reason for the reduced Lysine:CP ratio in the autoclaved DDGS is that only the concentration of Lysine, but not the concentration of CP, is reduced when heat damage occurs [8,23]. The Lysine:CP ratio also was reduced in soybean meal that was heat-damaged compared with unheated soybean meal [4]. The concentration of Lysine in the non-autoclaved DDGS was slightly greater than the average concentration of Lysine (0.76%) in 39 sources of DDGS [24]. Likewise, the Lysine:CP ratio was slightly greater in the unheated DDGS compared with the average value (2.77) reported previously [24]. However, the concentrations of Lysine and the Lysine:CP ratio in autoclaved DDGS are close to the average values reported by Stein and Shurson [24] for corn DDGS, which indicates that the heat damage caused by autoclaving simulated the varying degrees of heat damage caused by processing of DDGS in commercial production facilities.

The concentration of ADIN has been used as a predictor of heat damage in plant proteins and our results agree with these observations [25]. The changes observed between the concentrations of reducing sugars in the unheated DDGS and the 3 autoclaved DDGS were expected because, in the initial steps of the Maillard reactions, reducing sugars react with the ε-amino group of Lysine to form fructoselysine and maltuloselysine, respectively [26].

Digestibility of CP and AA

Values for the AID of CP and AA in the DDGS that was not autoclaved are in agreement with results from previous experiments [11,24]. Results of previous experiments also indicated that the digestibility of Lysine was more reduced by heat damage than the digestibility of other AA [4,22]. Values for the SID of CP and AA in the non-autoclaved DDGS were within the range of values for the SID of CP and AA in corn DDGS observed in previous experiments [24]. This observation is supported by the value for the Lysine:CP ratio (2.94) determined for

Table 5 Standardized ileal digestibility of crude protein (CP) and AA in distillers dried grains with solubles subjected to increasing levels of heat treatment by weanling pigs[1]

	Distillers dried grains with solubles				SEM	*P*-value[2]	
		Autoclaved at 130°C					
Item	Non-autoclaved	10 min	20 min	30 min		Linear	Quadratic
CP, %	77.9	72.1	66.1	68.5	2.0	< 0.01	0.25
Indispensable AA, %							
Arginine	87.9	81.3	81.0	78.5	1.7	< 0.01	0.11
Histidine	78.1	71.2	70.1	70.3	1.3	< 0.01	0.27
Isoleucine	77.3	71.1	71.4	68.2	1.4	< 0.01	0.03
Leucine	86.7	83.3	83.1	81.8	0.8	< 0.01	0.12
Lysine	66.8	54.9	55.3	51.9	1.9	< 0.01	0.04
Methionine	86.1	80.4	80.5	78.5	0.9	< 0.01	< 0.01
Phenylalanine	83.7	79.0	79.2	77.5	1.0	< 0.01	0.05
Threonine	70.2	64.0	63.1	61.3	1.6	< 0.01	0.33
Tryptophan	65.9	55.5	54.5	53.2	2.0	< 0.01	0.20
Valine	77.2	71.0	70.7	67.9	1.4	< 0.01	0.09
Mean	80.7	75.2	74.9	73.1	1.1	< 0.01	0.09
Dispensable AA, %							
Alanine	83.3	79.1	77.1	76.8	1.3	< 0.01	0.89
Aspartic acid	72.3	62.1	62.1	60.2	1.5	< 0.01	0.03
Cysteine	77.7	70.8	66.8	68.9	1.5	< 0.01	0.55
Glutamic acid	85.2	80.1	79.0	78.7	0.9	< 0.01	0.33
Glycine	73.4	66.0	58.1	61.7	4.6	< 0.01	0.39
Proline	76.9	72.4	71.3	70.5	1.3	< 0.01	0.48
Mean	78.1	71.7	69.0	69.4	1.5	< 0.01	0.91

[1]Data are means of 8 observations, except for the distillers dried grains with solubles diet that was autoclaved for 10 min (n = 7); Values for standardized ileal digestibility were calculated by correcting apparent ileal digestibility values for basal endogenous losses (g/kg of DMI), which were determined by feeding pigs a N-free diet: CP, 21.93; Arginine, 0.68; Histidine, 0.19; Isoleucine, 0.30; Leucine, 0.50; Lysine, 0.31; Methionine, 0.08; Phenylalanine, 0.30; Threonine, 0.66; Tryptophan, 0.15; Valine, 0.43; Alanine, 0.71; Aspartic acid, 0.77; Cysteine, 0.21; Glutamic acid, 0.98; Glycine, 2.08; and Serine, 0.62.
[2]Linear and quadratic effects of time of autoclaving.

corn DDGS in this experiment, which is in agreement with the average value (2.77) reported previously [24]. For the autoclaved DDGS, however, values for the SID of CP and AA were less than those reported by Stein and Shurson [24]. Reductions observed for the SID of Lysine with increasing time of autoclaving and the results from reducing sugars and the Lysine concentrations in autoclaved DDGS indicate that the conditions created by autoclaving of DDGS (i.e., heat, moisture, and pressure) were favorable to initiation of Maillard reactions, which renders Lysine unavailable to the animals and, therefore, reduces Lysine digestibility [4,27]. This indicates that heating alone may not be detrimental to the nutritional quality of DDGS, unless combined with a certain degree of moisture. The nutritional quality of soybean meal heat-processed in conventional oven was not different from that of conventional soybean meal, which is in agreement with the latter assumption [4]. It appears, however, that the degree of heat damage among autoclaved DDGS (i.e., 10, 20, or 30 min) was not different and this observation is supported by the fact that the SID of Lysine among these ingredients was not different. Reductions in the digestibility of AA other than Lysine have been attributed to the formation of cross-linkages between protein chains [28]. When these cross-links are formed, digestive enzymes such as trypsin have reduced access to the proteins, thus the ability of trypsin to hydrolyze peptide bonds is reduced. Exposure to excessive heat and pressure may also lead to AA racemization, thus converting L-AA to D-AA [29]. Proteolytic enzymes are unable to hydrolyze peptide bonds connecting D-AA and L-AA. Consequently, the digestibility is reduced not only for the D-AA, but also for the L-AA [30].

Coefficients of linear correlation and regression equations

The relatively good correlations between the concentration of SID AA and the concentration of each AA in DDGS agree with previous results [8]. Consequently,

Table 6 Coefficients of linear correlation between nutrient composition of distillers dried grains with solubles and the concentration (%) of standardized ileal digestible (SID) AA[1]

	Nutrient composition															
Item	Lys:CP	Reducing sugars	ADF	NDF	Lignin	ADIN[2]	Arg	His	Ile	Leu	Lys	Met	Phe	Thr	Trp	Val
SID Arg	0.80^{*}	0.50^{*}	-0.86^{*}	-0.60^{*}	-0.62^{*}	-0.87^{*}	0.84^{*}	-	-	-	-	-	-	-	-	-
SID His	0.72^{*}	0.06^{ns}	-0.81^{*}	-0.69^{*}	-0.48^{*}	-0.82^{*}	-	0.69^{*}	-	-	-	-	-	-	-	-
SID Ile	0.76^{*}	0.49^{*}	-0.81^{*}	-0.58^{*}	-0.55^{*}	-0.81^{*}	-	-	0.70^{*}	-	-	-	-	-	-	-
SID Leu	0.72^{*}	0.45^{**}	-0.81^{*}	-0.54^{*}	-0.65^{*}	-0.82^{*}	-	-	-	0.78^{*}	-	-	-	-	-	-
SID Lys	0.83^{*}	0.45^{**}	-0.91^{*}	-0.74^{*}	-0.54^{*}	-0.92^{*}	-	-	-	-	0.89^{*}	-	-	-	-	-
SID Met	0.78^{*}	0.40^{**}	-0.86^{*}	-0.73^{*}	-0.48^{*}	-0.87^{*}	-	-	-	-	-	0.75^{*}	-	-	-	-
SID Phe	0.78^{*}	0.47^{*}	-0.85^{*}	-0.63^{*}	-0.56^{*}	-0.86^{*}	-	-	-	-	-	-	0.83^{*}	-	-	-
SID Thr	0.58^{*}	0.25^{ns}	-0.70^{*}	-0.55^{*}	-0.56^{*}	-0.73^{*}	-	-	-	-	-	-	-	0.60^{*}	-	-
SID Trp	0.71^{*}	0.34^{ns}	-0.81^{*}	-0.70^{*}	-0.52^{*}	-0.83^{*}	-	-	-	-	-	-	-	-	0.84^{*}	-
SID Val	0.73^{*}	0.45^{**}	-0.78^{*}	-0.59^{*}	-0.50^{*}	-0.78^{*}	-	-	-	-	-	-	-	-	-	0.55^{*}

[1]*Arg*, Arginine; *His*, Histidine; *Ile*, Isoleucine; *Leu*, Leucine; *Lys*, Lysine; *Met*, Methionine; *Phe*, Phenylalanine; *Thr*, Threonine; *Trp*, Tryptophan; *Val*, Valine.
[2]*ADIN*, acid detergent insoluble nitrogen.
$^{*}= P < 0.01$; $^{**}= P < 0.05$; ns = not significant.

Table 7 Linear regression to predict the concentration (%) of standardized ileal digestible (SID) AA from the nutrient concentrations (%) in corn distillers dried grains with solubles (DDGS) fed to pigs[1]

	Intercept			Independent variable[2]				
Dependent variable	Estimate	SE	*P*-value	Variable	Estimate	SE	*P*-value	r^2
SID Arginine	−0.66	0.20	< 0.01	Arg	1.398	0.171	< 0.01	0.70
	1.16	0.02	< 0.01	ADIN	−0.506	0.053	< 0.01	0.76
SID Histidine	−0.62	0.22	< 0.01	His	1.639	0.322	< 0.01	0.47
	0.58	0.01	< 0.01	ADIN	−0.203	0.025	< 0.01	0.68
SID Isoleucine	−0.92	0.30	< 0.01	Ile	1.706	0.320	< 0.01	0.50
	0.15	0.08	0.08	Lys:CP	0.200	0.031	< 0.01	0.58
SID Leucine	−2.23	0.69	< 0.01	Leu	1.623	0.243	< 0.01	0.61
	2.61	0.03	< 0.01	ADIN	−0.554	0.071	< 0.01	0.67
SID Lysine	−0.46	0.08	< 0.01	Lys	1.214	0.115	< 0.01	0.79
	0.60	0.02	< 0.01	ADIN	−0.463	0.037	< 0.01	0.84
SID Methionine	−0.39	0.13	< 0.01	Met	1.53	0.252	< 0.01	0.56
	0.46	0.01	< 0.01	ADIN	−0.161	0.017	< 0.01	0.76
SID Phenylalanine	−1.11	0.26	< 0.01	Phe	1.726	0.217	< 0.01	0.68
	1.08	0.01	< 0.01	ADIN	−0.304	0.034	< 0.01	0.73
SID Threonine	−1.31	0.50	< 0.05	Thr	1.967	0.499	< 0.01	0.36
	0.75	0.02	< 0.01	ADIN	−0.247	0.043	< 0.01	0.54
SID Tryptophan	−0.24	0.04	< 0.01	Trp	1.767	0.219	< 0.01	0.70
	−0.04	0.03	0.19	Lys:CP	0.061	0.011	< 0.01	0.51
SID Valine	−0.88	0.50	0.08	Val	1.438	0.400	< 0.01	0.31
	0.24	0.11	0.04	Lys:CP	0.246	0.042	< 0.01	0.54

[1]n = 31 observations; for all models $P < 0.01$.
[2]*ADIN*, acid detergent insoluble nitrogen; *Arg*, Arginine; *His*, Histidine; *Ile*, Isoleucine; *Leu*, Leucine; *Lys*, Lysine; *Met*, Methionine; *Phe*, Phenylalanine; *Thr*, Threonine; *Trp*, Tryptophan; *Val*, Valine.

analyzed concentrations of most AA in DDGS may be used to predict the concentration of SID AA in DDGS although that is not the case for Threonine. It has been observed that heat damage of feed ingredients is associated with an increase in analyzed values for ADF, lignin, and ADIN, which results in a decrease in N digestibility in cattle and sheep [30]. The concentrations of ADF and ADIN is increased in dark-colored DDGS, which indicates a greater degree of heat damage in such ingredients [10], and likely a lower digestibility of AA. The negative correlation between the concentration of SID AA and the concentrations of ADF, lignin, or ADIN observed in this experiment support the latter observations.

The concentration of SID Lysine can accurately be predicted from the concentration of Lysine. This observation agrees with Kim et al. [8]. Prediction of the concentration of SID Lysine in this study was slightly improved if the concentration of ADIN was used in the model. This indicates that the concentration of analyzed ADIN may be used to predict the concentration of SID Lysine in DDGS.

Conclusions

Results of this experiment confirmed that the concentration and digestibility of AA in DDGS is reduced as a result of heat damage. The concentrations of Lysine and ADIN are good predictors of the concentration of most AA in heat-damaged DDGS, which indicates that such predictors may be used as indicators of the protein quality in DDGS. These equations, however, must be validated using different sources of DDGS.

Abbreviations

AA: Amino acids; ADF: Acid detergent fibre; ADIN: Acid detergent insoluble nitrogen; AID: Apparent ileal digestibility; BW: Body weight; CDS: Condensed solubles; CP: Crude protein; CV: Coefficient of variation; DDGS: Distillers dried grains with solubles; DM: Dry matter; HPLC: High performance liquid chromatography; ME: Metabolizable energy; NDF: Neutral detergent fibre; SID: Standardized ileal digestibility.

Competing interests

The authors declare that they have no competing interests.

Authors' contributions

All authors read and approved the final manuscript.

Acknowledgements

Financial support for this research from Enonik Industries AG, Rodenbacher Chaussee 4, 63457, Hanau, Germany, is appreciated.

Author details

[1]Department of Animal Sciences, University of Illinois, 1207 W Gregory Dr, Urbana, IL 61801, USA. [2]Evonik Industries AG, Rodenbacher Chaussee 4, Hanau 63457, Germany. [3]Evonik Degussa Corporation, 1701 Barrett Lakes Blvd NW, Kennesaw, GA 30144, USA.

References

1. Rosentrater KA, Ileleji K, Johnston DB: **Manufacturing of fuel ethanol and distillers grains – current and evolving process.** In *Distillers Grains Production, Properties, and Utilization.* Edited by Liu K, Rosentrater KA. Boca Raton, FL: CRC Press; 2012:73–102.
2. Pahm AA, Pedersen C, Stein HH: **Application of the reactive lysine procedure to estimate lysine digestibility in distillers dried grains with solubles fed to growing pigs.** *J Agric Food Chem* 2008, **56:**9441–9446.
3. Boucher SE, Pedersen C, Stein HH, Schwab CG: **Evaluation of the furosine and homoarginine methods for determining reactive Lys in rumen-undegraded protein.** *J Dairy Sci* 2009, **92:**3951–3958.
4. González-Vega JC, Kim BG, Htoo JK, Lemme A, Stein HH: **Amino acid digestibility in heated soybean meal fed to growing pigs.** *J Anim Sci* 2011, **89:**3617–3625.
5. Rutherfurd SM, Moughan PJ: **Development of a novel bioassay for determining the available Lys contents of foods and feedstuffs.** *Nutr Res Rev* 2007, **20:**3–16.
6. Fontaine J, Zimmer U, Moughan PJ, Rutherfurd SM: **Effect of heat damage in an autoclave on the reactive lysine contents of soy products and corn distillers dried grains with solubles. Use of the results to check on lysine damage in common qualities of these ingredients.** *J Agric Food Chem* 2007, **55:**10737–10743.
7. Cozannet P, Primot Y, Gady C, Métayer JP, Callu P, Lessire M, Skiba F, Noblet J: **Ileal digestibility of amino acids in wheat distillers dried grains with solubles for pigs.** *Anim Feed Sci Technol* 2010, **158:**177–186.
8. Kim BG, Kil DY, Zhang Y, Stein HH: **Concentrations of analyzed or reactive lysine, but not crude protein, may predict the concentration of digestible lysine in distillers dried grains with solubles (DDGS) fed to pigs.** *J Anim Sci* 2012, **90:**3798–3808.
9. Stein HH: **Feeding ethanol coproducts to swine.** In *Distillers Grains Production, Properties, and Utilization.* Edited by Liu K, Rosentrater KA. Boca Raton, FL: CRC Press; 2012:297–315.
10. Cromwell GL, Herkelman KL, Stahly TS: **Physical, chemical, and nutritional characteristics of distillers dried grains with solubles for chicks and pigs.** *J Anim Sci* 1993, **71:**679–686.
11. Fastinger ND, Mahan DC: **Determination of the ileal amino acid and energy digestibilities of corn distillers dried grains with solubles using grower-finisher pigs.** *J Anim Sci* 2006, **84:**1722–1728.
12. Stein HH, Shipley CF, Easter RA: **Technical note: a technique for inserting a T-cannula into the distal ileum of pregnant sows.** *J Anim Sci* 1998, **76:**1433–1436.
13. NRC: *Nutrient requirements of swine.* 10th edition. Washington, DC: Natl Acad Press; 1998.
14. Llames CR, Fontaine J: **Determination of amino-acids in feeds – collaborative study.** *J AOAC Int* 1994, **77:**1362–1402.
15. Directive C: **Establishing community methods for the determination of amino acids, crude oils and fats, and olaquindox in feeding stuff and amending directive 71/393/EEC, annex part a. Determination of amino acids.** *Official J European Communities* 1998, **L257:**14–23.
16. Directive C: **Establishing community methods for the determination of vitamin A, vitamin E and tryptophan, annex part C. Determination of Tryptophan.** *Official J European Communities* 2000, **L174:**45–50.
17. AOAC: *Official methods of analysis.* 18th edition. Gaithersburg, MD: Association of Official Analytical Chemists; 2007.
18. Holst DO: **Holst filtration apparatus for Van Soest detergent fiber analysis.** *J AOAC* 1973, **56:**1352–1356.
19. Sanderson P: **A new method of analysis of feedstuffs for the determination of crude oils and fats. 1986.** In *Recent Advances in Animal Nutrition.* Edited by Haresign W, Cole DJA. London, UK: Butterworths; 1986:77–81.
20. Dubois M, Gilles KA, Hamilton JK, Rebers PA, Smith F: **Colorimetric method for determination of sugars and related substances.** *Anal Chem* 1956, **28:**350–356.
21. Stein HH, Seve B, Fuller MF, Moughan PJ, de Lange CFM: **Invited review: Amino acid bioavailability and digestibility in pig feed ingredients: Terminology and application.** *J Anim Sci* 2007, **85:**172–180.
22. Martinez-Amezcua C, Parsons CM, Singh V, Srinivasan R, Murthy GS: **Nutritional characteristics of corn distillers dried grains with solubles as affected by the amounts of grains versus solubles and different processing techniques.** *Poult Sci* 2007, **86:**2624–2630.
23. Stein HH, Connot SP, Pedersen C: **Energy and nutrient digestibility in four sources of distillers dried grains with solubles produced from corn**

grown within a narrow geographical area and fed to growing pigs. *Asian-australas J Anim Sci* 2009, **22:**1016–1025.

24. Stein HH, Shurson GC: **Board-invited review: The use and application of distillers dried grains with solubles in swine diets.** *J Anim Sci* 2009, **87:**1292–1303.
25. Schroeder GE, Erasmus LJ, Leeuw KJ, Meissner HH: **The use of acid detergent insoluble nitrogen to predict digestibility of rumen undegradable protein of heat processed plant proteins.** *S Afr J Anim Sci* 1996, **26:**49–52.
26. Erbersdobler HF, Hupe A: **Determination of lysine damage and calculation of lysine bio-availability in several processed foods.** *Z Ernahrungswiss* 1991, **30:**46–49.
27. Nursten H: *The Maillard Reaction. Chemistry, Biochemistry, and Implications.* Cambridge, UK: Royal Society of Chemistry; 2005.
28. Zagon J, Dehne LI, Bögl KW: **D-amino acids in organisms and food.** *Nutr Res* 1994, **14:**445–463.
29. Finot PA: **The absorption and metabolism of modified amino acids in processed foods.** *J AOAC Int* 2005, **88:**894–902.
30. Broesder JT, Gunter SA, Krysl LJ, Judkins MB: **Digestion of ruminal masticate dried by three different methods.** *Anim Feed Sci Technol* 1992, **37:**323–331.

18

Effects of inorganic or organic selenium on immunoglobulins in swine

Ashley Gelderman and Jeffrey Clapper*

Abstract

A study was conducted to determine if Se source fed during gestation and lactation affects passive transfer of immunoglobulins. Sixty days prior to breeding (d −60), gilts were randomly assigned to one of three treatments prior to breeding and throughout gestation: control (Control, no supplemental Se; n = 8), inorganic Se (Inorganic Se, 0.3 ppm; n = 4) and organic Se (Organic Se, 0.3 ppm; n = 4). Blood was collected on d −60, 57 and 113 of gestation and on d 21 of lactation and milk was collected at d 0, 1, 7, 14, and 21 of lactation. Blood was collected from piglets at d 0, 1, 7, 14, and 21 of age. Gilts fed organic Se had greater ($P < 0.05$) circulating concentrations of Se than Inorganic and Control gilts. Regardless of treatment, circulating concentrations of Se were greatest ($P < 0.05$) at d −60 compared to all other days. Serum concentrations of IgG were greatest ($P < 0.05$) in gilts at d 57 of gestation compared to d 113. Serum concentrations of IgA were greatest ($P < 0.05$) on d 113 of gestation and d 21 of lactation compared to d −60 and 57. Serum concentrations of IgM were greater ($P < 0.05$) at d 57 compared to d −60. Inorganic gilts had greater ($P < 0.05$) colostral and milk concentrations of IgG and IgM than Organic or Control gilts. Circulating concentrations of Se in piglets were greatest ($P < 0.05$) at d 14 and 21 of age compared to all other days. Piglets from gilts supplemented with organic Se had greater ($P < 0.05$) circulating concentrations of Se on d 1 versus piglets from gilts supplemented with no additional Se. The immunoglobulin concentrations of IgG, IgA, and IgM were lowest ($P < 0.05$) on d 0 and then increased when compared to d 1. The addition of different sources of Se did not affect the immunoglobulin concentrations in the gilt or piglet.

Keywords: Immunoglobulin, Pig, Selenium

Background

Traditionally, Se is known to be toxic or harmful to animals in certain regions in the U.S., but Se deficiency is a more widespread problem. Selenium was not considered important until Eggert et al. [1] reported that Se is an essential nutrient in swine because it helped prevent hepatosis diaetetica. One of the main biochemical functions of Se is as a component of glutathione peroxidase [2]. Typically, Se is provided as a supplement in swine diets, either in the inorganic (sodium selenite) or organic (enriched Se yeast) forms. The current recommendation for the inclusion of Se in diets is 0.30 ppm [3].

Feeding organic Se compared to inorganic Se resulted in greater serum concentrations of Se in the females as well as progeny [4] and greater colostral concentrations of Se [5]. When feeding either inorganic or organic Se forms, there were no differences in total pigs born, litter average daily gain (ADG), or litter weight [6]. Furthermore, the addition of either inorganic or organic Se at 0.3 ppm reduced the number of stillborn piglets [7]. At birth, piglets had greater whole blood concentrations of Se when born to sows fed organic Se than piglets born to sows fed the control diet [7].

Colostrum and milk are essential for the survival of piglets as sources of digestible nutrients [8] and helps to establish passive immunity [9,10]. Increased IgG absorption and enhanced humoral immune function has been shown in diets with supplemental Se [11,12]. Research regarding the effects of adding Se to maternal diets to increase serum concentrations of IgG is equivocal [13-15]. Sows injected with Se had increased colostral concentrations of IgM compared to other treatment groups, indicating a basis for enhanced passive transfer of antibodies [16]. Beef cows supplemented with Se have also been shown to have increased colostral concentrations of IgG [17]. Thus, Se

* Correspondence: Jeffrey.Clapper@sdstate.edu
Department of Animal Science, South Dakota State University, Brookings 57007SD, USA

supplementation may prove useful in enhancing the immune profile in the newborn pig. Therefore, the objective of the current study was to determine if diets supplemented with an organic or inorganic source of Se would increase whole blood Se and serum concentrations of IgG, IgA, and IgM in gilts and their progeny.

Methods

Experimental design

Twenty-seven crossbred gilts (Yorkshire × Landrace × Hampshire; 130 ± 6.44 kg; approximately 240 d of age) were used in this experiment. Gilts were fed 1.81 kg/d of their respective diet beginning 60 d prior to breeding. Gilts were randomly assigned to one of the three dietary treatments: control (Control) diet with no supplemental selenium added (n = 9), an inorganic source of Se (Inorganic) with 0.3 ppm of Se added as sodium selenite (n = 9), and an organic source of Se (Organic) with 0.3 ppm of Se added as selenized yeast (n = 9). Diets were formulated to meet the 1998 NRC requirements for gestating swine (Table 1). Concentrations of Se in the corn and soybean meal used in the gestation and lactation diets were 0.02 ppm and 0.759 ppm, respectively. Diets contained 0.182 ppm, 0.469 ppm, and 0.419 ppm of Se for the control, inorganic source, and organic source diets, respectively. Gilts were individually housed in a 1.2 m × 1.8 m partially slatted pen in an environmentally controlled room with ad libitum access to water. Each gilt was fed 15 mg altrenogest (Matrix) each day for 15 d to synchronize estrus. Estrus detection was performed by exposing gilts to a mature boar twice daily for 20 min each, beginning the third day after cessation of the altrenogest treatment and continuing for 4 d. The first day the gilt stood immobile in the presence of the boar was designated as d 1 of the estrous cycle. Gilts were bred by artificial insemination 2 h after their initial display of standing estrus. Pregnancy was confirmed by ultrasound 28 d post breeding. Of the 27 females bred, 16 gilts became pregnant; Control (n = 8), Inorganic (n = 4), and Organic (n = 4).

Gilts were bled via jugular venipunture 60 d prior to breeding, d −60, (2 h after eating), d 57 and 113 of gestation, and d 21 of lactation. Blood was separated into two aliquots of approximately 2.5 mL each to be used for analysis of whole blood concentrations of Se and serum concentrations of immunoglobulins. Serum was collected by allowing blood samples to clot overnight at 4°C before centrifugation (1,500 × *g* for 30 min) and storage at −20°C. At 5 and 2 wk pre-farrowing, gilts were administered a 2 mL dose of Rhinogen BPE (Intervet, Inc., Millsboro, DE) intramuscularly. Gilts were transferred to their farrowing pens on d 107 of gestation where they had ad libitum access to water and fed 1.81 kg/d of their respective diets. Parturition was induced by administering 10 mg $PGF_{2\alpha}$, i.m. (Pfizer; New York, NY) on d 113 to ensure attendance at the start of farrowing. During lactation, diets (Table 2) were offered ad libitum and formulated to meet or exceed 1998 NRC recommendations for lactating sows. The concentration of Se in the lactation diets were 0.272 ppm, 0.523 ppm, and 0.547 ppm for the control, inorganic source, and organic source, respectively. Sows were milked at farrowing after birth of the first piglet (d 0) and on d 1, 7, 14, and 21 of lactation for determination of colostral and milk concentrations of Se and immunoglobulins. Prior to milking on d 1, 7, 14, and 21, piglets were removed for at least 1 h and dams were administered 10 mg of oxytocin i.m. (VetTech; Shippack, PA) to facilitate milk let down. Colostrum and milk samples were collected and immediately frozen at −20°C for subsequent analysis of Se and immunoglobulin concentrations.

To ensure that no pig had suckled prior to sampling, piglets were removed from their dam, dried off, and placed in plastic totes that had a covering of wood chips at the bottom and a heat lamp located above the tote. Three piglets from each litter were randomly selected and bled via

Table 1 Maternal diets formulated to meet or exceed 1998 NRC recommendations for lactating sows

	Lactation diets		
Ingredient, %	**No Se**	**Inorganic Se**	**Organic Se**
Corn	65.15	65.15	65.15
Soybean meal, 44%	27.50	27.50	27.50
Soybean oil	4.00	4.00	4.00
Dical, dynafos	1.60	1.60	1.60
Limestone	0.90	0.90	0.90
Salt	0.40	0.40	0.40
Vitamin premix[3]	0.30	0.30	0.30
Trace mineral	0.15[1]	0.15[2]	0.15[1]
Se[4]	-	-	0.05

[1]Trace mineral mix containing no sodium selenite.
[2]Trace mineral mix containing at least 200 ppm Se as sodium selenite.
[3]Vitamin premix contained the following quantities of vitamins and microminerals 4,545,455 IU vitamin A/kg, 681,818 IU vitamin D_3/kg, 39,182 IU vitamin E/kg, 1,818 mg vitamin B12/kg, 1,818 mg menadione/kg, 164 mg biotin/kg, 6,250 mg pyridoxine/kg, 4,091 mg riboflavin/kg, 1,364 mg thiamine/kg, 110,000 mg Zn/kg, 110,000 mg Fe/kg, 29,000 mg Mn/kg, 11,000 mg Cu/kg, 240 mg I/kg, 200 mg Se/kg.
[4] Sel-Plex® premix: Total selenium 600 ppm.

Table 2 Serum concentrations of IgG, IgA, IgM in gilts fed no additional Se supplementation (control), inorganic source of Se supplementation, or an organic source of Se supplementation

	Treatment groups			
Item	**Control (n = 8)**	**Inorganic Se (n = 4)**	**Organic Se (n = 4)**	***P*-value**
IgG, mg/mL	8.56 ± 1.16	9.48 ± 1.64	9.78 ± 1.65	0.805
IgA, mg/mL	0.25 ± 0.09	0.20 ± 0.13	0.37 ± 0.13	0.627
IgM, mg/mL	3.16 ± 0.42	3.04 ± 0.59	3.29 ± 0.59	0.958

Data are expressed as least square means ± SEM.

jugular venipuncture prior to suckling (d 0) and on d 1, 7, 14, and 21 for determination of whole blood Se and serum concentrations of immunoglobulin. Blood was separated into two aliquots to be used for analysis of whole blood concentrations of Se and serum concentrations of immunoglobulins. Serum was collected by allowing blood samples to clot overnight at 4°C before centrifugation (1,500 × *g* for 30 min) and storage at −20°C. Whole blood was frozen at −20°C for later analysis of circulating concentrations of immunoglobulins.

All procedures involving animals were approved by the Institutional Animal Care and Use Committee of South Dakota State University. All animals were housed and cared for in accordance with the Guide for the Care and Use of Animals in Agriculture Research (2010).

Immunoglobulin analysis

Serum concentrations of IgG, IgA, and IgM were quantified in piglet and sow serum by ELISA (Bethyl Laboratories, IgA E101-102, IgG E101-104, and IgM E101-100, Montgomery, TX). The assay was conducted in 96-well, high binding microtiter plates (NUNC-Immuno Plate, 446612, VWR International Batavia, IL). The assay for each immunoglobulin was conducted according to manufacturer recommendations. Standards were prepared according to manufacturer's instructions and pipetted into duplicate wells. Sow serum was diluted in sample/conjugate diluent (50 mmol/L Tris, 0.14 mol/L NaCl, 1% BSA, 0.05% Tween 20) to 1:120,000, 1:1,000, and 1:10,000 for IgG, IgA, and IgM, respectively. Sow colostrum and milk samples were diluted in sample/conjugate diluent to 1:250,000 (d 0), 1:100,000 (d 1), and 1:10,000 (d 7, 14, and 21) for IgG. Sow colostrum and milk samples were diluted to 1:100,000 (d 0), 1:25,000 (d 1), and 1:10,000 (d 7, 14, and 21) for analysis of IgA. Colostrum and milk samples for IgM were diluted to 1:25,000 (d 0) and 1:10,000 (d 1, 7, 14, and 21). Piglet serum was diluted to 1:40,000, 1:5,000, and 1:5,000 for IgG, IgA, and IgM, respectively.

Absorbance was read at 450 nm (Molecular Dynamics, Spectramax Plus 384). Intra assay CVs were 5.02%, 4.96%, and 3.94% for IgG, IgA, and IgM, respectively. The inter assay CVs were 13.70%, 16.18%, and 14.40% for IgG, IgA, and IgM, respectively. Sensitivity of the ELISA was 5.19 ng/mL, 12.01 ng/mL, and 12.23 ng/mL for IgG, IgA, and IgM, respectively.

Milk and colostrum samples

After thawing, colostrum samples were centrifuged at 9,700 × *g* at 4°C for 20 min and milk samples for 10 min. Skim milk was collected while the fat was discarded. Skim milk was then centrifuged at 41,000 × *g* at 4°C for 45 min for colostrum and 20 min for milk. The resulting fraction was saved and frozen at −20°C for analysis of immunoglobulin and selenium content while the casein fraction was discarded.

Whole blood selenium samples

Gilt and piglet whole blood Se concentrations in gilts and piglets samples were determined by fluorometric method [18] at the Olson Biochemistry Lab located on the campus of South Dakota State University.

Statistical analysis

Effect of treatment on serum concentrations of immunoglobulins, whole blood concentrations of Se, and milk and colostral concentrations of immunoglobulins were

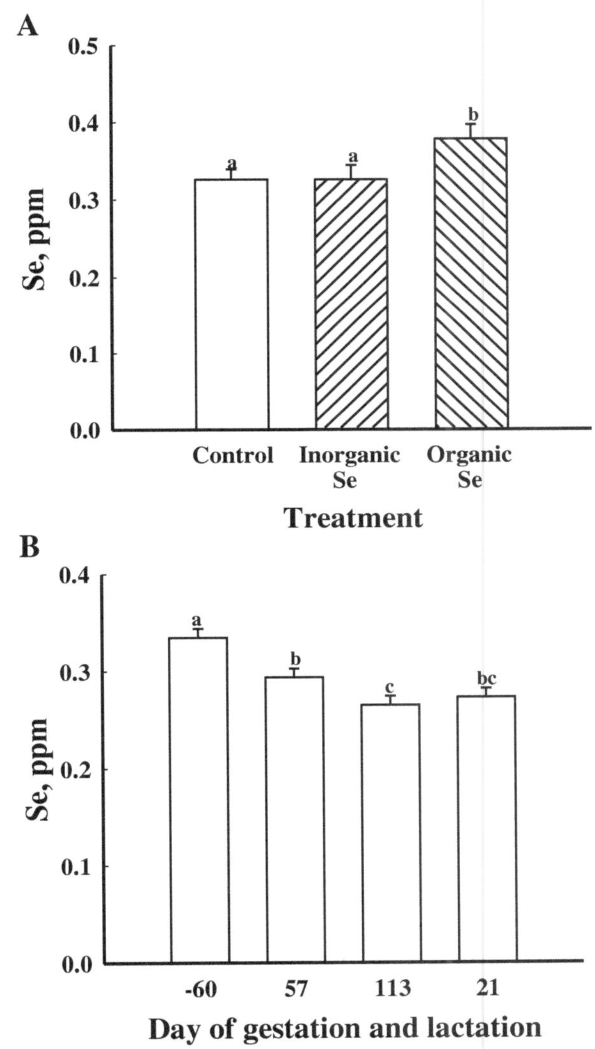

Figure 1 Mean whole blood concentrations of Se in gilts fed no additional Se (Control; n = 8), inorganic Se (n = 4), and organic Se (n = 4), (A) and (B) mean whole blood concentrations of Se in gilts (n = 16) at day −60, 57, and 113 of gestation and d 21 of lactation. Data are expressed as least-square means ± SEM. [abc]Means with different letters differ ($P < 0.05$).

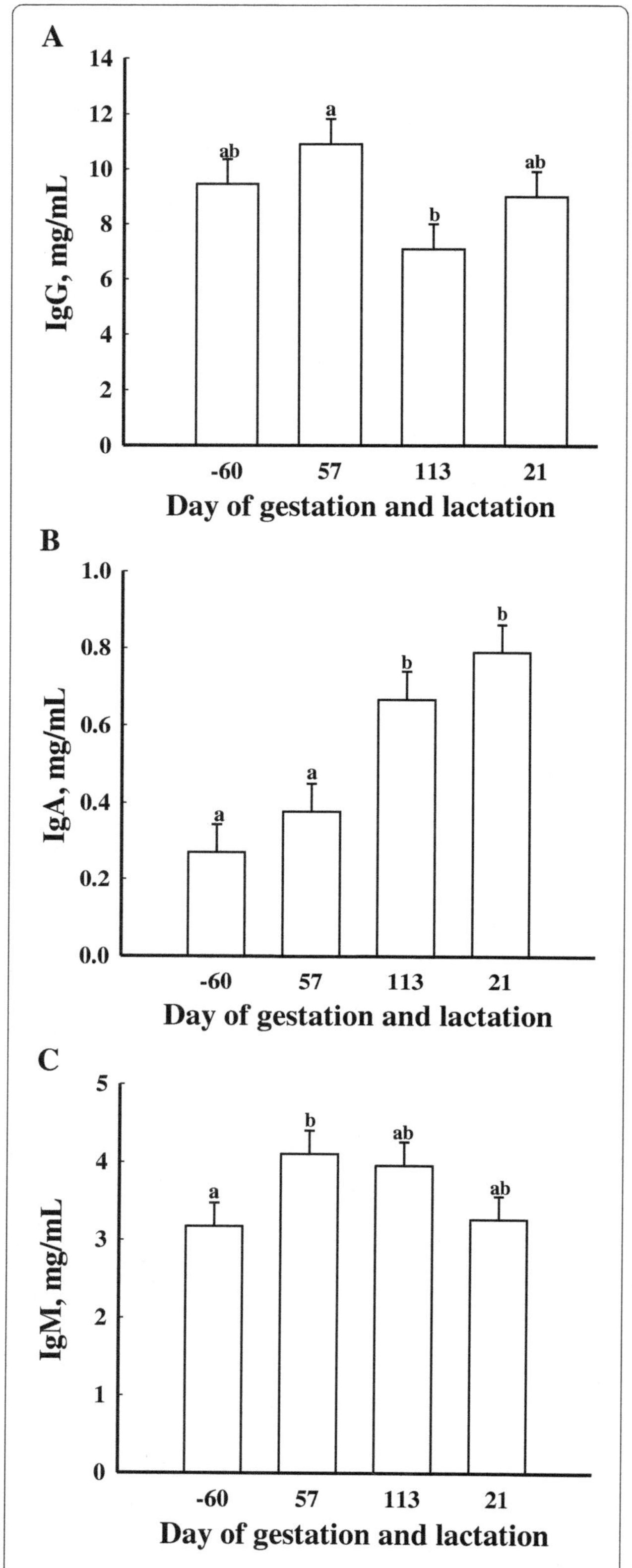

Figure 2 Mean serum concentrations of IgG (A), IgA (B), and IgM (C) in gilts (n = 16) at d −60, 57, and 113 of gestation and d 21 of lactation. Data are expressed as least-square means ± SEM. [ab]Means with different letters differ ($P < 0.05$).

analyzed by ANOVA for repeated measures in SAS by PROC MIXED [19]. The statistical model consisted of treatment, day, and their interactions. The effect of treatment was analyzed using litter within treatment as the error term, and the effects of day and any interaction were analyzed using the residual as the error term.

Results

There were no treatment × day interactions ($P > 0.05$) for whole blood concentrations of Se or serum concentrations of IgG, IgA and IgM among the gilts supplemented with an organic, inorganic or no supplemental Se. Gilts supplemented with an organic source of Se had greater ($P < 0.05$) whole blood concentrations of Se than gilts supplemented with an inorganic source or no additional supplementation of Se (Figure 1A). Regardless of treatment, whole blood concentrations of Se were greatest ($P < 0.05$) at d −60 then decreased by d 57 of gestation reaching its lowest concentration by d 113 of gestation ($P < 0.05$; Figure 1B).

Serum concentrations of IgG did not differ ($P > 0.05$) in gilts fed inorganic and organic sources of Se (Table 2). However, there was an effect of time on serum concentrations of IgG. Gilts at d 57 of gestation had greater ($P < 0.05$) serum concentrations of IgG than gilts at d 113 of gestation (Figure 2A).

There were no differences due to dietary treatment on serum concentrations of IgA ($P > 0.05$; Table 2). There was an effect of day of gestation on serum concentrations of IgA as d 113 of gestation and d 21 of lactation were greater ($P < 0.05$) compared to d −60 and d 57 of gestation (Figure 2B). There was no treatment × day effect for serum concentrations of IgA ($P > 0.05$).

There was no effect of dietary treatment ($P > 0.05$) on serum concentrations of IgM (Table 2). There was a trend for serum concentrations of IgM being the greatest at d 57 of gestation compared to d −60 ($P < 0.08$; Figure 2C).

There were no treatment × day interactions ($P > 0.05$) on colostral and milk concentrations of IgG, IgA and IgM among the gilts supplemented with an organic, inorganic or no supplemental Se. There was an overall effect of dietary treatment on colostral and milk concentrations of IgG ($P < 0.05$; Figure 3A). Gilts supplemented with an inorganic source of Se had greater ($P < 0.05$) colostral and milk concentrations of IgG compared to gilts supplemented with an organic source or no additional Se (Figure 3A).

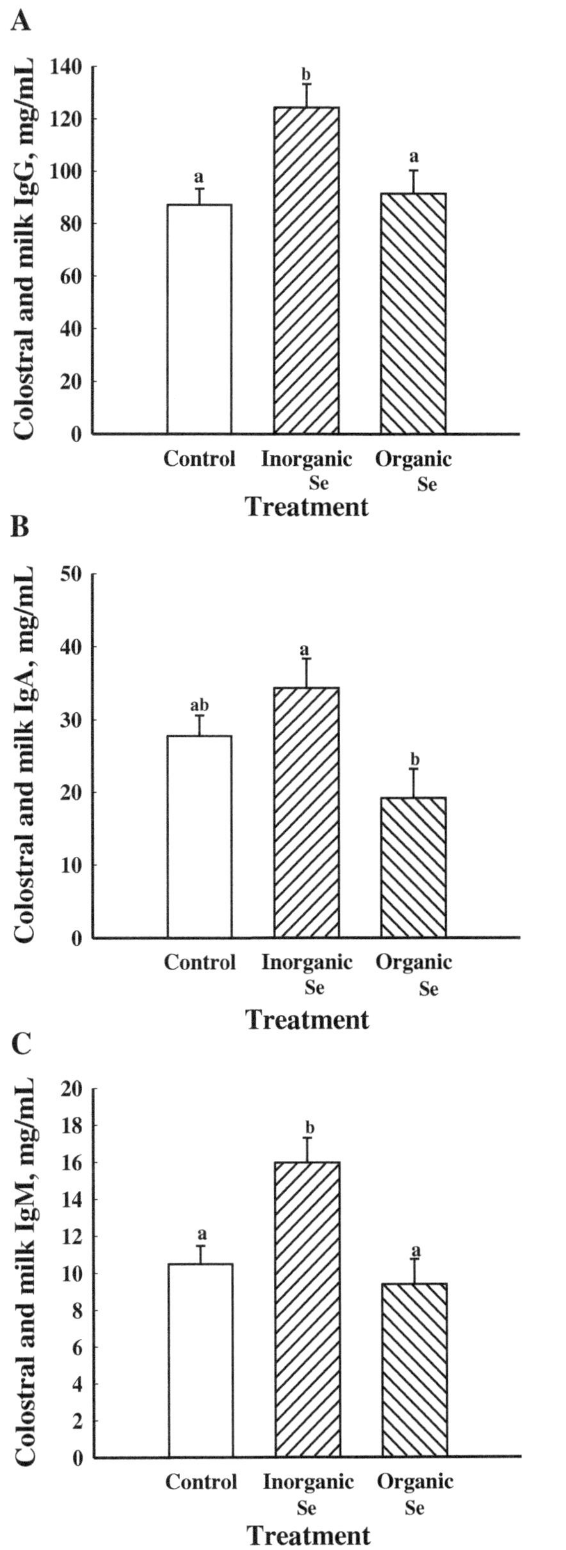

Figure 3 Mean colostral and milk concentrations of IgG (A), IgA (B), and IgM (C) in gilts fed no additional Se (Control; n = 8), inorganic Se (n = 4), and organic Se (n = 4). Data are expressed as least-square means ± SEM. [ab]Means with different letters differ ($P < 0.05$).

Colostral and milk concentrations of IgG were greatest at parturition ($P < 0.05$) then decreased by d 1 of lactation and remained low throughout the remainder of lactation (Figure 4A).

Gilts supplemented with an inorganic source of Se had greater ($P < 0.05$) colostral and milk concentrations of IgA than gilts supplemented with an organic source of Se (Figure 3B). Colostral and milk concentrations of IgA were greatest ($P < 0.05$) on the day of parturition (d 0) then decreased on d 1 and continued to be low throughout lactation (Figure 4B).

Colostral and milk concentrations of IgM were greatest ($P < 0.05$) in gilts supplemented with an inorganic source of Se compared to gilts supplemented with an organic source of Se or no additional Se (Figure 3C). Colostral and milk concentrations of IgM were greatest ($P < 0.05$) at parturition (d 0) then decreased throughout the remainder of lactation (Figure 4C).

Maternal dietary treatment affected ($P < 0.05$) the whole blood concentrations of Se in piglets (Figure 5A). Piglets born to gilts supplemented with an organic source of Se had greater ($P < 0.05$) whole blood concentrations of Se than piglets from gilts supplemented with no additional Se (Figure 5A). There was no effect of maternal dietary treatment on litter size ($P > 0.05$). Piglet whole blood concentrations of Se were greatest ($P < 0.05$) at d 14 and 21 of age compared to d 0, 1, and 7 (Figure 5B). Additionally, seven-day-old piglets had greater ($P < 0.05$) whole blood concentrations of Se compared to d 1 and d 0 piglets (Figure 5B).

No treatment or treatment × day effect was found for serum concentrations of IgG in piglets ($P > 0.05$). Serum concentrations of IgG in piglets differed by day ($P < 0.05$; Figure 6A). Serum concentrations of IgG in piglets on d 0 were decreased ($P < 0.05$) compared to all other days. Serum concentrations of IgG in piglets were greatest ($P < 0.05$) on d 1 compared to all other days. Seven day old piglets had greater ($P < 0.05$) serum concentrations of IgG than d 14 and 21 piglets.

Serum concentrations of IgA in piglets did differ by day ($P < 0.05$; Figure 6B) but not by treatment ($P > 0.05$). Serum concentrations of IgA were greatest ($P < 0.05$) on d 1 compared to all other days. Additionally, there was no treatment × day interaction ($P > 0.05$).

Serum concentrations of IgM were not affected by treatment and no day × treatment effect was found ($P > 0.05$). Serum concentrations of IgM in piglets did differ by day

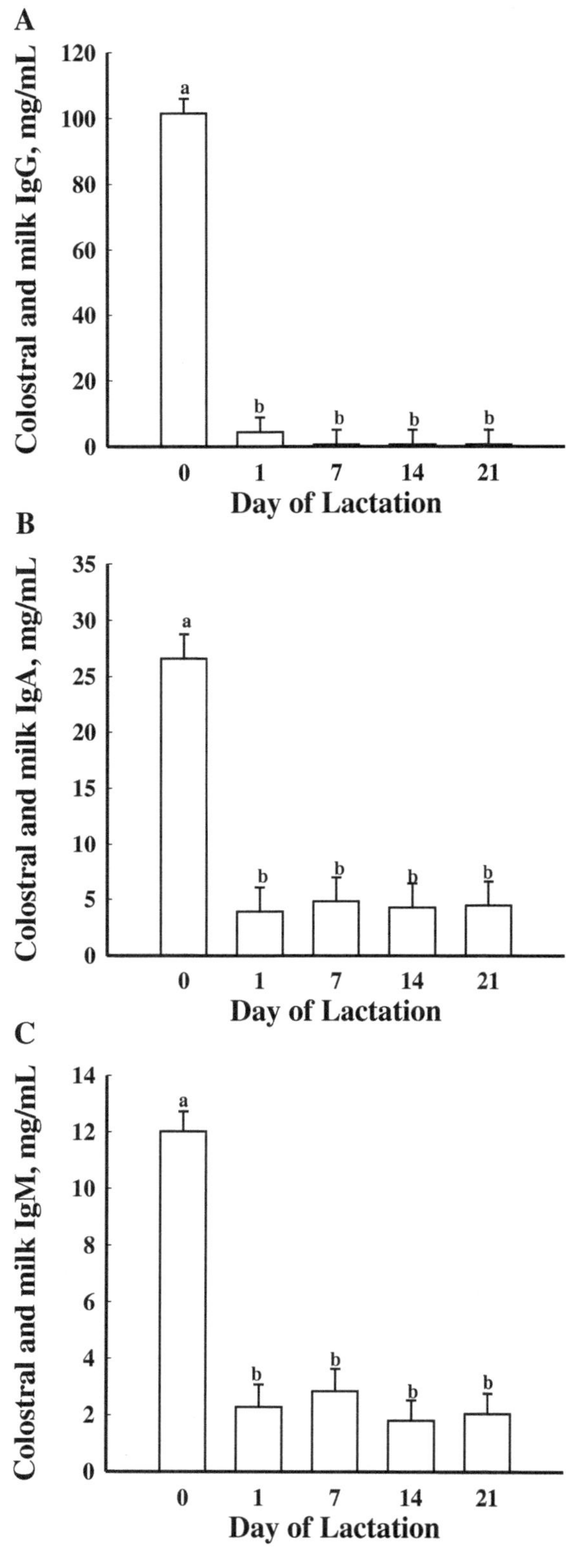

Figure 4 Mean colostral and milk concentrations of IgG (A), IgA (B), and IgM (C) in gilts (n = 16) on d 0, 1, 7, 14 and 21 of lactation. Day 0 represents the day of parturition. Data are expressed as least-square means ± SEM. [ab]Means with different letters differ ($P < 0.05$).

($P < 0.05$; Figure 6C). Serum concentrations of IgM in piglets on d 0 were decreased ($P < 0.05$) compared to all other days. Serum concentrations of IgM in piglets on d 1, 14, and 21 were greater ($P < 0.05$) than on d 0 and 7.

Discussion

The Se concentrations in the diets of the current study are greater than those reported in other studies [4,5,7]. In the current study, it was the intention to create maternal diets which were reflective of diets typically fed to gestating and lactating swine. Most pigs raised in the midwestern United States consume grains that are grown on soils which have moderate to high Se concentrations; therefore, the grains in this region are greater in Se compared to most other parts of the U.S. Also, the swine industry routinely adds Se to diets at 0.3 ppm as dictated by the FDA, regardless of the endogenous Se content of the grain.

In the current study, whole blood concentrations of Se were greater in gilts that were supplemented with organic Se. This is similar to the findings of Yoon and McMillian [7] who reported greater whole blood concentrations of Se at farrowing when females were fed organic Se. However, Mahan and Kim [4] observed no differences in blood concentrations of Se from females fed either organic or inorganic Se at 0.3 ppm. Whole blood concentrations of Se for gilts also decreased from the start of the trial to parturition regardless of the treatments in the current study. These results also correspond with studies done by Mahan and Peters [7] and Yoon and McMillan [7] as they all documented a decrease in whole blood concentrations of Se from the initial start of the trial at gestation to parturition. Mahan and Kim [4] hypothesized the decline in Se concentration from gestation to parturition could demonstrate a need for the female to transfer Se to developing fetuses or for Se to produce selenoproteins that could then be transferred to either fetal or mammary tissue.

The greater whole blood concentration of Se in the gilts from this study compared to those reported by Kim and Mahan [20] can be attributed to the greater dietary concentration of Se in the diet. When comparing the results of the current study to those of Kim and Mahan [20], both studies fed diets supplemented with 0.3 ppm. The current study reported gilt whole blood Se levels of 0.33 to 0.28 ppm throughout the course of the trial, which was greater than those reported by Kim and Mahan [20] and Yoon and McMillan [7]. While previous studies [7,20] supplemented Se at 0.3 ppm as inorganic

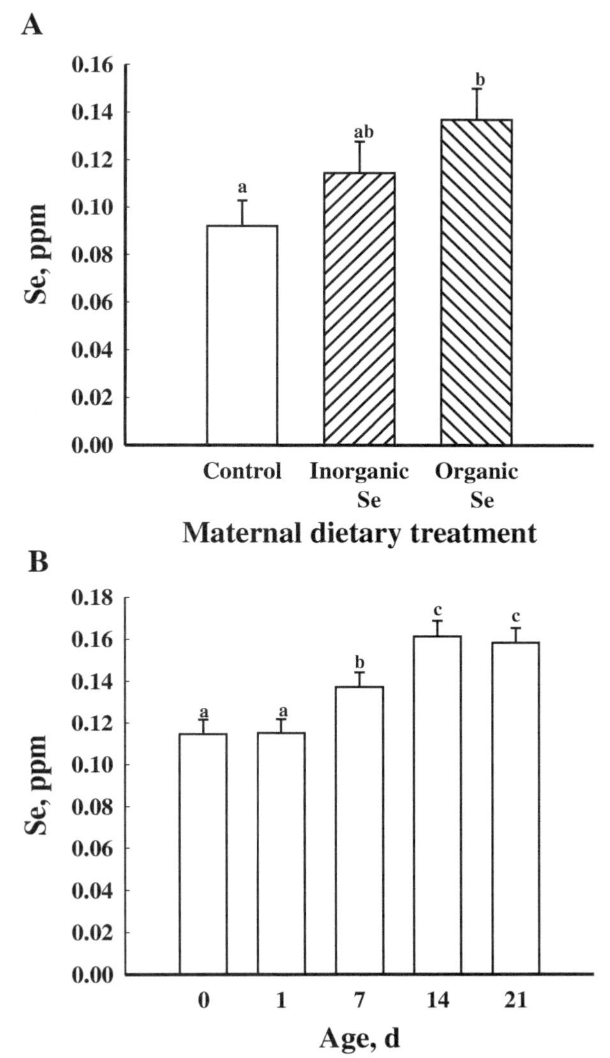

Figure 5 Mean whole blood concentrations of Se in piglets from gilts supplemented with no additional Se (Control; n = 8 litters), gilts supplemented with inorganic Se (n = 4 litters), and gilts supplemented with organic Se (n = 4 litters) (A) and mean whole blood concentrations of Se in piglets (n = 16 litters) at d 0, 1, 7, 14 and 21 of age (B). Day 0 represents the day of parturition. Data are expressed as least-square means ± SEM. [abc]Means with different letters differ ($P < 0.05$).

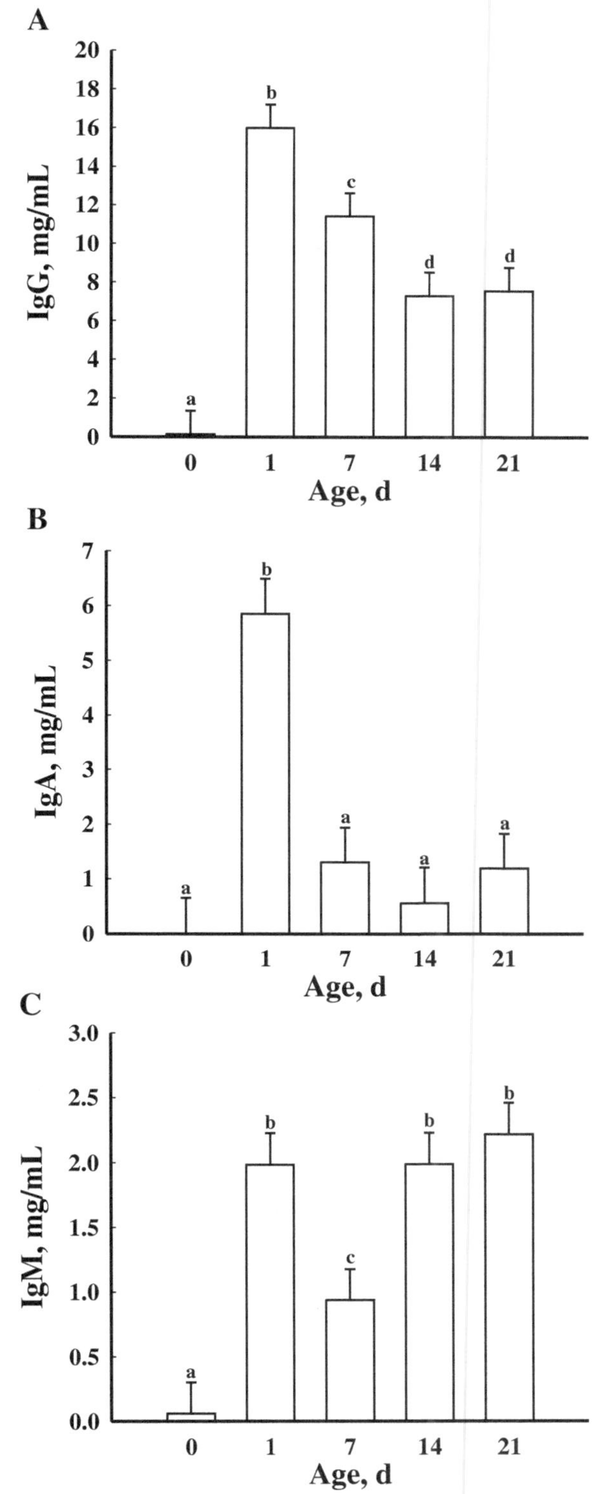

Figure 6 Mean serum concentrations of IgG (A), IgA (B), and IgM (C) in piglets (n = 16 litters) at d 0, 1, 7, 14 and 21 of age. Day 0 represents the day of parturition. Data are expressed as least-square means ± SEM. [abcd]Means with different letters differ ($P < 0.05$).

and organic Se, the discrepancy among studies for increased concentrations of Se may be attributed to gilts being fed their respective diets for 60 d before breeding as Yoon and McMillan [7] did not start feeding the treatment diets until 60 d prepartum.

Whole blood concentrations of Se were greatest for piglets born to gilts that received diets supplemented with organic Se in the current study. This is in agreement with the observations of Mahan and Kim [4] and Kim and Mahan [20] who reported greater piglet whole blood concentrations of Se from dams that were supplemented with organic Se. Though we did not have a day by treatment interaction effect, the whole blood concentrations of Se in piglets increased from d 1 of birth until weaning at d 21. Whole blood concentrations of Se were greatest on d 14 in piglets born to females supplemented with organic Se [5]. Yoon and McMillan [7] reported that piglets born to females receiving diets supplemented with organic Se had greater whole blood concentrations of Se than piglets born to females fed the control diet. The Yoon and McMillan [7] results are similar to the data reported in the current study as piglets from gilts who received diets supplemented with organic Se had greater whole blood concentrations of Se than piglets from gilts fed the control diet, but were not different than the piglets from gilts supplemented with inorganic Se. The current study did find greater values of Se in piglets, but it could be attributed to the maternal treatment diets being fed throughout the course of the pregnancy.

Altered thyroid hormone status may provide an explanation for the mechanism leading to an altered IgG absorption in dams that were supplemented with Se throughout gestation [21]. Fetal infusion of thyroid hormones resulted in decreased IgG absorption in lambs [22]. Selenium supplementation directly to calf colostrum increased serum concentrations of IgG in calves, but as the supplementation of Se increased the IgG absorption was impaired [11]. Kamada et al. [11] speculated that increased IgG absorption resulting from supplementation with Se was pharmacological and not nutritional. In the present study, the addition of either type of Se source had no effect on any of the serum immunoglobulins evaluated. Sows that were administered a Se injection on d 100 of gestation had decreased levels of serum immunoglobulins [16] compared to the serum immunoglobulins concentrations in the current study. The decreased serum concentrations of immunoglobulins could be attributed to only a single injection of Se or that the females used in the study had a more developed immune system [16].

Serum concentrations of IgG in piglets in the current study were greater than those reported by Yoon and McMillan [7]. Additionally, maternal dietary treatment had no effect on serum concentrations of IgG in piglets, which was also observed by Yoon and McMillan [7]. Increasing the concentration of circulating immunoglobulins in the neonatal pig is beneficial for long term piglet survival after weaning when the dam's milk is no longer available. Immunoglobulin G is the predominant immunoglobulin in colostrum and has a half-life of 14 d, while IgA and IgM have half-lives of 2.5 and 5 d, respectively [23]. A decrease in serum concentrations of IgG were observed over time. Serum concentrations of IgG were greatest on d 1 because of the high ingestion of colostral IgG from the gilts. Serum concentrations of IgM were greater on d 1 and then decreased by d 7 indicating that the IgM transferred from the colostrum had decreased in concentration, but concentrations of IgM were increased by d 14 and 21. This increase in concentrations of IgM may indicate that the piglet immune system is maturing.

The limited numbers of animals in the experimental groups in the present study probably precluded us from finding differences in many of the measured parameters. However, colostral and milk concentrations of immunoglobulins were increased in the sows treated with inorganic Se which could further protect the neonatal pig. When ewes were supplemented with supranutritional levels of inorganic or organic Se, colostral IgG concentrations increased but it was dependent upon the dosage of Se administered [24]. Additionally, Se supplementation may affect cytokine production as has been observed in the horse [25]. Horses receiving an inorganic source of Se had the greatest mean expression of IL-8 by stimulated neutrophils. Interleukin-8 can induce chemotaxis which is important for the recruitment of leukocytes to the site of infection as well as activation of neutrophils [26]. Thus, based upon our observations and the corollary work by others, further research examining cytokine profiles and/or dosage and source of Se administered to the pig is warranted.

Early weaning of piglets can reduce performance, which correlates to increased susceptibility to disease [27]. Piglets weaned at an older age may be more efficient at mounting an immune response because of their more developed immune system [28]. As weaning age increased, a reduction in plasma IgG concentration was observed and this reduction could reflect the stage of development of the innate immune system [28]. Swine producers are always examining ways to decrease piglet mortality. Weaning at an age when piglets are finally developing their own immune system can decrease mortality in finishing pigs Early weaning piglets to an isolated site can reduce the potential for disease transfer from the dam [29] and decrease immunological stress [30]. Early weaning to an isolated site improved growth and feed efficiency compared to piglets weaned to a conventional farrow-to-finish system [31]. Early weaning practices should be made to a clean environment as Blecha et al. [32] showed that early weaning compromises the cellular immunity of these younger animals. The practice of early weaning is stressful to the piglet as their immune system is not fully developed and considerations should be made as to what age the piglets weaning occurs.

Conclusion

The results of this study are similar to previously published reports, but the current study also demonstrates the effect of Se sources on passive transfer of immunoglobulins. The addition of inorganic or organic sources of Se in the maternal diet did not affect immunoglobulin concentration in the pig, but we were able to profile the concentrations of immunoglobulins over a 3 wk period in the piglet, which may help in determining when the piglet immune system is starting to mature.

Competing interests

None of the authors have any competing interest to declare.

Authors' contributions

All authors have made substantial contributions to: the research design, analysis or interpretation of data, and in drafting the paper. All authors read and approved the final manuscript.

References

1. Eggert RC, Patterson E, Akers WT, Stokstad EIR: **The role of vitamin E and selenium in the nutrtition of the pig [abstract].** *J Anim Sci* 1957, **16:**1037.
2. Rotruck JT, Pope AL, Ganther HE, Swanson AB, Hafeman DG, Hoekstra WG: **Selenium: biochemical role as a component of glutathione peroxidase.** *Science* 1973, **179:**588–590.
3. NRC: *Nutrient Requirements of Swine*. 10th edition. Washington, DC: National Academy Press; 1988:199.
4. Mahan DC, Kim YY: **Effect of inorganic or organic selenium at two dietary levels on reproductive performance and tissue selenium concentrations in first-parity gilts and their progeny.** *J Anim Sci* 1996, **74:**2711–2718.
5. Mahan DC: **Effect of organic and inorganic selenium sources and levels on sow colostrum and milk selenium content.** *J Anim Sci* 2000, **78:**100–105.
6. Mahan DC, Peters JC: **Long-term effects of dietary organic and inorganic selenium sources and levels on reproducing sows and their progeny.** *J Anim Sci* 2004, **82:**1343–1358.
7. Yoon I, McMillan E: **Comparative effects of organic and inorganic selenium on selenium transfer from sows to nursing pigs.** *J Anim Sci* 2006, **84:**1729–1733.
8. Lin C, Mahan DC, Wu G, Kim SW: **Protein digestibility of porcine colostrum by neonatal pigs.** *Livest Sci* 2009, **121:**182–186.
9. Miller ER, Harmon BG, Ullrey DE, Schmidt DA, Luecke RW, Hoefer JA: **Antibody absorption, retention and production by the baby pig.** *J Anim Sci* 1962, **21:**309–314.
10. Speer VC, Brown H, Quinn L, Catron DV: **The cessation of antibody absorption in the young pig.** *J Immunol* 1959, **83:**632–634.
11. Kamada H, Nonaka I, Ueda Y, Murai M: **Selenium addition to colostrum increases immunoglobulin G absorption by newborn calves.** *J Dairy Sci* 2007, **90:**5665–5670.
12. Knight DA, Tyznik WJ: **The effect of dietary selenium on humoral immunocompetence of ponies.** *J Anim Sci* 1990, **68:**1311–1317.
13. Boland TM, Keane N, Nowakowski P, Brophy PO, Crosby TF: **High mineral and vitamin E intake by pregnant ewes lowers colostral immunoglobulin G absorption by the lamb.** *J Anim Sci* 2005, **83:**871–878.
14. Lacetera N, Bernabucci U, Ronchi B, Nardone A: **Effects of selenium and vitamin E administration during a late stage of pregnancy on colostrum and milk production in dairy cows, and on passive immunity and growth of their offspring.** *Am J Vet Res* 1996, **57:**1776–1780.
15. Lacetera N, Bernabucci U, Ronchi B, Nardone A: **The effects of injectable sodium selenite on immune function and milk production in Sardinian sheep receiving adequate dietary selenium.** *Vet Res* 1999, **30:**363–370.
16. Hayek MG, Mitchell GE Jr, Harmon RJ, Stahly TS, Cromwell GL, Tucker RE, Barker KB: **Porcine immunoglobulin transfer after prepartum treatment with selenium or vitamin E.** *J Anim Sci* 1989, **67:**1299–1306.
17. Awadeh FT, Abdelrahman MM, Kincaid RL, Finley JW: **Effect of selenium supplements on the distribution of selenium among serum proteins in cattle.** *J Dairy Sci* 1998, **81:**1089–1094.
18. AOAC, Horwitz W, International A: *Official Methods of Analysis of the AOAC International*. Arlington, Va: The Association; 2000.
19. Littell RC, Henry PR, Ammerman CB: **Statistical analysis of repeated measures data using SAS procedures.** *J Anim Sci* 1998, **76:**1216–1231.
20. Kim YY, Mahan DC: **Prolonged feeding of high dietary levels of organic and inorganic selenium to gilts from 25 kg body weight through one parity.** *J Anim Sci* 2001, **79:**956–966.
21. Hammer CJ, Thorson JF, Meyer AM, Redmer DA, Luther JS, Neville TL, Reed JJ, Reynolds LP, Caton JS, Vonnahme KA: **Effects of maternal selenium supply and plane of nutrition during gestation on passive transfer of immunity and health in neonatal lambs.** *J Anim Sci* 2011, **89:**3690–3698.
22. Cabello G, Levieux D, Girardeau JP, Lefaivre J: **Intestinal K99+ escherichia coli adhesion and absorption of colostral IgG1 in the newborn lamb: effect of fetal infusion of thyroid hormones.** *Res Vet Sci* 1983, **35:**242–244.
23. Curtis J, Bourne FJ: **Half-lives of immunoglobulins IgG, IgA and IgM in the serum of new-born pigs.** *Immunology* 1973, **24:**147–155.
24. Stewart WC, Bobe G, Vorachek WR, Stang BV, Pirelli GJ, Mosher WD, Hall JA: **Organic and inorganic selenium: IV. passive transfer of immunoglobulin from ewe to lamb.** *J Anim Sci* 2013, **91:**1791–1800.
25. Montgomery JB, Wichtel JJ, Wichtel MG, McNiven MA, McClure JT, Markham F, Horohov DW: **Effects of selenium source on measures of selenium status and immune function in horses.** *Can J Vet Res* 2012, **76:**281–291.
26. Janeway CA, Travers P, Walport M, Shlomchik MJ (Eds): *Immunobiology*. 6th edition. New York, New York: Garland Science Publishing; 2005.
27. Pluske JR, Hampson DJ, Williams IH: **Factors influencing the structure and function of the small intestine in the weaned pig: a review.** *Livest Prod Sci* 1997, **51:**215–236.
28. Niekamp SR, Sutherland MA, Dahl GE, Salak-Johnson JL: **Immune responses of piglets to weaning stress: impacts of photoperiod.** *J Anim Sci* 2007, **85:**93–100.
29. Harris DL: **Alternative approaches to eliminating endemic diseases and improving performance of pigs.** *Vet Rec* 1988, **123:**422–423.
30. Johnson RW: **Explanation for why sick pigs neither eat well nor grow well.** In *Proceedings of Carolina Nutrtition Conference*. Research Triangle Park, NC: Carolina Feed Industry Association; 1997:49.
31. Williams NH, Stahly TS, Zimmerman DR: **Effect of chronic immune system activation on the rate, efficiency, and composition of growth and lysine needs of pigs fed from 6 to 27 kg.** *J Anim Sci* 1997, **75:**2463–2471.
32. Blecha F, Kelley KW: **Cold stress reduces the acquisition of colostral immunoglobulin in piglets.** *J Anim Sci* 1981, **52:**594–600.

19

Utilization of tea grounds as feedstuff for ruminant

Huili Wang and Chuncheng Xu*

Abstract

Researches on tea have been developed for decades, which prove that tea, especially green tea, has multiple functional components. With the rapid development of beverage industry, the resultant large amounts of tea grounds attract great attention. However, unreasonable utilization of tea grounds would lead to great waste and environmental pollution, especially in summer. In view of the high nutritive value and multiple functional components, tea grounds could be used as feedstuff. By now, researches of tea grounds as feedstuff are mainly on ruminant, as the utilization on other animals is limited to lower level due to high fiber content. Therefore, the following review will begin with a brief introduction of tea grounds and the possible utilization of tea grounds as feedstuff, and then elaborate on the application of ensiling and total mixed ration on ruminant. Apart from the fermentation quality, nutritive value is also provided to assess feasibilities of using tea grounds as feedstuff resources. Finally, a summary on the utilization situation and future direction of using tea grounds as feedstuff is provided in this review.

Keywords: Ensiling, Fermentation quality, Nutritive value, Tea grounds, Total mixed ration

Introduction

Tea is one of the most popular beverages in the world [1,2]. With the increasing awareness on health and nutrition issues in beverage market, consumption of tea drinks, especially green tea has been increasing significantly in recent years. Accompanied with the consumption of ready-made tea drinks in bottles, packs and cans, a large amount of tea grounds are released annually by beverage companies. Although only small amounts of tea grounds are converted into raw compost material, most are generally buried [3]. In view of the high nutritive value and multiple functional components, tea grounds could be used as feed resources or nutrient supplementary. However, unreasonable utilization of tea grounds could lead to great waste and environmental pollution, especially in summer. There is increasing demand for efficient use of food by-products due to economic and environmental concerns [4].

Nowadays in China, utilization of wet or dry tea grounds on the production of chickens and pigs has been reported, which indicates that the addition of tea grounds at proper amount could enhance or maintain the quality and output of the resultant meat and eggs [5,6]. However, high fiber content restricts the proportion of tea grounds in diet. While for ruminant, there is no such problem but how to make efficient utilization and obtain better animal performance. Therefore, researches published recently focus on the effective utilization of tea grounds on ruminant.

Chemical composition and feed characteristics

Tea grounds are mainly by-products of ready-made tea drinks, which are derived from tea leaves extracted by hot water to make tea drinks. Tea leaf has a useful content of amino acids, proteins, vitamins, tannins and polyphenols [7]. The catechins are present in higher quantities in green tea than in black or oolong tea, because of different processing of tea leaves after harvest [1]. It has been indicated that polyphenols especially the flavonoids exhibit anticarcinogenic, antimutagenic, and cardioprotective effects which are generally associated with their antioxidant (free radical scavenging and metal chelation) properties [2,8].

However, most of these nutrients remain in tea grounds. After extraction by hot water, green tea grounds (GTG) still contain a lot of protein, tannin, caffeine, beta-carotene and vitamin E [9], which might help to prevent disease if fed to ruminants. Total extractable tannin and condensed tannin (CT) contents are higher in GTG than in black tea and oolong tea grounds [10]. According to the

* Correspondence: xucc@cau.edu.cn
College of Engineering, China Agricultural University, Beijing 100083, P. R. China

previous studies [3,10-12], GTG usually contain 22-35% of crude protein (CP), 2.3-7.1% ether extract (EE), 24-37% acid detergent fiber (ADF) and 31-45% neutral detergent fiber (NDF) on dry matter (DM) basis. Whereas, barley tea grounds (BTG) contain 12-19% of CP, 2.2-3.4% EE, 15-25% ADF and 27-35% NDF on DM basis [13-15]. All these suggest that tea grounds have a potential as a feed resource.

However, the high moisture content makes it deteriorated easily after being released by beverage companies, especially in summer. Therefore, it must be used as soon as possible if fed directly. However, direct-fed exhibits poor palatability due to the tannin components existed in tea grounds. Furthermore, it can be used as dry feed due to its high nutrient, except that drying process would consume a lot of energy. Considering the above characteristics, ensiling could be one of the suitable ways to preserve high-moisture tea grounds.

Silage preparation and fermentation quality

Nutrient losses and proteolysis are inevitable during the ensiling process. They are always reduced or inhibited by a rapid decline in pH during the initial period to create a low pH environment that is unsuitable for the action of plant and microbial proteases [16,17].

Preservation by ensiling is highly dependent on lactic acid fermentation. Whereas, tea grounds usually contain about 10^6 aerobic bacteria, 10^3 to 10^4 mold and yeast in colony forming units/g of fresh matter, whereas lactic acid bacteria (LAB) could not be detected. Besides, water-soluble carbohydrates (WSC) are consistently at or below the detectable level [9,12]. Thereby, lactic acid fermentation scarcely occurs when untreated tea grounds are ensiled alone. In order to solve such problems, many researches have been carried out by adding additives or mixed ensiled with other feed resources.

Fermentation quality of tea grounds silage

Tea grounds are usually ensiled with additives or in mixed silage with other materials. Researches indicated that addition of *Lactobacillus plantarum* and commercial acremonium cellulase (AUS) in both GTG and BTG could improve the fermentation quality as indicated by lower pH and NH_3-N content and higher lactic acid content [9,12,18]. It is probably that the inoculation ensures sufficient LAB at the initial period, while AUS increased WSC content, which could be used by LAB to produce lactic acid. BTG silage could also be well preserved when treated by formic acid or sodium hydroxide (NaOH) which could inhibit even the fermentation of LAB, and therefore, exhibited low NH_3-N content, no lactic acid, acetic acid, propionic acid and butyric acid [18]. Wang et al. [3] evaluated the fermentation quality of GTG silage treated with LAB, enzyme, formic acid and formaldehyde, and suggested that LAB treatment had the best effect on improving fermentation quality and inhibiting protein degradation.

In addition, it has been found that the inclusion of high-quality materials in GTG silage may be an effective way to improve fermentation quality. GTG could be ensiled successfully without bacterial inoculants when mixed with materials containing sufficient sugars [19,20].

Besides, it is also indicated that the addition of tea grounds could enhance the lactic acid fermentation of sudangrass and oat silage, which contains adequate WSC content. However, the fermentation quality differs by variety of tea and addition ratio [10,21,22]. Apart from forage silage, GTG could also enhance the lactic acid fermentation of byproducts-mixed silage when there are insufficient materials for lactic acid production [23]. However, research reveals that neither GTG-associated LAB nor green tea polyphenols could account for the enhancement of lactic acid fermentation. It is probably that GTG supply some nutrients other than polyphenols, which are heat-stable and effective for LAB growth during ensiling [22]. Further research is still required to clarify the mechanisms of enhanced lactic acid fermentation by the addition of GTG.

Fermentation quality of total mixed ration (TMR) silage

Ensiling is suitable for preserving high-moisture tea grounds. However, there are problems with the use of tea grounds as feed, such as nutritional imbalance, poor palatability and poor preservation [24,25]. If ensiled with dry feeds as a TMR, the risk of effluent production would be minimized and the time for mixing prior to feeding could be omitted. In addition, unpalatable by-products could be incorporated into TMR if their odors and flavors could be altered by silage fermentation [4].

Noor et al. [26] reported that GTG could be used as an ingredient in TMR silage production for animal feeding. However, it is suggested that the fermentation period should not exceed 3 mo if ensiled in a flexible container bag silo, when some decrease in quality would occurred. Xu et al. [24,25] reported that the addition of GTG and BTG at 10%, 20% and 30% DM ratio of TMR could all be well preserved and exhibited low pH and NH_3-N content, and high lactic acid content. Suto et al. [27] recommended that the moisture content of TMR should be held at about 55%, when the TMR contains 10-30% of GTG on DM basis, as the lactic acid content tended to increase with the proportion of GTG increased and with the moisture content of the TMR decreased.

Besides, many researches have shown that the substitution of GTG for wet brewers' grains (BG) at ratios of up to 15% on DM basis of TMR could be well preserved with high lactic acid content, low pH and NH_3-N content [4,28]. While, the replacement of BTG and soybean meal mixture (7:3 on DM basis) for BG at ratios of up to 15% DM of TMR could increase lactic acid concentration, decrease pH, acetic acid and NH_3-N of the TMR silage [15].

Nutritive value of tea grounds silage and TMR silage

In addition to fermentation characteristics of silage, their nutritive value for ruminant also need to be evaluated, which is assessed by feed intake, digestibility, rumen fermentation, nitrogen (N) balance, blood component and milk production.

Nutritive value of tea grounds silage

Many researches have assessed the nutritive value of fresh tea grounds and tea grounds silage from different aspects.

The nutritive value of GTG is thought to be equivalent to that of BG. For GTG silage, digestibilities of CP and EE are 74.6% and 50.7% on DM basis, respectively; Besides, the estimated total digestible nutrients (TDN), digestible CP and digestible energy are 71.1%, 23.9% and 13.4 MJ/kg on DM basis, respectively [12]. For BTG silage, however, digestibilities of CP and organic cell wall are 26.8% and 30.2%, and the TDN, digestible CP, digestible energy are 65.1%, 4.8% and 13.0 MJ/kg on DM basis, respectively. It is suggested that the nutritive value of BTG silage is about 80% of that of barley [14,29]. Therefore, both GTG and BTG can be used as feed ingredient.

Xu et al. [18] reported that BTG silage treated with NaOH exhibited the highest degradability of DM and CP than that of the control and BTG, then it followed by the LAB + AUS treatment. However, formic acid treatment had no effect on the DM and CP disappearance of silage.

Many researches have indicated that well-fermented GTG silage contributes to preserve more tea catechins and antioxidative activity [12,20]. Nishino et al. [20] investigated the changes due to ensiling in tea catechins and antioxidative activity of wet GTG. Results indicated that ensiling significantly lowered antioxidative activity and decreased the contents of partial tea catechins in GTG silage, whereas, reductions of tea catechins were ameliorated and no marked changes were found in total phenols and antioxidative activity during ensiling when ensiled as a mixture with dried beet pulp. Inhibited degradation of tea catechins were also found in wet GTG silage treated with LAB and cell wall degrading enzymes [12].

Besides, it has been indicated that CT could reduce rumen forage protein degradation due to reversible binding to these proteins [30], suppress the breakdown of protein by rumen microorganisms [31] and decrease ruminal gas production [32]. It is consistent with the study of Nishino et al. [20] that addition of dried beet pulp increased gas production of GTG silage, which in other words suppressed gas production with the increase of GTG. However, the reduction of tea catechins during ensiling would not elicit an improvement in digestibility of GTG silage. It is consistent with the finding of Salawu et al. [31] that tannin protected proteolysis of proteins during silage fermentation but was digestible in the lower gut, and it also agreed with finding of Kondo et al. [21] that proteins in GTG seem to be stable during ensiling, but digestible post-ruminally.

Evidence also indicates that CT at levels of 20–45 g/kg DM can reduce rumen forage protein degradation and protect amino acids to increase the absorption in the small intestine of ruminants, while deteriorating intake and digestibility at >55 g/kg DM [20,30]. It may illustrate the following no detrimental effect on gas production with the increase of GTG, which probably due to the less amount of tannins than that is critical to suppress the activity of rumen bacteria. Kondo et al. [10] found that the addition of GTG to forage silage could increase gas production, whereas addition of oolong and black tea grounds suppressed gas production. It is agreed with Kondo et al. [23] that the addition of GTG to byproducts-mixture increased gas production.

Furthermore, many researches focused on the nutritive value of silage with GTG added in different ratio. Kondo et al. [21] studied the feeding value of oat silage with GTG added at a ratio of 0, 5% and 20% on fresh matter. No significant differences were found in DM intake and digestibility, except for the increased CP digestibility with the increment of GTG. In addition, N retention and ruminal NH_3-N were increased, while total volatile fatty acids (VFA) were reduced by addition of GTG. The high CP content, N digestibility and N retention indicated the potential of GTG as a protein supplement, and the addition of GTG could be up to 20% on fresh matter in oat silage.

Nutritive value of tea grounds TMR silage

Previous studies have shown that GTG can be used as protein supplements due to the high nutrition and low cost [9,12]. On the other hand, tannins exhibit contrasting effects on feed intake and digestibility when GTG applied in different ratio [30]. All these suggest that uncertain results may appear on the utilization of tea grounds. Thus, further researches are required to have a better understanding of the utilization of tea grounds.

Suto et al. [27] reported that the moisture content of TMR varied from 55% to 75% had small effects on the degradability and the effective degradability of TMR containing GTG at 10-30% on DM basis.

Besides, many researches have reported the feeding value of TMR silage with BTG in different application ratio. Xu et al. [25] reported that, digestibilities of DM, CP and TDN were significantly higher for TMR silages with 10% and 20% of BTG than that with 30% of BTG. Therefore, the ideal mixing proportion of BTG for TMR silage is 10% to 20% on DM basis. However, it was also reported that the possible proportion of replacing wet BG with BTG for TMR silage was suggested to be 10% or less of diet DM [15]. In the study, feed intake and digestibility for the 5% and 10% treatments were not different from the control. However, feed intake and digestibilities of EE and NDF were lower for the 15% treatment. In addition, the ruminal total VFA concentration was higher, while NH_3-N content was lower for the 15% treatment compared with the control.

Furthermore, progressive increase of GTG (0, 5%, 10% and 15% on DM of TMR) substituted for wet BG had no effect on voluntary feed intake of TMR silage, while digestibility was slightly lower than the control. No differences among treatments were observed in retention N, pH and total VFA concentration [4]. The study suggested that a high GTG level of 15% of diet DM can be recommended for silage based TMR. In contrast, feed intake and N retention of TMR silage were also not affected with the GTG increased up to 15% [28]. Compared with the control, digestibilities for the 5% and 10% treatments were not different, while that for 15% treatment was lower. With the increase in GTG, N intake did not differ, but fecal N increased, while urinary N and plasma urea N content were decreased. However, ruminal NH_3-N in the 15% treatment was lower than that of the control. Therefore, the possible mixing proportion of GTG for TMR silages can be 10% of the diet DM.

Kondo et al. [11] found that the DM intake of cows fed TMR with GTG silage was slightly but not significantly increased with the increment of GTG, while ensiled GTG contained high amounts of lactic acid, acetic acid and CT. It is inconsistent with the negative correlation between DM intake of silage and the acetic acid and lactic acid concentrations [33], or the decreased feed intake in ruminants owning to high tannin [34]. It is supposed that the inclusion of GTG silage at 5% DM in TMR were relatively low to show the negative impact of tannin on feed intake. Furthermore, GTG silage incorporated into TMR at a ratio of 5% on DM and 10% on CP basis had no detrimental effects on the performance of lactating cows. Therefore, the addition of GTG silage is suggested to be no less than 5% on DM basis of TMR, which is also consistent with the results of Eruden et al. [35] that addition of GTG silage at 5% of diet DM did not affect the performance of midlactating cows. However, Eruden et al. [36] reported that, considering that the feed intake was not affected, the upper limit of mixing ratio for GTG silage is expected to be 8 and 15% in TMR with timothy hay and maize silage as main roughage, respectively.

Eruden et al. [37,38] also studied the addition of GTG silage at 10% or 20% of diet DM. Compared with the control, no significant effects were found on feed intake and digestibility in both 10% and 20% tea group, except that the CP intake was increased and digestibilities of ADF and NDF were decreased in the 20% group. Besides, addition of GTG at 10% diet DM had no significant effect on ruminal fermentation, plasma metabolites, as well as milk yield and components. All the above indicated that GTG silage could be added up to 20% of diet DM. Furthermore, Nishida et al. [39] suggested that feeding diets containing 20% of GTG silage had no negative impact on ruminal fermentation, but increased the plasma antioxidative activity and vitamin E concentration.

Conclusions

Ensiling is suitable for preserving tea grounds. Both ensiled with additives and mixed ensiled could be well preserved. Besides, adjusting moisture with dry feeds as a TMR is also effective to incorporate GTG as feedstuff. According to the previous researches, the addition of tea grounds in mixed silage could be up to 20% on fresh matter and the possible mixing proportion of tea grounds for TMR silage is suggested to be less than 10% or 10% to 20% on DM basis. Whereas, the inclusion of tea grounds silage for TMR is inconsistent and differs by the variety of main roughage, ranging from 5% to 15% or more on DM basis.

Apart from the possible addition ratios in silage and TMR, researches also provide feasibilities of using tea grounds as protein resources or substitution of other feed materials. The characteristics of high protein content and low cost will certainly relieve the dependence to some extent on imported protein products. However, there are also problems that GTG are always mass-produced and beverage companies are always scattered at suburbs. Therefore, it is necessary to work out a reasonable plan to reduce the unnecessary cost.

Furthermore, in view of the controversies on tea catechins and CT, it is necessary to have a further understanding of the functional ingredients in tea grounds and their association with animal performance, so as to make efficient use and obtain better performance. Besides, efforts should be made to reduce the high fiber content so as to reduce the limitation on other animals, as well as to assess the potential risk of tea grounds, such as pesticide residues.

Abbreviations

ADF: Acid detergent fiber; BG: Brewers' grains; BTG: Barley tea grounds; CP: Crude protein; CT: Condensed tannin; DM: Dry matter; EE: Ether extract; GTG: Green tea grounds; LAB: Lactic acid bacteria; N: Nitrogen; NDF: Neutral detergent fiber; TDN: Total digestible nutrients; TMR: Total mixed ration; VFA: Volatile fatty acids; WSC: Water-soluble carbohydrates.

Competing interests

The authors declare that they have no competing interests.

Authors' contributions

HLW collected and sorted the related information and drafted the manuscript. CCX conceived the study, participated in its design and arrangement of the review, and helped to draft the manuscript. Both authors have read and approved the manuscript.

Authors' information

Huili Wang is a doctor candidate of College of Engineering, China Agricultural University, with expertise in the utilization and evaluation of non-conventional feed resources. Chuncheng Xu, Ph.D., is a professor of Animal Nutrition and Science in China Agricultural University, with expertise in research and development of non-conventional feed ingredients and nutrition evaluation for ruminants.

References

1. Graham HN: **Green tea composition, consumption, and polyphenol chemistry.** *Prev Med* 1992, **21**:334–350.
2. Khokhar S, Magnusdottir S: **Total phenol, catechin, and caffeine contents of teas commonly consumed in the United Kingdom.** *J Agric Food Chem* 2002, **50**:565–570.
3. Wang RR, Wang HL, Liu X, Xu CC: **Effects of different additives on fermentation characteristics and protein degradation of green tea grounds silage.** *Asian-Australas J Anim Sci* 2011, **24**:616–622.
4. Xu C, Cai Y, Moriya N, Ogawa M: **Nutritive value for ruminants of green tea grounds as a replacement of brewers' grains in totally mixed ration silage.** *Anim Feed Sci Technol* 2007, **138**:228–238.
5. Ma T: **Application prospect of tea polyphenols in feed.** *Anim Breed Feed* 2008, **6**:92–94 (In Chinese).
6. Gao F, Tian K, Wang J: **A study on the feeding value of quick-melting tea residue: 1. the residue's feeding effect on fattening pigs.** *J Hunan Agr Univ* 1998, **24**:465–467 (In Chinese).
7. Yamamoto T, Juneja LR, Chu DC, Kim M: *Chemistry and application of green tea.* Boca Raton, Florida: CRC Press; 1997.
8. Dufresne CJ, Farnworth ER: **A review of latest research findings on the health promotion properties of tea.** *J Nutr Biochem* 2001, **12**:404–421.
9. Cai Y, Masuda N, Fujita Y, Kawamoto H, Ando S: **Development of a new method for preparation and conservation of tea grounds silage.** *Anim Sci J* 2001, **72**:536–541.
10. Kondo M, Kita K, Yokota H: **Effects of tea leaf waste of green tea, oolong tea, and black tea addition on sudangrass silage quality and *in vitro* gas production.** *J Sci Food Agric* 2004, **84**:721–727.
11. Kondo M, Nakano M, Kaneko A, Agata H, Kita K, Yokota HO: **Ensiled green tea waste as partial replacement for soybean meal and alfalfa hay in lactating cows.** *Asian-Australas J Anim Sci* 2004, **17**:960–966.
12. Xu C, Cai Y, Fujita Y, Kawamoto H, Sato T, Masuda N: **Chemical composition and nutritive value of tea grounds silage treated with lactic acid bacteria and *Acremonium* cellulase.** *Anim Sci J* 2003, **74**:355–361.
13. Cai Y, Fujita Y, Sato T, Masuda N, Nishida T, Ogawa M: **Preparation and conservation of barley tea grounds silage and its fermentation quality.** *Anim Sci J* 2002, **73**:283–289.
14. Enishi O, Tsukahara N, Kajikawa H, Terada F: **Analysis of *in situ* ruminal disappearance and nutritive value of barley water residue.** *Anim Sci J* 2000, **71**:252–257.
15. Xu C, Cai Y, Fukasawa M, Matsuyama H, Moriya N: **The effect of replacing brewers' grains with barley tea grounds in total mixed ration silage on feed intake, digestibility and ruminal fermentation in wethers.** *Anim Sci J* 2008, **79**:575–581.
16. McDonald P, Henderson AR, Heron SJE: *The Biochemistry of Silage.* 2nd edition. Marlow, UK: Chalcombe Publications; 1991.
17. Rooke JA, Hatfield RD: **Biochemistry of ensiling.** In *Silage Science and Technology.* Edited by Buxton DR, Muck RE, Harrison HJ. Madison, Wisconsin: American Society of Agronomy; 2003:95–139.
18. Xu C, Cai Y, Zhang H, Fukasawa M, Moriya N: **Ensiling and subsequent ruminal degradation characteristics of barley tea grounds treated with contrasting additives.** *Anim Feed Sci Technol* 2008, **141**:368–374.
19. Cai Y, Fujita Y, Xu C, Ogawa M, Sato T, Masuda N: **Mixed silage preparation of green tea grounds and corn and its fermentation quality.** *Anim Sci J* 2003, **74**:203–211.
20. Nishino N, Kawai T, Kondo M: **Changes during ensilage in fermentation products, tea catechins, antioxidative activity and *in vitro* gas production of green tea waste stored with or without dried beet pulp.** *J Sci Food Agric* 2007, **87**:1639–1644.
21. Kondo M, Kita K, Yokota H: **Feeding value to goats of whole-crop oat ensiled with green tea waste.** *Anim Feed Sci Technol* 2004, **113**:71–81.
22. Kondo M, Naoki N, Kazumi K, Yokota H: **Enhanced lactic acid fermentation of silage by the addition of green tea waste.** *J Sci Food Agric* 2004, **84**:728–734.
23. Kondo M, Kita K, Yokota HO: **Evaluation of fermentation characteristics and nutritive value of green tea waste ensiled with byproducts mixture for ruminants.** *Asian-Australas J Anim Sci* 2006, **19**:533–540.
24. Xu C, Cai Y, Kida T, Matsuo M, Kawamoto H, Murai M: **Silage preparation of total mixed ration with green tea grounds and its fermentation quality and nutritive value.** *Grassl Sci* 2004, **50**:40–46.
25. Xu C, Cai Y, Murai M: **Fermentation quality and nutritive value of total mixed ration silage with barley tea grounds.** *Anim Sci J* 2004, **75**:185–191.
26. Noor RM, Cai Y, Uegaki R, Matsuyama H, Xu C, Yoshida N, Tando Y, Miwa T: **Fermentation quality of TMR silage with green tea grounds ensiled in flexible container bag silo.** In *XXI International Grassland Congress and VIII International Rangeland Congress; July 2008.* Edited by Xie H, Huang J. Hohhot, China: Guangdng People's Publishing House; 2008:733.
27. Suto R, Horiguchi K, Takahashi T, Toyokawa K: **Effect of mixing proportion of green tea waste and moisture content on the fermentation quality and the rate of *in situ* degradation of TMR silage.** *JPN J Grassl Sci* 2007, **53**:127–132.
28. Xu C, Cai Y, Moriya N, Eruden B, Hosoda K, Matsuyama H: **Influence of replacing brewers' grains with green tea grounds on feed intake, digestibility and ruminal fermentation characteristics of wethers.** *Anim Sci J* 2008, **79**:226–233.
29. Xu C, Cai Y, Fujita Y, Kawamoto H, Sato T, Masuda N: **Silage preparation of barley tea grounds and their nutritive value.** *Anim Sci J* 2003, **74**:343–348.
30. Min B, Barry T, Attwood G, Mcnabb W: **The effect of condensed tannins on the nutrition and health of ruminants fed fresh temperate forages: a review.** *Anim Feed Sci Technol* 2003, **106**:3–19.
31. Salawu M, Acamovic T, Stewart C, Hvelplund T, Weisbjerg MR: **The use of tannins as silage additives: effects on silage composition and mobile bag disappearance of dry matter and protein.** *Anim Feed Sci Technol* 1999, **82**:243–259.
32. Makkar HP, Blümmel M, Becker K: ***In vitro* effects of and interactions between tannins and saponins and fate of tannins in the rumen.** *J Sci Food Agric* 1995, **69**:481–493.
33. Jones G, Larsen R, Lanning N: **Prediction of silage digestibility and intake by chemical analyses or in vitro fermentation techniques.** *J Dairy Sci* 1980, **63**:579–586.
34. Silanikove N, Nitsan Z, Perevolotsky A: **Effect of a daily supplementation of polyethylene glycol on intake and digestion of tannin-containing leaves (*Ceratonia siliqua*) by sheep.** *J Agric Food Chem* 1994, **42**:2844–2847.
35. Eruden B, Nishida T, Matsuyama H, Hosoda K, Shioya S: **Effect of the addition of various levels of green tea grounds silage at on the feed intake and milk production in lactating dairy cows.** *Anim Sci J* 2005, **76**:295–301.
36. Eruden B, Nishida T, Matsuyama H, Hosoda K, Shioya S, Nonaka K, Yamada A: **Effect of main roughages on the palatability by lactating dairy cows of total mixed rations with green tea waste silage.** *JPN J Grassl Sci* 2006, **52**:232–236.
37. Eruden B, Nishida T, Hosoda K, Shioya S, Cai Y: **Effects of green tea grounds silage on digestibility, rumen fermentation, and blood components in lactating dairy cows.** *Anim Sci J* 2003, **74**:483–490.

38. Eruden B, Nishida T, Hosoda K, Shioya S: **Nutritive value of green tea grounds silage and influence of polyethylene glycol on nitrogen metabolism in steers.** *Anim Sci J* 2004, **75:**559–566.
39. Nishida T, Eruden B, Hosoda K, Matsuyama H, Nakagawa K, Miyazawa T, Shioya S: **Effects of green tea (*Camellia sinensis*) waste silage and polyethylene glycol on ruminal fermentation and blood components in cattle.** *Asian-Australas J Anim Sci* 2006, **19:**1728–1736.

20

Restricted nutrient intake does not alter serum-mediated measures of implant response in cell culture

Christopher D Reinhardt[1*], Tiffany L Lee[2], Daniel U Thomson[2], Laman K Mamedova[1] and Barry J Bradford[1]

Abstract

Background: During nutritional stress, reduced intake may reduce the efficacy of anabolic implants. This study was conducted to evaluate basic cellular responses to a growth promotant implant at two intake levels.

Methods: Sixteen crossbred steers (293 ± 19.3 kg) were used to evaluate the impact of anabolic implants in either an adequate or a restricted nutritional state. Steers were trained to individual Calan gates, and then randomly assigned to 1 of 4 treatments in a 2 × 2 factorial arrangement. Treatments consisted of: presence or absence of an anabolic growth implant (Revalor-XS, 200 mg TBA and 40 mg estradiol; IMPLANT or CONTROL) and a moderate energy, pelleted, starting cattle diet fed at either 2.0 × or 1.0 × maintenance energy (NE_M) requirements (HIGH or LOW). Serum (d 0, 14, and 28) was used for application to bovine muscle satellite cells. After treatment with the serum (20% of total media) from the trial cattle, the satellite cells were incubated for 72 h. Protein abundance of myosin heavy chain (MHC), phosphorylated extracellular signal-related kinase (phospho-ERK), and phosphorylated mammalian target of rapamycin (phospho-mTOR) were analyzed to determine the effects of implant, intake, and their interaction (applied via the serum).

Results: Intake had no effect on MHC ($P = 0.85$) but IMPLANT increased ($P < 0.01$) MHC abundance vs. CONTROL. Implant status, intake status, and the interaction had no effect on the abundance of phospho-ERK ($P \geq 0.23$). Implanting increased phospho-mTOR ($P < 0.01$) but there was no effect ($P \geq 0.51$) of intake or intake × implant.

Conclusions: The nearly complete lack of interaction between implant and nutritional status indicates that the signaling molecules measured herein respond to implants and nutritional status independently. Furthermore, results suggest that the muscle hypertrophic effects of anabolic implants may not be mediated by circulating IGF-1.

Keywords: Beef cattle, Implant, Muscle, Myosin heavy chain, Nutrient restriction, Satellite cells

Background

Anabolic implants increase average daily gains 15 to 20%, and improve feed efficiency by 6 to 14% in feedlot cattle [1]. However, when animals first arrive in a feedlot many are in a state of stress and have not yet become accustomed to the environment, feeding routine, other animals, and other factors, especially if they have traveled long distances or have recently been weaned. These cattle are considered at high risk to develop bovine respiratory disease (BRD). Many of these high-risk cattle have reduced intakes for up to three weeks after arrival at the feedlot [2].

Stress and infection can lead to a hypermetabolic state, characterized by weight loss, increased resting metabolic rate, and muscle protein catabolism [3]. In cattle infected with BRD, decreased feed intake, weight loss, and decreased body condition are often observed [3]. Although ruminants may have a greater capacity to store digesta because of their larger digestive tracts, this weight loss and decrease in body condition may be comparable to the hypermetabolic state that develops in non-ruminants [3]. Munson et al. [4] reported that delaying implant administration in high-risk calves did not alter feedlot performance or the risk of respiratory disease, suggesting that stress brought on by the

* Correspondence: cdr3@ksu.edu
[1]Department of Animal Sciences and Industry, 232 Weber Hall, Kansas State University, Manhattan, KS 66506, USA
Full list of author information is available at the end of the article

transition to the feedlot did not diminish the anabolic signals from implanting early in the finishing phase.

Anabolic implants impact growth rate at least in part by stimulating muscle hypertrophy [5]. Androgenic and estrogenic steroids increase *in vitro* bovine satellite cell proliferation, and these effects are mediated in part by the extracellular signal-regulated kinase (ERK) and phosphatidylinositide 3-kinase/AKT/mammalian target of rapamycin (mTOR) signaling pathways [6]. Activation of these pathways, however, is dependent on adequate nutritional status, which is sensed through a complex array of signals [7,8]. Therefore, animals in a depleted nutritional state, accompanied by alterations in hormonal signals, may not have the expected muscle growth response to anabolic implants.

The objectives of this study were to examine the effects of anabolic implants on measures of nutrient balance, metabolic status, and growth factors in animals consuming nutrients either adequate or inadequate to support growth. Moreover, effects of serum-derived signals on *in vitro* proliferation of bovine muscle satellite cells were also assessed.

Materials and methods

Animals

All procedures conducted in the present study were approved by the Kansas State University Institutional Animal Care and Use Committee (Protocols 2745.0 and 2745.1).

Sixteen crossbred steers (293 ± 19.3 kg) were purchased locally and transported to Kansas State University. They were weighed, tagged, and vaccinated with a 5-way respiratory viral vaccine and a 7-way clostridial vaccine. The steers were trained to individual Calan® gates over a period of 4 wk.

After training was complete, the steers were randomly assigned to 1 of 4 treatments in a 2 × 2 factorial arrangement. Main effect treatments were as follows: administration or absence of an anabolic implant (Revalor-XS, 200 mg TBA and 40 mg estradiol 17-β [Merck Animal Health, Whitehouse Station, NJ]; IMPLANT or CONTROL, respectively) and either 2.0 × or 1.0 × maintenance energy (NE_M) requirements of a moderate energy, pelleted, starting cattle diet (HIGH or LOW, respectively; Table 1). Energy content of the diet was calculated from ingredient composition and tabular estimates of individual ingredient energy values [9].

Steers were weighed on d 0, 14, and 28 of the trial. At each of these time points, maintenance energy requirements were calculated for each steer [10], and the diet was fed based these calculations and the treatment assignment for each animal.

Biochemical assays

All assays were conducted in duplicate. Satellite cell isolation was conducted as described by Johnson et al. [5].

Table 1 Ingredient and nutrient composition (dry matter basis) of the diet fed during the experimental period

Ingredient, %	
Alfalfa meal	33.00
Corn gluten feed, dry	18.18
Wheat middlings	14.63
Distillers grains, dried	11.52
Cracked corn	10.94
Cottonseed hulls	7.75
Molasses, liquid	2.00
Limestone	1.85
Salt	0.10
Bovatec 91[1]	0.025
Vitamin A 650	0.006
Zn SO_4	0.0055
Nutrient composition (calculated)[2]	
CP, %	15.31
Ca, %	0.56
P, %	1.43
NE_M, Mcal/kg	1.44
NE_G, Mcal/kg	0.85

[1]The final diet contained 49.89 mg lasalocid sodium per kg.
[2]Nutrient composition calculated based on ingredient composition in the final diet using tabular estimates of nutrient values (NRC, 1996).

Using sterile techniques, 500 g of the semimembranosus muscle was dissected from 2 steers which were not part of the present study (18 mo of age). After removal of connective tissue, the muscle was passed through a sterile meat grinder. The ground muscle was incubated with 0.1% pronase in Earl's Balanced Salt Solution (EBSS) for 1 h at 37°C with frequent mixing. After incubation the mixture was centrifuged at 1,500 × *g* for 5 min, the pellet was suspended in phosphate buffered saline (PBS; Ca- and Mg-free PBS; Life Technologies, Grand Island, NY), and the suspension was centrifuged at 500 × *g* for 10 min. The supernatant was centrifuged at 1,500 × *g* for 10 min to pellet the mononucleated cells. The PBS wash and differential centrifugation were repeated 2 more times. Cell viability was over 90%. The resulting mononucleated cell preparation was suspended, pooled (density 3×10^4 cells/mL) in cold DMEM containing 10% fetal bovine serum (FBS) and 10% dimethylsulfoxide and frozen in liquid nitrogen. Cells were counted using a hemocytometer and 0.4% Trypan Blue solution.

Blood samples (serum and plasma) were collected from each animal at d 0, 14, and 28 of the trial and used in determining levels of total IGF-1, and additional serum samples were collected from each animal for application to satellite cells [11-13] which had been previously isolated. Total IGF-1 concentration was determined in plasma samples using a commercial ELISA kit (Octeia

IGF-1, catalog #AC27F1, Immunodiagnostic Systems Inc., Scottsdale, AZ). The kit includes reagents for pretreatment of samples to release IGF-1 from binding proteins [14]. Frozen satellite cells were re-suspended in DMEM supplemented with 5% FBS and 1% penicillin-streptomycin mixture (containing 10,000 IU/mL penicillin and 10,000 μg/mL streptomycin). Cells were passaged once, then at 70% confluence, cells were incubated in DMEM for 72 h with serum (20% of total media) taken from individual animals on d 0, 14, or 28 and with an additional control containing 20% FBS and no serum from trial cattle. Halting incubation at 72 h prevents over-confluency of the cells, which can lead to excessive differentiation; since these are skeletal muscle cells and would contract or twitch after differentiation, excessive differentiation would lead the cell layers to come off the cell culture plate. All media contained 1% penicillin-streptomycin mixture, but FBS was not added to wells containing treatment serum.

Protein abundance of myosin heavy chain (MHC; d 0, 14, and 28), phosphorylated ERK (phospho-ERK; d 0 and 28), and phosphorylated mTOR (phospho-mTOR; d 0 and 28) was analyzed in differentiated satellite cells to determine the effects of implant, intake, and their interaction (applied via the serum). MHC was used as a marker of myotube formation, and phospho-ERK and phospho-mTOR are regulators of cell proliferation.

Western blot analysis

Cultured cells were lysed at 4°C with RIPA lysis buffer (50 mmol/L Tris–HCl, pH 7.5, 150 mmol/L NaCl, 1% NP-40, 0.5% sodium deoxycholate, and 0.% SDS; Santa Cruz Biotechnology, Santa Cruz, CA) containing X1 broad-spectrum protease inhibitor cocktail (Protease inhibitor cocktail I; Calbiochem, Gibbstown, NJ) and centrifuged at 15,000 × *g* for 10 min at 4°C. Protein concentration was measured using a Coomassie protein assay kit (Pierce, Rockford, IL) according to Bradford [15]. The proteins in the supernatant were denatured with SDS buffer containing dithiothreitol for 5 min at 95°C. Forty micrograms of protein were loaded in each lane, and proteins were separated using SDS-PAGE on a 4 – 12% TRIS-glycine gel, then transferred to a nitrocellulose membrane (iBlot; Invitrogen, Carlsbad, CA). Membranes were blocked in Tris buffer (pH 7.4) with 5% dry milk powder for 2 h at room temperature and then incubated with the following antibodies: a rabbit anti-MYH (H-300) polyclonal antibody raised against amino acids 1641–1940 of myosin heavy chain 3 (sc-20641; Santa Cruz Biotechnology, Santa Cruz, CA); rabbit anti-phospho-mTOR (Ser 2481) polyclonal antibody (Cell Signaling Technology, Beverly, MA), and rabbit anti-phospho-ERK (Tyr 204; sc-7976, Santa Cruz Biotechnology) overnight at 4°C. All primary antibodies were diluted 1:1,000. After washing, membranes were incubated for 1 h at room temperature with a secondary antibody (anti-rabbit IgG, sc-2301; Santa Cruz Biotechnology) diluted 10,000-fold in Tris buffer (pH 7.4). Immunodetection was performed by chemiluminescence (West-Dura; Thermo Scientific, Waltham, MA), and band images were visualized using a photodocumentation system (ChemiDoc-It Imaging System; UVP Inc., Upland, CA). Densitometry analysis was carried out using Image J (http://rsbweb.nih.gov/ij/).

Statistical analysis

Animal was the experimental unit for all measures. Data were analyzed using the MIXED procedure of SAS® (v. 9.3; SAS Institute, Cary, NC); IGF-1 and MHC were analyzed using repeated measures and d 0 values were included in the model as a covariate; phospho-ERK and phospho-mTOR were analyzed as a completely randomized design but included d 0 values as a covariate. The independent variables included the main and interactive effects of implant status and intake level. Effects were considered significant with a protected F-test with $\alpha < 0.05$; effects were considered a trend if $P < 0.10$.

Results and discussion

Level of feeding achieved different growth rates (0.76 vs. 0.02 kg/d for 2.0 × vs. 1.0 × maintenance intake, respectively; $P < 0.01$); implant status did not affect 28-d growth rates (0.36 vs. 0.41 kg/d for implanted vs. control, respectively; $P = 0.38$). There was no interaction between level of intake and implant status for growth rate ($P = 0.38$).

There was no intake × implant for serum IGF-1 levels ($P = 0.56$). High-intake steers had greater IGF-1 levels ($P = 0.02$) vs. low-intake steers (Figure 1). Elsasser et al. [16] reported that greater energy intake increases circulating IGF-1 levels in growing steers, and Hayden et al. [17] reported reduced IGF-1 values in steers when intake was restricted by 23%.

The IGF-1 concentrations in implanted steers were not different from those in control cattle in this study. Bryant et al. [18,19] reported no effects of implants on IGF-1 in feedlot heifers until d 42 after implant administration; however, Johnson et al. [18,19] and Reinhardt [20] reported that circulating IGF-1increased in implanted steers by d 21 and d 28, respectively, compared to non-implanted steers. Anderson et al. [21] reported that exogenous estradiol supplementation in post-menopausal women increased circulating growth hormone 2.1-fold but actually decreased circulating IGF-1 by 27%.

When serum was applied to the cultured satellite cells from the semimembranosus muscle of non-study steers, level of intake had no effect ($P = 0.85$; Figure 2) on MHC abundance but implant increased MHC abundance ($P < 0.01$), with no interaction observed between intake level and implant status ($P = 0.23$). In this study,

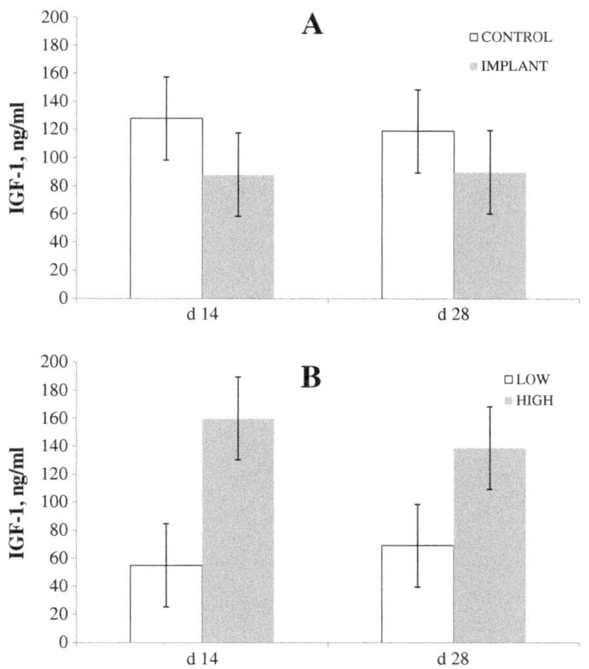

Figure 1 Effects of implant or intake level on plasma urea nitrogen. Plasma urea nitrogen concentration of steers (Chart **A**) either implanted (solid line) or not implanted (dashed line) with a long-acting TBA:E_2 implant; and (Chart **B**) fed a common diet at either 2.0 (solid line; HIGH) or 1.0 (dashed line; LOW) × maintenance NE requirements. (Effect of intake $P = 0.56$; implant $P = 0.62$; intake × implant $P = 0.99$; day × intake $P < 0.01$; day × implant $P = 0.33$; day × intake × implant $P = 0.12$. Pooled SE for treatment means was 0.261. [a,b]means without a common superscript differ ($P < 0.05$).

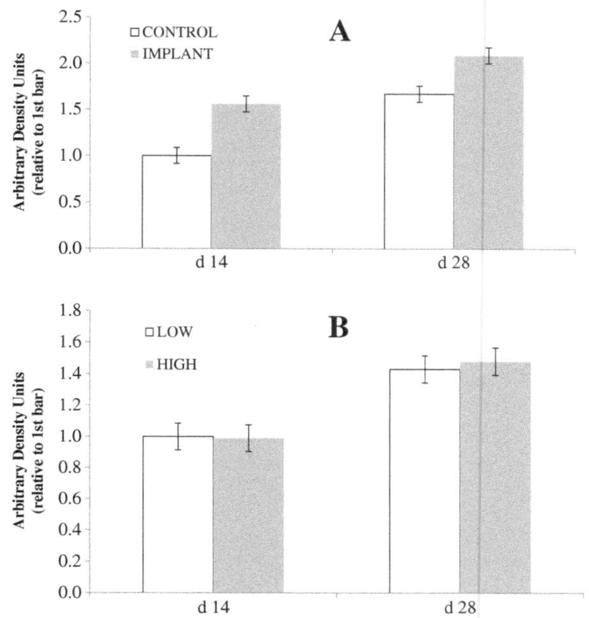

Figure 2 Effects of implant and intake level on myosin heavy chain abundance in cultured myocytes. Myocytes from a non-study steer were treated with serum (20% of total media) from study steers drawn on d 0, 14, and 28, and myosin heavy chain abundance was measured. In chart **A**, steers were either implanted (gray bars) or not implanted (open bars) with a long-acting TBA:E_2 implant; in chart **B**, steers were fed a common diet at either 2.0 (gray bars; HIGH) or 1.0 (white bars; LOW) × maintenance NE requirements. (Effect of day $P < 0.01$; intake $P = 0.85$; implant $P < 0.01$; intake × implant $P = 0.23$; day × intake $P = 0.88$; day × implant $P < 0.01$; day × intake × implant $P = 0.44$; Error bars represent pooled SEM for treatment means = 0.103).

MHC was used as a marker of myotube formation, reflecting satellite cell proliferation and differentation. Baxa et al. [22] found that implanting increased MHC type IIa mRNA *in vivo*. In a smaller study, Baxa [23] found that implanting with a combined estradiol-17β (E_2)/trenbolone acetate (TBA) led to numerical increases in MHC type I and IIa mRNA. Chung et al. [24] found no effect of E_2/TBA implants on MHC type I, IIa or IIx mRNA, but did find that MHC IIa increased over time which was also seen in the present study (effect of day $P < 0.01$). Implanting with E_2/TBA increases muscle fiber diameter [25,26], due to an initial increase in DNA transcription followed by an increase in nuclei within the muscle fiber which support hypertrophy, but ultimately due to an increase in total protein produced by the cell [27], of which MHC is a key constituent.

Extracellular signal-regulated kinase plays a role in regulating insulin- and potentially IGF-I-mediated glucose uptake by skeletal muscle cells [28], and certain fatty acids reduce muscle cell proliferation via ERK [29]. Implant status did not affect the phospho-ERK abundance ($P = 0.46$; Figure 3). Intake had no effect ($P = 0.65$) on phospho-ERK, and there was no interaction between the two main effects ($P = 0.41$). Because ERK is phosphorylated (activated) by IGF-I signaling [6,30], the lack of phospho-ERK response to implant in the present study is understandable, given the lack of effect on circulating IGF-1. However, intake level also failed to influence phospho-ERK despite the fact that HIGH increased serum IGF-1 compared to LOW.

The protein kinase mTOR plays an essential part in transcription, translation, muscle cell size, and protein stability [31,32]. To help regulate these processes, mTOR integrates information about nutrient status, energy status, and many growth factors and may be a central component in the regulation of cellular processes that determine cell growth and proliferation, and even ultimate overall animal size [33]. Phosphorylation of mTOR can occur at both Ser 2448 [34] and Ser 2481 [35]. We investigated Ser 2481 phosphorylation of mTOR. This site is responsive to nutrient supply and Ser 2481 phosphorylation status is directly correlated with activity of mTOR complex 1, which is responsible for promoting protein synthesis through activation of S6 kinase 1 and eukaryotic translation initiation factor 4E-binding protein 1 [35]. There was no interaction

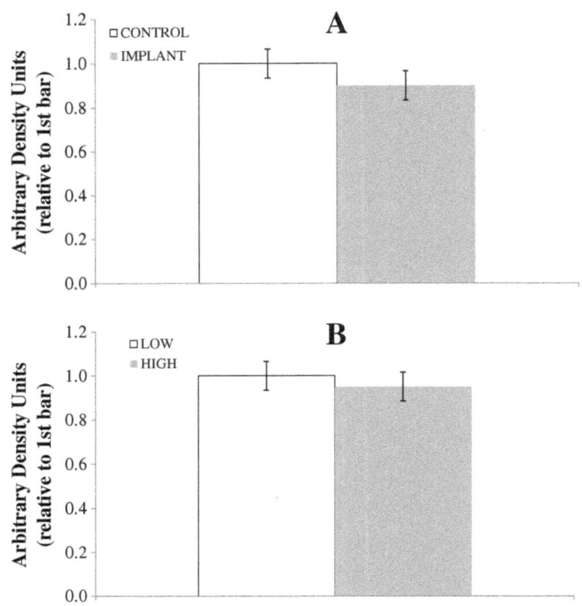

Figure 3 Effect of implant and intake level on phosphorylated extracellular signal-related kinases (phosphor-ERK). Myocytes from a non-study steer were treated with serum (20% of total media) from study steers drawn on d 0 and 28, and phospho-ERK abundance was measured. Steers were (Chart **A**) either implanted (gray bars) or not implanted (white bars) with a long-acting TBA:E_2 implant and (Chart **B**) fed a common diet at either 2.0 (gray bars; HIGH) or 1.0 × (white bars; LOW) maintenance NE requirements. (Effect of intake $P = 0.65$; implant $P = 0.47$; intake × implant $P = 0.41$). Pooled SE for treatment means was 0.065.

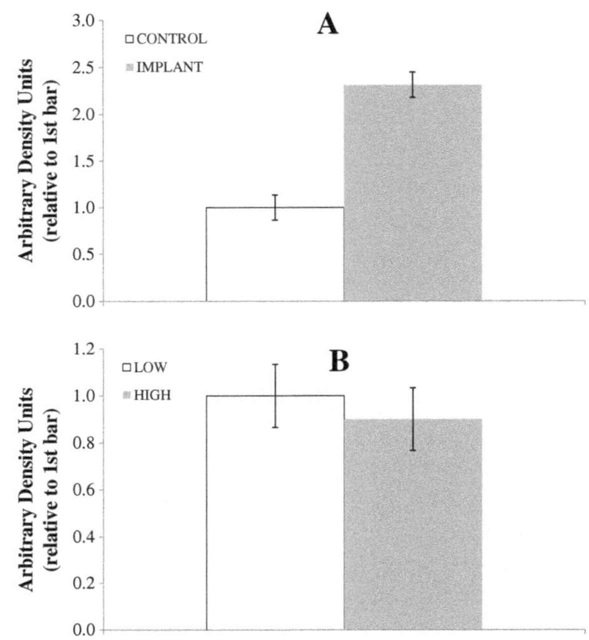

Figure 4 Effects of implant and intake level on phosphorylated mammalian target of rapamycin (phospho-mTOR) abundance in cultured myocytes. Myocytes from a non-study steer were treated with serum (20% of total media) from study steers drawn on d 0 and 28. In chart **A**, steers were either implanted (gray bars; IMPLANT) or not implanted (open bars; CONTROL) with a long-acting TBA:E_2 implant; in chart **B**, steers were fed a common diet at either 2.0 (gray bars; HIGH) or 1.0 (white bars; LOW) × maintenance NE requirements. (Effect of intake $P = 0.51$; implant $P < 0.01$; intake × implant $P = 0.88$. Error bars represent pooled SEM = 0.134).

between intake and implant status ($P = 0.88$; Figure 4) for phospho-mTOR, and intake had no effect ($P = 0.51$); however, serum from implanted steers ($P < 0.01$) stimulated greater phospho-mTOR levels vs. serum from CONTROL steers. Phosphorylated mTOR normally increases when nutrition is adequate for growth and decreases when the nutrient environment is not favorable for growth [31,36]. Although we did not see this response, it is important to recognize that our experimental design only partially subjected satellite cells to the milleu that muscle tissue of LOW animals would have seen. Our protocol provided cells with essentially unlimited nutrient supply via the DMEM that comprised 80% of the media, thereby negating effects of intake level on nutrient supply to the satellite cells. The lack of intake effect on phospho-mTOR, therefore, suggests only that endocrine signals (or compounds not supplied in DMEM) that differed by intake failed to alter satellite cell mTOR signaling. Yang et al. [32] reported that mTOR is highly sensitive to upregulation by IGF-1 and that the stress caused by low nutrient availability can inhibit mTOR signaling. In the present study, although we saw no effect of implant on IGF-1, we did measure a strong response of implant on phospho-mTOR. We saw a significant effect of increased intake level on IGF-1 but no corresponding effect of high nutrient intake stimulating phospho-mTOR.

Johnson et al. [18,19] reported increased mitogenic activity of serum from implanted vs. control steers when applied directly to satellite cell cultures, but that study also reported increased levels of circulating IGF-I in the implanted steers. Results from the present study suggest that elevated IGF-I is not the sole serum-mediated factor which increases signalling for increased mitogenic activity *in vitro*, given that serum from nutrient abundant and nutrient restricted calves, which differed greatly in IGF-I concentration, did not alter MHC or phospho-mTOR. The phospho-mTOR and MHC response to implant, in the absence of a circulating IGF-I response, could be explained by a number of additional, non-somatomedin-mediated pathways associated with growth regulation. Binding proteins specific for growth hormone and some, although not all, for IGF-I increase in plasma in response to elevated nutrient availability independent of treatment with exogenous growth hormone [37], and the same study indicated that elevated nutrient status actually reduced growth hormone in plasma. Concentration of IGF-I mRNA is elevated in longissimus muscle cells from animals previously administered exogenous E_2 [38,39], and muscle mRNA concentrations for the adipogenic regulators PPARγ and stearoyl CoA desaturase are reduced after administration of exogenous TBA [24]. In addition, both E_2 and TBA, which were used in combination in the present study, stimulate protein synthesis and reduce protein degradation in a dose-dependent manner when directly applied to fused satellite cell cultures [40,41]. This action is at least partially modulated through stimulation of muscle IGF-I mRNA by E_2 and TBA [42].

Treatment of proliferating bovine satellite cells cultures with either E_2 or TBA resulted in increases in IGF-1 mRNA as well as increased ^{3}H-thymidine incorporation rate [42]. Direct treatment of fused bovine satellite cell cultures with E_2 increases protein synthesis and decreases protein degradation rates, and blocking of estradiol binding suppresses estradiol-induced modifications in protein synthesis and degradation [40]. Testosterone also increases satellite cell activation in rats [43]. These documented effects indicate a direct mechanism for both estradiol and TBA on satellite cell proliferation and protein synthesis.

Implanting with a combination of E_2 and TBA increases circulating estradiol and TBA, but only exogenous E_2 increases circulating growth hormone [44-46]. Circulating estradiol is correlated to circulating growth hormone in humans [47], and exogenous E_2 supplementation increased growth hormone secretion in postmenopausal women [48] and ovariectamized ewes [49]. Sotiropoulos et al. [50] reported that growth hormone stimulates myoblast fusion with myotubes. Ge et al. [51] reported that growth hormone did not stimulate myoblast fusion or proliferation, but growth hormone did increase protein synthesis.

Hepatocyte growth factor increases satellite cell activation [52-54] and fibroblast growth factor enhances proliferation of satellite cells in culture [55,56]. Although hepatocyte growth factor [57-59] and fibroblast growth factor [60,61] are elevated *in vitro* in response to treatment with E_2, liver and muscle mRNA for hepatocyte growth factor are not affected *in vivo* after implanting cattle with E_2 and TBA [62].

The satellite cell results, where serum from implanted cattle activated mTOR signaling, make it tempting to speculate that mTOR is a more critical mediator of muscle hypertrophy in response to implants than ERK. Certainly, mTOR has been shown to be a key regulator of protein synthesis [8] and is activated by testosterone treatment [63]. The primary upstream regulators of mTOR are AMP-activated protein kinase (AMPK), AKT, and ERK. Blockade of either the AKT and ERK signalling pathways ablated the mitogenic response to trenbalone acetate in bovine satellite cells, suggesting that at least basal activation of both of these pathways is required [6]. Conversely, activation of AMPK, which is triggered by nutrient deficiency, impairs mTOR phosphorylation , and this pathway has been implicated in loss of muscle mass in nutrient-restricted cows [36]. Given that implant status did not enhance phospho-ERK abundance in this study, we propose that basal ERK signaling was likely necessary to permit activation of mTOR, but that activation of the AKT pathway was probably more directly involved in stimulating mTOR signaling. This is also consistent with the unique roles of ERK and mTOR signaling pathways. Although ERK does contribute to mTOR activation, it is probably more important as a regulator of cell proliferation [64]. Proliferation of satellite cells is important for maintainance of future hypertrophic potential *in vivo*, but myotube formation is driven by cell differentation and activation of protein synthesis, which is more closely related to the function of mTOR.

Conclusions

No interactions were found between nutrient restriction and implant status in muscle satellite cell culture treated with serum from calves either implanted or not implanted and fed at either 1× or 2 × maintenance. This suggests that the effects of exogenous anabolic agents and nutrient status on growth regulation, at least insofar as it is regulated by the molecules measured in this study, are independent of one another. Serum from implanted cattle promoted increased abundance of the active phospho-mTOR regulatory protein as well as myotube formation, measured by MHC abundance, in bovine satellite cells. These finding suggest the presence of circulating factors in implanted cattle that influence cell signals and promote hypertrophy of muscle tissue.

Competing interests

The authors declare that they have no competing interests.

Authors' contributions

CR is the corresponding author; TL gathered all samples, participated in laboratory procedures, conducted statistical analysis, and wrote the primary draft of the manuscript; DT created the concept and design of the study; LM coordinated all laboratory analyses; BB was responsible for determining the appropriate analytical methods to be conducted. All authors have read and approved the final manuscript.

Author details

[1]Department of Animal Sciences and Industry, 232 Weber Hall, Kansas State University, Manhattan, KS 66506, USA. [2]Department of Clinical Sciences, A-111 Mosier Hall, Kansas State University, Manhattan, KS 66506, USA.

References

1. Duckett SK, Andrae JG: **Implant strategies in an integrated beef production system.** *J Anim Sci* 2001, **79**(E. Suppl):E110–E117.
2. Hutcheson DP, Cole NA: **Management of transit-stress syndrome in cattle: nutritional and environmental effects.** *J Anim Sci* 1986, **62**:555–560.

3. Cole NA, Delaney DD, Cummins JM, Hutcheson DP: **Nitrogen metabolism of calves inoculated with bovine adenovirus-3 or with infectious bovine rhinotracheitis virus.** *Am J Vet Res* 1986, **47**:1160–1164.
4. Munson RD, Thomson DU, Reinhardt CD: **Effects of delayed steroid implanting on health, performance, and carcass quality in high health risk, auction market-sourced feedlot steers.** *J Anim Sci* 2012, **90**:4037–4041.
5. Johnson BJ, Halstead JN, White ME, Hathaway MR, DiConstanzo A, Dayton WR: **Activation state of muscle satellite cells isolated from steers implanted with a combined trenbolone acetate and estradiol implant.** *J Anim Sci* 1998, **76**:2779–2786.
6. Kamanga-Sollo E, White ME, Hathaway MR, Chung KY, Johnson WR, Dayton WR: **Roles of IGF-I and the estrogen, androgen and IGF-I receptors in estradiol-17β- and trenbolone acetate-stimulated proliferation of cultured bovine satellite cells.** *Domest Anim Endocrinol* 2008, **35**:88–97.
7. Lewis MI, Bodine SC, Kamangar N, Xu X, Da X, Fournier M: **Effect of severe short-term malnutrition on diaphragm muscle signal transduction pathways influencing protein turnover.** *J Appl Physiol* 2006, **100**:1799–1806.
8. Howell J, Manning M: **mTOR couples cellular nutrient sensing to organismal metabolic homeostasis.** *Trends Endocrinol Metab* 2011, **22**:94–102.
9. NRC: *Nutrient requirements of beef cattle. 7th rev. ed.* Washington, D.C: Natl. Acad. Press; 1996.
10. NRC: *Nutrient requirements of beef cattle. 6th rev. ed.* Washington, D.C: Natl. Acad. Press; 1984.
11. Loomis RJ, Marshall LA, Johnston PV: **Sera fatty acid effects on cultured rat splenocytes.** *J Nutr* 1983, **113**:1292–1298.
12. Jewell DE, Drewry MM, Martin RJ, Hausman GJ: **Effect of sera from control and overfed rats on preadipocyte growth in culture.** *J Nutr* 1988, **118**:803–808.
13. Reecy JM, Williams JE, Kerley MS, MacDonald RS, Thornton WH Jr, Wallace LM: **Abomasal casein infusion enhances the mitogenic activity of serum from protein-restricted steers.** *J Nutr* 1994, **124**:67–77.
14. Krebs A, Wallaschofski H, Spilcke Liss E, Kohlmann T, Brabant G, Völzke H, Nauck M: **Five commercially available insulin-like growth factor I (IGF-I) assays in comparison to the former Nichols Advantage IGF-I in a growth hormone treated population.** *Clin Chem Lab Med* 2008, **46**:1776–1783.
15. Bradford MM: **A rapid and sensitive method for the quantitation of microgram quantities of protein utilizing the principle of protein-dye binding.** *Anal Biochem* 1976, **72**:248–254.
16. Elsasser TH, Rumsey TS, Hammond AC: **Influence of diet on basal and growth hormone-stimulated plasma concentrations of IGF-1 in beef cattle.** *J Anim Sci* 1989, **67**:128–141.
17. Hayden JM, Williams JE, Collier RJ: **Plasma growth hormone, insulin-like growth factor, insulin, and thyroid hormone association with body protein and fat accretion in steers undergoing compensatory gain after dietary energy restriction.** *J Anim Sci* 1993, **71**:3327–3338.
18. Bryant TC, Engle TE, Galyean ML, Wagner JJ, Tatum JD, Anthony RV, Laudert SB: **Effects of ractopamine and trenbolone acetate implants with or without estradiol on growth performance, carcass characteristics, adipogenic enzyme activity, and blood metabolites in feedlot steers and heifers.** *J Anim Sci* 2010, **88**:4102–4119.
19. Johnson BJ, Hathaway MR, Anderson PT, Meiske JC, Dayton WR: **Stimulation of circulating insulin-like growth factor I (IGF-I) and insulin-like growth factor binding proteins (IGFBP) due to administration of a combined trenbolone acetate and estradiol implant in feedlot cattle.** *J Anim Sci* 1996, **74**:372–379.
20. Reinhardt CD: *Metabolic indices for growth: Endocrine profile of steers on different nutritional and growth regulation regimes*, Master's thesis. College Station, TX: Texas A&M University; 1991.
21. Anderson SM, Shah N, Evans WS, Patrie JT, Bowers CY, Veldhuis JD: **Short-term estradiol supplementation augments growth hormone (GH) secretory responsiveness to dose-varying GH-releasing peptide infusions in healthy postmenopausal women.** *J Clin Endocrinol Metab* 2001, **86**:551–560.
22. Baxa TJ, Hutcheson JP, Miller MF, Brooks JC, Nichols WT, Streeter MN, Yates DA, Johnson BJ: **Additive effects of a steroidal implant and zilpaterol hydrochloride on feedlot performance, carcass characteristics, and skeletal muscle messenger ribonucleic acid abundance in finishing steers.** *J Anim Sci* 2010, **88**:330–337.
23. Baxa TJ: *Effect of zilpaterol hydrochloride and steroid implantation on yearling steer feedlot performance, carcass characteristics, and skeletal muscle gene expression*, Master's Thesis. Manhattan, KS: Kansas State University; 2004.
24. Chung KY, Baxa TJ, Parr SL, Luqué LD, Johnson BJ: **Administration of estradiol, trenbolone acetate, and trenbolone acetate/estradiol implants alters adipogenic and myogenic gene expression in bovine skeletal muscle.** *J Anim Sci* 2012, **90**:1421–1427.
25. Gonzalez JM, Carter JN, Johnson DD, Ouellette SE, Johnson SE: **Effect of ractopamine-hydrochloride and trenbolone acetate on longissimus muscle fiber area, diameter, and satellite cell numbers in cull beef cows.** *J Anim Sci* 2007, **85**:1893–1901.
26. Kellermeier JD, Tittor AW, Brooks JC, Galyean ML, Yates DA, Hutcheson JP, Nichols WT, Streeter MN, Johnson BJ, Miller MF: **Effects of zilpaterol hydrochloride with or without an estrogen-trenbolone acetate terminal implant on carcass traits, retail cutout, tenderness, and muscle fiber diameter in finishing steers.** *J Anim Sci* 2009, **87**:3702–3711.
27. Johnson BJ, Reinhardt CD: **Growth promotants for beef production: Anabolic steroids: Performance responses and Mode of action.** In *Current Veterinary Therapy: Food Animal Practice 5th Volume.* Edited by Anderson DE, Rings DM. St. Louis, MO: Saunders/Elsevier, Inc; 2009:643–651.
28. Hwang SL, Jeong YT, Li X, Kim YD, Yue L, Chang YC, Lee IK, Chang HW: **Inhibitory cross-talk between the AMPK and ERK pathways mediates endoplasmic reticulum stress-induced insulin resistance in skeletal muscle.** *Br J Pharmacol* 2013. doi:10.1111/bph.12124. [Epub ahead of print].
29. Peng Y, Zheng Y, Zhang Y, Zhao J, Chang F, Lu T, Zhang R, Li Q, Hu X, Li N: **Different effects of omega-3 fatty acids on the cell cycle in C2C12 myoblast proliferation.** *Mol Cell Biochem* 2012, **367**:165–173.
30. Fuentes EN, Bjornsson BT, Valdes JA, Einarsdottir IE, Lorca B, Alvarez M, Molina A: **IGF-I/PI3K/AkT and IGF-I/MAPK/ERK pathways *in vivo* in skeletal muscle are regulated by nutrition and contribute to somatic growth in the fine flounder.** *Am J Physiol* 2011, **300**:R1532–R1542.
31. Corradetti MN, Guan KL: **Upstream of the mammalian target of rapamycin: do all roads pass through mTOR?** *Oncogene* 2006, **25**:6347–6360.
32. Yang X, Yang C, Farberman A, Rideout TC, de Lange CFM, France J, Fan MZ: **The mammalian target of rapamycin-signaling pathway in regulating metabolism and growth.** *J Anim Sci* 2008, **86**:E36–E50.
33. Sorbassov DD, Ali SM, Sabatini DM: **Growing roles for the mTOR pathway.** *Curr Opin Cell Biol* 2005, **17**:596–603.
34. Nave BT, Ouwens M, Withers DJ, Alessi DR, Shepherd PR: **Mammalian target of rapamycin is a direct target for protein kinase B: identification of a convergence point for opposing effects of insulin and amino-acid deficiency on protein translation.** *Biochem J* 1999, **344**(Pt 2):427–431.
35. Soliman GA, Acosta-Jaquez HA, Dunlop EA, Ekim B, Maj NE, Tee AR, Fingar DC: **mTOR Ser-2481 autophosphorylation monitors mTORC-specific catalytic activity and clarifies rapamycin mechanism of action.** *J Biol Chem* 2010, **285**:7866–7879.
36. Du M, Zhu MJ, Means WJ, Hess BW, Ford SP: **Nutrient restriction differentially modulates the mammalian target of rapamycin signaling and the ubiquitin-proteasome system in skeletal muscle of cows and their fetuses.** *J Anim Sci* 2005, **83**:117–123.
37. Vestergaard M, Purup S, Frystyk J, Løvendahl P, Sørensen MT, Riis PM, Flint DJ, Sejrsen K: **Effects of growth hormone and feeding level on endocrine measurements, hormone receptors, muscle growth and performance of prepubertal heifers.** *J Anim Sci* 2003, **81**:2189–2198.
38. Pampusch MS, Johnson BJ, White ME, Hathaway MR, Dunn JD, Waylan AT, Dayton WR: **Time course of changes in growth factor mRNA levels in muscle of steroid-implanted and non-implanted steers.** *J Anim Sci* 2003, **81**:2733–2740.
39. Pampusch MS, White ME, Hathaway MR, Baxa TJ, Chung KY, Parr SL, Johnson BJ, Weber WJ, Dayton WR: **Effects of implants of trenbolone acetate, estradiol, or both, on muscle insulin-like growth factor-I, insulin-like growth factor-I receptor, estrogen receptor-a, and androgen receptor messenger ribonucleic acid levels in feedlot steers.** *J Anim Sci* 2008, **86**:3418–3423.
40. Kamanga-Sollo E, White ME, Hathaway MR, Weber WJ, Dayton WR: **Effect of Estradiol-17 beta on protein synthesis and degradation rates in fused bovine satellite cell cultures.** *Domest Anim Endocrinol* 2010, **39**:54–62.
41. Kamanga-Sollo E, White ME, Hathaway MR, Weber WJ, Dayton WR: **Effect of trenbolone acetate on protein synthesis and degradation rates in fused bovine satellite cell cultures.** *Domest Anim Endocrinol* 2011, **40**:60–66.
42. Kamanga-Sollo E, Pampusch MS, Xi G, White ME, Hathaway MR, Dayton WR: **IGF-I mRNA levels in bovine satellite cell cultures: effects of fusion and anabolic steroid treatment.** *J Cell Physiol* 2004, **201**:181–189.

43. Joubert Y, Tobin C: **Testosterone Treatment Results in Quiescent Satellite Cells Being Activated and Recruited into Cell Cycle in Rat Levator Ani Muscle.** *Devel Biol* 1995, **169**(1):286–294.
44. Breier BH, Gluckman PD, Bass JJ: **Influence of nutritional status and oeshadiol-17B on plasma growth hormone, insulin-like growth factors-I and -II and the response to exogenous growth hormone in young steers.** *J Endocrinol* 1988, **118**:243.
45. Hayden JM, Bergen WG, Merkel RA: **Skeletal muscle protein metabolism and serum growth hormone, insulin, and cortisol concentrations in growing steers implanted with estradiol-17 beta, trenbolone acetate, or estradiol-17 beta plus trenbolone acetate.** *J Anim Sci* 1992, **70**:2109–2019.
46. Hongerholt DD, Crooker BA, Wheaton JE, Carlson KM, Jorgenson DM: **Effects of a growth hormone-releasing factor analogue and an estradiol-trenbolone acetate implant on somatotropin, insulin-like growth factor I, and metabolite profiles in growing Hereford steers.** *J Anim Sci* 1992, **70**:1439–1448.
47. Milewicz T, Krzysiek J, Sztefko K, Radowicki S, Krzyczkowska-Sendrakowska M: **17 beta-estradiol regulation of human growth hormone (hGH), insulin-like growth factor-I (IGF-I) and insulin-like growth factor binding protein-3 (IGFBP-3) axis in hypoestrogenic, hypergonadotropic women.** *Endokrynol Pol* 2005, **56**:876–882.
48. Dawson-Hughes B, Stern D, Goldman J, Reichlin S: **Regulation of growth hormone and somatomedin-C secretion in postmenopausal women: effect of physiological estrogen replacement.** *J Clin Endocrinol Metab.* 1986, **63**:424–432.
49. Scanlan N, Skinner DC: **Estradiol modulation of growth hormone secretion in the ewe: No growth hormone-releasing hormone neurons and few somatotropes express estradiol receptor α.** *Biol Reprod* 2002, **66**:1267–1273.
50. Sotiropoulos A, Ohanna M, Kedzia C, Menon RK, Kopchick JJ, Kelly PA, Pende M: **Growth hormone promotes skeletal muscle cell fusion independent of insulin-like growth factor 1 up-regulation.** *Proc Natl Acad Sci U S A* 2006, **103**:7315–7320.
51. Ge X, Yu J, Jiang H: **Growth hormone stimulates protein synthesis in bovine skeletal muscle cells without altering insulin-like growth factor-I mRNA expression.** *J Anim Sci* 2012, **90**:1126–1133.
52. Allen RE, Sheehan SM, Taylor RG, Kendall TL, Rice GM: **Hepatocyte growth factor activates quiescent skeletal muscle satellite cells *in vitro*.** *J Cell Physiol* 1995, **165**:307–312.
53. Tatsumi R, Anderson JE, Nevoret CJ, Halevy O, Allen RE: **HGF/SF is present in normal adult skeletal muscle and is capable of activating satellite cells.** *Devel Biol* 1998, **194**:114–128.
54. Sheehan SM, Tatsumi R, Temm-Grove CJ, Allen RE: **HGF is an autocrine growth factor for skeletal muscle satellite cells *in vitro*.** *Muscle Nerve* 2000, **23**:239–245.
55. Gospodarowicz D, Moran J, Braun D, Birdwell C: **Clonal growth of bovine vascular endothelial cells: fibroblast growth factor as a survival agent.** *Proc Natl Acad Sci U S A* 1976, **73**:4120–4124.
56. Rusnati M, Dell'Era P, Urbinati C, Tanghetti E, Massardi ML, Nagamine Y, Monti E, Presta M: **A distinct basic fibroblast growth factor (FGF-2)/FGF receptor interaction distinguishes urokinase-type plasminogen activator induction from mitogenicity in endothelial cells.** *Mol Biol Cell* 1996, **7**:369–381.
57. Taylor KM, Chen C, Gray CA, Bazer FW, Spencer TE: **Expression of messenger ribonucleic acids for fibroblast growth factors 7 and 10, hepatocyte growth factor, and insulin-like growth factors and their receptors in the neonatal ovine uterus.** *Biol Reprod* 2001, **64**:1236–1246.
58. Coleman KD, Wright JA, Ghosh M, Wira CR, Fahey JV: **Estradiol modulation of hepatocyte growth factor by stromal fibroblasts in the female reproductive tract.** *Fertil Steril* 2009, **92**:1107–1109.
59. Coleman KD, Ghosh M, Crist SG, Wright JA, Rossoll RM, Wira CR, Fahey JV: **Modulation of hepatocyte growth factor secretion in human female reproductive tract stromal fibroblasts by poly (I:C) and estradiol.** *Am J Reprod Immunol* 2012, **67**:44–53.
60. Frey RS, Johnson BJ, Hathaway MR, White ME, Dayton WR: **Growth factor responsiveness of primary satellite cell cultures from steers implanted with trenbolone acetate and estradiol-17b.** *Basic Appl Myol* 1995, **5**:71–79.
61. Wing LY, Chuang PC, Wu MH, Chen HM, Tsai SJ: **Expression and mitogenic effect of fibroblast growth factor-9 in human endometriotic implant is regulated by aberrant production of estrogen.** *J Clin Endocrinol Metab* 2003, **88**:5547–5554.
62. White ME, Johnson BJ, Hathaway MR, Dayton WR: **Growth factor messenger RNA levels in muscle and liver of steroid-implanted and nonimplanted steers.** *J Anim Sci* 2003, **81**:965–972.
63. Wu Y, Bauman WA, Blitzer RD, Cardozo C: **Testosterone-induced hypertrophy of L6 myoblasts is dependent upon Erk and mTOR.** *Biochem Biophys Res Commun* 2010, **400**:679–683.
64. Jones NC, Fedorov YV, Rosenthal RS, Olwin BB: **ERK1/2 is required for myoblast proliferation but is dispensable for muscle gene expression and cell fusion.** *J Cell Physiol* 2001, **186**:104–115.

21

Effect of dietary phosphorus content on milk production and phosphorus excretion in dairy cows

Chong Wang[1,2], Zhen Liu[1], Diming Wang[1], Jianxin Liu[1*], Hongyun Liu[1] and Zhiguo Wu[3]

Abstract

Background: Phosphorus (P) supplementation is costly and can result in excess P excretion. This study investigated the effects of reducing dietary P on milk production and P excretion in dairy cows over a full lactation.

Method: Forty-five multiparous Holstein dairy cows were divided into 15 blocks according to expected calving date and previous milk yield, and assigned randomly to one of the three dietary treatments: 0.37, 0.47, and 0.57% P (DM basis); these P levels represent the NRC recommendations, Chinese recommendations, and the amount of dietary P commonly fed by Chinese dairy farmers, respectively. Average daily feed intake was calculated from monthly data on feed offered and refused. Milk yields of individual cows were recorded weekly, and milk samples were taken for analysis of protein, fat, solids-not-fat, lactose, and somatic cell count. Blood samples were collected on days −6, −3, 0, 3, 6 relative to calving, and then monthly throughout lactation, and analyzed for P and Ca concentrations. Spot samples of feces and urine were collected for 3 consecutive d during weeks 12, 24, and 36, and P concentrations were analyzed. Reproduction and health data were recorded.

Results: Dietary P did not affect dry matter intake or milk yield ($P > 0.10$). Milk fat content was slightly higher in cows fed 0.37% P than in cows fed 0.47% P ($P = 0.05$). Serum concentrations of P and Ca did not reflect dietary P content ($P > 0.10$). Fecal and urinary P both declined linearly ($P < 0.05$) as dietary P decreased from 0.57 to 0.37%. Fecal P content was 25% less when dietary P was 0.37% compared to 0.57%. Health events and reproductive performance were not associated with dietary P content ($P > 0.05$).

Conclusions: Lowering dietary P from 0.57 to 0.37% did not negatively affect milk production, but did significantly reduce P excretion into environment.

Keywords: Dairy cows, Milk production, Phosphorus excretion, Phosphorus requirement

Background

Phosphorus (P) is an essential nutrient for animal production and reproduction, and dairy producers often feed inorganic supplements to increase the P content of the diet. However, P supplementation is costly and can result in excess P excretion, contributing to the eutrophication of waterways. A study conducted by Wu et al. [1] showed that dietary P supplementation is not necessary for most dairy herds in the US. The latest NRC [2] recommendation reduced the P requirements for lactating cows from previously recommended values [3]. These requirements range from 0.32% (DM basis) for cows milking 25 kg/d to 0.38% for cows milking 54 kg/d.

Cows in China produce approximately 4,000 kg/yr of milk, compared to the 9,000 kg/yr produced by dairy cattle in the US [4]. In East China, milk production is higher, averaging 7,200 kg/yr [5]. The P requirement recommended by the Chinese feeding standard is 0.45% for this level of production, higher than the 0.37% currently recommended by the NRC [2]. A recent survey of Chinese dairy farms [4] found that as much as 0.55% P was fed on most of the farms contacted. These findings suggest that Chinese dairy cows are currently fed too much P. Therefore, the purpose of this study was to investigate whether

* Correspondence: liujx@zju.edu.cn
[1]Institute of Dairy Science, Zhejiang University, Hangzhou, P. R. China
Full list of author information is available at the end of the article

reducing dietary P over a full lactation affects the lactation performance of dairy cows in China.

Materials and methods

Animals, diets and experimental design

The use of animals was approved by the Animal Care Committee of Zhejiang University, Hangzhou, China. The experiment was conducted on a farm with a 1,000-cow herd fed TMR. Forty-five multiparous Holstein dairy cows (609.0 ± 23.0 kg body weight) were selected and divided into 15 blocks of three cows each according to expected calving date and milk yield during the previous lactation (6,381.0 ± 145.0 kg), and then assigned randomly within block to one of the following dietary treatments: 0.37 (**LP**), 0.47 (**MP**), and 0.57% P (**HP**) (Table 1). These P levels were obtained by altering the calcium phosphate and dicalcium phosphate content of the diets (Table 1). The amount of P in the LP diet was close to the level recommended by the NRC, the MP diet contained P at a level similar to the Chinese feeding standard recommendations [6], and the HP diet represented the amount of P commonly fed by milk producers in China [4]. All diets contained 45% forage, primarily corn silage and grass hay. The cows were fed their assigned diets for the duration of their lactation. Diet formulations (as-fed basis) were adjusted weekly to compensate for changes in the DM content of the ingredients.

Table 1 Ingredients and chemical composition of diets (DM basis)

Items	Dietary treatment[1]		
	LP	MP	HP
Ingredient,%			
Ground corn grain	22.0	21.9	21.8
Wheat bran	4.1	4.1	4.1
Soybean meal	4.2	4.2	4.2
Sesame meal	3.2	3.1	3.1
Cotton meal	3.0	2.9	2.9
Cotton seed	2.1	2.1	2.1
Calcium carbonate	0.9	0.6	0.2
Salt	0.4	0.4	0.4
Dicalcium phosphate	0.0	0.5	1.1
Corn silage	20.1	20.7	20.7
Grass hay	16.9	16.9	16.9
Alfalfa hay	7.6	7.6	7.7
DDGS	8.3	8.3	8.3
Apple pulp	6.0	6.0	6.0
Premix[2]	0.5	0.5	0.5
Chemical composition[3]			
Dry matter,%	57.9	57.9	57.9
P,%	0.37	0.47	0.57
Ca,%	0.74	0.74	0.74
CP,%	15.2	15.2	15.2
NDF,%	32.6	32.8	32.9
NEL, Mcal/kg	1.51	1.51	1.51

[1]LP = low P; MP = medium P; and HP = high P.
[2]Formulated to provide (per kg of DM) 1,000,000 IU of vitamin A, 200,000 IU of vitamin D, 1,250 IU of vitamin E, 14,000 mg of Zn, 100 mg of Se, 180 mg of I, 3,000 mg of Fe, 40 mg of Co, 3,000 mg of Mn, and 3,000 mg of Cu.
[3]All values were the actually determined averages from the monthly TMR analyses with the exception of the NEL, which was estimated using tabular values in the Chinese standard [6].

The experiment was conducted from January 2009 until February 2010. Cows were housed in a tie-stall barn with free access to water, offered TMR three times daily ad libitum (5 to 10% refusal), and milked at 0630 h, 1430 h, and 1930 h. Average daily feed intake was calculated from monthly data on feed offered to and refused by individual cows. After completion of a 40 wk lactation cows were dried off or removed from the experiment. As a normal herd management practice, some cows were dried off after less than 40 wk of lactation in order to insure an 8-wk dry period before the next lactation, or were removed from the experiment when their milk yield dropped to 9 kg/d or when they developed significant health problems. During the dry period following the trial, all cows were fed a standard dry cow diet with a P level of 0.40%.

Sampling, measurement, and analyses

Feed ingredients and TMR were sampled monthly (every 4 wk). Samples were dried at 60°C for 48 h, ground and passed through a Wiley mill with 1-mm screen, and analyzed for neutral detergent fiber (NDF) [7], DM, calcium (**Ca**), crude protein (CP) [8], and P [9]. A certified commercial P solution (VHG Labs, Inc., Manchester, NH) was used as a calibration standard. Accuracy of the analysis was assured by using additional commercial standards (Standard Reference Material 1570a-spinach leaves, and 8436-durum wheat flour; National Institute of Standards and Technology, Gaithersburg, MD) and by interlaboratory comparison of reference samples, which revealed differences of <5%. Chemical analyses were based on DM measurements made at 105°C.

Milk yields of individual cows were recorded weekly. At the same time, daily milk samples were taken in amounts proportional to yield (4: 3: 3, composite) and analyzed for protein, fat, solids-not-fat (SNF), lactose and somatic cell count (SCC).

Blood samples (10 mL) were collected 3 h after feeding on days −6, −3, 0, 3, 6 relative to calving, and then monthly throughout lactation. Blood samples were centrifuged at 2,200 × g for 15 min, and serum P and Ca concentrations were analyzed by Nanjing Jiancheng Bioengineering Institute (Nanjing, China) according to AOAC procedures [8].

Table 2 Milk production of cows fed diets differing in P content

Items[1]	Dietary treatments[2]			SEM	P^3				
	LP	MP	HP		T	t	T*t	L	Q
DMI, kg/d	22.4	22.0	22.4	0.77	0.12	0.03	0.43	0.69	0.65
Milk yield, kg/d	21.5	20.7	22.0	1.03	0.63	0.01	0.03	0.66	0.61
3.5% FCM, kg/d	20.8	19.7	21.5	0.62	0.09	0.01	0.47	0.54	0.60
Milk protein,%	3.33	3.30	3.28	0.05	0.77	0.01	0.02	0.79	0.68
Milk fat,%	3.71	3.41	3.61	0.09	0.05	0.01	0.11	0.36	0.15
Milk lactose,%	4.61	4.69	4.66	0.03	0.46	0.03	0.29	0.55	0.71
Milk SNF,%	8.77	8.74	8.72	0.06	0.25	0.03	0.33	0.59	0.55
SCC, 10^3/mL	302	350	231	44.9	0.05	0.58	0.29	0.35	0.03

[1]DMI = Dry matter intake; FCM = Fat corrected milk; SNF = Solid not fat; SCC = Somatic cell count.
[2]LP = low P; MP = medium P; and HP = high P.
[3]T, effect of treatment; t, effect of time; T*t, treatment by time interaction; L, linear effect; Q, quadratic effect.

Spot samples of feces and urine were collected three times during the experiment (on 3 consecutive d during weeks 12, 24, and 36) as described by Wu et al. [1]. Both fecal and urinary P levels were analyzed using the same method as for feed analysis.

Reproduction and health data were recorded. Visual detection of estrus was performed by the farm technician while the cows were in the holding area before milking, and was based on standing, mounting, and mucus discharge. Cows were inseminated at their first estrus after 52 d postpartum and at every estrus thereafter until conception occurred. Pregnancy was confirmed by rectal palpation. Conception rate at the first AI was the percentage of cows that conceived on first AI. Pregnancy rate was defined as the number of cows confirmed pregnant divided by the number of cows bred. Cows were identified as nonbreeders when they failed to become pregnant by 230 d in milk. These cows were not used in the calculation of pregnancy rates.

Statistical analysis

Measured milk yield, milk component percentages, and DMI were reduced to monthly averages. These averages, and the blood Ca and P concentrations, between-diet differences were assessed by repeated measures analysis using Proc MIXED in SAS (SAS institute Inc., Cary, NC); treatment, time, treatment × time, and block were included as effects in the model. An AR (1) covariance structure was utilized. Milk production during the previous lactation and serum Ca and P concentrations at −6 d relative to calving were included as covariates in the models for analysis of milk yield and serum variables, respectively. Other data were analyzed using the general linear models procedure in SAS. Treatments were also tested for linear and quadratic effects by orthogonal polynomial contrasts. Categorical data were analyzed for treatment effects using Proc FREQ (SAS) with a chi-square and Fisher's exact test.

Probability values of $P < 0.05$ were used to define statistically significant results, with statistical trends being defined at $P < 0.10$. Statistics related to block are excluded from discussion because of lack of significance of the term.

Results and discussion

Diet composition and feed intake

Treatment diets were formulated to differ only in P content, which was 0.37, 0.47, or 0.57% (DM basis, Table 1). The other ingredients of the diets were identical in their nutrient content throughout the experiment; therefore, all diets had the same CP, NDF, and Ca ($P > 0.05$).

Feed intake did not differ among treatment groups (Table 2, Figure 1). Valk and Sebek [10] reported that feeding cows a 0.24% P diet reduced DMI during the dry period following the first year of the study, and during the second lactation. However, most published studies suggest that dietary P content does not affect DMI [10,11] unless the amount is so low (<0.25%) that the activity of rumen microbes is impaired [12]. Clearly,

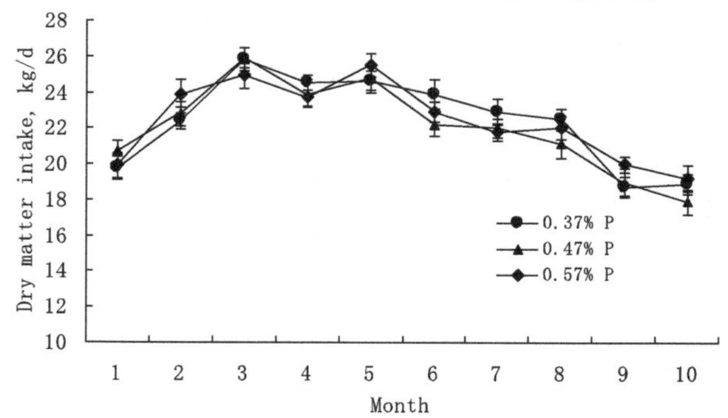

Figure 1 Change in dry matter intake of cows fed diets containing 0.37, 0.47, or 0.57% of P over a full lactation period. Means for the respective treatments over the entire lactation were 22.4, 22.0, and 22.4 kg/d (P = 0.12), with no treatment by sampling period interaction (P= 0.43). Values are expressed as mean ± SE.

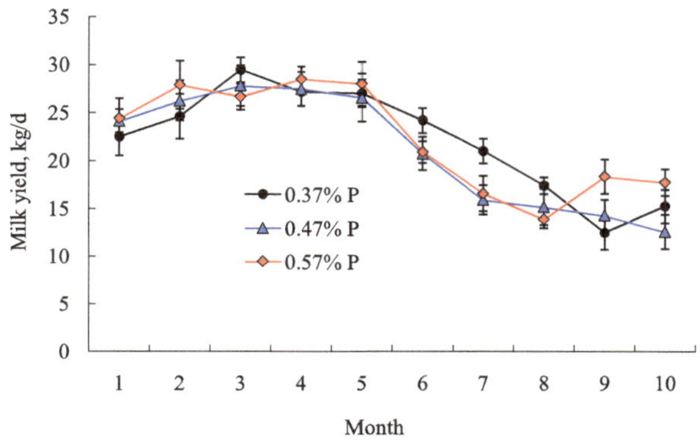

Figure 2 Change in milk yield of cows fed diets containing 0.37, 0.47, or 0.57% of P over a full lactation period. Means for the respective treatments over the entire lactation were 21.5, 20.7, and 22.0 kg/d (P=0.63), respectively, with a treatment by sampling period interaction (P = 0.03). Values are expressed as mean ± SE.

our LP diet (0.37% P) was not so low in P that it reduced DMI.

Lactation performance

Eight cows (2, 3, and 3 in the LP, MP, and HP groups, respectively) were culled prior to completing lactation. To obtain 280-day lactation information for these cows, the milk yields from the last five weekly averages were used to extrapolate estimates for the missing weeks by linear regression.

The lactation curves in Figure 2 show that during the first 5 mo milk yield was similar ($P > 0.10$) among treatment groups. During months 6, 7, and 8, the average milk yield was higher in the LP group than in the other 2 groups (20.9 vs. 17.3 and 17.2 kg/d, $P < 0.05$). This relationship was reversed during the following 2 mo (treatment by time interaction, $P < 0.05$). The lower milk yield of the cows fed the two higher P amounts during months 6 to 8 was unexpected, but similar observations were reported by Carstairs et al. [13], who showed that cows receiving 35% more P than required produced less milk than those fed according to the NRC recommendation [3]. Average daily milk yield did not significantly differ among treatment groups (21.5, 20.7, and 22.0 kg for the LP, MP, and HP groups, respectively; $P > 0.05$). Similarly, Wu and Satter [14] reported no difference in milk production over a complete lactation when cows were fed low (0.31 to 0.38%) or high (0.44 to 0.48%) dietary P. Lopez et al. [15] also reported that milk production was the same for cows fed 0.33 or 0.46% P.

Milk protein content was 3.33, 3.30, and 3.28% for the LP, MP, and HP groups, respectively (Table 2), and there were no significant differences among treatments ($P > 0.1$). Milk fat was lowest in the MP group ($P = 0.05$); overall, however, no linear or quadratic effect was observed. Similarly, there were no significant differences in milk lactose, fat-corrected milk (FCM), SNF, or SCC ($P > 0.05$). These observations are consistent with previous reports in the literature [16], which indicate that varying dietary P from 0.37 to 0.57% does not affect milk composition.

Serum Ca and P

Mean serum Ca (2.05 to 2.06 mmol/L) and P (1.21 to 1.28 mmol/L) concentrations were similar among treatment groups before calving, when all animals were receiving the same diet. Serum Ca and P remained similar (2.00 to 2.04 mmol/L for Ca and 1.20 to 1.25 mmol/L for P) after calving for the entire lactation period, even though the diets differed in P content (Table 3, $P > 0.1$). Peterson et al. [17] reported that serum P concentrations ranged from 2.09 to 2.17 mmol/L in dairy cows fed diets containing 0.21 to 0.44% P. Knowlton et al. [18] found that serum P concentrations increased from 1.26 to 1.83 mmol/L when dietary P content increased from 0.34 to 0.67%. Dietary Ca concentration did not affect P balance in a 20 wk experiment in which the serum P concentrations ranged from 1.65 to 1.74 mmol/L [19]. The serum concentrations of P that we found in this study were consistent with results reported by Liu [4], who found serum P concentrations of 1.23 to 1.29 mmol/L in cows in East China. Although serum P concentrations do decline if cows receive inadequate P in their diet [1,18], Valk and Sebek [10] suggested that 0.28% dietary P is sufficient for cows producing 9,000 kg/yr of

Table 3 Serum Ca and P of cows fed diets differing in P content

Items	Dietary treatments[1]			SEM	P^2				
	LP	MP	HP		Trt	t	T*t	L	Q
Serum Ca, mmol/L									
Prepartum	2.05	2.06	2.06	0.27	0.92	0.28	0.78	0.56	0.57
Postpartum	2.03	2.04	2.00	0.13	0.87	0.32	0.66	0.55	0.61
Serum P, mmol/L									
Prepartum	1.21	1.22	1.28	0.04	0.34	0.12	0.84	0.49	0.60
Postpartum	1.25	1.23	1.20	0.07	0.60	0.22	0.74	0.43	0.58

[1]LP = low P; MP = medium P; and HP = high P.
[2]T, effect of treatment; t, effect of time; T*t, treatment by time interaction; L, linear effect; Q, quadratic effect.

milk. This is substantially less than 0.37%, the lowest amount used in the present study, and our data suggest that the cows receiving 0.37% dietary P were not deficient in P. This observation is supported by the serum Ca data, which showed no difference among treatment groups. In a study in sheep, Breves et al. [20] found that serum Ca increases during P deficiency.

Fecal and urinary P excretion

The excretion of P during peak, middle, and late lactation reflected dietary P concentrations ($P < 0.05$, Table 4, Figure 3). Fecal and urinary P both declined linearly as dietary P decreased from 0.57 to 0.37%. These results suggest that reducing the amount of P fed to dairy cows in China from current levels to an amount consistent with NRC recommendations would reduce P excretion by 40% without affecting milk production. Tallam et al. [21] found that fecal P content was 29% lower in cows fed 0.35% P than in cows fed 0.47% P. Similar reductions in fecal P were found when dietary P was reduced [22,23]; such a reduction would have a significant impact on the environment.

Health problems and reproductive performance

There was no association between health events and dietary P content (Table 5). Reproductive performance also did not differ among treatment groups ($P > 0.05$; Table 5). Although, the present study did not use a sufficient number of animals to draw definite conclusions about dietary P content and animal health or reproductive performance, these data may be useful when combined with results from similar experiments. Several other studies have also showed no effect of dietary P on reproduction [14,21,24]. Noller et al. [25] reported similar reproductive performance in cows fed 0.22 or 0.32% P. Call et al. [11] reported that feeding diets containing 0.24, 0.32, or 0.42% P in their 12-month trial did not affect reproductive performance. In many parts of the world, dairy producers feed high dietary P to cows with the belief that P can improve reproductive performance [4]; however, our results indicate that reducing dietary P to the amounts recommended by the NRC [2] or the Chinese feeding standard [6] would not affect animal health or reproductive performance.

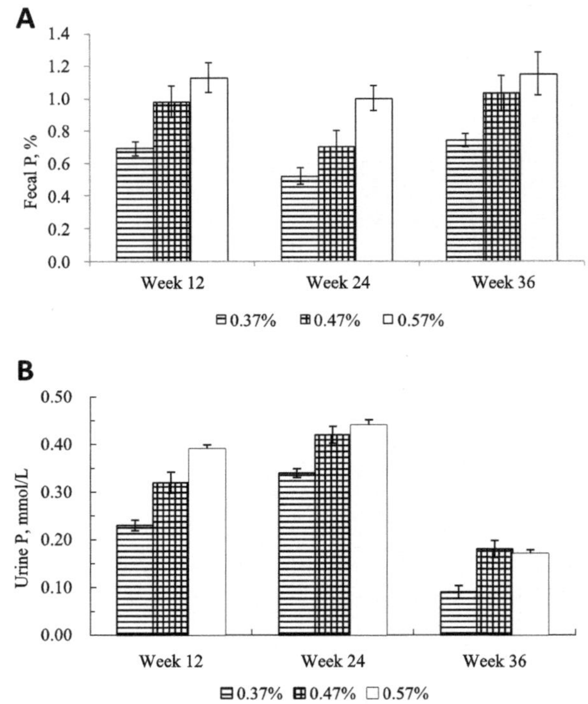

Figure 3 Concentration of fecal P (A) and urinary P (B) in lactation dairy cows fed diets containing 0.37, 0.47, or 0.57% of P measured during different stages of lactation. Values are expressed as mean ± SE.

Table 4 Fecal and urinary P concentrations at different stages of lactation in cows fed diets differing in P content

Item	Dietary treatment[1]			SEM	P^2		
	LP	MP	HP		T	L	Q
Fecal P,% of DM							
Peak lactation	0.69	0.98	1.13	0.031	0.01	0.02	0.45
Middle lactation	0.52	0.70	1.00	0.102	0.03	0.04	0.39
Late lactation	0.74	1.03	1.15	0.085	0.02	0.02	0.26
Urine P, mmol/L							
Peak	0.23	0.32	0.39	0.011	0.01	0.01	0.44
Middle	0.34	0.42	0.44	0.018	0.01	0.03	0.29
Late	0.36	0.44	0.50	0.008	0.01	0.04	0.30

[1]LP = low P; MP = medium P; and HP = high P.
[2]T, effect of treatment; L, linear effect; Q, quadratic effect.

Table 5 Health events and reproductive performance of cows fed diets differing in P content

Item	Dietary treatments[1]			P
	LP	MP	HP	
Mastitis	3	2	2	0.63
Foot rot	3	4	3	0.72
Retained placenta	4	3	3	0.57
Displaced abomasum	3	2	1	0.64
Udder edema	2	3	3	0.56
Ketosis	3	2	2	0.63
Dystocia	1	0	1	0.55
Total health problems	19	16	15	0.43
Proportion of cows that conceived,%	80.2	88.5	70.5	0.54
Proportion of cows pregnant,%	75.9	77.5	72.0	0.66
Weight of calves, kg	35.4	37.1	35.8	0.64

[1]LP = low P; MP = medium P; and HP = high P.

Conclusions

Lowering the dietary P content from 0.57 to 0.37% (NRC recommendation) did not negatively affect milk production in Chinese dairy cows, but did significantly reduce P excretion. The results of this study suggest that a dietary P content of 0.37% is sufficient for cows producing 21.5 kg/d of milk during lactation. Depending on the feed ingredients used, this concentration of P can be obtained without the addition of inorganic P supplements to the feed.

Competing interests

The authors declare that they have no competing interests.

Authors' contributions

JXL and ZG Wu conceived the study and designed the experiment. CW, ZL, and HYL carried out the statistical analysis and drafted the manuscript. CW, ZL, and DMW performed the animal experiments. All authors read and approved the final manuscript.

Acknowledgements

This work was supported by the earmarked fund for China Agriculture Research System (CARS-37), the National Key Technologies R & D Program of China (2012BAD12B02), and the Special Fund for Agro-scientific Research in the Public Interest (No. 201303143). The authors gratefully thank all staff of the Hangzhou Zhengxing Animal Industries for their inputs to this study, and Dr. Ferguson at the University of Pennsylvania for advice on the experimental design.

Author details

[1]Institute of Dairy Science, Zhejiang University, Hangzhou, P. R. China. [2]College of Animal Science and Technology, Zhejiang Agriculture and Forestry University, Lin'an, Hangzhou, P. R. China. [3]School of Veterinary Medicine, University of Pennsylvania, 382 West Street Road, 19348 Kennett Square, USA.

References

1. Wu Z, Satter LD, Blohowiak AJ, Stauffacher RH, Wilson JH: **Milk production, estimated phosphorus excretion, and bone characteristics of dairy cows fed different amounts of phosphorus for two or three years.** *J Dairy Sci* 2001, **84:**1738–1748.
2. NRC: *Nutrient Requirements of Dairy Cattle.* 7 revth edition. Washington, DC: National Research Council; 2001. Nat. Acad. Sci.
3. NRC: *Nutrient Requirements of Dairy Cattle.* 6 revth edition. Washington, DC: National Research Council; 1989. Nat. Acad. Sci.
4. Liu Z: *Effect of Dietary Phosphorus Amount on Milk Production and Phosphorus Excretion of Dairy Cows in China.* Master thesis, Zhejiang University, College of Animal Sciences; 2010.
5. Wang C, Liu JX, Zhai SW, Lai JL, Wu YM: **Effect of rumen degradable protein to rumen undegradable protein ratio on nitrogen conversion of lactating dairy cow.** *Acta Agric Scand A Anim Sci* 2008, **58**(2):100–103.
6. Ministry of Agriculture (MOA), PRC: *Feeding Standard of Dairy Cattle (NY/T 34 – 2004).* P.R.China: Ministry of Agriculture; 2004.
7. Van Soest PJ, Bobertson JB, Lewis BA: **Methods of dietary fiber, neutral detergent fiber, and nonstarch polysaccharides in relation to animal nutrition.** *J Dairy Sci* 1991, **74:**3583–3597.
8. AOAC: *Official Methods of Analysis,* Volume 1. 15th edition. Arlington, VA: Assoc. Offic. Anal. Chem; 1990.
9. Combs DK, Satter LD: **Determination of markers in digesta and feces by direct current plasma emission spectroscopy.** *J Dairy Sci* 1992, **75:**2176–2183.
10. Valk H, Sebek LBJ: **Influence of long term feeding of limited amounts of phosphorus on dry matter intake, milk production and body weight of dairy cows.** *J Dairy Sci* 1999, **82:**2157–2163.
11. Call JW, Butcher JE, Shupe JL, Lamb RC, Boman RL, Olson AE: **Clinical effects of low dietary phosphorus concentrations in feed given to lactating dairy cows.** *Am J Vet Res* 1987, **48:**133–136.
12. Ternouth JH: **Phosphorus and beef production in Northern Australia.** *Trop Grasslands* 1990, **24:**159–169.
13. Carstairs JA, Neitzel RR, Emery RS: **Energy and phosphorus status as factors affecting postpartum performance and health of dairy cows.** *J Dairy Sci* 1981, **64:**34–41.
14. Wu Z, Satter LD: **Milk production and reproductive performance of dairy cows fed two concentrations of phosphorus for two years.** *J Dairy Sci* 2000, **83:**1052–1063.
15. Lopez H, Kanitz FD, Moreira VR, Wiltbank MC, Satter LD: **Effect of dietary phosphorus on performance of lactating dairy cows: milk production and cow health.** *J Dairy Sci* 2004, **87:**139–146.
16. Wu Z, Tallam SK, Ishler VA, Archibald DD: **Utilization of phosphorus in lactating cows fed varying amounts of phosphorus and forage.** *J Dairy Sci* 2003, **86:**3300–3308.
17. Peterson AB, Orth MW, Goff JP, Beede DK: **Periparturient responses of multiparous Holstein cows fed different dietary phosphorus concentrations prepartum.** *J Dairy Sci* 2005, **88:**3582–3594.
18. Knowlton KF, Herbein JH: **Phosphorus partitioning during early lactation in dairy cows fed diets varying in phosphorus content.** *J Dairy Sci* 2002, **85:**1227–1236.
19. Taylor MS, Knowlton KF, McGilliard ML, Swecker WS, Ferguson JD, Wu Z, Hanigan MD: **Dietary calcium has little effect on mineral balance and bone mineral metabolism through twenty weeks of lactation in Holstein cows.** *J Dairy Sci* 2009, **92:**223–237.
20. Breves G, Ross R, Holler H: **Dietary phosphorus depletion in sheep: effects on plasma inorganic phosphorus, calcium, 1, 25-$(OH)_2$-Vit. D_2 and alkaline phosphatase and on gastrointestinal P and Ca balances.** *J Agric Sci (Camb)* 1985, **105:**623–629.
21. Tallam SK, Ealy AD, Bryan KA, Wu Z: **Ovarian activity and reproductive performance of dairy cows fed different amounts of phosphorus.** *J Dairy Sci* 2005, **88:**3609–3618.
22. Dou Z, Knowlton KF, Kohn RA, Wu Z, Satter LD, Zhang GY, Toth JD, Ferguson JD: **Phosphorus characteristics of dairy feces affected by diets.** *J Environ Qual* 2002, **31:**2058–2065.
23. Wu Z: **Utilization of phosphorus in lactating cows fed varying amounts of phosphorus and sources of fiber.** *J Dairy Sci* 2005, **88:**2850–2859.
24. Lopez H, Wu Z, Satter LD, Wiltbank MC: **Effect of dietary phosphorus concentration on estrous behavior of lactating dairy cows.** *Theriogenol* 2004, **61:**437–445.
25. Noller CH, Castro AG, Wheeler WE, Hill DL, Moeller NJ: **Effect of phosphorus supplementation on growth rate, blood minerals, and conception rate of dairy heifers.** *J Dairy Sci* 1977, **60:**1932–1940.

22

Effect of variety and drying method on the nutritive value of corn for growing pigs

Quanfeng Li†, Meng Shi†, Chuanxin Shi, Dewen Liu, Xiangshu Piao, Defa Li and Changhua Lai*

Abstract

Background: This experiment was conducted to determine the nutritive value of corn from the north of China for growing pigs. The experiment examined corn variety (LS1, LS2, LS3 and LS4) grown in one location, drying method (sun dried and artificially dried) and different drying temperatures. Corn harvested at 20-25% moisture was dried to about 12% moisture by sun drying and artificially drying at 80, 100, or 120°C in a fluidized bed dryer. Ninety-six barrows (average BW of 33.4 ± 2.7 kg) were housed in individual metabolism crates to facilitate separate collection of feces and urine. A five-day collection period followed a seven-day diet acclimation period.

Results: The results indicated that variety significantly influenced ($P < 0.01$) the 1,000 kernel weight of corn but not the bulk weight. Variety also influenced the available energy content (digestible energy of dry matter, $P < 0.01$; metabolisable energy of dry matter, $P < 0.01$) and digestibility of organic matter ($P < 0.01$), as well as dry matter ($P < 0.01$) and gross energy (GE) content ($P < 0.02$). The drying method of corn significantly influenced the 1,000 kernel weight ($P < 0.01$), bulk weight ($P < 0.01$) and digestibility of ether extract (EE) ($P < 0.01$). No effect of drying temperature on the digestibility of organic matter, dry matter (DM), crude protein (CP), neutral detergent fiber (NDF), acid detergent fiber (ADF) and gross energy was observed, but gelatinization ($P < 0.05$) and test weight ($P < 0.01$) decreased with an increase in temperature.

Conclusions: Variety has a significant impact on the nutritive value of corn for growing pigs, and greater attention needs to be paid to these influences in the assignment of the nutritive value of corn given to growing pigs.

Keywords: Corn, Digestible energy, Drying method, Pigs, Variety

Background

Corn growers are usually confronted with difficulty in the safe storage of their grain crop because of a high moisture content (22-30%) at the time of harvest. This condition is especially pronounced in the north of China where the weather conditions are unfavorable for natural field drying. So, the purpose of drying is to lower the moisture content in grain to the safe storage content of about 14%. Therefore, most of the corn is artificially dried at a temperature of 120°C in Jilin, Liaoning, and Heilongjiang province, which are the biggest corn producing areas in China. In order to decrease the damage to corn from a high temperature, the low temperature drying process, which occurs at 80°C is used. However, corn in the Shandong and Henan provinces, which are in the east of China, was dried in the sun.

As expected, there were many studies on corn variety in feeds, but the varieties studied were high-oil [1,2], high-lysine [3], low-phytate [4] and so on. There varieties are very special, and the planting area is not large. The varieties of corn used in feeds still are conventional varieties. However, there is a lack of congruent information on the variation in nutritive value as affected by conventional variety, drying method and drying temperature of corn. Therefore, the objectives of this experiment were to measure the influence of conventional variety on the nutritive value of corn and to compare estimates of the effects of drying method on the nutritive value of corn. An additional objective was to evaluate varying temperature during drying on the nutritive value of corn in growing pigs.

* Correspondence: Laichanghua999@163.com
†Equal contributors
State Key Laboratory of Animal Nutrition, Ministry of Agriculture Feed Industry Centre, China Agricultural University, Beijing 100193, China

Materials and methods

The parts of this experiment that involved animals were conducted in the Metabolism Laboratory of the Ministry of Agriculture Feed Industry Centre (China Agricultural University, Beijing, China). The protocol for the experiment was reviewed and approved by the Institutional Animal Care and Use Committee at China Agricultural University.

Selection and preparation of the corn samples

Four corn variety samples (LS1, LS2, LS3 and LS4) were obtained from one location in the north of Hebei province. The four varieties were very widespread, and the planting area was very large. The varieties of hardness were classified as soft (LS1), hard (LS3), and of intermediate hardness (LS2; LS4). Corn was hand-harvested at 20-25% moisture and was spread on a tarp on the ground for a maximum of one week prior to threshing and drying.

Grain samples from each variety were divided into four subplots of 200 kg each. One of the subplots was placed on an airfield and allowed to dry to about 12% moisture content after threshing. The remaining three subplots were dried in a fluidized bed dryer at 80, 100 or 120°C. Final moisture contents were about 12% (ranging from9.4 to 14.7%). At each drying temperature, the varieties were dried in a random order. Approximate drying times were 60, 50 and 40 min, respectively, for temperatures of 80, 100 and 120°C. When samples were removed from the dryer, they were cooled at room temperature before packaging in bags.

Animals, housing, and experimental design

Ninety-six barrows (initial body weight of 33.4 ± 2.7 kg) were used in this experiment. The pigs were Duroc × Landrace × Yorkshire crossbreeds. Pigs were individually housed in stainless steel metabolism cages (1.4 m × 0.45 m × 0.6 m). A feeder and a nipple drinker were installed in each pen. The crates were located in an environmentally controlled room with a temperature of 22 ± 1°C. Pigs were allotted to one of sixteen diets according to a completely randomized design, and each diet was measured with six pigs.

Diets, feeding, and sample collection

Sixteen diets were formulated to contain 96.8% of one of each of the corn samples and 3.2% minerals and vitamins (Table 1). Corn was assumed to be the only source of energy in the diet, as the slight contribution of energy from the vitamin and mineral premix was assumed to be negligible. Vitamins, salt, and minerals were included in all diets to meet or exceed the estimated requirements for growing pigs [5].

Feed was provided twice daily as a mash, at 0800 and 1700 h. Water was freely available for each pig. During the period of adjustment to the metabolism crates and diets, average daily feed intake was gradually increased until it was estimated to supply 4% of the BW determined at the initiation of each adaptation period. During the collection period, all fresh fecal samples were collected as often as possible throughout the day from all pigs and stored at −20°C. The collection and sample preparation of feces and urine were conducted according to the methods described by Song et al. [2].

Table 1 Composition of the experimental diets (as-fed basis) fed to growing pigs for comparison of the energy digestibility between different corn samples

Ingredient, %	Experimental diet
Corn	96.8
Antioxidant[1]	0.1
Calcium phosphate	1.7
Limestone	0.6
Salt	0.3
Vitamin and mineral premix[2]	0.5

[1]Santoquin MAX composite antioxidant, contained no less than 10% Ethoxyquin, no less than 3% ButylatedHydroxytoluene (BHT) and Citric acid, provided by Novus International, Inc.
[2]Premix provided the following per kg of complete diet for growing pigs: vitamin A, 5,512 IU; vitamin D_3, 2,200 IU; vitamin E, 30 IU; vitamin K_3, 2.2 mg; vitamin B_{12}, 27.6 µg; riboflavin, 4 mg; pantothenic acid, 14 mg; niacin, 30 mg; choline chloride, 400 mg; folacin, 0.7 mg; thiamine 1.5 mg; pyridoxine 3 mg; biotin, 44 µg; Mn, 40 mg (MnO); Fe, 75 mg ($FeSO_4 \cdot H_2O$); Zn, 75 mg (ZnO); Cu, 100 mg ($CuSO_4 \cdot 5H_2O$); I, 0.3 mg (KI); Se, 0.3 mg (Na_2SeO_3).

Chemical analyses

At the conclusion of the experiment, the fecal samples were dried at 60°C in a forced-air oven for 72 h. After drying, all samples were finely ground, and ground samples of diets and feces were analyzed for DM [6], ether extract (EE) [7], and ash [6]. Kjeldahl N was determined according to the method of Thiex et al. [8]. The diets and feces samples were analyzed for NDF and ADF, and the content of NDF and ADF were determined using filter bags and fiber analyzer equipment (Fiber Analyzer, Ankom Technology, Macedon, NY) following a modification of the procedure of Van Soest et al. [9]. The gelatinized starch of corn samples was determined by enzymatic hydrolysis as described by Xiong et al. [10]. Urine samples (4 mL) were injected into 2 filter papers in a special crucible and then dried for 8 h in a 65°C drying oven prior to determination of the energy content. The gross energy (GE) of urine was measured by injecting 4 mL of sample into 2 filter papers in a special crucible and then dried for 8 h in a 65°C drying oven to determine the energy content. The GE of feces and diets were measured using an automatic adiabatic oxygen bomb calorimeter (Parr 6400 Calorimeter, Moline, IL). The 1,000 kernel weight (g/1,000 seeds) was measured in each sample of test corn by first cleaning it of all foreign materials and then counting 1,000 seeds. The particle size were determined according to the method ASAE S319.4 (2008) [11].

Calculations and statistical analyses

The digestible energy (DE), metabolisable energy (ME) and apparent total tract digestibility (ATTD) of nutrients in the 16 feed samples were measured and then later converted to reflect the digestibility of the individual corn samples. The small portion of the experimental diets that consisted of minerals and vitamins (3.2%) was assumed to have a negligible contribution to the digestibility of GE.

The DE and ME contents of the 16 diets were calculated using Eq:DE = (GEi – GEf)/Ft, ME = (GEi – GEf – GEu)/Ft, where DE is the DE content in diets (kcal/kg of DM), GEi is the total GE intake (kcal of DM), GEf is the GE content in feces (kcal of DM), Ft is the total feed intake (kg of DM), ME is the ME content in diets (kcal/kg of DM), and GEu is the GE content in urine (kcal of DM). The apparent total tract digestibility (ATTD, %) for energy, organic matter (OM), DM, EE, ADF, NDF and CP was calculated using Eq: ATTD = [(Fi – Ff)/Fi] × 100, where Fi is the total intake (kcal or g) of respective component in the collection period, and Ff is the total fecal output (kcal or g) of respective component originating from the feed that was given during the collection period. Data were analyzed as a randomized complete block design by using the PROC MIXED procedure (SAS Inst. Inc., Cary, NC) with pig as the experimental unit. Main effects of variety and drying method interactions were tested. Orthogonal polynomial contrasts were used to detect linear and quadratic responses to drying temperature. Means were separated using the LSMeans procedure and the PDIFF option of SAS. Pig was the experimental unit for all calculations, and an alpha level of 0.05 was used to assess the significance of difference between means.

Results

The mean physical characterization and chemical composition of the four corn varieties are presented in Table 2. The bulk weight of corn ranged from 716.1 to 765.8 g/L, with a mean value of 744.6 g/L (CV 3.0%). The 1,000 kernel weight was highly variable, and the content ranged from 282.2 to 356.6 g with a mean value of 318.4 g (CV 10.1%).

Drying the corn samples using different temperatures affected certain characteristics of the grain kernels. In a blind taste-comparison test, five individuals were given these samples of corn to taste. Test subjects found that a progressive brown darkening and parched corn odor and taste occurred in corn dried at 120°C. Even though the same hammer mill and screen were used in all cases in the experiment, the particle size of corn decreased as drying temperatures increased, except for the variety of LS1. The EE, NDF, ADF, CP and ash contents were not significantly affected by drying temperature (Table 2). However, the DM content of corn increased linearly as drying temperature increased (linear, $P < 0.01$; quadratic, $P < 0.03$).The difference between two drying methods was significant with regard to bulk weight and 1,000 kernel weight ($P < 0.01$). However, increasing the drying temperature decreased the bulk weight (linear, $P < 0.01$; quadratic, $P < 0.01$). There was no difference between varieties and drying methods for starch gelatinization (Table 3).

The DE and ME content of the 16 test corn samples and digestibility of nutrients are presented in Table 4. The interaction term between variety and drying method was not significant. The LS1 variety had a higher digestible energy than LS2 ($P < 0.05$); while no differences among LS2, LS3 and LS4 were observed. The ATTD for OM, DM and GE in LS1 were greater ($P < 0.05$) than in all other corn varieties while no differences among LS2, LS3 and LS4 were observed. Acid detergent fiber and NDF digestibility coefficients varied from 34.6 to 51.3% and 45.2 to 51.0%, respectively, for the four varieties. The EE and CP fractions were digested to a much greater degree than the structural carbohydrates. Drying method did not affect the digestible energy content and digestibility of nutrients, with the exception of DE (as fed basis), ME (as fed basis) and the ATTD of EE.

In the 16 corn samples, fibrous compounds had a negative correlation with DE and ME content, while the correlation of GE with DE and ME content was positive (Table 5). The NDF content had the highest correlation of any characteristic (chemical or physical) with DE and ME content ($r = -0.42$; $P < 0.01$; $r = -0.43$; $P < 0.01$), followed by ADF ($r = -0.38$; $P < 0.01$; $r = -0.35$; $P < 0.01$) and GE ($r = 0.32$; $P < 0.05$; $r = 0.39$; $P < 0.05$). Interestingly, in the 16 corn samples with a bulk weight between 662.0 and 765.9 g/L, bulk weight was not correlated with DE and ME content ($r = -0.39$; $P > 0.05$; $r = -0.26$; $P > 0.05$) (Table 5).

Discussion

Relationships of physical characteristics and energy content

LS4 had the highest test weight and 1,000 kernel weight among the four varieties, but the results indicate that there is no significant correlation between physical characteristics (1,000 kernel weight and bulk weight) and available energy value in these grains. In the past, several attempts have been made to link the nutritional value of cereal grains to physical characteristics, such as density and kernel weight. Many researchers [12-14] concluded that available energy in growing pigs was not related to

Table 2 Physical and chemical characterization of the corn samples[a]

Variety	Drying method	1,000 kernel weight, g	Bulk weight g/L	DM, %	EE, %	NDF, %	ADF, %	CP, %	Ash, %	GE, MJ/kg	Starch gelatinization, %	Particle size, μm
LS1	Sundried	330.4	716.2	85.35	3.17	9.15	2.11	7.79	1.25	15.88	12.34	538.09 ± 1.11
	80°C	330.7	699.5	88.00	3.52	9.05	1.86	7.84	1.21	16.27	17.26	549.78 ± 1.12
	100°C	317.0	662.0	90.63	3.51	9.03	2.20	7.80	1.35	16.92	18.60	619.47 ± 1.11
	120°C	313.3	682.3	90.24	3.43	8.82	1.98	7.73	1.22	16.81	12.20	578.91 ± 1.12
LS2	Sundried	282.2	758.6	85.14	3.15	10.28	2.24	7.34	1.12	15.83	14.50	520.46 ± 1.10
	80°C	269.5	716.9	88.01	3.74	10.03	2.12	7.38	1.20	16.45	19.92	501.04 ± 1.12
	100°C	279.5	692.3	88.21	3.57	9.98	1.90	7.81	1.12	16.31	15.61	481.23 ± 1.09
	120°C	273.9	670.5	89.77	3.78	10.05	2.16	7.28	1.19	16.60	13.46	506.32 ± 1.11
LS3	Sundried	304.6	738.2	85.93	3.21	10.49	2.27	7.11	1.17	16.00	15.62	541.38 ± 1.12
	80°C	294.8	701.9	88.45	3.18	10.27	2.25	7.50	1.11	16.50	15.82	525.26 ± 1.12
	100°C	296.3	678.8	88.93	3.50	10.65	2.37	7.25	1.20	16.45	19.67	512.40 ± 1.12
	120°C	286.2	664.5	89.33	3.49	10.04	2.17	7.70	1.28	16.57	13.19	379.59 ± 1.14
LS4	Sundried	356.6	765.9	86.30	3.04	9.85	2.13	8.60	1.11	15.96	13.82	518.24 ± 1.11
	80°C	317.4	692.8	88.53	2.85	9.42	2.15	8.91	1.17	16.37	13.73	511.25 ± 1.13
	100°C	309.7	681.5	89.10	2.81	10.16	2.57	8.32	1.27	16.48	13.54	496.71 ± 1.12
	120°C	316.5	655.5	89.50	2.68	10.28	2.14	8.94	1.29	16.55	12.57	413.96 ± 1.10

[a]All data are the results of a chemical analysis conducted in duplicate.

density in wheat samples. Likewise, in poultry, there were no correlations between ME and bushel weight or weight per 1,000 kernels in wheat samples [15,16]. In China, bulk weight is the most important marker of grade index for most feed companies. However, physical parameters such as bulk weight and 1,000 kernel weight cannot be used to estimate nutritional value of corn accurately for pigs.

The particle size of corn samples decreased with drying temperature, except for LS1. Drying temperature has substantial effects on breakage susceptibility, stress cracking, and dry-milling quality [17]. Increasing drying temperature increases breakage susceptibility [18,19]. The reason why particle size of LS1 was not affected by drying temperature may be that the LS1 was the soft corn hybrid.

Effects of variety on the nutritive value of corn

Although special varieties of corn, such as high-oil, high-lysine and NutriDense, were not used in this experiment, the results of this study indicate that corn genetics significantly influence the 1,000 kernel weight and available energy content. The DE and ME are the most important parameters used in characterizing the nutritional value of cereal grains for livestock. Although the variety of LS1 was not a NutriDense corn, the DE

Table 3 Effect of variety and drying method on the physical characteristics and starch gelatinization of corns for growing pigs

Item	Variety				Sun dried	Artificially dried °C			SEM	*P*-value[1]			
										Variety	Drying method[2]	Air temperature	
	LS1	LS2	LS3	LS4		80	100	120				Linear	Quadratic
Bulk weight, g	690.0	695.8	709.5	698.92	744.6	702.7	678.6	668.1	10.37	0.58	0.01	0.01	0.01
1,000 kernel weight, g/kg	322.8[a]	295.4[b]	276.2[c]	325.0[a]	318.4	303.1	300.6	297.4	4.70	0.01	0.01	0.71	0.94
Starch gelatinization of DM, %	17.03	18.24	18.09	15.19	16.42	18.91	18.88	14.33	1.35	0.48	0.37	0.06	0.06
DM, %	88.36	87.78	88.16	88.56	85.68	88.25	89.22	89.71	0.84	0.62	0.01	0.01	0.03
GE, MJ/kg	16.34	16.30	16.38	16.47	15.92	16.40	16.54	16.63	0.18	0.57	0.01	0.08	0.22
Particle size, μm	485.04	502.26	489.66	571.56	529.54	521.83	527.45	469.70	49.24	0.10	0.43	0.27	0.42

[1]Interaction between variety and drying method was not significant ($P > 0.10$).
[2]Contrast for Artificial dried vs. sun dried.
[a-c]Means followed by the same letter within each row are not significantly different from each other ($P > 0.05$).

Table 4 Effect of variety and drying method on the nutritive value of corns for growing pigs

Item	Variety				Sun dried	Artificially dried °C			SEM	P-value[1]			
										Variety	Drying method[2]	Air temperature	
	LS1	LS2	LS3	LS4		80	100	120				Linear	Quadratic
DE of DM, MJ/kg	16.14[a]	15.86[b]	15.86[b]	15.82[b]	15.82	16.03	15.92	15.91	0.06	0.01	0.06	0.23	0.45
ME of DM, MJ/kg	15.87[a]	15.51[b]	15.61[b]	15.51[b]	15.57	15.76	15.63	15.61	0.06	0.01	0.06	0.19	0.40
Digestibility coefficients, %													
Organic matter	91.56[a]	90.10[b]	90.21[b]	90.03[b]	90.24	90.77	90.48	90.41	0.27	0.01	0.31	0.45	0.73
Dry matter	90.03[a]	88.54[b]	88.70[b]	88.62[b]	88.66	89.43	88.96	88.85	0.30	0.01	0.23	0.20	0.41
Energy	89.58[a]	88.11[b]	88.00[b]	88.24[b]	88.02	89.04	88.47	88.41	0.34	0.02	0.13	0.21	0.40
NDF	57.01	48.20	45.78	45.24	52.28	50.20	46.33	47.43	3.35	0.09	0.27	0.62	0.78
ADF	51.32	41.94	34.65	34.58	48.54	40.92	36.27	36.76	4.47	0.06	0.06	0.57	0.79
Crude protein	80.81	78.56	76.76	78.56	78.99	79.37	77.17	79.16	1.18	0.15	0.75	0.91	0.38
Ether extract	61.13	65.25	64.07	63.71	59.18	66.46	66.83	67.68	1.39	0.31	0.01	0.53	0.82

[1]Interaction between variety and drying method was not significant ($P > 0.10$).
[2]Contrast for Artificial dried vs. sun dried.
[a-b]Means followed by the same letter within each row are not significantly different from each other ($P > 0.05$).

content in LS1 was 2% greater than in LS4 corn; this is very valuable trait for feed companies. The reason for this observation was the ATTD of nutrients in the LS1 were greater than the other. It is likely that natural differences among varieties of corn exist, as has been demonstrated for corn and other cereal grains [14,20,21] and for sweet lupins [22].

Effects of drying method and drying temperatures on the nutritive value of corn

The EE, NDF, ADF, CP, and ash contents were not significantly affected by drying temperature. The results are in agreement with those reported by Costa et al. [23]. A brown discoloration and a parched corn odor were also observed by Coates et al. [22] in corn dried at temperatures of 138 and 160°C. These characteristics suggest that a Maillard reaction has occurred in the grain kernel [24]. The parched odor of the corns dried at 120°C might have originated in similar fashion as the flavors from Maillard reaction described by Hathaway et al. [25] and Coates et al. [23]. In this study, drying method affected the bulk weight and 1,000 kernel weight. Other researchers have found that bulk weight and 1,000 kernel weight of corn generally decreases with increasing drying temperatures [16,19,26]. A factor that may affect bulk weight and 1,000 kernel weight is artificial drying, which removes moisture within the kernel without changing the kernel size.

With the exception of ATTD of EE, the available energy content (DE of DM; ME of DM) and ATTD of nutrients were not significantly affected by drying method. These results are in agreement with those reported by Costa et al. [23]. However, the ATTD of NDF and ADF for sundried corn samples were not significantly higher

Table 5 Correlation coefficients between chemical and physical characteristics and energy values of corn

Item	DE	ME	Bulk weight	1,000 kernel weight	GE	Ash	EE	NDF	ADF	CP
DE	1.00									
ME	0.95**	1.00								
Bulk weight	−0.39	−0.26	1.00							
1,000 kernel weight	0.12	0.07	0.21	1.00						
GE	0.32*	0.39*	0.17	−0.21	1.00					
Ash	0.32	0.27	−0.47	0.21	0.17	1.00				
EE	0.24	0.39	−0.05	−0.42	0.37	0.01	1.00			
NDF	−0.42*	−0.43*	0.27	−0.28	0.03	−0.09	−0.11	1.00		
ADF	−0.38*	−0.35*	0.20	−0.08	−0.02	0.10	−0.33	0.80**	1.00	
CP	0.21	0.04	−0.22	0.45	−0.26	0.21	−0.70	−0.54	−0.28	1.00

*, **, $P < 0.05$, $P < 0.01$, respectively.

than for artificially dried samples. The ATTD of nutrients and available energy contents were not significantly different among corn dried at different temperatures. The percent of starch gelatinization did not increase with drying temperature. Our findings are not in agreement with increases in starch gelatinization percentage reported by Costa et al. [16]. The reason may be that the drying time, moisture, and processing condition in this study were not suited for starch gelatinization. Wood [26] showed that a high moisture content (30-50%) is required for efficient gelatinization. Furthermore, when limited amounts of water are present, more heating time is needed to complete drying [27]. Chiang and Johnson [28] and Della Valle et al. [29] reported that decreasing the retention time of the sample during processing decreased starch gelatinization. There was no extrusion process used in the present study, and the heating time was shorter than that of Costa et al. [16].

Conclusions

In conclusion, variety had a significant impact on the nutritive value of corn for growing pigs. Heat treatment had some effects on the physical characteristics of corn, but this did not affect the nutritive value of corn for swine. These data suggest that greater accountability of these factors is required when assessing the nutritive value of corn for use in pig feed.

Abbreviations

DE: Digestible energy; ME: Metabolizable energy; CP: Crude protein; NDF: Neutral detergent fibre; ADF: Acid detergent fibre; ATTD: Apparent total tract digestibility; EE: Ether extract; GE: Gross energy; DM: Dry matter; RSD: Residual standard deviation.

Competing interests

The authors declare that they have no competing interests.

Authors' contributions

QFL carried out the experiment trial, performed the statistics and drafted the manuscript. SM and XSP participated in design of the study. DWL and SH participated animal trial. DFL and CHL conceived the study, and participated in its design and coordination. All authors read and approved the final manuscript.

Acknowledgements

This research was financially supported by the National Key Technology R & D Program in the 11th Five Year Plan of China (2006BAD12B01) and Special Public Sector Fund in Agriculture (200903006). Thanks to Chuanxin Shi, Zhongchao Li and Zhaoyu Liu for their practical support.

References

1. Adeola O, Bajjalieh NL: **Energy concentration of high-oil corn varieties for pigs.** *J Anim Sci* 1997, **75**:430–436.
2. Song GL, Li DF, Piao XS, Chi F, Yang WJ: **Apparent ileal digestibility of amino acids and the digestible and metabolizable energy content of high-oil corn varieties and its effects on growth performance of pigs.** *Arch Anim Nutr* 2003, **57**:297–306.
3. O'Quinn PR, Nelssen JL, Goodband RD, Knabe DA, Woodworth JC, Tokach MD, Lohrmann TT: **Nutritional value of a transgenic high-lysine, high-oil corn for young pigs.** In *Report of progress (Kansas State University. Agricultural Experiment Station and Cooperative Extension Service)*; 2010:116–121.
4. Linneen SK, DeRouchey JM, Goodband RD, Tokach MD, Dritz SS, Nelssen JL, Snow JL: **Evaluation of NutriDense low-phytate corn and added fat in growing and finishing swine diets.** *J Anim Sci* 2008, **86**:1556–1561.
5. NRC: *Nutrient requirements of swine 10th rev. ed.* Washington DC, USA: National Academy Press; 1998.
6. AOAC: *Official methods of analysis.* 17th edition. Arlington, VA, USA: Association of Official Analytical Chemists; 2000.
7. Thiex NJ, Anderson S, Gildemeister B: **Crude fat, diethyl ester extraction, in feed, cereal grain, and forage (Randall/Soxtec/submersion method): collaborative study.** *J AOAC Int* 2003, **86**:888–898.
8. Thiex NJ, Manson H, Anderson S, Persson JA: **Determination of crude protein in animal feed, forage, grain, and oilseeds by using block digestion with copper catalyst and steam distillation into boric acid: collaborative study.** *J AOAC Int* 2002, **85**:309–317.
9. Van Soest PJ, Robertson JB, Lewis BA: **Methods for dietary fiber and non-starch polysaccharides in relation to animal nutrition.** *J Dairy Sci* 1991, **74**:3568–3597.
10. Xiong YQ, Bartle SJ, Preston RL: **Improved enzymatic method to measure processing effects and starch availability in sorghum grain.** *J Anim Sci* 1990, **68**:3861–3870.
11. ASAE: *Method of determining and expressing fineness of feed materials by sieving. ASAE standard S319.4.* St. Joseph, MI: American Society of Agricultural and Biological Engineers; 2008.
12. Bhatty RS, Christison GI, Sosulski FW, Harvey BL, Hughes GR, Berdahl JD: **Relationships of various physical and chemical characters to digestible energy in wheat and barley cultivars.** *Can J Anim Sci* 1974, **54**:419–427.
13. Christison GI, Bell JM: **An assessment of bulk weight and other simple criteria for predicting the digestible energy values of feed grains.** *Can J Plant Sci* 1975, **55**:515–528.
14. Anderson DM, Bell JM: **The digestibility by pigs of dry matter, energy, protein and amino acids in wheat cultivars.I. Glenlea, Neepawa, Inia-66, Pitic-62 and Twin, compared with Bonanza barley.** *Can J Plant Sci* 1983, **63**:385–391.
15. Stewart AH, Acamovic T, Taylor AG, Fraser H: **An evaluation of wheat specific weight as a determinant of nutritive value for pigs and poultry.** *Anim Sci* 1997, 66–66.
16. Costa PM, Jensen AH, Harmon BG, Norton HW: **Effects of roasting and roasting temperatures on the nutritive value of high-moisture corn for swine.** *J Anim Sci* 1977, **44**:234–241.
17. Kirleis AW, Stroshine RL: **Effects of hardness and drying air temperature on breakage susceptibility and dry-milling characteristics of yellow dent corn.** *Cereal Chem* 1990, **67**:523–528.
18. Peplinski AJ, Brekke OL, Griffin EL, Hall G, Hill LD: **Corn quality as influenced by harvest and drying conditions.** *Cereal Foods World* 1975, **20**:145.
19. Gunaskaran S, Paulsen MR: **Breakage resistance of corn as a function of drying rates.** *Trans ASAE* 1985, **28**:2071–2093.
20. Pedersen C, Boersma MG, Stein HH: **Energy and nutrient digestibility in NutriDense corn and other cereal grains fed to growing pigs.** *J Anim Sci* 2007, **85**:2473–2483.
21. Fairbairn SL, Patience JF, Classen HL, Zijlstra RT: **The energy content of barley fed to growing pigs: characterizing the nature of its variability and developing prediction equations for its estimation.** *J Anim Sci* 1999, **77**:1502–1512.
22. Kim JC, Mullan BP, Heo JM, Hernandez A, Pluske JR: **Variation in digestible energy content of Australian sweet lupins (Lupinus angustifolius L.) and the development of prediction equations for its estimation.** *J Anim Sci* 2009, **87**:2565–2573.
23. Costa PM, Jensen AH, Harmon BG, Norton HW: **The effects of roasting and roasting temperatures on the nutritive value of corn for swine.** *J Anim Sci* 1976, **42**:365–374.
24. Moran ET Jr, Summers JD: **Heat processing of wheat germ meal and its effect on utilization and protein quality for the growing chick: toasting and autoclaving.** *Cereal Chem* 1968, **45**:304–310.
25. Hathaway IL, Yung FD, Kiesselbach TA: **The effect of drying temperature upon the nutritive value and commercial grade of corn.** *J Anim Sci* 1952, **11**:430–440.
26. Hall GE: **Test weight changes of shelled corn during drying.** *Trans ASAE* 1972, **15**:320.

27. Sun Z, Yang W, Siebenmorgen T, Stelwagen A, Cnossen A: **Thermomechanical transition of rice kernels.** *Cereal Chem* 2002, **79**:349–353.
28. Chiang BY, Johnson JA: **Measurement of total and gelatinized starch by glucoamylase ando-toluidine reagent.** *Cereal Chem* 1977, **54**:429–435.
29. Della Valle G, Tayeb J, Melcion JP: **Relationship of extrusion variables with pressure and temperature during twin screw extrusion cooking of starch.** *J Food Eng* 1987, **6**:423–444.

23

Effects of different sources of carbohydrates on intake, digestibility, chewing, and performance of Holstein dairy cows

Simin Poorkasegaran[1] and Asadollah Teimouri Yansari[1,2*]

Abstract

To investigate the effects of different sources of carbohydrates on intake, digestibility, chewing, and performance, nine lactating Holstein dairy cows (day in milk= 100±21 d; body weight=645.7 ± 26.5 kg) were allotted to a 3 × 3 Latin square design at three 23-d periods. The three treatments included 34.91% (B), 18.87% (BC), and 18.86% (BB) barley that in treatment B was partially replaced with only corn or corn plus beet pulp in treatments BC and BB, respectively. The concentration of starch and neutral detergent soluble carbohydrate varied (22.2, 20.2, and 14.5; 13.6, 15.9, and 20.1% of DM in treatments B, BC, and BB, respectively). Cows in treatment BB showed a higher DMI and improved digestibility of DM, NDF, and EE compared with treatments B or BC. Ruminal pH was higher in cows fed on BB (6.83) compared with those that received B or BC treatments (6.62 and 6.73, respectively). A lower proportion of propionate accompanied the higher pH in the BB group; however, a greater proportion of acetate and acetate: propionate ratio was observed compared with cows fed either on the B or BC diet. Moreover, cows fed on the BB diet showed the lowest ruminal passage rate and longest ruminal and total retention time. Eating time did not differ among treatments, rumination time was greater among cows fed on the BB diet compared with the others, whereas total chewing activity was greater than those fed on BC, but similar to those fed on B. The treatments showed no effect on milk yield. Partially replacing barley with corn or beet pulp resulted in an increase in milk fat and a lower protein concentration. Changing dietary NFC with that of a different degradability thus altered intake, chewing activity, ruminal environment, retention time or passage rate, and lactation performance. The results of this study showed that beet pulp with a higher NDF and a detergent-soluble carbohydrate or pectin established a more consistent ruminal mat than barley and corn, thus resulting in higher mean retention time and chewing activity, whereas no changes in 3.5% FCM and milk fat were observed.

Keywords: Chewing activity, Dairy cow, Detergent-soluble carbohydrate, Fiber, Neutral detergent fiber, Physically effective fiber, Ruminal characteristics

Background

Three components affect the balance of carbohydrates in the diets of lactating dairy cattle. The first two are the nature of the NDF and the nature of non-fiber carbohydrates (NFC), and the third component is one of the following: NDF content, NFC content, or NDF:NFC ratio [1-3]. Carbohydrates consisting of NDF and NFC fractions make up 65%–75% of rations. However, issues have been associated with the use of an NDF because this component ranges from 27% to 37% in dietary DM, whereas NFC content ranges from 45% to 35% [3,4]. The NFC has been represented as a single value for feeds or diets, but the type of carbohydrates in this fraction largely varies. For example, the NFC in corn grain is mostly starch (65%–70% of DM), beet and citrus pulp provides sugar (12%–40% of DM) and neutral detergent-soluble carbohydrate (NDSC) (25%–44% of DM), and sugar (mono- and oligosaccharides) predominates in molasses [5]. In addition, some of the variation in the results of the effect of the type and amount of NFC on high-producing lactating cows may be related to: 1) the

* Correspondence: astymori@yahoo.com
[1]Department of Animal Science, Animal Science and Aquaculture Faculty, Sari Agricultural and Natural Resource University, Sari, Mazandaran, Iran
[2]Department of Animal Science, Educational Center of Tajan, GhaemShahar, Mazandaran, Iran

effects of rapidly degradable starch on ruminal digestion of fiber; 2) the level to which NFC replaces fiber in the ration; 3) the site of starch digestion; 4) the level of DMI and the physiological state of the animal; and 5) the conservation and processing methods of NFC sources used to alter the rate and the extent of NFC digestion [4-6]. However, the optimal concentration of NFC in dairy cow diets has not been well defined. The lack of clear recommendations may be at least due to the wide compositional and nutritional variation in the sources of NFC, the kinetics of ruminal digestion, and the yields of metabolizable nutrients [4]. Therefore, improving the understanding of the digestion and metabolism of various NFC types that are fed to dairy cattle can improve animal performance and profitability while maintaining health.

The physically effective NDF (peNDF) incorporates information on particle length, NDF content of the diet, and impact of replacing forage fiber with NFC based on the use of small particles of byproducts and measuring the animal's chewing activity [7,8]. However, this concept does not account for different sources of NFC or differences in fermentability among various feeds. The nature of NFC differs among feeds and diets. In evaluating the effectiveness of the NDF approach, the effect of removing or adding NFC has been entirely attributed to NDF [1,2,9]. Based on this simple assumption, the effectiveness of NDF from all sources is thus based on the replacement of NFC. Therefore, understanding the nutritional value of each NFC type may aid in the use of available feeds, including byproducts, to efficiently formulate dairy cow diets that would enhance production and health. However, extensive information on the optimal level of starch and digestible fiber in rations for dairy cows is warranted to facilitate in the design of strategies to prevent ruminal acidosis and the efficient conversion of energy into milk. The objective of this trial was to compare the effects of three different carbohydrate sources, namely barely, corn, and beet pulp (BP), on stimulating intake, digestibility, chewing activity, ruminal characteristics, and early lactation in Holstein dairy cows.

Materials and methods

Animals and diets

The experiment was conducted at the Dairy Farm of the Agricultural Education and Research Centre of Jehad-e-Keshavarzi, Sari, Iran. Nine multiparous mid-lactation Holstein dairy cows (BW = 645.7 ± 26.5 kg; DIM = 100 ± 21 d) were subjected to a 3 × 3 Latin square design. The experiment consisted of three 23-d periods (adaptation, 14 d; sample collection, 7 d; and measurement of chewing activity, 2 d). The three treatments containing 34.91% (B), 18.87% (BC), and 18.86% (BB) barley as DM were used. The barley in treatment B was partially replaced with only corn or corn plus beet pulp in treatments BC and BB, respectively (Table 1). Barley grain, corn grain, and BP were prepared by milling normal BP using a 2-mm screen pore size miller. Diets were similar in chemical (16% CP and 6.65 MJ/kg of NEL) and physical (geometric mean and peNDF) characteristics

Table 1 Ingredients and chemical composition of three total mixed rations containing three different sources of non-fiber carbohydrates

Item	Treatments[1]			SEM	*P* value
	B	BC	BB		
Ingredients, % of DM[2]					
Alfalfa hay	23.41	23.69	23.57		
Wheat straw	15.53	15.84	15.8		
Barley grain	34.91	18.87	18.86		
Corn grain	...	15.89	...		
Beet pulp	...	...	15.89		
Cotton seed meal	11.19	8.88	8.90		
Soybean meal	3.01	2.99	3.01		
Wheat bran	8.01	9.68	9.69		
Sunflower meal	2.69	2.99	3.01		
Calcium carbonate	0.43	0.4	0.5		
Calcium phosphate	0.25	0.20	0.20		
Salt	0.10	0.10	0.10		
Vitamin supplement[3]	0.25	0.25	0.25		
Urea	0.22	0.22	0.22		
Chemical compositions, % of DM					
CP	16.2	15.9	15.9	0.232	0.321
NDF	36.2	36.9	37.7	0.864	0.079
ADF	23.1	22.6	24.6	0.541	0.092
NFC	38.9	38.9	37.9	0.682	0.061
Starch	22.2^a	20.2^a	14.5^b	0.792	0.001
Sugar	3.1	2.8	3.3	0.432	0.213
NDSC[4]	13.6^c	15.9^b	20.1^a	0.232	0.001
EE	2.3	2.4	2.4	0.219	0.212
Ash	6.4	5.9	6.1	0.672	0.431
NEl[5], Mcal/kg of DM	1.58	1.59	1.58	0.232	0.452

$^{a, b, c}$Means within a row with different subscripts differ ($P < 0.05$).

[1]Treatments were diets containing barley + basal diet (B), barley and corn + basal diet (BC), and barley and beet pulp + basal diet (BB).

[2]Alfalfa, wheat straw, barley grain, corn, beet pulp, cotton seed meal, soy bean meal, wheat bran, and sun flower meal (DM basis) were chemically composed of 89.1%, 50.9%, 18.7%, and 15.3%; 91.2, 75.2, 10.4, and 3.4; 93.2, 22.1, 55.4, and 10.5; 96.7, 11.0, 78.3, and 8.5; 93.4, 44.4, 41.0, and 9.0; 90.7, 45.2, 16.3, and 27.3; 90.3, 17.8, 24.7, and 42.5; 90.3, 52.3, 24.4, and 16.5; and 92.0, 55.7, 17.3, and 16.1 for OM, NDF, NFC, and CP, respectively.

[3]Contained 40% NaCl, 25% Dyna mate, 10% K, 7% Mg, 12% S, 1,000 mg Fe/kg, 2% $ZnSO_4.H_2O$, 2% $MnSO_4.4H_2O$, 0.01% $CoSO_4.6H_2O$, 0.009% Na_2SeO_3, 0.012% ethylenediamine dihydroiodide, 0.8% $CuSO_4.5H_2O$, 680,000 IU/kg of Vitamin A, 160,000 IU/kg of Vitamin D, and 2,000 IU/kg of Vitamin E.

[4]NDSC = Neutral detergent soluble carbohydrate.

[5]NEl was calculated based on NRC [4].

(Table 2). Cows were housed in tie-stalls and fed *ad libitum* twice daily at 0900 and 2100 h with TMR, allowing for at least 10% residual (as-fed basis). Water and mineralized salt stone were available for cows during the experiment.

Particle length and effectiveness fiber

Feed particle size and distribution were determined by dry sieving in four replicates using two sets of a Penn State particle separator [10,11]. The NDF content of all materials retained on each sieve of the Penn State particle separator were measured [12]. Using three systems, the physical effective factor (**pef**) of the TMRs was determined. According to Mertens [13], Lammers et al. [10] and Kononoff [11], the pef was determined based on the proportion of DM retained on the 1.18-m sieve (**pef_m**), on two 19- and 8-mm sieves (**$pef_{>8}$**), and three 19-, 8-, and 1.18-mm sieves (**$pef_{>1.18}$**), respectively. The $peNDF_m$, $peNDF_{>8}$, and $peNDF_{>1.18}$ were calculated by multiplying the NDF content of each portion on each sieve with the pef_m, $pef_{>8}$, and $pef_{>1.18}$, respectively (Table 2).

Table 2 Physical characteristics of three total mixed rations containing three different sources of non-fiber carbohydrate

Item	Treatments[1]			SEM	P value
	B	BC	BB		
Particle size distribution (material retain on each sieve, % of DM)					
19 mm	9.7	10.1	9.8	0.732	0.213
8 mm	22.2	22.1	20.4	0.912	0.133
1.18 mm	34.4	34.2	36.9	1.648	0.422
Pan	33.7	33.6	32.9	0.641	0.455
Geometric mean, mm	3.85	3.92	3.80	0.692	0.087
Standard deviation of geometric mean	1.20	1.24	1.34	0.106	0.093
Physical effective factor					
pef_m[2]	66.2	66.3	66.9	0.509	0.101
$pef_{>8}$[3]	34.4	34.6	31.3	0.322	0.288
$pef_{>1.18}$[4]	66.3	66.4	67.1	0.694	0.083
Physical effective NDF, % of DM					
$peNDF_m$	24	24	26	1.13	0.063
$peNDF_{>8}$	12	13	12	1.03	0.054
$peNDF_{>1.18}$	24	24	26	1.14	0.063

[1]Treatments were diets containing barley + basal diet (B), barley and corn + basal diet (BC), and barley and beet pulp + basal diet (BB).
[2]pef_m = physically effective factor, based on DM retained on sieve 1.18 mm [13].
[3]$pef_{>8}$ = physically effective factor, determined as the proportion of DM retained on sieves of the original version of Penn State particle separator [10].
[4]$pef_{>1.18}$ = physically effective factor, determined as the proportion of DM retained on sieves of the new version of Penn State particle separator [11].
[5]The peNDF was calculated by multiplying NDF content of each portion on each sieve on each pef [13].

Body weight, intake, and apparent total-tract nutrient digestibility

The BW of cows was determined each wk. The DMI of the cows was measured daily (Table 3). Daily samples of TMRs, residuals and. total feces were collected daily for 5 d of each period at 0700 h on d 14 to 0700 h on d 20, weighed, and d 5 samples were composited and subsampled. The samples were dried at 55°C for 48 h and ground using a Wiley mill (1-mm screen), analyzed for DM, Kjeldahl N, ether extract, OM, ash at 605°C for 3 h [14], NDF, and ADF [12; with α-amylase being added for concentrates during NDF extraction; sodium sulfite was not added]. Neutral detergent fiber was expressed without residual ash. Non-fibrous carbohydrate was calculated using the following equation: 100 - % CP +%

Table 3 Body weight, intake, and digestibility of nutrients of three total mixed rations containing three different sources of non-fiber carbohydrates

Item	Treatments[1]			SEM	P value
	B	BC	BB		
BW, kg	600.63	591.84	583.25	11.212	0.140
Intake, kg/d					
DM	21.14[b]	21.02[b]	22.32[a]	0.530	0.004
NDF	7.65[b]	7.76[b]	8.57[a]	0.021	0.001
$peNDF_m$[2]	5.10[b]	5.14[b]	5.74[a]	0.034	0.025
$peNDF_{>8}$[3]	2.63	2.68	2.68	0.231	0.075
$peNDF_{>1.18}$[4]	5.07[b]	5.15[b]	5.75[a]	0.023	0.024
CP	3.42	3.34	3.55	0.456	0.133
NFC	8.22	8.18	8.46	0.345	0.309
Starch	4.69[a]	4.25[a]	3.24[b]	0.431	0.002
Sugar	0.66	0.59	0.73	0.084	0.064
NDSC[5]	2.88[c]	3.34[b]	4.49[a]	0.104	0.001
EE	0.49	0.50	0.54	0.121	0.087
OM	19.79	19.78	20.96	0.961	0.342
Digestibility, %					
DM	64.58[b]	65.01[b]	66.91[a]	0.824	0.004
NDF	61.58[b]	62.38[b]	68.82[a]	1.028	0.004
ADF	51.67	52.21	52.33	0.769	0.100
CP	77.33	77.42	76.89	0.571	0.532
NFC	83.36[a]	80.37[b]	80.90[b]	1.424	0.032
EE	62.44[b]	66.52[ab]	67.33[a]	0.277	0.024
OM	64.68	64.61	63.53	1.143	0.200

[a, b, c]Means within a row with different subscripts differ ($P < 0.05$).
[1]Treatments were diets containing barley + basal diet (B), barley and corn + basal diet (BC), and barley and beet pulp + basal diet (BB).
[2]$peNDF_m$ = physically effective NDF based on DM retained on sieve 1.18 mm [13].
[3]$peNDF_{>8}$ = physically effective NDF, determined as the proportion of DM retained on sieves of the original version of Penn State particle separator [10].
[4]$peNDF_{>1.18}$ = physically effective NDF determined as the proportion of DM retained on sieves of the new version of Penn State particle separator [11].
[5]NDSC = Neutral detergent soluble carbohydrate.

NDF +% ash +% EE [4]. The TMR composites were analyzed for soluble sugars using sucrose as the standard and for starch [5]. The proportion of NDSC was computed as NFC (%) using the following equation: [(Sugar (%) + Starch (%)] [5]. Using the chemical components of TMRs and feces, the intake and digestibility of nutrients were calculated. Intake (kg/d and% of DMI) of $peNDF_m$, $peNDF'_{>8}$ and $peNDF_{>1.18}$ were calculated for each treatment (Table 3).

Rumen fermentation and kinetics

On d 21 of each period, ruminal fluid samples (50 mL) from the ventral sac were collected 3 h after the morning feeding using the rumenocentesis technique. Ruminal pH was measured immediately after ruminal fluid collection using a pH meter (model 632, Metrohm, Herisau, Switzerland), and samples were frozen at −20°C. The concentration of ammonia nitrogen (NH_3-N) was measured using a Kjeltec Auto Analyzer (Model 1030, Tecator Co. Sweden). Ruminal fluids were acidified by mixing with 2.5 mL of 6 mol/L HCl and then frozen for further analysis of VFA and NH_3-N. Ruminal fluid was centrifuged at 25,000 × *g* for 20 min before measurement of VFA by gas chromatography (4% carbomax 80/120 BDA column, Supelco Inc., Bellefonte, PA; Autosystem XL gas chromatograph, Perkin Elmer Inc., Norwalk, CT) [15].

Digestive kinetics were measured using Cr-mordanted NDF alfalfa that was prepared by repeatedly soaking a mixture of alfalfa hay in dilute neutral detergent and rinsing until the NDF content of the material exceeded 80%. The fiber was then dried at 55°C. Chromium (Cr)-mordanted alfalfa NDF was prepared as previously described by Uden et al. [16], as the single-dose marker for solid passage rate measurement. Cr-mordanted fiber was prepared by mordanting alfalfa NDF ground through a 5-mm screen using a Wiley mill. Each cow was orally fed with a mix of 2 kg of concentrate and 250 g of marker before morning feeding on the 15th d of each period. Fecal grab samples were collected at 0, 6, 10, 12, 14, 18, 22, 26, 30, 36, 42, 48, 54, 60, 72, 84, 96, 120, and 144 h after dosing from the rectum to determine the passage rate, ruminal, and total mean retention time, and time delay (transit time) of the marker. Samples were dry-ashed, and fecal Cr concentrations were determined by direct current plasma emission spectroscopy [14]. Fecal Cr excretion curves were fitted to the double compartment model, as represented by the following two exponential constants and a time delay [17]:

$$Y = Ae_1^{k(t\text{-}TT)} - Ae_2^{\text{-}k}(t\text{-}TT), k_1 = k_2 \text{ for } t^3T; Y = 0 \text{ for } t = TT;$$

in which Y = marker concentration (ppm); A = scale parameter; k_1 = ruminal passage rate (%/h); k_2 = lower digestive tract passage rate (%/h); t = sampling time post dosing (h); and TT = transit time or time delay of marker. The total mean retention time was calculated as the sum of ruminal mean retention time ($1/k_1$), and for the lower digestive tract, mean retention time ($1/k_2$) plus the transit time. Parameters were estimated by NLIN regression using the PROC NLIN (iterative Marquardt method) of SAS® [18]. The estimated parameters were analyzed according to a previously described experimental design.

Eating, ruminating, and chewing times

Eating and ruminating activities were monitored visually for all cows in each treatment over a 24-h period for 2 d and at d 22 to 23 of each period. Eating and ruminating activities were noted at 5-min intervals and each activity was assumed to persist for the entire interval. A period of rumination was defined as at least 5 min of ruminating activity, followed by at least 5 min without ruminating activity. Total time spent on chewing was calculated as the total time spent eating and ruminating [8].

Milk production

Milk yield was recorded daily during the course of the experiment. On d 18 to 23 of each period, milk samples were collected each morning and evening. Approximately 100 mL of each milk sample were composited and analyzed for total protein, fat, and lactose [14].

Statistical analysis

The data on particle size were analyzed as a completely randomized design for the effects of the diets and the two methods of particle size measurement using the REML variance component and PROC MIXED of SAS® [18]. Mean separation was determined using the PDIFF procedure and significance was declared at $P < 0.05$. The geometric mean and its standard deviation were calculated using ASAE S424.1 [7].

Using PROC MIXED of SAS® [18], the experimental data were analyzed in a 3 × 3 replicated Latin square design using following model:

$$Y_{ijkl} = \mu + T_i + S_j + Cow_{k(j)} + period + e_{ijkl}$$

in which Y_{ijkl} is the dependent variable; μ is overall mean; T_i is fixed effect of the treatments (i = 1, 2, and 3 for diets containing B, BC, and BB); S_j, random effect of square (j = 1, 2, and 3); $Cow_{k(j)}$, effect of cow within square; $period_l$ within square, and e_{ijkl} is residual error. Means were separated using Duncan's multiple range tests, with an alpha level of 0.05.

Results

Nutrient composition, particle size distribution, and physical effectiveness of different treatments

Except for starch (P = 0.001) and NDSC (P = 0.001) content, the three TMRs showed similar chemical compositions and physical characteristics (Tables 1 and 2). Concentrations of starch were similar for treatments B and BC and both were greater than that observed with treatment BB (22.2, 20.2, and 15.9%, respectively). In addition, the concentration of NDSC increased with partial replacement of barley by corn and to a greater extent by partial replacement with BP (13.6, 15.9, and 20.1% for treatments B, BC, and BB, respectively). Distribution of particles, expressed as the percentage of total mass, retained on the 19, 8, 1.18-mm sieves for the three TMRs were also similar and averaged 9.7, 10.1, and 9.8; 22.2, 22.1, and 20.4; 34.4, 34.2, and 36.9%, respectively. The values of pef_m, $pef_{>8}$, and $pef_{>1.18}$ of the treatments were different among the systems. However, regardless of the methods and evaluation system, as well as the geometric mean and its standard deviation, their values were similar within each system. The peNDF values (% of DM) were similar in all treatments (Table 2).

Intakes and apparent total-tract nutrient digestibility

Dry matter intake (P = 0.004), as well as NDF (P = 0.001), $peNDF_m$ (P = 0.025), and $peNDF_{>1.18}$ (P = 0.024) intake were significantly different, although the intake of CP, NFC, EE, and ash were similar among treatments. Replacing barley with BP increased DMI, but DMI in treatments B and BC were similar (Table 3). The nature of NFC affected DMI, as well as the production and composition of milk. Positive intake responses to feeding on BP were observed. Cows in the treatment BB group consumed more NDF, $peNDF_m$, and $peNDF_{>1.18}$ compared with those in treatments B and BC. The intake of starch (P = 0.002), sugars, and NDSC (P = 0.001) varied among the different NFC treatments. Cows fed on treatments B and BC showed a greater intake of starch compared with treatment BB. Cows assigned to the B diet showed the smallest sugar and NDSC intakes. NDF intake did not vary among treatments. However, there were some unexpected differences in the total NFC intake, ash, and CP. Digestibility of DM (P = 0.004), NDF (P = 0.004), NFC (P = 0.032), and EE (P = 0.03) varied among treatments, although digestibility of ADF, CP, and OM remained the same (Table 3). The highest DM, NDF, and EE and significantly lower NFC digestibility were observed in treatment BB.

Rumen fermentation and kinetics

Lower pH levels were observed in cows that received treatments B and BC compared with those in treatment BB (P = 0.003). In addition, the NH_3-N concentration varied among treatments (P = 0.035). The NH_3-N concentrations were significantly lower in treatments B and BC than in treatment BB. The total VFA concentration in treatment BC was similar to those observed in treatments B and BB. Acetate concentration levels were also significantly different ($P < 0.001$) among treatments. The lowest values of acetate and acetate: propionate ratio was observed in treatment B, whereas the highest were observed in treatment BB. The butyrate concentration remained the same among treatments. The lowest values for propionate, isobutytate, valerate, and isovalerate were observed in treatment BB, whereas the highest were in treatment B.

The ruminal particulate passage rate was highest and lowest in treatment BC and BB, respectively. The ruminal and total mean retention time were higher in treatments BB than in treatments B and BC. Contrary to the passage rate, treatments BB and BC showed the highest and lowest ruminal mean retention times, respectively. However, total mean retention time in treatment BB was higher than in treatments B and BC (Table 4). The time delay of the marker was significantly lower in treatment B than in BC (Table 4).

Eating, ruminating, and chewing times

No differences in eating time were observed among treatments, although rumination time ($P \leq 0.001$) and total chewing activity (P <0.001) varied (Table 5). Rumination time and total chewing activity were lower and higher in BC compared with that observed in B and BB. Treatment BB showed the highest rumination time and total chewing activity. Eating time (min/kg) per NDF intake, $peNDF_m$, and $peNDF_{>8}$ were higher in treatment B compared with treatments BC and BB. However, no significant differences with regard to eating time per NDF intake, $peNDF_m$, and $peNDF_{>8}$ were observed among treatments B and BB. Rumination time (min/kg) per DMI, NDF intake, $peNDF_m$, $peNDF_{>8}$, $peNDF_{>1.18}$, and NFC were higher in treatments BB and BB compared with treatment BC. However, no significant differences on rumination time per DMI, NDF intake, $peNDF_m$, $peNDF_{>8}$, $peNDF_{>1.18}$, and NFC were observed among treatments BB and BB. In addition, total chewing activity (min/kg) per DMI, NDF intake, $peNDF_m$, $peNDF_{>8}$, $peNDF_{>1.18}$, and NFC were higher in treatment BB compared with treatments BC and BB. However, no differences on total chewing activity per $peNDF_m$, $peNDF_{>8}$, $peNDF_{>1.18}$ were observed between treatments BC and BB, and the total chewing activity per DMI, NDF and NFC intake was lower in treatment BC compared with treatment BB. The eating time, rumination time, and total chewing activity per intake starch was higher in treatment BB than in treatments B and BC, whereas that of treatments B and BC were similar. Also, the eating time, rumination time, and total chewing activity per

Table 4 Ruminal metabolites and kinetic of digestion of three total mixed rations containing three different sources of non-fiber carbohydrate

Item	Treatments[1]			SEM	*P* value
	B	BC	BB		
pH	6.62^{c}	6.73^{b}	6.83^{a}	0.002	0.003
NH_3-N[2], mg/dL	6.76^{b}	6.68^{b}	7.71^{a}	0.111	0.035
VFA concentration					
Total, mmol/L	125.44^{a}	121.21^{b}	125.01^{a}	0.650	<0.001
Acetate, mol/100 mol	58.21^{c}	60.21^{b}	63.05^{a}	0.435	<0.001
Propionate, mol/100 mol	26.75^{a}	25.45^{b}	23.04^{c}	0.121	<0.001
Butyrate, mol/100 mol	6.48	6.33	6.28	0.104	0.075
Isobutytate, mol/100 mol	2.39^{a}	2.29^{a}	2.11^{b}	0.093	0.023
Valerate, mol/100 mol	3.47^{a}	3.35^{ab}	3.25^{b}	0.096	<0.001
Isovalerate, mol/100 mol	2.49^{a}	2.37^{ab}	2.27^{b}	0.073	0.014
Acetate: propionate	2.16^{c}	2.37^{b}	2.75^{a}	0.053	<0.001
Ruminal passage rate, %/h	4.36^{b}	5.46^{a}	4.06^{c}	0.097	0.012
Ruminal mean retention time, h	23.48^{b}	18.32^{c}	24.63^{a}	0.871	0.002
Total mean retention time, h	54.36^{b}	52.64^{b}	61.35^{a}	1.214	0.003
Time delay, h	9.25^{b}	10.24^{a}	9.86^{ab}	0.444	0.021

[a, b, c]Means within a row with different subscripts differ ($P < 0.05$).
[1]Treatments were diets containing barley + basal diet (B), barley and corn + basal diet (BC), and barley and beet pulp + basal diet (BB).
[2]NH_3-N = ammonia nitrogen concentration.

intake NDSC was lower in treatment BB than in treatments B and BC, and was significantly lower in treatment BC than in treatment B (Table 5).

Milk production

Although milk yield was not altered, replacing barley with corn and BP in the diet resulted in an increase in milk fat [yield ($P = 0.013$)] and percentage ($P < 0.001$) and a decrease in milk protein (yield ($P = 0.021$) and percentage [($P = 0.001$); Table 6)], respectively.

Discussion

Nutrient composition, particle size distribution, and physical effectiveness

The $pef_{>8}$ values were lowest for TMRs, although the pef_m and $pef_{>1.18}$ values were higher and very close. In contrast to the $peNDF_m$ and $peNDF_{>1.18}$, the $peNDF_{>8}$ values were similar among treatments. In this experiment, $peNDF_m$ and $peNDF_{>1.18}$ were higher than 21% of the DM that Mertens [13] suggested to adequately stimulate chewing activity that was required to maintain an average ruminal pH of greater than 6.0. The peNDF content of experimental rations was adequate and the differences among treatments were attributable to the effects of NFC sources or its nature. The concentration of starch was similar in the B and BC treatments and higher than that observed in the BB treatment, but the NDSC significantly varied among treatments. These results thus confirm that intake of starch and NDSC varied with different NFC treatments. We hypothesized that the use different digestion products, which are the NFC complement of the diet, would alter feed intake, ruminal digestion kinetic, chewing activity, and the production and composition of milk.

Intake and apparent total-tract nutrient digestibility

The intake of $peNDF_m$ and $peNDF_{>1.18}$ did not follow the same trend as the pef content of the diets, although this was similar to that of NDF; however, this increased in treatment BB. Treatment BB, which contained more NSDC, resulted in a significant increase in DMI. These results have been confirmed by several previous studies. Substitution of sugars and NDSC with starch has increased [19-22] or did not change [23] the intake in dairy cattle. Increases in dietary concentrations of NDSC at the expense of starch also increased [20,24], decreased [22,25], or did not affect DMI [24]. Changes in the passage rate or fiber digestibility using various NFC sources may partial explain these changes [6]. A higher intake in treatment BB could have been expected based on the high NDSC and elevated starch content of barley and corn compared with treatments B and BC.

In this experiment, intake might have been regulated by the metabolism of propionate in the liver, and ruminal propionate production generally increases with a higher rate of starch fermentation. The amount of starch intake was 4.69, 4.25, and 3.24 kg/d, and ruminal molar percent of propionate in total VFA was 26.75, 25.45, and 23.04 mol/100 mol

Table 5 Chewing activity of cows fed on three total mixed rations of three total mixed rations containing three different sources of non-fiber carbohydrates

Items	Treatments[1]			SEM	*P* value
	B	BC	BB		
Eating, min/d	257.8[a]	232.8[b]	249.4[ab]	10.272	0.045
Rumination, min/d	338.9[b]	286.1[c]	356.1[a]	7.430	<0.001
Total chewing activity, min/d	596.7[a]	518.9[b]	605.5[a]	10.482	<0.001
Chewing behaviour per different nutrients, min/kg					
Eating					
DMI	12.19	11.08	11.174	0.552	0.200
NDF	33.70[a]	30.00[b]	29.10[b]	0.972	0.004
$peNDF_m$[2]	50.55[a]	45.29[b]	43.45[b]	2.430	0.033
$peNDF_{>8}$[3]	97.93[a]	86.75[b]	92.97[ab]	4.431	0.002
$peNDF_{>1.18}$[4]	50.85	45.20	43.37	4.294	0.063
NFC	31.36	28.46	29.42	1.154	0.088
Starch	54.97[b]	54.78[b]	76.98[a]	4.452	<0.001
NDSC[5]	89.52[a]	69.70[b]	55.55[c]	3.762	<0.001
Rumination					
DMI	16.03[a]	13.61[b]	15.95[a]	0.350	<0.001
NDF	44.30[a]	36.87[b]	41.55[a]	1.561	<0.001
$peNDF_m$	66.45[a]	55.66[b]	62.04[a]	1.902	0.003
$peNDF_{>8}$	128.74[a]	106.61[b]	132.74[a]	5.242	0.009
$peNDF_{>1.18}$	66.84[a]	55.55[b]	61.93[a]	1.615	0.002
NFC	41.23[a]	34.98[b]	42.09[a]	0.812	<0.001
Starch	72.26[b]	67.32[b]	109.91[a]	6.692	<0.001
NDSC	117.67[a]	85.66[b]	79.31[c]	3.833	<0.001
Total chewing activity					
DMI	28.23[a]	24.69[c]	26.13[b]	0.671	<0.001
NDF	78.00[a]	66.87[c]	68.06[b]	2.031	<0.001
$peNDF_m$	117.00[a]	100.95[b]	101.61[b]	3.292	<0.001
$peNDF_{>8}$	216.41[a]	193.35[b]	225.71[a]	15.320	<0.001
$peNDF_{>1.18}$	117.69[a]	100.76[b]	101.43[b]	3.462	0.006
NFC	72.59[a]	63.43[c]	68.94[b]	1.752	<0.001
Starch	127.23[b]	122.09[b]	186.88[a]	7.860	<0.001
NDSC	207.19[a]	155.36[b]	134.86[c]	6.744	<0.001

[a, b, c]Means within a row with different subscripts differ ($P < 0.05$).
[1]Treatments were diets containing barley + basal diet (B), barley and corn + basal diet (BC), and barley and beet pulp + basal diet (BB).
[2]$peNDF_m$ = physically effective NDF, based on DM retained on sieve 1.18 mm [13].
[3]$peNDF_{>8}$ = physically effective NDF, determined as the proportion of DM retained on sieves of the original version of Penn State particle separator [10].
[4]$peNDF_{>1.18}$ = physically effective NDF, determined as the proportion of DM retained on sieves of the new version of Penn State particle separator [11].
[5]NDSC = Neutral detergent soluble carbohydrate.

in treatments B, BC, and BB, respectively, showing that starch intake decreased with a higher rate of starch intake (Table 4). Therefore, it seems that propionate metabolism in the liver caused the reduction in DMI in treatments B and BC compared with BB because hepatic propionate oxidation probably decreased with the supplementation of BP [25]. In addition, ruminal pH might be implicated in the control of feed intake, particularly in satiety, because VFAs were more rapidly absorbed in the undissociated form at lower pH. In addition, substituting a high-fiber byproduct such as BP for grain may have resulted in the reduction of energy content of the ration or the limitation of intake by the physical filling effects of the diet.

The NFC sources such as BP have a similar range of NDF content as forages, although its particle length was generally smaller. In several experiments, these sources

Table 6 Milk production and composition of cows fed on three total mixed rations containing three different sources of non-fiber carbohydrates

Item	Treatments[1]			SEM	P value
	B	BC	BB		
Yield of milk and milk components, kg/d					
Milk	25.11	25.50	25.83	0.653	0.062
3.5% FCM	21.69	22.43	22.69	0.645	0.070
Fat	0.86[b]	0.90[a]	0.91[a]	0.033	0.013
Protein	0.81[a]	0.80[b]	0.80[b]	0.025	0.021
Lactose	1.17	1.20	1.16	0.020	0.062
Total solid	3.47	3.42	3.57	0.060	0.076
Milk composition, %					
Fat	3.43[b]	3.55[b]	3.63[a]	0.087	<0.001
Protein	3.23[a]	3.15[b]	3.10[b]	0.030	0.001
Lactose	4.67	4.69	4.66	0.045	0.058
Total solid	11.93	11.89	12.01	0.098	0.060

[a, b]Means within a row with different subscripts differ ($P < 0.05$).
[1]Treatments were diets containing barley + basal diet (B), barley and corn + basal diet (BC), and barley and beet pulp + basal diet (BB).

were substituted with forage or grain in diets and then the DMI was measured [3,5,6,22]. When DMI is regulated by ruminal distension, its substitution with forage in diets results in an increase in DMI, whereas its substitution with grain decreases DMI [2,4,13]. However, non-forage fiber sources showed inconsistent effects on DMI when substituted with either forage or grain in the diets of lactating cows. In the current experiment, the physical characteristics of the diets were similar; therefore, it seems that these do not differ in their relative filling effects. In treatment BB, the DMI was not regulated by ruminal distension [3,25]. The greater concentration of propionate in the ruminal liquid in treatment B compared with treatments BC and BB suggests that stimulated propionate receptors in the ruminal region may have controlled satiety and DMI, because propionate is more hypophagic than acetate, resulting in a lower DMI in treatment B versus treatments BC and BB [19,26].

Partially replacing barley with BP in the dairy cow diet resulted in an increase in total digestibility of DM, NDF, and EE, and a decrease in the digestibility of NFC. However, partially replacing barley with corn in the dairy cow diets showed minimal effects on total digestibility. Unfortunately, total digestibility of starch was not measured in this study. In the present study, decreased fiber digestibility in treatment B using barley was consistent with the reduction in rumination time, ruminal pH, and acetate to propionate ratio. The consumption of diets rich in NDSC imparted positive effects on ruminal digestion. In addition, the results of this experiment might be influenced by differences in the digestible NDF of the treatment diet and not simply the source of NFC, as implied in the objectives of this study.

Differences in digestible NDF rations or starch, NDSC, and sugar content were associated with several parameters, especially acetate and propionate concentrations and their ratio in the rumen, as well as milk fat and protein content. These were significantly different among treatments ($P < 0.001$) and treatment BB showed the highest intake of potentially digestible fiber or NDSC. Previous studies have confirmed that animal intake response to supplemented starch- or pectin-rich feeds remained the same, although changes in digestibilities have been observed. Valk et al. [22] reported that cows fed on corn consumed more DM and produced more milk than cows fed on BP. However, DMI and digestibility of NDF were higher in cows fed on BP compared with those fed on corn meal.

The extent and rate of degradation is extremely variable among grains, byproducts, and different sources of NFC [6,27]. Variation in ruminal degradation of DM and OM across feeds has been reported to range from 29% to 90%, and 29% to 67%, respectively [27]. Total tract digestibility of corn ranges from 91.2% to 98.9%, depending on the processing method and grain type, with ground corn averaging 93.5% [28]. Ruminal degradation of starch in ground corn and ground barley ranged from 51% to 93%, and 42% to 91%, respectively [28]. In addition, ruminal degradation of DM and NDF in BP was 78% [29] and 94.1% [30], respectively. The potentially digestible fraction leaves the rumen either by degradation or by passage to the lower tract. Furthermore, feed characteristics and ruminal fractional passage rates would most likely vary; these tend to be higher with fine, non-forage fiber sources due to differences in ruminal retention time, which may be influenced by ruminal consistency [6]. The inclusion of BP in treatment BB resulted in a decrease in particulate passage rates, an increase in ruminal and total mean retention time, and a positive effect on ruminal and total digestion.

Rumen fermentation and kinetics

Variables of ruminal pH such as the outflow rate of digesta from the reticulorumen or ruminal indicate that retention time influences total fiber degradation. Rapidly fermenting carbohydrates such as sugars, NDSC, and some starches have the potential to rapidly decrease ruminal pH by virtue of the sheer mass of organic acids produced in this digestive region within a relatively short period of time. The ruminal pH corresponded to a higher rate of degradation and fermentation of barley grain that produced the highest concentration of total VFA and propionate, lower acetate, and acetate: propionate ratio in treatment B compared with the other treatments (Table 4).

Beet pulp, which is rich in pectin or NDSC, has been shown to yield relatively high amounts of acetic acid, thus resulting in an increase in the acetate to propionate ratio and less butyrate and lactate compared with other carbohydrate sources [31]. In addition, using twelve Friesian cows and sixteen heifers, Lees et al. [21] reported that in wk 9 and 16, the ruminal VFA proportions of cows fed on BP were higher for acetate and butyrate and lower for propionate compared with cows fed on corn. The molar proportion of acetate in the current study and in previous investigations was probably altered by the tendency of BP to increase fiber intake and digestion. In contrast, Van Vuuren et al. [32] did not detect any differences in ruminal fluid pH or VFA concentrations in either experiment involving cows fed on corn products or BP and soybean hulls.

Changing the dietary source of NFC or NDSC and starch content of diet rations of lactating dairy cows has been shown to alter the ruminal environment and fermentation products. The effects, although not always consistent, have been observed on the proportions of VFA, the concentration of rumen NH_3-N, and the extent of fiber digestion with feeding starch, sugars, or NDSC-rich feeds [6]. The application of treatment BC resulted in a higher acidic pH, lower total VFA and proportion of propionate, and increased acetate concentration and acetate:propionate ratio in the ruminal fluid. Replacing BP with barley resulted in a higher acidic pH, increased acetate concentration in ruminal fluid, without any change in total VFA, and decreased proportion of propionate, valerate, isobutyrate, and isovalerate. Therefore, compared with barley, BP created more favorable conditions for microbial utilization of other carbohydrates in the rumen, at least in part due to the neutral pH. Decreasing daily mean ruminal pH, which was associated with an increase in ruminal outflow rate of digesta, resulted in a decrease in ruminal NDF digestibility and total tract digestibility [2,8,9,33]. Zebeli et al. [9] showed a decrease in total tract NDF digestibility by increasing the outflow rate of particulate digesta, which was associated with a decrease in ruminal mat consistency. It might be possible that a low digesta passage rate and high consistency in the ruminal mat due to inclusion of BP in the treatment BB was associated with a higher water holding capacity, greater entrapment of small particles, decreased outflow rate of solid digesta, and increased ruminal NDF digestibility in dairy cows.

The ruminal particulate passage rate was highest and lowest in treatments BC and BB, respectively. Because corn has a higher specific gravity than barley and BB, inclusion of corn in the diet resulted in an increase in ruminal passage rate and a decrease in rumen mean retention time (Table 4), and consequently, it decreased rumination time and total chewing activity. In contrast, the higher total chewing activity in treatment BB might have resulted from the lower pH that was attributable to a high degradability of barley, which may have decreased the rate and extent of NDF digestion. In treatment BB, the BP did not result in a reduction in ruminal pH despite its high rate of NDF digestion because of the production of a higher concentration of acetate and lower propionate. However, it has a low functional specific gravity and a large particle size [34], which could decrease its passage rate and increase its ruminal mean retention time and consequently, rumination and total chewing activity.

Two characteristics that affect the passage of digesta are density and particle size. In addition, there is a clear negative relationship between particle density and ruminal mean retention time. Particles with a specific gravity of lower than the critical size (1.18 mm) quickly pass out of the reticulorumen [8]. The specific gravity for BP was 1.152, whereas that of barley and corn grain were higher than 1.237 [34]. Therefore, inclusion of BP in treatment BB decreased particulate passage rates and increased ruminal and total mean retention time. Particles with a density range of 1.2 to 1.5 showed the highest rate of passage in cattle and sheep [3,8,9]. Grain and its byproducts (except for BP and soy hulls] showed a higher rate of passage from the rumen than forages because of the higher functional specific gravity and a particle size much smaller than that associated with stimulation of rumination. Therefore, most grains would have a faster rate of passage from the rumen than forages and other byproducts.

In treatment B, negative associative effects resulting from a lower pH due to a high degradability of barley would probably decrease the rate of digestion. Beet pulp did not reduce ruminal pH despite its high rate of NDF digestion [3]; however, it has a low functional specific gravity and a large particle size [34], which could decrease its passage rate and increase its ruminal mean retention time. Therefore, BP should have a very high ruminal digestibility [29], potentially supporting levels of ruminal microbial protein synthesis and milk production, similar to that observed in corn [31]. However, generalization of these results under different feeding conditions would be an oversimplification because different dietary treatments can interact with each other, thus affecting passage rate.

Eating, ruminating, and chewing times

Although the effect of forage particle size on chewing behavior has been substantially investigated, information on the effects of NFC, starch, or NDSC content of ration, especially when barley grain partially replaced with corn and BP, is limited. Despite the absence of differences in eating time, the rumination time and total chewing activity significantly varied among treatments. Prolonged total chewing time due to the increase in dietary peNDF resulted from an increase in ruminating time rather than an increase in eating time.

Treatment BB showed the highest rumination time and total chewing activity. Total daily chewing time varied from 8.6, 9.9, and 10.1 h, with ruminating time at 5.6, 4.8, and 5.9 h in treatments B, BC, and BB, respectively. These values were within limits previously reported for dairy cows [9,13], although the mean chewing time per kg DM ingested was less than 30 min, contrary to the recommendations of Mertens [13], who proposed values equal to or exceeding 30 min/kg DM intake as suitable for limiting the risk of digestive disorders. Eating time per day and per kilogram of DMI increased with the supplementation of BP. As diet NDF intake increased with added BP in the current study, a decrease in time eating per kilogram of NDF was observed. However, no effect of BP substitution for corn on time eating per day or per kilogram of DMI was reported [2,13]. Tafaj et al. [2] and Zebeli et al. [9] have previously shown that the fiber content (NDF or forage NDF) and nonfiber carbohydrates in the diet affected daily chewing and rumination activities. Forage content remained the same across treatments; therefore, the nature of NFC or starch and NDSC content impacted the rate of feed intake, ruminal mean retention time, rumination time, and total chewing activity.

Milk production

Although milk yield was not altered, replacing corn and BP with barley in diets resulted in an increase in milk fat (yield and percentage), and a decrease in protein concentration (yield and percentage; Table 6). Dietary BP can increase [27], decrease [22], or support milk yield, which have also been observed in cows fed on corn or barley [22,35]. In addition, intake may be improved [21], decreased [22,35], or remain the same [27,35] using BP. Various dietary carbohydrates can result in changes in ruminal fermentation and generate VFAs patterns that influence the hormonal status of the animal and potentially affect milk production and composition.

The increase in fat secretion in treatments BC and BB has been shown to elevate both acetate and acetate: propionate ratio in the rumen. However, treatments BC and BB did not alter ruminal concentrations of either the total VFA or butyrate. Mansfield et al. [31] previously showed a higher milk fat concentration (3.82 vs. 3.64), whereas milk protein (3.01% vs. 2.90%) and yield were lower in cows fed on the BP diet compared with the corn diet. The fat content of milk was enhanced in treatment BB, which can be directly attributed to the increase in fiber intake and digestion in cows and the molar proportion of acetate in the rumen. Consistent with the current experiment, milk fat increased in cows fed on the BP diet as a replacement for grain [22,31] due to the higher fiber intake and greater acetate concentrations in the rumen. It seems that the higher milk fat observed in treatment BB was due to the increase in intake of digestible NDF and NDSC, ruminal mean retention time, rumination time, total chewing activity, ruminal acetate concentration, and ruminal pH.

A decrease in milk CP content and yield was observed when corn and BP were substituted with barley. This finding was confirmed by Mansfield et al. [31], but was in contrast to the findings of Valk et al. [22]. In contrast to the current experiment, a lower intake of CP, lower NH_3-N concentrations in the rumen, and lower CP degradability for cows fed on BP was observed, compared with that in cows fed on corn [20] or barley [33]. The consumption of carbohydrates such as NDSC that are more extensively fermented in the rumen may improve the utilization of the NPN through the stimulation of microbial protein synthesis. In contrast, when NH_3-N concentrations in the rumen cannot meet microbial requirements, the fermentation of fiber is reduced and the animal responds with a lower intake and digestibility. Total VFA concentration in treatments BC and BB were observed to decrease. The lower total rumen VFA is an indicator of lower ruminal fermentation of fiber that limited milk protein yield. Huhtanen [27], Rooke et al. [36], and Chcster-Jones et al. [20] previously showed that BP supports VFA concentrations in the rumen similar to those for corn and barley grain, whereas VFA concentrations decreased in some situations [6]. In addition, NFC sources that were used in the current experiment varied in their DM rate and extent degradability, support of microbial yield, and supply of protein to the small intestine because microbial CP yield is influenced by carbohydrate and nitrogen sources, rate of carbohydrate fermentation, bacterial growth rate, ruminal dilution rate, and pH [22].

Conclusion

The results of this study indicate that changing dietary NFC or starch and NDSC that possess a different level of degradability may alter intake, ruminal mat consistency, chewing activity, ruminal environment, and retention time or passage rate, and lactation performance. However, studies on the optimal concentrations of NFC from different sources, in combination with other dietary components, are warranted to monitor the lactation performance of dairy cows.

Competing interests

The authors declare that they have no competing interests.

Authors' contributions

All authors have made substantial contributions to: the research design, analysis or interpretation of data, and in drafting the paper. Both authors read and approved the final manuscript.

References

1. Grant RJ: **Interactions among forages and non-forage fiber sources.** *J Dairy Sci* 1997, **80:**1438–1446.
2. Tafaj M, Zebeli Q, Baes C, Steingass H, Drochner W: **A meta-analysis examining effects of particle size of total mixed rations on intake, rumen digestion and milk production in high yielding dairy cows in early lactation.** *Anim Feed Sci Technol* 2007, **138:**137–161.
3. Varga AG: *Soluble carbohydrates for lactating dairy cows*, Proceeding of Tri-State Dairy Nutrition Conference. Columbus: The Ohio State University; 2003:59–73.
4. National Research Council (NRC): *Nutrient Requirements of Dairy Cattle, 7th rev.* Washington, DC: Natl. Acad. Sci; 2001.
5. Hall MB, Hoover WH, Jennings JP, Webster TKM: **A method for partitioning neutral detergent-soluble carbohydrates.** *J Sci Food Agric* 1999, **79:**2079–2086.
6. Larson CC: *The effects of non-fiber carbohydrate sources and protein degradability on performance of Holstein cows*, Ph.D. Thesis. University of Florida: Department of Animal Sci; 2003.
7. American Society of Agricultural Engineers: *Method of determining and expressing particle size of chopped forage (S424.1).* 70th edition. St. Joseph, MI: Am Society Agric Engineering; 2002:619–921.
8. Teimouri Yansari A, Primohammadi R: **Effect of particle size of alfalfa hay and reconstitution with water on intake, digestion, and milk production in Holstein dairy cows.** *Animal* 2009, **3:**218–227.
9. Zebeli Q, Dijkstra J, Tafaj M, Steingass H, Ametaj BN, Drochner W: **Modeling the adequacy of dietary fiber in dairy cows based on the responses of ruminal pH and milk fat production to composition of the diet.** *J Dairy Sci* 2008, **91:**2046–2066.
10. Lammers BP, Buckmaster DR, Heinrichs AJ: **A simple method for the analysis of particle sizes of forages and total mixed rations.** *J Dairy Sci* 1996, **79:**922–928.
11. Kononoff PJ: *The effect of ration particle size on dairy cows in early lactation*, Ph.D. Thesis. The Pennsylvania State Univ: Department of Animal Sci; 2002.
12. Van Soest PJ, Robertson JB, Lewis BA: **Methods for dietary fiber, neutral detergent fiber, and non-starch polysaccharide in relation to animal nutrition.** *J Dairy Sci* 1991, **74:**3583–3597.
13. Mertens DR: **Creating a system for meeting the fiber requirements of dairy cows.** *J Dairy Sci* 1997, **80:**1463–1481.
14. Association of Official Analytical Chemists: *Official Methods of Analysis*, Volume 1. 17th edition. Arlington, VA: AOAC; 2002:1–75.
15. Merchen NR, Firkins JL, Berger LL: **Effect of intake and forage level on ruminal turn over rates, bacterial protein synthesis and duodenal amino acid flows in sheep.** *J Anim Sci* 1986, **62:**216–224.
16. Uden P, Colucci E, Van Soest PJ: **Investigation of chromium, cerium, and cobalt as markers in digesta rate of passage studies.** *J Sci Food Agric* 1980, **31:**625–632.
17. Grovum WL, Williams VJ: **Rate of passage of digesta in sheep 4: Passage of markers through the alimentary tract and the biological relevance of rate-constants derived from changes in concentration of markers in feces.** *Br J Nutr* 1973, **30:**231–242.
18. User's Guide SAS: *Statistics, version 8.2.* Cary, NC: SAS Inst. Inc; 1998.
19. Broderick GA, Luchini ND, Radloff WJ, Varga GA, Ishler VA: *Effect of replacing dietary starch with sucrose on milk production in lactating dairy cows.* U.S: Dairy Forage Research Center 2000–2001 Research Report; 2002:116–118.
20. Chester-Jones H, Stern MD, Metwally HM, Linn JG, Ziegler DM: **Effects of dietary protein-energy interrelationships on Holstein steer performance and ruminal bacterial fermentation in continuous culture.** *J Anim Sci* 1991, **69:**4956–4966. 69.
21. Lees JA, Oldham ID, Haresign W, Gamsworthy PC: **The effect of patterns of rumen fermentation on the response by dairy cows to dietary protein concentration.** *Br J Nutr* 1990, **63:**177–187.
22. Valk H, KleinPoelhuis HW, Wentink HJ: **Effects of fibrous and starchy carbohydrates in concentrates as supplements in a herbage-based diet for high yielding dairy cows.** *Neth J Agric Sci* 1990, **38:**475–486.
23. Friggens NC, Emmans GC, Robertson S, Chamberlain DG, Whittemore CT, Oldham JD: **The lactational responses of dairy cows to the amount of feed and to the source of carbohydrate energy.** *J Dairy Sci* 1995, **78:**1734–1744.
24. Leiva E, Hall MB, Van Horn HH: **Performance of dairy cattle fed citrus pulp or corn products as sources of neutral detergent-soluble carbohydrates.** *J Dairy Sci* 2000, **83:**2866–2875.
25. Allen MS: **Effects of diet on short-term regulation of feed intake by lactating dairy cattle.** *J Dairy Sci* 2000, **83:**1598–1624.
26. Broderick GA, Mertens DR, Simons R: **Efficacy of carbohydrate sources for milk production by cows fed diets based on alfalfa silage.** *J Dairy Sci* 2002, **8:**1767–1776.
27. Huhtanen P: **The effect of dietary inclusion of barley, unmolassed sugar beet pulp and molasses on milk production, digestibility and digesta passage in dairy cows given silage based diet.** *J Sci Food Agric* 1987, **59:**l01–l111.
28. Huntington GB: **Starch utilization by ruminants: from basics to the bunk.** *J Anim Sci* 1997, **75:**852–867.
29. Torrent J, Johnson DE, Kujawa MA: **Co-product fiber digestibility: kinetic and in vivo assessment.** *J Anim Sci* 1994, **72:**790–795.
30. DePeters EJ, Fadel JG, Arosemena A: **Digestion kinetics of neutral detergent fiber and chemical composition within some selected by-product feedstuffs.** *Anim Feed Sci Technol* 1991, **67:**127–140.
31. Mansfield HR, Stern MD, Otterby DE: **Effects of beet pulp and animal by-products on milk yield and in vitro fermentation by rumen microorganisms.** *J Dairy Sci* 1994, **77:**205–216.
32. Van Vuuren AM, Van Der Koelen CJ, Vroons-De Bruin J: **Ryegrass versus corn starch or beet pulp fiber diet effects of digestion and intestinal amino acids in dairy cows.** *J Dairy Sci* 1993, **76:**2692–2700.
33. Teimouri Yansari A, Valizadeh R, Naserian A, Christensen DA, Yu P, Eftekhari Shahroodi F: **Effects of alfalfa particle size and specific gravity on chewing activity, digestibility, and performance of Holstein dairy cows.** *J Dairy Sci* 2004, **87:**3912–3924.
34. Bhatti SA, Firkins JL: **Kinetics of hydration and functional specific gravity of fibrous feed by-products.** *J Anim Sci* 1995, **73:**1449–1458.
35. DeVisser H, Van der Togh PL, Tamminga S: **Structural and nonstructural carbohydrates in concentrate supplements of silage-based dairy cow rations. 1. Feed intake and milk production.** *Neth J Agric Sci* 1990, **38:**487–497.
36. Rooke JA, Rymer C, Maya FM, Armstrong DG: **Effect of including barley or molassed sugar beet feed in grass silage diets on their digestion by cattle and sheep.** *J Sci Food Agric* 1992, **48:**475–489.

Effect of dietary defatted diatom biomass on egg production and quality of laying hens

Xiangjun Leng[1,2], Kun-Nan Hsu[1], Richard E Austic[1] and Xin' gen Lei[1*]

Abstract

Background: This study was to determine if feeding laying hens with defatted diatom microalgal biomass (DFA) from biofuel production affected their egg production and health status.

Methods: Five replicates of 5 individually caged ISA Babcock White leghorn hens were fed 4 diets, including a corn-soybean meal control diet, a diet containing 7.5% DFA substituting for soybean meal, and diets containing 7.5% or 15% DFA substituting for corn and soybean meal. Body weights, feed intake, feed conversion ratio (FCR), rate of egg production, egg size, egg mass, and several characteristics of eggs were determined at 4 and 8 wk. Venous blood was sampled at 4 and 8 wk for measurement of 5 biomarkers of health.

Results: The 15% DFA diet decreased ($P < 0.05$) feed intake, egg production, and plasma uric acid concentrations as compared with the control diet, but increased ($P < 0.05$) egg albumen weight and height compared with the 7.5% DFA diets. The two levels of DFA produced dose-dependent ($P < 0.05$) changes in three color measures of egg yolk, without affecting four hen plasma biochemical indicators of health.

Conclusions: Feeding laying hens with 7.5% DFA in the corn-soybean meal diet for 8 wk had no adverse effect on their health, egg production, or egg quality, but 15% inclusion reduced feed intake, egg production, and efficiency of feed utilization.

Keywords: Defatted diatom algae, Egg, Health, Laying hen, Soybean

Background

Although fossil fuels are the major source of energy for heating, transportation, manufacturing, and the generation of electricity, these fuels are non-renewable. Therefore, the search for renewable energy sources has become a key challenge of this century. Microalgae are recognized as one of the oldest living micro-organisms, and are seen as a very attractive source of renewable biofuels, especially biodiesel. Many species of microalgae contain large amounts of lipids that are suitable for the production of biofuels [1]. Microalgae are the natural food source for many important aquaculture species such as molluscs, shrimps and fish [2]. Several species of microalgae have been reported to be acceptable for inclusion in diets for swine, rabbits, broiler chickens, laying hens, and ruminant animals [3,4]. Soybean meal is a by-product of the extraction of oil from soybeans. With high protein content, it is rich in essential amino acids and serves as the main protein source for food-producing animals. Due to the increased human demand for soybean products, soybean meal is becoming more expensive and limited in supply [5,6]. Therefore, it is important to explore other high protein feedstuffs for the sustainable development of animal production.

Microalgae are a rich source of protein, essential fatty acids, vitamins and minerals [3]. After lipid removal, the residual biomass contains even higher concentrations of protein and other nutrients. Microalgae are good sources of long chain polyunsaturated fatty acids (PUFA) and have been used to enrich diets with omega-3 PUFA [7-9]. Diatoms are a class of microalgae that characteristically accumulate amorphous silicon in their membranes, resulting in distinct external features called frustules [10]. The defatted diatom biomass contains protein, residual fat, carbohydrates, silicon and a large mineral fraction that includes calcium, phosphorus, sodium, potassium, chloride, and several other nutritionally significant minerals

* Correspondence: XL20@cornell.edu
[1]Department of Animal Science, Cornell University, Ithaca, NY 14853, USA
Full list of author information is available at the end of the article

[3,11-13]. At the present time, a diatom species, *Staurosira sp*, is under investigation as a lipid source for biofuel production. Apparently, economics of that attempt would be remarkably enhanced if its defatted biomass is used as a feedstuff. However, no such study has been reported on feeding values of the defatted biomass of *Staurosira sp* or any diatom species in diets for poultry or swine. The frustules structure and high ash content in the defatted biomass could be a practical constraint for feed application. With a limited amount of defatted *Staurosira sp* microalgal biomass available for testing, we conducted the present laying hen experiment for 8 wk. Our objectives were to determine if the defatted biomass could partly replace soybean meal or a combination of corn and soybean meal in a practical diet; and how the replacements affected feed intake, health status, and egg production of the hens, and biochemical and biophysical properties of their eggs including pigmentation of yolk.

Table 1 Chemical composition of the defatted diatom microalgal biomass[1]

Item	Content	Item	Content
ME, Kcal/g	1.32	Met	0.33
Protein	19.1	Cys	0.32
Fat	3.3	Gly	0.96
Fiber	14.7	Ser	0.76
Ash	44.9	His	0.30
Moisture	6.9	Ile	0.78
Ca	2.78	Leu	1.33
P	0.76	Phe	0.86
Na	3.94	Tyr	0.57
K	1.66	Thr	0.88
Cl	6.34	Trp	0.18
Arg	0.93	Val	0.98
Lys	0.83		

[1]All items except ME are % of biomass (as fed basis).

Materials and methods

Animals, dietary treatments, and management

ISA Babcock White Leghorn laying hens (n = 100, 47 wk old, *Gallus gallus domesticus*), with an initial body weight of 1.57 ± 0.20 kg, were randomly assigned to 4 dietary treatments. There were 5 replicates for each treatment and each replicate consisted of 5 individually-caged hens. The cages with a size of 0.44 m high × 0.30 m wide × 0.45 m deep were equipped with nipple drinkers and trough feeders. The hen-house was provided with 16 h light per day and the hens were given free access to feed and water. The duration of the experiment was 8 wk. The animal protocol was approved by the Institutional Animal Care and Use Committee at Cornell University.

Defatted *Staurosira sp* microalgal biomass (DFA) in the form of powder was generated from the research on algal cultures for biofuel production (Cellana, Kailua-Kona, HI). Proximate and mineral analyses of the defatted alga were done by Dairy One, Inc. (Ithaca, NY) and amino acid analyses were performed by Experiment Station Chemical Laboratories (University of Missouri, Columbia, MO). The composition of the defatted alga is presented in Table 1. The four experimental diets included a corn-soybean meal control diet, the diet with 7.5% DFA and the 5 most limiting amino acids (Lys, Met, Ile, Thr, Trp) substituting for 7.5% soybean meal (7.5% algae-SBM), and two diets with DFA substituting for 7.5% (7.5% algae-SBM/Corn) or 15% (15% algae-SBM/Corn) soybean meal and corn (about 1:3.1). The same source of algae was used for all diets. The DFA had a similar crude protein level and amino acids composition as the mixture of soybean meal and corn with a ratio of 1:3.1. All diets were formulated to be isocaloric (2.80 Mcal of ME/kg), and iso-Ca, and iso-P by adjusting vegetable oil, dicalcium phosphate and limestone. The DFA contained many other minerals, but the dietary levels were not adjusted because the objective of the experiment was to test the acceptability of the algae, with its high ash content, as a feed ingredient. The compositions and nutrient concentrations of the diets are presented in Table 2.

Sample collection and procedures

Body weights of the laying hens were recorded at the start and the end of the experiment. Eggs were collected daily and egg production was calculated on a hen-day basis. Feed intake and egg production was recorded weekly by replicates to calculate feed conversion ratio (FCR). Eggs produced on the last 3 d of the 4th wk and 8th wk were individually weighed and analyzed for their interior and exterior quality. Eggs were examined for shell quality by measuring egg shell thickness, breaking strength, and egg specific gravity. Shell thickness, without shell membrane, was measured by micrometer with a mean value of measurements at 3 locations on the eggs (near the equator). The strength of egg shells was determined as the compression pressure necessary to crack the shell when the egg was placed horizontally between the plates of an Instron model 5969 (Instron, Norwood, MA). The specific gravity of eggs was determined by the buoyancy of eggs in salt solutions of varying density. Egg components, including albumen, yolk and shell, were weighed separately. Haugh units, a measure of the height of the albumen of eggs broken out on a flat surface were determined by the use of a micrometer [14]. Yolk color, measured as L^*-, a^*- and b^*-values, was determined with a Macbeth Color Eye (Macbeth Division of Kollmorgen Instruments Corp. Newburgh, NY). The L* value represents lightness (negative towards black, positive towards white), the a* value red–greenness (negative towards

Table 2 Compositions and nutrient concentrations of experimental diets

Ingredient,% as fed	Control	7.5% Algae -SBM	7.5% Algae -SBM/Corn	15% Algae -SBM/Corn
Corn (yellow)	65.65	65.28	59.45	52.55
Vegetable oils	0.55	1.55	2.30	4.30
Soybean meal	22.50	15.00	20.50	18.50
(48.5% crude protein)				
Dicalcium phosphate	1.50	1.40	1.35	1.20
Limestone	8.80	8.40	8.40	7.95
Salt	0.50	0.00	0.00	0.00
Vitamin mix[1]	0.25	0.25	0.25	0.25
Mineral mix[2]	0.15	0.15	0.15	0.15
DL-methionine	0.10	0.10	0.10	0.10
Defatted algae	0.00	7.50	7.50	15.00
Amino acids premix[3]	0.00	0.37	0.00	0.00
Total	100.00	100.00	100.00	100.00
Nutritional composition (% except for ME)				
ME, Kcal/g	2.80	2.80	2.80	2.80
Protein	16.49	14.26	16.43	16.30
Fat	3.27	4.73	5.31	7.58
Fiber	2.32	3.12	3.21	4.08
Ca	3.75	3.76	3.76	3.76
P	0.60	0.59	0.60	0.60
Avail. P	0.38	0.40	0.40	0.42
Arg	1.03	0.86	1.03	1.02
Lys	0.84	0.84	0.85	0.85
Met	0.36	0.36	0.38	0.39
Cys	0.28	0.23	0.26	0.23
Gly	0.68	0.61	0.70	0.73
Ser	0.80	0.68	0.80	0.79
His	0.44	0.36	0.42	0.40
Ile	0.67	0.67	0.68	0.69
Leu	1.50	1.33	1.48	1.46
Phe	0.78	0.67	0.78	0.78
Tyr	0.64	0.55	0.64	0.64
Thr	0.61	0.61	0.63	0.65
Trp	0.21	0.21	0.21	0.22
Val	0.76	0.68	0.78	0.79

[1]Provided (in IU/kg of diet): vitamin A, 6,500; vitamin D_3, 3,500; vitamin E, 25 and (in mg/kg of diet): riboflavin, 25; d-calcium pantothenate, 25; nicotinic acid, 150; cyanocobalamin, 0.011; choline chloride, 1,250; biotin, 1.0; folic acid, 2.5; thiamine hydrochloride, 7.0; pyridoxine hydrochloride, 25.0; menadione sodium bisulfite, 5.0; and ethoxyquin, 66.
[2]Provided (in mg//kg of diet): $CuSO_4 5H_2O$, 31.42; KI, 0.046; $FeSO_4 7H_2O$, 224.0; $MnSO_4 H_2O$, 61.54; Na_2SeO_3, 0.13; ZnO, 43.56; $Na_2MoO_4 2H_2O$, 1.26.
[3]Provided 1.4 g Lys, 0.1 g Met, 0.9 g Ile, 0.6 g Thr, 0.3 g Trp per kilogram of diet.

green, positive towards red) and the b* value the blue–yellow color scale (negative towards blue, positive towards yellow).

Uric acid, alanine transaminase (ALT), cholesterol and glucose in plasma were determined with Uric acid liquid stable reagent (Infinity TM, Fisher Diagnostics, Fisher Scientific Company, LLC, Middletown, NY), ALT Liquid stable reagent (Thermo Electron Corporation, Pittsburgh, PA), Cholesterol reagent (Wako Pure Chemical Industries, Ltd, Osaka, Japan), and Glucose Assay Kit (Sigma, St Louis, MO), respectively. Alkaline phosphatase (AKP) was determined by the method of Bowers and McComb [15]. Yolk cholesterol was extracted by the method of Folch *et al.* [16] and measured by the HPLC method of Beyer and Jensen [17].

Statistical analysis

The experimental data were presented as mean ± standard error, using SPSS17.0 for one-way analysis of variance, combined with Duncan's method for multiple comparisons. The significance level for differences was $P < 0.05$.

Results

Body weight, feed intake, egg production rate, and biochemical indices of hens

There were no significant differences in body weight among treatment groups at the beginning or the end of the experiment, despite an average loss of 20 to 70 g/hen during the experiment across all groups (Table 3). In the first 4 wk, the inclusion of defatted algae tended to decrease feed intake, egg production rate, egg mass, and FCR. In the second 4 wk period, hens fed 15% DFA had lower ($P < 0.05$) feed intake, egg production rate, egg mass, and higher ($P < 0.05$) FCR than those of the control group. During the whole 8-wk period, hens fed 15%, but not 7.5% DFA had lower ($P < 0.05$) egg production rate (−12%), daily feed intake (− 9 g/hen/d) and higher ($P < 0.05$) FCR (+0.09 g feed/g egg) than those fed the control diet (Table 3). There were no significant differences in plasma AKP, ALT, cholesterol, or glucose among dietary groups at the end of study (the 8th wk) ($P > 0.05$) (Table 4). However, hens fed 15% DFA had lower ($P <$ 0.05) plasma uric acid level than that of control group.

Egg quality and yolk color

At the end of 4th and 8th wk, there were no significant differences between the control and the DFA diets in egg quality indexes (Table 5), including egg weight, Haugh unit, yolk weight, shell weight, albumen weight, and shell thickness. However, the 15% Algae-SBM/Corn diet had higher ($P < 0.05$) egg albumen weight and height than the 7.5% Algae-SBM and 7.5% Algae-SBM/Corn diets at the 8th wk. Additionally, the DFA diets

Table 3 Effect of defatted diatom microalgal biomass on body weight, feed intake, egg production rate, egg mass and FCR

Item	Control	7.5% Algae -SBM	7.5% Algae -SBM/Corn	15% Algae -SBM/Corn
Body weight, kg				
Wk 0	1.53 ± 0.16	1.60 ± 0.19	1.58 ± 0.18	1.58 ± 0.20
Wk 8	1.47 ± 0.08	1.58 ± 0.08	1.53 ± 0.11	1.51 ± 0.10
Daily feed intake, g/hen				
Wk 1-4	100.1 ± 4.9	96.9 ± 3.9	95.9 ± 6.7	92.6 ± 11.2
Wk 5-8	96.0 ± 8.3^{a}	96.0 ± 3.6^{a}	95.0 ± 6.9^{a}	85.5 ± 4.5^{b}
Wk 1-8	98.1 ± 6.0^{a}	96.4 ± 3.4^{ab}	95.4 ± 5.9^{ab}	89.1 ± 5.9^{b}
Egg production rate, %[1]				
Wk 1-4	86.7 ± 3.4	83.7 ± 6.3	84.7 ± 6.6	75.1 ± 16.4
Wk 5-8	81.3 ± 11.2^{a}	85.6 ± 5.0^{a}	83.9 ± 8.7^{a}	69.0 ± 6.6^{b}
Wk 1-8	84.0 ± 7.1^{a}	84.6 ± 4.7^{a}	84.3 ± 7.1^{a}	72.1 ± 10.1^{b}
Egg mass, g/hen/d				
Wk 1-4	52.02 ± 2.02	49.79 ± 3.68	50.21 ± 3.23	45.63 ± 8.65
Wk 5-8	48.77 ± 5.02^{a}	49.73 ± 2.23^{a}	49.71 ± 4.34^{a}	41.43 ± 4.01^{b}
Wk 1-8	50.40 ± 3.87^{a}	49.74 ± 2.93^{a}	49.96 ± 3.36^{a}	43.53 ± 5.71^{b}
FCR[2]				
Wk 1-4	1.92 ± 0.03	1.95 ± 0.05	1.91 ± 0.06	2.02 ± 0.11
Wk 5-8	1.97 ± 0.08^{a}	1.93 ± 0.04^{a}	1.92 ± 0.05^{a}	2.06 ± 0.05^{b}
Wk 1-8	1.94 ± 0.06^{a}	1.94 ± 0.04^{a}	1.92 ± 0.05^{a}	2.05 ± 0.09^{b}

^{a,b}Means in the same row without a letter in common are different ($P < 0.05$).
[1]Egg production = 100 [number of eggs laid ÷ (number of hens x number of days)].
[2]FCR = feed intake (g) / egg mass (g).

produced no difference from the control diet in egg specific gravity, egg shell breaking strength or yolk cholesterol at the end of study (the 8th wk). Including DFA in the diets affected ($P < 0.05$) yolk color in a dose-dependent fashion (Table 6). The L* (lightness) value and + b* (yellowness) value of yolk decreased ($P < 0.05$) with the increasing DFA of the diet. At the 8th wk, but not at the 4th wk, the values for lightness and yellowness were lower ($P < 0.05$) for 7.5% algae-SBM/Corn than for 7.5% algae-SBM. The + a* (redness) values of yolk were elevated ($P < 0.05$) by all three DFA diets at the end of 4th wk, but by only the two 7.5% DFA diets at the 8th wk.

Discussion

Results of the present study indicate that 7.5% DFA of *Staurosira sp* could replace 7.5% soybean meal or a combination of corn and soybean meal in diets for laying hens, without adverse effects on their body weights, feed intake, FCR, health status, or egg mass, egg production rate and quality. However, the inclusion of 15% DFA decreased hen feed intake by 11% and egg production rate by 12%. Past studies on inclusions of algae in diets for poultry have shown that algae can be used safely at dietary levels of 5% to 10% [18-24]. Thus, our finding on 7.5% DFA as a tolerable level for hens just falls into the

Table 4 Effect of defatted diatom microalgal biomass on plasma biochemical indices of hens at the 8th wk

Item	Control	7.5% Algae -SBM	7.5% Algae -SBM/Corn	15% Algae -SBM/Corn
AKP, U/mL	111.5 ± 29.0	146.4 ± 34.1	155.4 ± 44.2	133.8 ± 31.1
ALT, U/L	24.1 ± 6.5	24.3 ± 9.1	33.5 ± 15.8	29.5 ± 7.4
Cholesterol, mg/dL	117.5 ± 8.5	117.4 ± 12.3	112.6 ± 22.5	106.0 ± 15.6
Glucose, mg/dL	204.9 ± 4.6	210.7 ± 13.4	209.1 ± 9.4	198.6 ± 7.0
Uric acid, mg/dL	10.7 ± 1.9^{a}	10.2 ± 2.7^{ab}	8.6 ± 2.4^{ab}	7.6 ± 1.3^{b}

^{a,b}Means in the same row without a letter in common are different ($P < 0.05$).

Table 5 Effect of defatted diatom microalgal biomass on egg quality at the 4th and 8th wk

Item	Control	7.5% Algae -SBM	7.5% Algae -SBM/Corn	15% Algae -SBM/Corn
At the end of the 4th wk				
Egg weight, g	60.00 ± 3.93	59.49 ± 5.25	59.28 ± 3.79	60.69 ± 3.61
Haugh unit	86.6 ± 4.0	84.6 ± 6.4	85.6 ± 7.1	88.6 ± 5.7
Yolk weight, g	15.79 ± 1.42	16.35 ± 1.42	15.77 ± 1.22	15.92 ± 0.82
Egg shell weight, g	5.29 ± 0.57	5.70 ± 2.11	5.37 ± 0.39	5.60 ± 0.54
Albumen weight, g	38.81 ± 2.96	38.21 ± 3.89	38.13 ± 2.83	39.00 ± 3.24
Egg shell thickness, mm	0.34 ± 0.03	0.34 ± 0.03	0.36 ± 0.03	0.36 ± 0.03
At the end of the 8th wk				
Egg weight, g	59.99 ± 2.51	58.09 ± 1.49	59.25 ± 1.75	60.05 ± 1.26
Haugh unit	82.6 ± 3.1	82.0 ± 3.3	81.6 ± 2.0	84.6 ± 1.4
Yolk weight, g	16.11 ± 0.71	16.25 ± 0.55	16.08 ± 0.48	15.90 ± 0.32
Egg shell weight, g	5.47 ± 0.26	5.28 ± 0.25	5.38 ± 0.25	5.49 ± 0.35
Albumen weight, g	38.44 ± 2.17^{ab}	36.78 ± 1.40^{a}	37.87 ± 1.45^{ab}	39.24 ± 1.15^{b}
Albumen height, mm	6.92 ± 0.43^{ab}	6.73 ± 0.47^{ab}	6.71 ± 0.24^{a}	7.23 ± 0.23^{b}
Egg shell thickness, mm	0.35 ± 0.02	0.35 ± 0.01	0.34 ± 0.03	0.35 ± 0.02
Egg specific gravity	1.082 ± 0.004	1.080 ± 0.001	1.080 ± 0.005	1.080 ± 0.002
Egg shell strength, N	34.4 ± 2.2	34.4 ± 2.0	32.6 ± 4.6	35. 3 ± 3.6
Yolk cholesterol, mg/g	14.20 ± 0.66	14.06 ± 0.34	13.50 ± 0.66	13.69 ± 0.71

^{a,b}Means in the same row without a letter in common are different ($P < 0.05$).

middle of these studies. The decreases in feed intake and egg production rate in the 15% DFA group were likely attributed to the high ash (44.9%) and sodium chloride (10.3%) concentrations of the DFA used in the study. Although we removed salt and decreased limestone and dicalcium phosphate supplementation in the DFA diets, the 15% inclusion rate rendered the diet still salty and probably less palatable than the control diet. As expected, hens fed this diet were observed to drink more water and have bulky wet droppings. While the hens seemed to tolerate 7.5% DFA in the diet well for 8 wk, effects of long-term ingestion of high ash and salt on the health and egg production of these animals and on moisture and bulk of their excreta and the subsequent manure disposal should be evaluated in future research.

Inclusion of 7.5 and 15% DFA resulted in dose-dependent changes of egg yolk color, but did not significantly affect egg weights, egg shell weight, egg shell thickness, egg breaking strength, albumen firmness (Haugh units), or yolk cholesterol concentrations. While the latter outcome is consistent with other reports involving chickens [21,22,24-27] and Japanese quail [22], in which one or more of these traits were reported, the former is likely due to the presence of various carotenoids in DFA. Although algae powder appears as a bluish-green color, it contains high levels of carotenoids such as

Table 6 Effect of defatted algae on yolk color

Item	Control	7.5% Algae -SBM	7.5% Algae -SBM/Corn	15% Algae -SBM/Corn
At the end of the 4th wk				
L*(lightness)	54.806 ± 0.541^{a}	52.502 ± 0.602^{b}	51.943 ± 0.650^{b}	50.328 ± 0.878^{c}
+a*(redness)	11.576 ± 0.209^{a}	13.438 ± 0.305^{b}	13.316 ± 1.015^{b}	12.939 ± 0.329^{b}
+b*(yellowness)	35.660 ± 0.314^{a}	34.522 ± 0.320^{b}	34.013 ± 0.383^{b}	33.038 ± 0.558^{c}
At the end of the 8th wk				
L*(lightness)	54.758 ± 0.430^{a}	52.718 ± 0.393^{b}	51.690 ± 1.014^{c}	50.732 ± 0.762^{d}
+a*(redness)	11.084 ± 0.317^{a}	12.573 ± 0.288^{b}	12.518 ± 0.889^{b}	11.356 ± 0.607^{a}
+b*(yellowness)	35.479 ± 0.238^{a}	34.499 ± 0.149^{b}	33.697 ± 0.561^{c}	32.856 ± 0.239^{d}

^{a,b,c,d}Means in the same row without a letter in common are different ($P < 0.05$).

β-carotene, lutein and zeaxanthin [28]. The red carotenoid astaxanthin is present in some algae [2]. Because poultry and other animals cannot synthesize carotenoids *de novo*, the color of broiler skin and leg and egg yolk can be enhanced by including algae in feeds [3]. In a study using a color meter to measure yolk pigmentation [29], 2.4% or 4.8% of a commercial preparation of the marine algae, *Schizochytrium*, was fed to Leghorn hens over 4 wk. The yolk redness (+a* value) was increased, and the yellowness (+b* value) and lightness (+L* value) were decreased in a dose-responsive manner to the algae. In another study, 10% and 20% algae (*Nannochloropsis oculata*) in diet increased the yolk +a* value, and decreased the +L* value [30]. Consistently, the present study showed DFA dose-dependent increases in the + a value and decreases in the +b and +L* values of yolk. These color changes are supposed to be highly correlated with the Roche color fan that is used to measure the pigmentation in egg yolks and poultry skin by visual examination. Increases in the numerical color fan score have been observed in various studies on the use of algae in diets of laying hens [9,24,27,30,31]. Depending on the amount of algae or lipid extract of algae incorporated into wheat- and sorghum-based feeds, scores ranged from about 7 to 9, which are considered acceptable for table eggs, to 12 to 15 which are in the range suitable for manufactured food products such as egg noodles and bakery items. Increases in yolk pigmentation were reported to occur within 24 h and to reach plateau at 7 to 8 d [9,31]. In the present study, we found that 7.5 or 15% DFA shifted yolk pigmentation score from 7.9 in the control group to 9.3 to 9.5 (data not shown). This shift implies that DFA may be used as a source of pigments for egg yolk formation to improve its consumers' acceptance.

Several genera of algae have been analyzed for protein quality based on protein digestibility, biological value, and net protein utilization [3,4]. Although the results indicate that protein quality varies among algae, many species have relatively high protein content and excellent protein quality by these measures. In the present experiment, DFA was added to the diet in replacement of an equivalent weight of soybean meal or a mixture of soybean and corn. The DFA contained 19% crude protein in contrast to 48% crude protein in soybean meal. Substitution of one-third of the soybean meal with DFA necessarily means that the resulting diets had less protein than the control diet. Several amino acids were added to the 7.5% Algae-SBM diet in an attempt to meet essential amino acid requirements. This diet supported production responses that were similar to those of the control diet. Only uric acid, among the plasma biomarkers, was affected by the DFA treatments. The lower plasma uric acid concentration in the hens fed 15% defatted diatom suggests that less amino acid catabolism had occurred in this group of hens as compared to the control groups. A simple explanation for this observation would be that the 11% lower daily feed (and protein) consumption resulted in fewer amino acids available for catabolism. Alternatively or additionally, a low digestibility of the protein in the 15% DFA diet may have contributed to reduced absorption of amino acids, thereby resulting in lower amino acid catabolism. Additional research would be necessary to determine the actual cause of the reduction in plasma uric acid concentration.

Due to the limited amount of DFA available for the present study, our feeding experiment lasted for only 8 wk. However, all responses of feed intake, egg production, egg quality, and plasma biomarkers were consistent that DFA could serve at a dietary level of 7.5% over this period as a source of protein, energy, calcium, phosphorus, and sodium chloride for laying hens. The question is if the period of 8 wk was sufficient for the objectives of this experiment including determining effects of the DFA nutrients on egg production and composition. Physiologically, the formation of egg yolk occurs in a process that lasts about 9 d. Egg albumen, eggshell membranes, and eggshell are usually secreted within slightly more than 24 h. Experimentally, changes in yolk pigmentation occurred within 8 d after the addition of algae to the diet [9,31]. Changes in egg production occurred within 2 to 3 mo of the feeding protein- or amino acid-deficient diets to laying hens [32,33]. Concentrations of selenium or fat-soluble vitamins in eggs were affected by diet within days to a few weeks [34-36]. Clearly, many nutritional effects on egg production and composition could be detected within 8 wk. In fact, we observed significant effects of DFA on egg yolk color measures after only 4-wk feeding. Metabolically, animals should be most responsive during the initial phase to negative impacts of DFA as longer time feeding may give them chance to adapt. Nevertheless, it is prudent that a feeding trial of several months or a full year would be necessary to evaluate long-term nutritional values of DFA during a full egg production cycle.

Competing interests

The authors declare that they have no competing interests.

Authors' contributions

XGL and REA conceived the study, supervised the trial, and revised the manuscript. XL, K-NH, and REA raised the animals and carried out the laboratory analysis. XL and REA drafted the initial manuscript, and XGL helped revise and finalize the manuscript. All authors read and approved the final manuscript.

Acknowledgements

This study was supported in part by a USDA/DOE Biomass R&D Initiative grant. The defatted microalgal biomass was provided by Cellana, Kailua-Kona, HI, USA. We thank Dr. James Bartsch in the Department of Biological and Environmental Engineering and Dr. David Barbano in the Department of Food Science for instruction on the use, and access to the Instron and

Macbeth COLOR EYE, respectively, for measurement of egg breaking strength and egg yolk color.

Author details

[1]Department of Animal Science, Cornell University, Ithaca, NY 14853, USA. [2]College of Aquaculture and Life Science, Shanghai Ocean University, Shanghai 201306, China.

References

1. Gouveia L, Oliveira AC: **Microalgae as raw material for biofuels production.** *J Ind Microbiol Biotechnol* 2009, **36:**269–274.
2. Spolaore P, Joannis-Cassan C, Duran E, Isambert A: **Commercial applications of microalgae.** *J Biosci Bioeng* 2006, **101**(2):87–96.
3. Becker W: **Microalgae in Human and Animal Nutrition.** In *Handbook of Microalgal Culture: Biotechnology and Applied Phycology.* Edited by Richmond A. Oxford: Blackwell Science; 2004:312–351.
4. Becker EW: **Micro-algae as a source of protein.** *Biotechnol Adv* 2007, **25:**207–210.
5. U.S. Department of Agriculture, Economic Research Service: *Economics, Statistics, and Market Information System.* Washington, DC: Oil Crops, Situation and Outlook Yearbook; 2011.
6. Mielke T: **Major challenges ahead: world soybean supply and demand outlook for 2012/13.** *Int News on Fats, Oils and Related Materials (INFORM)* 2012, **23**(7):468–470.
7. Herber SM, van Elswyk ME: **Dietary marine algae promotes efficient deposition of n-3 fatty acids for the production of enriched shell eggs.** *Poult Sci* 1996, **75:**1501–1507.
8. Barclay W, Abril R, Abril P, Weaver C, Ashford A: **Production of docosahexaenoic acid from microalgae and its benefits for use in animal feeds.** In *The return of ω3 fatty acids into the food supply. I. Land-based animal food products and their health effects.* Edited by Simopoulos AP. Basel, Switzerland: Karger: World Rev Nutr Diet; 1988, **83:**61-76.
9. Nitsan Z, Mokady S, Sukenik A: **Enrichment of poultry products with ω3 fatty acids by dietary supplementation with the alga *Nannochloropsis* and mantur oil.** *J Agric Food Chem* 1999, **47:**5127–5132.
10. Martin-Jézéquel V, Hildebrand M, Brezinski MA: **Silicon metabolism in diatoms: implications for growth.** *J Phycol* 2000, **36:**821–840.
11. Harrison G, Morel FMM: **Response of the marine diatom, *Thalassiosira weissflogii,* to iron stress.** *Limnol Oceanogr* 1986, **3**(5):989–997.
12. Taraldsvik M, Myklestad SM: **The effect of pH on growth rate, biochemical composition and extracellular carbohydrate production of the marine diatom, *Skeletonema costatum*.** *Eur J Phycol* 2000, **35:**189–194.
13. Pahl SL, Lewis DM, Chen F, King KD: **Growth dynamics and the proximate biochemical composition and fatty acid profile of the heterotrophically grown diatom *Cyclotella cryptica*.** *J Appl Phychol* 2010, **22:**165–171.
14. USDA: *U.S. Department of Agriculture, Consumer and Marketing Service.* Washington, DC: Egg grading manual. Agriculture Handbook No. 75; 1969.
15. Bowers GN Jr, McComb RB: **A continuous spectrophotometric method for measuring the activity of serum alkaline phosphatase.** *Clin Chem* 1966, **12:**70–89.
16. Folch J, Lees M, Stanley GHS: **A simple method for the isolation and purification of total lipides from animal tissues.** *J Biol Chem* 1957, **226:**497–509.
17. Beyer RS, Jensen LS: **Overestimation of the cholesterol content of eggs.** *J Agric Food Chem* 1989, **37:**917–920.
18. Combs GF: **Algae (*Chlorella*) as a source of nutrients for the chick.** *Science* 1952, **116:**453–454.
19. Grau CR, Klein NW: **Sewage-grown algae as a feedstuff for chicks.** *Poult Sci* 1957, **36:**1046–1051.
20. Yoshida M, Hoshii H: **Nutritive value of new type of *Chlorella* for poultry feed.** *Jpn Poult Sci* 1982, **19:**56–58.
21. Lipstein B, Hurwitz S: **The nutritional value of sewage-grown samples of *Chlorella* and *Micractinium* in broiler diets.** *Poult Sci* 1983, **62:**1254–1260.
22. Ross E, Dominy W: **The nutritional value of dehydrated, blue-green algae (*Spirulina platensis*) for poultry.** *Poult Sci* 1990, **69:**794–800.
23. Venkataraman LV, Somasekaran T, Becker EW: **Replacement value of blue-green alga (*Spirulina platensis*) for fishmeal and a vitamin-mineral premix for broiler chicks.** *Br Poult Sci* 1994, **35:**373–381.
24. Halle I, Janczyk P, Freyer G, Souffrant WB: **Effect of microalgae *Chlorella vulgaris* on laying hen performance.** *Arch Zootech* 2009, **12**(2):5–13.
25. Lipstein B, Hurwitz S, Bornstein S: **The nutritional value of algae for poultry. Dried Chlorella in layer diets.** *Br Poult Sci* 1980, **21:**23–27.
26. Lipstein B, Hurwitz S: **The nutritional value of sewage-grown, alum flocculated *Micractinium* algae in broiler and layer diets.** *Poult Sci* 1981, **60:**2628–2638.
27. Sauveur B, Zybko A, Colas B: **Protéines alimentaires et qualité de l'oeuf. I. Effet de quelques protéines sur la qualité interne de l'oeuf et les propriétes fonctionnelles.** *Ann Zootech* 1979, **28:**271–295.
28. Miki W, Yamaguchi K, Konosu S: **Carotenoid composition of Spirulina maxima.** *Bull Jpn Soc Sci Fish* 1986, **52:**1225–1227.
29. Herber-McNeill SM, van Elswyk ME: **Dietary marine algae maintains egg consumer acceptability while enhancing yolk color.** *Poult Sci* 1998, **77:**493–496.
30. Fredriksson S, Elwinger K, Pickova J: **Fatty acid and carotenoid composition of egg yolk as an effect of microalgae addition to feed formula for laying hens.** *Food Chem* 2006, **99:**530–537.
31. Zahroojian N, Morave H, Shivazad M: **Comparison of marine algae (*Spirulina platensis*) and synthetic pigment in enhancing egg yolk colour of laying hens.** *Br Poult Sci* 2011, **52**(5):584–588.
32. Summers JD, Atkinson JL, Spratt D: **Supplementation of a low protein diet in an attempt to optimize egg mass output.** *Can J Anim Sci* 1991, **71:**211–220.
33. Harms RH, Russell GB: **A re-evaluation of the methionine requirement of the commercial layer.** *J Appl Anim Res* 1996, **9:**141–151.
34. Latshaw JD, Osman M: **Distribution of selenium in egg white and yolk after feeding natural and synthetic selenium compounds.** *Poult Sci* 1975, **54:**1244–1252.
35. Suzuki Y, Okamoto M: **Production of hen's eggs rich in vitamin K.** *Nutr Res* 1997, **17**(10):1607–1615.
36. Mattila P, Rokka T, Konko K, Valaja J, Rossow L, Ryhanen EL: **Effect of cholecalciferol enriched hen feed on egg quality.** *J Agric Food Chem* 2003, **51:**283–287.

Metabolizable energy, nitrogen balance, and ileal digestibility of amino acids in quality protein maize for pigs

Gerardo Mariscal-Landín[1*], Tércia Cesária Reis de Souza[2] and Ericka Ramírez Rodríguez[1]

Abstract

Background: To compare the nutritional value and digestibility of five quality protein maize (QPM) hybrids to that of white and yellow maize, two experiments were carried out in growing pigs. In experiment 1, the energy metabolizability and the nitrogen balance of growing pigs fed one of five QPM hybrid diets were compared against those of pigs fed white or yellow maize. In experiment 2, the apparent and standardized ileal digestibility (AID and SID, respectively) of proteins and amino acids from the five QPM hybrids were compared against those obtained from pigs fed white and yellow maize. In both experiments, the comparisons were conducted using contrasts.

Results: The dry matter and nitrogen intakes were higher in the pigs fed the QPM hybrids ($P < 0.05$) than in the pigs fed white or yellow maize. Energy digestibility ($P < 0.001$) and metabolizability ($P < 0.01$) were higher in the pigs fed the white and yellow maize diets than in those fed the QPM diets. The AID of lysine was higher ($P < 0.01$) in the QPM diets than in the white and yellow maize. The AIDs of leucine, isoleucine, valine, phenylalanine, and methionine were lower in the QPM diets than those of maize (white and yellow) (all $P < 0.05$). Maize (white and yellow) had greater SIDs of leucine, isoleucine, valine, phenylalanine, glutamic acid, serine, alanine, tyrosine, and proline ($P < 0.05$).

Conclusions: Based on these results, it was concluded that QPM had a lower metabolizable energy content and a higher amount of digestible lysine than normal maize.

Keywords: Amino acids, Energy balance, Maize, Nitrogen balance, Pigs

Background

Maize (*Zea mays*) is the most widely harvested cultivar in the world, and it is often used as the principal source of energy for pigs. It is also an important source of protein and amino acids in finishing pigs [1]. There are varieties of maize, other than yellow maize, that contain different nutrient concentrations [2]. Owing to these differences in nutrient density and/or composition, new types of maize such as quality protein maize (**QPM**) may offer nutritional advantages over the conventional yellow maize varieties.

Maize protein is deficient in the amino acids, lysine and tryptophan, which limits its value for monogastric animals [3]. Mertz et al. [4] first reported the mutant maize called Opaque 2 in 1963, which has a higher content of these amino acids. However, maize Opaque 2 had a soft endosperm that made it susceptible to pests and crop storage problems, for which its production was ceased. Subsequent conventional breeding efforts by the International Maize and Wheat Improvement Centre (CIMMYT) generated numerous cultivars with improved agronomic characteristics, collectively referred to as QPM [5], a type of maize with a hard endosperm rich in lysine and tryptophan due to a change in the ratio of 19- and 22-kD α zeins, an increment of 27 kD γ zein [6], and a non zein protein called elongation factor 1α (**EF-1 α**) [7]. The substantial reduction in synthesis of α-zeins results in smaller, less numerous protein bodies and a concomitant increase in non-zein endosperm proteins [3].

The metabolizable energy (**ME**) and amino acid (**AA**) digestibility of yellow maize have been extensively investigated and summarized in previous publications [8,9]. This allows for the accurate formulation of yellow maize-based diets for pigs to meet ME and digestible AA

* Correspondence: mariscal.gerardo@inifap.gob.mx
[1]CENID Fisiología, Instituto Nacional de Investigaciones Forestales Agrícolas y Pecuarias, km 1 Carretera a Ajuchitlán, 76280 Querétaro, México
Full list of author information is available at the end of the article

requirements. However, there is currently limited information about the nutritive value of QPM.

Therefore, the first aim of the current study was to compare the energy metabolizability and nitrogen balance of QPM to those of yellow and white maize, and the second aim was to compare the apparent and standardized ileal digestibility (**AID** and **SID**, respectively) of protein and AAs in QPM hybrids to those of yellow and white maize.

Material and methods

This study was approved by the Scientific Associate Technical Group Committee of CENID Physiology. Two experiments were conducted at the experimental farm of CENID-Physiology (INIFAP, México). The experimental animals were treated according to the guidelines of the International Guiding Principles for Biomedical Research Involving Animals [10] and the Official Mexican Standard for production, care and use of laboratory animals [11].

Raw materials

The following QPM hybrids were evaluated: 538Ta, 537Ta, 537Ig, 334Ce, and QPM1. Evaluation was also conducted on one white and one yellow maize obtained from a commercial supplier in the state of Guanajuato, Mexico. The chemical composition of these materials is shown in Table 1.

Energy metabolizability and nitrogen balance (Experiment 1)

Four consecutive groups of seven Landrace x Duroc pigs with a mean weight of 61.2 ± 2.6 kg were used (28 pigs in total, four replicates by treatment). The pigs were placed in individual metabolic cages equipped with a self-feeder and a low-pressure drinking nipple connected to a watering system that controlled the water supply. Screens were placed under the floors, which allowed for total collection of feces and urine. The room temperature ranged from 19–22°C.

The pigs were fed twice daily at 0800 h and 1800 h. The experimental diets (Table 2) were prepared with one QPM or one maize (white or yellow) as the sole protein and energy source. The diets contained calcium carbonate, dicalcium phosphate, salt, and premixed vitamins and minerals. Chromic oxide (3 g/kg of feed) was included as an indigestible marker. The feed intake of the pigs was 2.5 times their digestible energy (**DE**) requirement for maintenance (460 kJ/kg $BW^{0.75}$ [12]) as recommended by Adeola [13] for pigs weighing >50 kg. The water container in each metabolic cage was filled just before each meal to restrict water intake to 2.5 L/kg of dry matter (**DM**) intake [13].

The experimental period lasted for 10 d for each consecutive group: 5 d for adaptation and 5 d for the total collection of feces and urine. The feces were frozen and kept at −20°C. At the end of the experimental period, the feces were defrosted and homogenized to obtain 10% of the weight as a final sample for lyophilizing. Urine was collected via funnels underneath the urine collection tray. This collection system included a glass wool mat to avoid contamination with feed or feces. To reduce urine pH and avoid NH_3 volatilization, 40 mL of HCl 6 mol/L were added to each urine container twice a day. The urine was removed twice a day, weighed, and filtered again through four layers of cheesecloth into a clean container. Then a 5% aliquot was taken and kept at −20°C until analysis. Three urine subsamples of 20 mL per pig were lyophilized to measure energy [14].

Ileal digestibility (Experiment 2)

Seven Landrace x Duroc pigs with a mean weight of 62.3 ± 4.9 kg at the time of data collection were used. When the pigs weighed 45 kg, a T-cannula was fitted at the terminal ileum [15]. After surgery, the pigs were placed in individual metabolism cages that included a self-feeder and a low-pressure drinking nipple. The room temperature ranged from 19–22°C.

The post-operative period lasted for 21 d. During this period, the pigs were fed a grower diet (160 g of CP/kg) twice daily at 0800 h and 1800 h. The amount fed was increased 100 g/d until the level of intake was the same as that prior to surgery.

During the experimental period, the pigs received one of the experimental diets (Table 3). The diets were formulated using maize as the sole source of dietary protein. To avoid the effect of the level of the dietary protein on protein apparent digestibility [16], maize starch was used to ensure that the experimental diets consisted of the same protein level despite the different protein levels of the maize. All experimental diets contained calcium carbonate, dicalcium phosphate, and salt. To reduce dust, 20 g/kg of maize oil was included. Vitamins and minerals exceeded the NRC [8] requirements. Chromic oxide (3 g/kg of feed) was included as an indigestible marker. The feed intake of the pigs was 2.5 times their digestible energy (DE) requirement for maintenance, 460 kJ/kg $BW^{0.75}$ [12]. The animals had free access to water.

The four experimental periods lasted seven d each: 5 d for adaptation and 2 d for digesta collection. Ileal digesta were collected continuously over a period of 10 h (0800 h–1800 h) using plastic bags (11 cm × 5 cm) that contained 10 mL of a 0.2 mol/L HCl solution to inhibit bacterial activity and were attached to the cannula using a rubber band. When the collecting bags were full, the ileal digesta was transferred to a container and frozen at −20°C until lyophilisation.

Table 1 Chemical composition of maize (g/kg)[1]

	QPM[2]					Maize	
Nutrient, g /kg	QPM1	334Ce	537Ta	537Ig	538Ta	White	Yellow
Dry Matter	910.7	911.4	893.0	909.0	911.2	911.0	910.8
Protein	85.0	84.0	86.0	85.0	87.0	80.0	75.0
NDF	114.0	91.0	102.0	84.0	89.0	104.0	94.0
ADF	48.0	39.0	40.0	38.0	35.0	34.0	27.0
Ether Extract	34.0	34.0	47.0	39.0	52.0	39.0	37.0
Ash	17.0	14.0	16.0	17.0	17.0	11.0	12.0
Gross Energy, MJ/kg	16.7	16.6	16.4	16.0	16.6	16.5	16.5
Amino acids							
Alanine	5.1	4.9	4.9	5.0	4.2	5.9	5.6
Arginine	5.9	3.5	5.3	5.3	4.9	3.8	3.6
Aspartic acid	6.7	4.6	6.5	6.4	5.4	5.4	5.3
Cystine	3.5	3.5	2.4	3.5	3.4	3.0	3.3
Glutamic acid	14.1	12.6	13.6	14.4	11.4	15.3	14.2
Glycine	4.1	2.9	3.9	4.0	3.5	3.1	3.0
Histidine	3.5	2.7	3.4	2.5	2.6	2.5	2.1
Isoleucine	2.6	2.0	2.5	2.4	2.1	2.6	2.5
Leucine	7.3	7.7	7.1	7.3	5.8	9.6	9.1
Lysine	3.1	2.7	2.9	2.7	2.7	2.0	2.0
Methionine	1.8	1.8	2.2	2.3	2.3	1.5	2.1
Phenylalanine	3.9	3.1	3.4	3.9	2.9	4.2	3.9
Proline	8.4	7.4	9.4	8.0	7.9	9.2	8.9
Serine	3.7	3.1	3.6	3.9	3.1	3.7	3.5
Threonine	3.3	2.4	3.0	3.5	2.7	2.8	2.6
Tyrosine	2.6	2.2	2.4	2.2	2.1	2.6	2.5
Valine	4.7	3.1	5.4	4.4	4.6	3.9	4.3

[1]As fed basis.
[2]Quality Protein Maize.

At the end of the experiment, narcosis was induced by CO_2 inhalation, followed by euthanasia by exsanguination. A post-mortem examination for fistula along the length of the small intestine was performed to verify its integrity.

Chemical analysis

The raw materials, diets, digesta, and feces were ground using a laboratory mill (Arthur H. Thomas Co., Philadelphia, PA) to pass through a 0.5 mm mesh sieve. DM and nitrogen analysis methods, 934.01 and 976.05 [17], respectively, were performed. Chromium analysis was performed as described by Fenton and Fenton [18]. Gross energy was estimated using an adiabatic bomb calorimeter (model 1281, Parr, Moline, IL). Samples from the raw materials, diets, and digesta were hydrolyzed at 110°C for 24 h in 6 mol/L HCl to use in AA analysis method 994.12 [17]. For methionine and cystine analyses, oxidation with performic acid was carried out before acid hydrolysis [17]. Tryptophan was not estimated. AA analysis was performed using reverse phase HPLC (1100 HPLC Hewlett Packard), according to Henderson *et al.* [19]. Nitrogen in the liquid urine was estimated according to AOAC [17] method 976.05. Energy in the lyophilized urine was estimated according to Le Bellego [14].

Calculations

The AID or apparent total tract digestibility (**ATTD**) was estimated using the equation proposed by Fan and Sauer [20]:

$$AID = [1-((ID \times AF)/(AD \times IF))] \times 100$$

where AID = apparent digestibility (ileal or total tract) of a nutrient in the diet, **ID** = concentration of the marker in the diet (mg/kg DM), **AF** = concentration of the nutrient in ileal digesta or feces (mg/kg DM),

Table 2 Composition of the experimental diets (g/kg): experiment 1[1]

Ingredient, g/kg	QPM[2]					Maize	
	QPM1	334Ce	537Ta	537Ig	538Ta	White	Yellow
QPM1	967.5						
334Ce		967.5					
537Ta			967.5				
537Ig				967.5			
538Ta					967.5		
White						967.5	
Yellow							967.5
Calcium Phosphate	18.5	18.5	18.5	18.5	18.5	18.5	18.5
Calcium Carbonate	5.5	5.5	5.5	5.5	5.5	5.5	5.5
Salt	3.5	3.5	3.5	3.5	3.5	3.5	3.5
Minerals[3]	1.0	1.0	1.0	1.0	1.0	1.0	1.0
Vitamins[4]	1.0	1.0	1.0	1.0	1.0	1.0	1.0
Chromic oxide	3.0	3.0	3.0	3.0	3.0	3.0	3.0
Chemical composition[5]							
Dry Matter	906.6	903.3	891.1	904.6	901.4	905.6	904.8
Protein	82.2	81.3	83.2	82.2	84.2	77.4	72.6
Gross Energy, MJ/kg	16.5	16.2	16.1	15.9	16.1	16.0	16.0

[1]As fed basis.
[2]Quality Protein Maize.
[3]Furnished by kg of feed: Cl 1.65 g, Na 0.87 g, Cu 7.7 mg, Fe 89.25 mg, Mn 19.98 mg, Se 0.087 mg, I 0.053 mg.
[4]Furnished by kg of feed: Vitamin A 6,600 IU, D 660 IU, E 100 IU, Choline 350 mg, Niacin 54 mg, Pantothenic acid 13.15 mg, Riboflavin 2.2 mg, B_{12} 36 μg.
[5]Analyzed values, on an as fed basis.

AD = concentration of the nutrient in the diet (mg/kg DM), and **IF** = concentration of the marker in the ileal digesta or feces (mg/kg DM).

The SID was estimated using the formula proposed by Furuya and Kaji [21]:

$$SID = [AID + (Endogenous/Dietary\ \ Content)] \times 100$$

where Endogenous = endogenous losses of a nutrient in mg/kg DM intake, and Dietary content = amount of nutrient consumed in mg/kg DM intake. The calculations were performed using endogenous values reported by Mariscal-Landin and Reis de Souza [22].

The ME was obtained using the formula proposed by Adeola [13]:

$$ME = [(GE–FE–UE)/GE] \times 100$$

where ME = metabolizable energy in MJ/d, **GE** = gross energy intake in MJ/d, **FE** = fecal energy output in MJ/d, and **UE** = urine energy output in MJ/d.

Statistical analyses

Experiment 1

Data were analyzed as a randomized complete block design [23] with four blocks of seven animals each and using the GLM procedure in SAS v9.2 [24]. The variables were the DM amount, nitrogen intake, energy intake, the amount of nitrogen and energy excreted in feces and urine, the apparent total tract digestibility of DM, nitrogen, and energy, nitrogen balance, and energy metabolizability. An alpha value of 0.05 was used to assess significance.

Experiment 2

Data were analyzed using a Latin square with additional columns, or a 4 × 7 "Youden square" [23], which included seven animals, seven treatments, and four experimental periods. The experimental variables were the protein and amino acids AID and SID of the maize. Data were analyzed using the GLM procedure in SAS v9.2 [24]. An alpha value of 0.05 was used to assess significance.

In both experiments, the means were compared using Duncan's method and the QPM (QPM1, 334Ce, 537Ta, 537Ig, or 538Ta) and normal maize (white or yellow) were compared using contrasts [23].

Results

Energy metabolizability and nitrogen balance

Energy metabolizability, QPM vs normal maize

Energy intake was similar ($P > 0.05$) between the treatments: 29.5 MJ/d of QPM or normal maize. Apparent

Table 3 Composition of the experimental diets (g/kg): experiment 2[1]

Ingredient, g/kg	QPM[2]					Maize	
	QPM1	334Ce	537Ta	537Ig	538Ta	White	Yellow
QPM1	845.5						
334Ce		845.5					
537Ta			825.5				
537Ig				835.5			
538Ta					816.5		
White						887.5	
Yellow							947.5
Maize starch	102.0	102.0	122.0	112.0	131.5	60.0	
Calcium Phosphate	18.5	18.5	18.5	18.5	18.5	18.5	18.5
Calcium Carbonate	5.5	5.5	5.5	5.5	5.5	5.5	5.5
Maize oil	20.0	20.0	20.0	20.0	20.0	20.0	20.0
Salt	3.5	3.5	3.5	3.5	3.5	3.5	3.5
Minerals[3]	1.0	1.0	1.0	1.0	1.0	1.0	1.0
Vitamins[4]	1.0	1.0	1.0	1.0	1.0	1.0	1.0
Chromic oxide	3.0	3.0	3.0	3.0	3.0	3.0	3.0
Chemical composition[5]							
Dry matter	905.6	901.3	894.1	864.2	899.2	906.5	902.8
Protein	71.1	70.8	70.6	71.0	71.3	70.5	71.6
Gross Energy, MJ/kg	15.8	15.7	15.6	15.2	15.7	15.7	15.6
Alanine	4.70	4.81	4.45	4.32	4.01	5.55	6.08
Arginine	5.36	3.47	4.86	4.57	4.70	3.59	3.93
Aspartic acid	6.10	4.54	5.83	5.59	5.04	5.04	5.69
Cystine	3.22	3.47	2.19	3.13	3.25	2.82	3.53
Glutamic acid	12.87	12.47	12.38	12.62	10.85	14.43	15.41
Glycine	3.71	2.85	3.56	3.56	3.33	2.90	3.24
Histidine	3.22	2.67	3.08	2.54	2.48	2.39	2.36
Isoleucine	2.39	1.96	2.19	1.61	1.96	2.39	2.65
Leucine	6.60	7.57	6.39	6.10	5.55	9.05	9.81
Lysine	2.80	2.67	2.59	2.54	2.48	1.88	2.16
Methionine	1.65	1.16	2.02	1.10	1.45	1.45	2.26
Phenylalanine	3.55	3.12	3.08	2.80	2.82	3.93	4.22
Proline	6.35	6.14	8.50	4.91	7.52	8.71	9.62
Serine	3.38	3.03	3.24	3.47	2.90	3.50	3.73
Threonine	2.97	2.40	2.75	3.39	2.56	2.65	2.85
Tyrosine	2.39	2.23	2.19	2.12	1.96	2.48	2.65
Valine	4.29	3.03	4.86	2.71	4.36	3.59	4.71

[1]As fed basis.
[2]Quality Protein Maize.
[3]Furnished by kg of feed: Cl 1.65 g, Na 0.87 g, Cu 7.7 mg, Fe 89.25 mg, Mn 19.98 mg, Se 0.087 mg, I 0.053 mg.
[4]Furnished by kg of feed: Vitamin A 6,600 IU, D 660 IU, E 100 IU, Choline 350 mg, Niacin 54 mg, Pantothenic acid 13.15 mg, Riboflavin 2.2 mg, B_{12} 36 μg.
[5]Analyzed values, on an as fed basis.

total tract digestibility was lower in the pigs fed QPM (89.8) than in those fed normal maize (91.0, $P < 0.001$). While UE was similar (0.58 MJ/d, $P > 0.05$) between all the treatments, the metabolizability of energy was lower in the pigs fed QPM (87.9) than in those fed normal maize (88.9) (Table 4).

Table 4 Energy metabolizability and nitrogen balance: experiment 1[1]

Traits	QPM[2]					Maize			Contrast		
	QPM1	334Ce	537Ta	537Ig	538Ta	White	Yellow	SEM[3]	QPM	Normal	Prob[4]
Dry Matter, kg/d											
Intake	1.67^{a}	1.66^{a}	1.66^{a}	1.68^{a}	1.64^{a}	1.66^{a}	1.55^{b}	0.010	1.66	1.60	0.05
Energy, MJ/d											
Intake	29,9^{a}	29.7^{a}	29.7^{a}	30.2^{a}	29.5^{a}	29.7^{a}	27.7^{b}	0.17	29.8	28.7	0.63
In feces	3.1^{a}	3.2^{a}	3.0^{a}	2.9^{ab}	2.9^{ab}	2.6^{b}	2.6^{b}	0.04	3.0	2.6	0.001
Digestible	26.8^{a}	26.5^{a}	26.7^{a}	27.3^{a}	26.5^{a}	27.1^{a}	25.1^{b}	0.16	26.7	26.1	0.01
In Urine	0.50	0.58	0.53	0.60	0.62	0.60	0.61	0.013	0.57	0.60	0.21
Metabolizable	26.3^{a}	25.9^{a}	26.2^{a}	26.7^{a}	25.9^{a}	26.5^{a}	24.5^{b}	0.16	26.2	25.5	0.01
Nitrogen, g/d											
Intake	22.4^{b}	24.6^{a}	24.5^{a}	24.6^{a}	24.4^{a}	22.0^{b}	22.1^{b}	0.13	24.0	22.0	0.001
In feces	3.6	4.1	3.8	3.7	3.8	3.4	3.5	0.08	3.82	3.46	0.32
Digestible	18.7^{b}	20.5^{a}	20.6^{a}	20.4^{a}	20.5^{a}	18.5^{b}	18.6^{b}	0.35	20.1	18.6	0.001
In Urine	10.2^{a}	12.6^{b}	13.2^{b}	11.9^{b}	12.0^{b}	12.1^{b}	12.2^{b}	0.21	11.97	12.2	0.74
Retained	8.5^{a}	7.9^{ab}	7.5^{ab}	8.5^{a}	8.5^{a}	6.5^{ab}	6.4^{b}	0.24	8.18	6.44	0.19
Dry matter, MJ/kg											
Digestible energy	16.11^{cd}	15.96^{d}	16.12^{bcd}	16.29^{ab}	16.18^{abc}	16.34^{a}	16.22^{abc}	0.007	16.13	16.28	0.01
Metabolizable energy	15.59^{ab}	15.38^{b}	15.59^{ab}	15.70^{a}	15.55^{ab}	15.74^{a}	15.61^{ab}	0.009	15.56	15.68	0.09
Coefficient											
DM Digestibility	88.7^{c}	88.2^{c}	88.9^{c}	90.1^{ab}	89.2^{bc}	90.8^{a}	90.2^{a}	0.12	89.0	90.5	0.001
E Digestibility	89.6^{bc}	89.1^{c}	89.8^{bc}	90.4^{ab}	90.1^{bc}	91.3^{a}	90.6^{ab}	0.14	89.8	91.0	0.001
Metabolizability	87.9^{b}	87.2^{b}	88.0^{b}	88.4^{ab}	88.0^{b}	89.3^{a}	88.4^{ab}	0.15	87.9	88.9	0.01
N Digestibility	83.7	83.3	84.3	84.7	84.3	84.3	84.2	0.30	84.1	84.3	0.72
N retention as percentage of											
Intake[5]	38.2^{a}	32.1^{ab}	30.5^{ab}	35.2^{ab}	35.0^{ab}	29.6^{ab}	28.8^{b}	1.01	34.2	29.2	0.28
Absorbed[6]	45.7^{a}	38.5^{ab}	36.1^{ab}	41.6^{ab}	41.3^{ab}	35.0^{b}	34.2^{b}	1.14	40.3	34.6	0.20

[1]All data are reported on a DM basis.
[2]Quality Protein Maize.
[3]Standard error of the mean.
[4]Contrast Probability.
[5](N retained/N intake)*100.
[6](N retained/N absorbed)*100.
[abc]different letters in the same line differ ($P < 0.05$).

Energy metabolizability, means comparison

Energy intake was lower in the pigs fed yellow maize (27.7 MJ/d) than in the pigs fed white maize (29.8 MJ/d, $P < 0.05$). Apparent total tract digestibility was higher in the pigs fed white maize (91.3) and lower in the pigs fed 334Ce QPM (89.1, $P < 0.05$). The metabolizability of energy was higher in the pigs fed white maize (89.3) than in the pigs fed QPM1 (87.9), except for those fed QPM 537Ig (88.4; Table 4).

Nitrogen balance, QPM vs normal maize

The DM intake was higher in the pigs fed QPM, compared to those fed normal maize (1.66 vs 1.60 kg/d, $P < 0.05$), and the DM total tract digestibility was lower in the QPM pigs, compared to those fed normal maize (89.0 vs 90.5, $P < 0.001$). The daily nitrogen intake was higher in the pigs fed QPM (24.0 g/d) than in the pigs fed normal maize (22.0 g/d, $P < 0.001$); however, nitrogen digestibility was similar between the treatments (84.1, $P > 0.05$). Consequently, the digestible nitrogen intake was higher in the pigs fed QPM. Urinary nitrogen excretion was similar between treatments (12.0 g/d, $P > 0.05$). Nitrogen retention as a proportion of nitrogen intake or nitrogen absorption was also similar ($P > 0.05$) between the treatments (Table 4).

Nitrogen balance, means comparison

Although the nitrogen intake in the pigs fed QPM1 was lower (22.4 g/d) than in the pigs fed the other hybrid QPMs (24.5 g/d, $P < 0.001$) and similar to the pigs fed normal maize (22.0 g/d), the pigs fed QPM1 retained more nitrogen as a percentage of nitrogen intake (38.2%) or nitrogen absorption (45.7%) than the pigs fed yellow maize (28.8% and 34.2%, respectively, $P < 0.05$; Table 4).

Ileal digestibility

Apparent ileal digestibility, QPM vs normal maize

The DM digestibility was lower in the pigs fed QPM (78.7) than in those fed normal maize (80.0, $P < 0.05$). The CP digestibility was similar in all of the diets (mean 73.0, $P > 0.05$).

Lysine digestibility was higher in the pigs fed QPM than in the pigs fed normal maize ($P < 0.05$; Table 5). The digestibility of leucine, isoleucine, phenylalanine, glutamic acid, alanine, tyrosine, proline, valine, serine, and methionine was lower in the QPM pigs than in those fed normal maize (all $P < 0.05$; Table 5).

Apparent ileal digestibility, means comparison

In general, QPM 537Ig had the lowest digestibility, except for that for arginine, threonine, serine, and glycine; for these AAs, QPM 334Ce had the lowest digestibility (Table 5). The white maize diet demonstrated lower AID of the sulphur AAs than the yellow maize diet. The digestibility of leucine, alanine, and tyrosine was lower in all of the QPM diets than in the normal maize diets (Table 5).

Standardized ileal digestibility, QPM vs normal maize

The digestibility of CP, lysine, arginine, histidine, methionine, threonine, aspartic acid, glycine, and cystine was similar between QPM and normal maize (all $P > 0.05$; Table 6). The digestibility of glutamic acid, tyrosine, leucine, isoleucine, phenylalanine, alanine, valine, serine, and proline was lower in the QPM diets than in the normal maize diets (all $P < 0.05$; Table 6).

Standardized ileal digestibility, means comparison

The SID of QPM 537Ig was consistently the lowest among the QPMs. The digestibility of glutamic acid and tyrosine was lower in the QPM diets than in the normal maize diets (Table 6). Methionine was less digestible in the white maize diet than in the yellow maize diet (Table 6).

Discussion

Energy metabolizability and nitrogen balance

The total tract digestibility of energy in the QPM diets was 1.3% lower, on average, than the normal maize diets.

Table 5 Apparent ileal digestibility of the maize: experiment 2

Traits	QPM[1]					Maize		SEM[2]	Contrast		Prob[3]
	QPM1	334Ce	537Ta	537Ig	538Ta	White	Yellow		QPM	Normal	
Dry Matter	77.5^{c}	77.6^{c}	81.9^{a}	77.7^{c}	78.6^{bc}	80.9^{ab}	79.1^{abc}	0.33	78.7	80.0	0.05
Protein	72.0	74.3	73.4	71.4	74.3	71.8	74.1	0.32	73.1	72.9	0.77
Amino acids											
Alanine	73.2^{b}	72.7^{b}	71.6^{b}	69.0^{b}	70.6^{b}	79.8^{a}	80.1^{a}	0.67	71.4	79.9	0.001
Arginine	82.1^{a}	73.8^{b}	82.3^{a}	81.3^{a}	85.5^{a}	80.0^{a}	82.7^{a}	0.72	81.0	81.4	0.79
Aspartic acid	75.2	68.4	75.9	71.5	71.9	75.4	75.6	0.61	72.6	75.5	0.06
Cystine	72.9^{c}	78.9^{ab}	66.9^{d}	74.9^{bc}	78.0^{abc}	75.1^{bc}	80.9^{a}	0.56	74.3	78.0	0.07
Glutamic acid	83.3^{b}	83.1^{b}	83.1^{b}	81.5^{b}	82.1^{b}	86.9^{a}	86.9^{a}	0.29	82.6	86.9	0.001
Glycine	55.1^{ab}	49.8^{b}	58.9^{a}	54.6^{ab}	51.9^{ab}	55.8^{ab}	57.0^{ab}	0.91	54.0	56.4	0.83
Histidine	87.2^{ab}	83.8^{bc}	88.3^{a}	82.6^{c}	85.4^{abc}	86.1^{abc}	84.8^{abc}	0.45	85.4	85.4	0.93
Isoleucine	75.4^{ab}	69.9^{b}	74.3^{ab}	60.2^{c}	71.8^{b}	79.4^{a}	79.2^{a}	0.66	70.3	79.3	0.001
Leucine	82.4^{b}	83.2^{b}	83.1^{b}	80.5^{b}	82.1^{b}	89.0^{a}	89.4^{a}	0.43	82.3	89.2	0.001
Lysine	77.2^{a}	76.7^{a}	77.1^{a}	73.8^{ab}	75.1^{ab}	69.0^{b}	72.4^{ab}	0.72	76.0	70.7	0.01
Methionine	80.5^{b}	74.2^{c}	85.5^{a}	70.7^{d}	75.5^{c}	76.8^{c}	84.3^{a}	0.36	77.3	80.5	0.05
Phenylalanine	82.7^{b}	80.6^{bc}	81.9^{b}	77.3^{c}	82.6^{b}	88.0^{a}	87.6^{a}	0.46	81.0	87.7	0.001
Proline	51.2^{cd}	61.6^{bc}	72.5^{ab}	45.8^{d}	61.5^{bc}	71.6^{ab}	80.3^{a}	1.63	58.5	75.9	0.001
Serine	71.0^{ab}	67.8^{b}	72.0^{ab}	70.4^{ab}	69.4^{b}	76.8^{a}	76.6^{a}	0.70	70.1	76.6	0.01
Threonine	60.4^{ab}	54.6^{b}	61.6^{ab}	64.1^{a}	60.1^{b}	64.8^{a}	62.4^{ab}	0.96	60.1	63.6	0.39
Tyrosine	77.9^{b}	76.8^{b}	77.7^{b}	73.8^{b}	74.7^{b}	83.7^{a}	83.9^{a}	0.57	76.2	83.8	0.001
Valine	77.6^{a}	70.4^{b}	81.7^{a}	63.3^{c}	80.9^{a}	78.5^{a}	82.8^{a}	0.18	74.8	80.6	0.01

[1]Quality Protein Maize.
[2]Standard error of the mean.
[3]Probability of contrast.
[abcd]different letters in the same line differ ($P < 0.05$).

Table 6 Standardized ileal digestibility of the maize: experiment 2

Traits	QPM[1]					Maize			Contrast		
	QPM1	334Ce	537Ta	537Ig	538Ta	White	Yellow	SEM[2]	QPM	Normal	Prob[3]
Protein	86.7	88.0	88.2	85.7	88.2	86.9	88.1	0.32	87.4	87.5	0.98
Amino acids											
Alanine	80.7^{ab}	80.0^{bc}	79.4^{c}	77.1^{c}	79.4^{c}	86.1^{a}	85.8^{ab}	0.66	79.3	86.0	0.01
Arginine	87.9^{ab}	82.8^{b}	88.8^{ab}	88.2^{ab}	92.2^{a}	88.8^{ab}	90.8^{a}	0.72	88.0	89.8	0.65
Aspartic acid	85.2	81.9	86.4	82.4	84.0	87.5	86.4	0.61	84.0	86.9	0.06
Cystine	77.1^{cd}	82.8^{ab}	73.1^{c}	79.2^{bc}	82.1^{abc}	79.9^{abc}	88.7^{a}	0.56	78.9	82.3	0.10
Glutamic acid	88.7^{b}	88.7^{b}	88.7^{b}	87.1^{b}	88.6^{b}	91.7^{a}	91.4a	0.28	88.3	91.6	0.001
Glycine	73.8	74.2	78.3	74.2	72.7	79.7	78.4	0.91	74.6	79.0	0.26
Histidine	92.0^{ab}	89.5^{ab}	93.3^{a}	88.6^{b}	91.5^{ab}	92.5^{ab}	91.3^{ab}	0.46	91.0	91.9	0.39
Isoleucine	86.3^{ab}	83.3^{b}	86.3^{ab}	76.4^{c}	85.0^{ab}	90.3^{a}	89.0^{ab}	0.66	83.4	89.7	0.01
Leucine	89.2^{bc}	89.1^{bc}	90.2^{abc}	87.9^{c}	90.2^{abc}	94.0^{a}	94.0^{a}	0.44	89.3	94.0	0.01
Lysine	88.1	88.2	88.9	85.9	87.5	85.3	86.6	0.72	87.7	85.9	0.18
Methionine	84.4^{b}	79.7^{b}	88.7^{a}	76.4^{d}	79.8^{c}	81.1^{c}	87.1^{ab}	0.36	81.8	84.1	0.30
Phenylalanine	89.5^{bcd}	88.4^{cd}	89.8^{bcd}	86.0^{d}	91.3^{abc}	94.2^{a}	93.3^{ab}	0.46	89.0	93.7	0.01
Proline	70.2^{c}	81.2^{abc}	86.7^{ab}	70.3^{c}	77.5^{bc}	85.4^{ab}	92.8^{a}	1.63	77.2	89.1	0.02
Serine	84.6^{ab}	82.9^{b}	86.2^{ab}	83.6^{ab}	85.2^{ab}	89.9^{a}	88.9^{ab}	0.70	84.5	89.4	0.03
Threonine	77.7	76.0	80.2	79.2	80.2	84.1	80.4	0.95	78.6	82.3	0.35
Tyrosine	83.2^{b}	82.5^{b}	83.4^{b}	79.8^{b}	81.1^{b}	88.8^{a}	88.7^{a}	0.57	82.0	89.7	0.001
Valine	86.0^{ab}	82.3^{b}	89.1^{a}	76.6^{c}	89.1^{a}	88.5^{a}	90.4^{a}	0.58	84.6	89.5	0.02

[1]Quality Protein Maize.
[2]Standard error of the mean.
[3]Probability.
[abcd]different letters in the same line differ ($P < 0.05$).

While urinary energy was similar between the diets, metabolizability was also lower in the QPM diets than in the white maize diet.

The starch type of QPM may explain its lower digestibility in the current study; amylose is negatively correlated with average daily gain [25]. Furthermore, the starch of waxy sorghums, which are low in amylose, is more digestible than the starch of non-waxy sorghums [26]; this has also been demonstrated with maize starch in ducks [27]. Although QPM has a vitreous endosperm phenotype, it is rich in a no-crystalline amylopectin that forms bonds with γ zein (27-kDa) [28].

The negative effect of fiber on energy digestibility [29,30] could provide another feasible explanation; dietary fiber is less digestible than other nutrients such as starch, sugars, fat, or protein (<50% vs 80-100%) [31]. Moreover, corn fiber is essentially insoluble [32]; QPM had a higher **ADF** content than the normal maize (40 g/kg QPM vs 30 g/kg maize), and this could have resulted in considerable effects on digestibility given that maize fiber, in general, is poorly digested by growing pigs [31]. Moreover, the ME:DE ratio in the current study was 0.98, which is similar to the 0.96 estimated by Noblet and van Milgen [31].

Generally, 50% of the nitrogen that is absorbed is retained in the body, and the other 50% is excreted in urine [29]. The retention of nitrogen in the current study was lower (37.5%), which could be attributed to a diet of low protein quality. Maize protein is deficient in lysine and tryptophan [3], and it is well-known that nitrogen retention in growing pigs is related to the lysine level in the diet as lysine is the first limiting AA [33]. Lysine digestibility was higher in the pigs fed QPM1 maize, and these same pigs retained 26% more nitrogen (1.7 g/d). These results are consistent with previous reports based on animal studies [1,34,35], as well as in humans where the consumption of QPM by children resulted in a 12% increase in weight [34] and a 9% increase in height and weight [36].

Ileal digestibility

The average AA profile of the proteins in the QPM was different from that of yellow and white maize. QPM had more lysine (45%), arginine (37%), histidine (31%), glycine (23%), methionine (19%), threonine (13%), aspartic acid (13%), valine (10%), and cystine (7%) than white and yellow maize. In contrast, QPM had less leucine (−23%), alanine (−14%), phenylalanine (−13%), glutamic acid (−8%), tyrosine (−8%), and proline (−7%) than white and yellow maize. Other studies have also reported that QPM is rich in lysine [5,37,38], while the low leucine and proline content is associated with a decrease in zein protein [6].

Dietary protein content affects apparent digestibility [16,39,40]. To avoid this effect, the experimental diets were iso-nitrogenous, resulting in similar protein digestibility in all of the diets. However, the differences in amino acid content may explain the differences in digestibility. It has been previously reported that the high lysine content in QPM results in a higher AID [34,35], as was found in the current study. Similarly, the high leucine, phenylalanine, glutamic acid, alanine, tyrosine, and proline content of normal maize (white and yellow) could explain the high AID that was observed. Moreover, the low AID observed for threonine may be caused by its richness within the endogenous protein [22,41,42].

Although the use of the estimate from one endogenous protein in each experiment has been recommended to estimate the SID [39,40], it is also true that the SID has been estimated from previously published AID data and corrected with an endogenous protein that was estimated later [9,43,44] or previously [45,46]. This supports the use of a "standard" endogenous protein to correct the AID. The SID removes the effect of nutrient level on the digestibility value by correcting for basal endogenous losses [39,43,47]. The SID of lysine was similar in all of the maize diets. Additionally, when SID was estimated, threonine reached similar values to those of the other amino acids; this may be related to its richness within the endogenous protein. The SID coefficients estimated for maize in the present work were similar to those reported in previous studies [8,9,43]. However, no values for SID have been reported previously for QPM.

Conclusions

The energy furnished by QPM was used less efficiently (metabolizability) than the energy furnished by normal maize. The AID of lysine was higher in the QPM than in the normal maize; however, the SID of lysine was similar between QPM and normal maize. The current study provides additional data about the nutrient composition, AA digestibility, and nitrogen utilization of QPM.

Abbreviations

CIMMYT: International maize and wheat improvement centre; QPM: Quality protein maize; EF-1 α: Elongation factor 1α; DM: Dry matter; DE: Digestible energy; ME: Metabolizable energy; AA: Amino acid; ADF: Acid detergent fiber; AID: Apparent ileal digestibility; SID: Standardized ileal digestibility; ATTD: Apparent total tract digestibility.

Competing interests

The authors declare that they have no competing interests.

Authors' contributions

GML conceived and designed the study; ERR carried out the lab analysis; TCRS contributed to data analysis; GML and TCRS wrote the manuscript. All authors read and approved the final manuscript.

Acknowledgements

To National Research Institute for Forestry, Crops and Livestock for the financial support given to this research project.

Author details

[1]CENID Fisiología, Instituto Nacional de Investigaciones Forestales Agrícolas y Pecuarias, km 1 Carretera a Ajuchitlán, 76280 Querétaro, México. [2]Facultad de Ciencias Naturales, Universidad Autónoma de Querétaro, México, Av. de las Ciencias s/n, 76230 Querétaro, México.

References

1. Cantarelli VS, Fialho ET, de Sousa RV, de Freitas RTF, Lima JAF: **Composição química, vitreosidade e digestibilidade de diferentes híbridos de milho para suínos.** *Ciênc Agrotec* 2007, **31**:860–864.
2. Snow JL, Stein HH, Ku PK, Trottier NL: **Amino acid digestibility and nitrogen utilization of high oil, high lysine, and waxy maize fed to growing pigs.** *Anim Feed Sci Technol* 2004, **113**:113–126.
3. Jia M, Wu H, Clay K, Jung R, Larkins B, Gibbon B: **Identification and characterization of lysine-rich proteins and starch biosynthesis genes in the opaque2 mutant by transcriptional and proteomic analysis.** *BMC Plant Biology* 2013, **13**:60.
4. Mertz ET, Bates LS, Nelson OE: **Mutant gene that changes protein composition and increases lysine content of maize endosperm.** *Science* 1964, **145**:279–280.
5. Nuss ET, Tanumihardjo SA: **Quality protein maize for Africa: closing the protein inadequacy gap in vulnerable populations.** *Adv Nutr* 2011, **2**:217–224.
6. Wallace JC, Lopes MA, Paiva E, Larkins BA: **New methods for extraction and quantitation of zeins reveal a high content of γ-zein in modified opaque-2 maize.** *Plant Physiol* 1990, **92**:191–196.
7. Habben JE, Moro GL, Hunter BG, Hamaker BR, Larkins BA: **Elongation factor 1α concentration is highly correlated with the lysine content of maize endosperm.** *PNAS* 1995, **92**:8640–8644.
8. NRC: *Nutrient requirements of swine.* 10th edition. Washington, DC: The National Academy Press; 1998.
9. INRA: *Tables de composition et de valeur nutritive des matières premières destinées aux animaux d'élevage. Porcs, volailles, bovins, ovins, caprins, lapins, chevaux, poissons.* Paris, France: Institut National de la Recherche Agronomique; 2002.
10. International guiding principles for biomedical research involving animals: *The development of science-based guidelines for laboratory animal care - NCBI Bookshelf.* http://cioms.ch/publications/guidelines/1985_texts_of_guidelines.htm.
11. Diario Oficial de la Federación: **Especificaciones técnicas para la producción, cuidado y uso de los animales de laboratorio. Norma Oficial Mexicana NOM-062-ZOO-1999.** *México, D.F. Diario Oficial de la Federación, Miércoles 2 de Agosto* 2001.
12. INRA: *L'alimentation des animaux monogastriques: porc, lapin, volailles.* Paris, France: Institut National de la Recherche Agronomique; 1984.
13. Adeola O: **Digestion and balance technique in pigs.** In *Swine Nutrition.* 2nd edition. Edited by Lewis AJ, Southern LL. Boca Raton, USA: CRC Press; 2001:903–916.
14. Le Bellego L, van Milgen J, Dubois S, Noblet J: **Energy utilization of low-protein diets in growing pigs.** *J Anim Sci* 2001, **79**:1259–1271.
15. de Souza RTC, Mar BB, Mariscal LG: **Canulación de cerdos posdestete para pruebas de digestibilidad ileal: Desarrollo de una metodología.** *Téc Pecu Méx* 2000, **38**:143–150.
16. Fan MZ, Sauer WC, Hardin RT, Lien KA: **Determination of apparent ileal amino acid digestibility in pigs: effect of dietary amino acid level.** *J Anim Sci* 1994, **72**:2851–2859.
17. AOAC: *Official Methods of Analysis.* 17th edition. Arlington, VA. USA: Assoc. Offic. Anal. Chem; 2000.

18. Fenton TW, Fenton M: **An improved procedure for determination of chromic oxide in feed and feces.** *Can J Anim Sci* 1979, **59:**631–634.
19. Henderson JH, Ricker RD, Bidlingmeyer BA, Woodward C: **Rapid, accurate and reproducible HPLC analysis of amino acids. Amino acid analysis using Zorbax Eclipse AAA columns and the Agilent 1100 HPLC.** *Agilent Technologies* 2000, **Part No5980-1193E 2000:10**. Agilent technologies home page at: www.agilent.com/chem/supplies.
20. Fan MZ, Sauer WC: **Determination of apparent ileal amino acid digestibility in barley and canola meal for pigs with the direct, difference, and regression methods.** *J Anim Sci* 1995, **73:**2364–2374.
21. Furuya S, Kaji Y: **Estimation of the true ileal digestibility of amino acids and nitrogen from their apparent values for growing pigs.** *Anim Feed Sci Technol* 1989, **26:**271–285.
22. Mariscal-Landín G, de Souza RTC: **Endogenous ileal losses of nitrogen and amino acids in pigs and piglets fed graded levels of casein.** *Arch Anim Nutr* 2006, **60:**454–466.
23. Steel RGD, Torrie JH: *Principles and procedures of statistics. A Biometrical approach.* 2nd edition. New York: McGraw-Hill; 1980.
24. SAS version 9.2: *Statistical Analysis Systems Institute User's guide.* 92nd edition. : SAS Institute Inc; 2008.
25. Moore SM, Stalder KJ, Beitz DC, Stahl CH, Fithian WA, Bregendahl K: **The correlation of chemical and physical corn kernel traits with growth performance and carcass characteristics in pigs.** *J Anim Sci* 2008, **86:**592–601.
26. Wong JH, Marx DB, Wilson JD, Buchanan BB, Lemaux PG, Pedersen JF: **Principal component analysis and biochemical characterization of protein and starch reveal primary targets for improving sorghum grain.** *Plant Sci* 2010, **179:**598–611.
27. Zhou Z, Wan HF, Li Y, Chen W, Qi ZL, Peng P, Peng J: **The influence of the amylopectin/amylose ratio in samples of corn on the true metabolizable energy value for ducks.** *Anim Feed Sci Technol* 2010, **157:**99–103.
28. Gibbon BC, Wang X, Larkins BA: **Altered starch structure is associated with endosperm modification in Quality Protein Maize.** *PNAS* 2003, **100:**15329–15334.
29. Kil D, Kim B, Stein H: **Feed energy evaluation for growing pigs.** *Asian-Aust J Anim Sci* 2013, **26:**1205–1217.
30. Szabó C, Halas V: **Shortcomings of the energy evaluation systems in pigs: a review.** *Agric Conspec Sci* 2013, **78:**153–158.
31. Noblet J, van Milgen J: **Energy value of pig feeds: effect of pig body weight and energy evaluation system.** *J Anim Sci* 2004, **82**(E-Suppl):E229–E238.
32. Cowieson AJ: **Factors that affect the nutritional value of maize for broilers.** *Anim Feed Sci Technol* 2005, **119:**293–305.
33. Susenbeth A, Dickel T, Diekenhorst A, Höhler D: **The effect of energy intake, genotype, and body weight on protein retention in pigs when dietary lysine is the first-limiting factor.** *J Anim Sci* 1999, **77:**2985–2989.
34. Burgoon KG, Hansen JA, Knabe DA, Bockolt AJ: **Nutritional value of quality protein maize for starter and finisher swine.** *J Anim Sci* 1992, **70:**811–817.
35. Sullivan JS, Knabe DA, Bockholt AJ, Gregg EJ: **Nutritional value of quality protein maize and food corn for starter and growth pigs.** *J Anim Sci* 1989, **67:**1285–1292.
36. Gunaratna NS, de Groote H, Nestel P, Pixley KV, McCabe GP: **A meta-analysis of community-based studies on quality protein maize.** *Food Policy* 2010, **35:**202–210.
37. Azevedo RA, Arruda P: **High-lysine maize: the key discoveries that have made it possible.** *Amino Acids* 2010, **39:**979–989.
38. Zarkadas CG, Yu JZ, Hamilton RI, Pattison PL, Rose NGW: **Comparison between the protein quality of northern adapted cultivars of common maize and quality protein maize.** *J Agric Food Chem* 1995, **43:**84–93.
39. Stein HH, Sève B, Fuller MF, Moughan PJ, de Lange CFM: **Invited review: amino acid bioavailability and digestibility in pig feed ingredients: terminology and application.** *J Anim Sci* 2007, **85:**172–180.
40. Stein HH, Fuller MF, Moughan PJ, Sève B, Moshentin R, Jansman AJM, Fernandez JA, de Lange CFM: **Deffinition of apparent, true and standardized ileal digestibility of amino acids in pigs.** *Livest Sci* 2007, **109:**282–285.
41. de Souza RTC, Aguilera BA, Mariscal-Landín G: **Estimation of endogenous protein and amino acid ileal losses in weaned piglets by regression analysis using diets with graded levels of casein.** *J Anim Sci Biotechnol* 2013, **4:**36.
42. Mariscal-Landín G, Sève B, Collèaux Y, LeBreton Y: **Endogenous amino nitrogen collected from pigs with end to end ileorectal anastomosis is affected by the method of estimation and altered by dietary fiber.** *J Nutr* 1995, **125:**136–146.
43. Yin Y-L, Li T-J, Huang R-L, Liu Z-Q, Kong XF, Chu W-Y, Tan B-E, Deng D, Kang P, Yin F-G: **Evaluating standardized ileal digestibility in growing pigs.** *Anim Feed Sci Technol* 2008, **140:**385–401.
44. Pedersen C, Boisen S: **Establishement of tabulated values for standardized ileal digestibility of crude protein and essential amino acids in common feedstuffs for pigs.** *Acta Agric Scand Sect A Anim Sci* 2002, **52:**121–140.
45. Urbaityte R, Mosenthin R, Eklund M, Piepho HP, Rademacher M: **Determination of standardized ileal crude protein and amino acid digestibilities in protein supplements for piglets.** *Animal* 2009, **3:**1696–1705.
46. Yin Y, Gurung NK, Jeaurond E, Sharpe P, de Lange C: **Digestible energy and amino acid contents in Canadian varieties of sorghum, pearl millet, high-oil corn, high-oil-high-protein corn and regular corn samples for growing pigs.** *Can J Anim Sci* 2002, **82:**385–391.
47. Furuya S, Kaji Y: **Additivity of the apparent and true digestible amino acid supply in barley, maize, wheat or soya bean based diets for growing pigs.** *Anim Feed Sci Technol* 1991, **32:**321–331.

Dairy sheep production research at the University of Wisconsin-Madison, USA – a review

David L Thomas[1*], Yves M Berger[2], Brett C McKusick[1,3] and Claire M Mikolayunas[1,4]

Abstract

Commercial milking of sheep is a new agricultural industry in the United States starting approximately 30 yr ago. The industry is still small, but it is growing. The majority of the sheep milk is used in the production of specialty cheeses. The United States is the major importer of sheep milk cheeses with 50 to 60% of annual world exports coming to the United States during the past 20 yr. Therefore, there is considerable growth potential for the industry in the United States. The only dairy sheep research flock in North America is located at the Spooner Agricultural Research Station of the University of Wisconsin-Madison. The research program started in 1993 and has been multifaceted; dealing with several areas important to commercial dairy sheep farmers. The East Friesian and Lacaune dairy breeds were compared and introduced to the industry through the research program. Both dairy breeds produced significantly more milk than traditional meat-wool breeds found in the U.S., but the two breeds differed in their production traits. East Friesian-cross ewes produced more lambs and slightly more milk than Lacaune-cross ewes whereas Lacaune-cross ewes produced milk with a higher percentage of fat and protein than East Friesian-cross ewes. Lactation physiology studies have shown that ewes with active corpora lutea have increased milk yields, oxytocin release during milking is required to obtain normal fat percentages in the milk, large udder cisterns of dairy ewes can allow for increased milking intervals, and short daylengths during late pregnancy results in increased milk yield. In the nutrition area, legume-grass pastures and forages with a higher percentage of legume will result in increased milk production. Grazing ewes respond to additional supplementation with increased milk yield, but it is important to match the supplement to the quality of the grazing. Ewes on high quality legume-grass pastures that are high in rumen degradable protein respond with increased milk production to supplements high in energy and/or high in rumen undegraded protein.

Keywords: Dairy sheep, East Friesian, Grazing, Lacaune, Lactation physiology, Nitrogen efficiency, RDP, RUP, Supplementation

Brief history of the Spooner Agricultural Research Station

The Spooner Agricultural Research Station is the site of the only dairy sheep research flock in North America. The station is located in northwestern Wisconsin (N 45.8, W 91.9) and is the oldest of the 12 research stations operated throughout Wisconsin by the College of Agricultural and Life Sciences of the University of Wisconsin-Madison. The station was established in 1909 with a donation of 32 hectares of land to the University of Wisconsin by the city of Spooner. The station currently occupies 157 hectares and has active research programs in field crops, pasture, horticulture, and sheep production. The environment is temperate, exhibiting four distinct seasons during the year. The average monthly high temperature occurs in July (27°C) and the average low monthly temperature occurs in January (-17°C). Average annual precipitation (rain or snow) is 76 cm and it occurs throughout the year with the greatest amount in August (106 mm) and the least amount in January (18 mm).

Sheep were added to the station in 1936 after an outbreak of brucellosis in the small dairy cow herd resulted in disposal of the cattle. The original ewes were commercial western whiteface, and these were graded-up to Shropshire. With changes in breed popularity and needs for research, the flock over time was composed of Shropshire, Suffolk, Targhee, Finnsheep-Targhee, Romanov-Targhee,

* Correspondence: dlthomas@wisc.edu
[1]University of Wisconsin-Madison, Department of Animal Sciences, Madison, Wisconsin, USA
Full list of author information is available at the end of the article

Dorset-cross, and the present dairy breeding of East Friesian (EF) and Lacaune (LA).

The first dairy sheep research program in North America was established by Dr. William Boylan at the University of Minnesota in 1984 [1], but this program ceased in about 1996 with his retirement. The dairy sheep program at the Spooner Station had its start in the summer of 1993 with the importation of two ½ EF rams from the flock of Hani Gasser, Chase, British Columbia, Canada. Construction on the sheep milking parlor started in April 1995. The milking system is a double-12, high-line, Casse system with a pit for the milkers (DeLaval®). The initial parlor had 6 milking units requiring manual cleaning. These original milking units were increased to 12 with in-place cleaning of machines. The first ewes were milked starting in April 1996, and with the collection of the first milk yield data shortly after, the dairy sheep research program began. From the years just prior to the introduction of dairy sheep through to the present time, the ewe flock has been maintained at approximately 300 breeding ewes. At the present time, the program at the Spooner Station is the only dairy sheep research program in North America.

Dairy sheep production research

Summaries of the results of some studies conducted at the station on dairy sheep production follow. The studies selected for summarization are those whose results may have the greatest opportunity for the improvement of the efficiency of dairy sheep production. In addition, several studies have been conducted in the Department of Food Science and the Center for Dairy Research, University of Wisconsin-Madison on various processing aspects of sheep milk, and these are not summarized here. Results of most of the processing studies can be found in the Journal of Dairy Science by searching for the main authors of W. Wendorff and J. Jaeggi.

The successful completion of these studies was due in large part to the dedicated staff at the Spooner Agricultural Research Station. We (D. Thomas and Y. Berger) also had the opportunity during the past several yr to work with two excellent graduate students who are co-authors of this paper, Brett McKusick and Claire Mikolayunas, who earned their M.S. and Ph.D. degrees in Animal Sciences through these projects. They not only conducted the work, but also conceived the hypotheses and designed the experiments for most of the following studies. Without their efforts, there would be much less to present in this review.

Breed comparisons

Low percentage East Friesian-crosses compared to Dorset-crosses

The first dairy sheep genetics available in North America for commercial dairy sheep production were on the farm of Hani Gasser, Chase, British Columbia, Canada who had imported frozen semen of EF rams from Switzerland and used it to inseminate his Rideau ewes. The Rideau is a composite breed developed by the Centre for Food and Animal Research (formerly Animal Research Institute), at Ottawa, Ontario, Canada [2]. The Rideau breed is estimated to be 40% Finnsheep, 20% Suffolk, 14% EF, 9% Shropshire, and 8% Dorset breeding with less than 1% additional contribution from each of the following breeds: North Country Cheviot, Leicester, Romnelet, and Corriedale. From Mr. Gasser, we purchased two 1/2 EF × 1/2 Rideau rams in 1993, one 3/4 EF × 1/4 Rideau ram in 1994, and one 7/8 EF × 1/8 Rideau ram in 1995. Three different Swiss EF rams sired the four EF-cross rams that we purchased.

Crossbred ewes of 1/2 Dorset × 1/4 (Romanov or Finnsheep) × 1/4 Targhee breeding (commercial ewes) were randomly assigned to either an EF-cross ram or polled Dorset ram in a single-sire mating pen during the late summers or autumns of the four years from 1993 to 1996. The Dorset rams were purchased from Wisconsin breeders from rams consigned to the Wisconsin Ram Test Station. Most female lambs resulting from these matings were retained as replacements and mated to either Dorset or EF-cross rams. The mating system resulted in the production of EF-cross lambs and ewes of 12.5 to 50% EF breeding and Dorset-cross lambs and ewes of 75 or 87.5% Dorset breeding and 0.0% EF breeding.

Growth data were available for 420 lambs from EF-cross sires and 216 lambs from Dorset sires on commercial ewes and an additional 546 lambs from EF-cross dams and 150 lambs from Dorset-cross dams. Reproduction data were collected from 338 matings of EF-cross ewes and 146 matings of Dorset-cross ewes. Milk production data was reported from 246 EF-cross lactations and 76 Dorset-cross lactations collected in 1996 and 1997 from one- and two-year-old ewes [3].

East Friesian-cross lambs had greater ($P < 0.05$) birth, weaning, and postweaning weights than Dorset-cross lambs. When lambing at 1 and 2 yr of age, EF-cross ewes gave birth to 0.27 more ($P < 0.05$) lambs per ewe lambing, reared 0.15 more ($P < 0.05$) lambs per ewe mated, had 33.5 more ($P < 0.05$) d in lactation, and produced 1.9 times more ($P < 0.05$) milk and more ($P < 0.05$) weight of milk fat (+2.2 kg) and milk protein (+2.2 kg) than Dorset-cross ewes (Table 1). The EF-cross ewes produced milk with a lower ($P < 0.05$) percentage fat and protein compared to Dorset-cross ewes which was expected given the negative phenotypic and genetic correlations between milk yield and percentage milk fat and protein that have been reported by several authors (e.g., [4]). The EF-cross ewes and lambs in this study were of 12.5 to 50.0% EF breeding and provided a strong endorsement for the use of dairy sheep genetics over domestic meat/

Table 1 Lactation performance[1] of young East Friesian-cross and Dorset-cross ewes (1996 and 1997)

Trait	Breeding of ewe	
	Dorset-cross[3]	East Friesian-cross[2]
Number of lactations	76	246
Lactation length, d	92.7^{a}	126.2^{b}
Milk yield, kg	56.9^{a}	109.1^{b}
Fat,%	5.5^{a}	5.0^{b}
Fat yield, kg	3.3^{a}	5.5^{b}
Protein,%	5.4^{a}	5.0^{b}
Protein yield, kg	3.2^{a}	5.4^{b}
Somatic cell count, $\log_{10}$	4.99	5.02

[1]Ewes were milked starting approximately 30 d after parturition.
[2]75% or 87.5% Dorset breeding and 0.0% East Friesian breeding.
[3]12.5% to 50% East Friesian breeding.
[a,b]Means within a row with no superscript in common are different ($P < 0.05$).

wool genetics for commercial sheep dairies in North America.

Crossbred performance of East Friesian and Lacaune breeds

As purebred rams and semen of the EF breed became available after 1995, 14 different EF rams were used in the flock. In addition, the first LA genetics in the U.S. was imported by the Spooner Station from the U.K. (semen from 3 rams) and Canada (3 rams) in 1998. Since very few, if any, additional EF and LA rams or semen from EF rams have been imported into the U.S. beyond these 14 EF and 6 LA rams, the Spooner flock is well-representative of the EF and LA genetics currently present in the U.S.

An analysis of growth, reproduction, and milk production data collected from 1999 through 2004 from lambs and ewes born between 1999 and 2004 and sired by either purebred EF or LA rams were reported [5]. Records were available on 1,749 lambs for growth, 942 exposures of 483 ewes for reproduction, and 796 lactations from 402 ewes. Lambs sired by EF rams had greater ($P < 0.05$) 30-d weights than lambs sired by LA rams but there were no differences between sire breeds for birth or 150-d weights (Table 2). Ewes sired by EF rams gave birth to 0.16 more ($P < 0.05$) lambs per ewe lambing than did ewes sired by LA rams (1.85 vs. 1.69, respectively) (Table 2). Ewes sired by EF rams produced 14.6 kg more (not statistically significant) milk per lactation than did ewes sired by LA rams. However, the greater ($P < 0.05$) percentage of both fat and protein of milk produced by LA-sired ewes compared to EF-sired ewes resulted in very similar amounts of fat and protein produced in a lactation by ewes sired by the two breeds.

A more sophisticated analysis on all records collected from 1996 through 2005 was conducted by J. Casellas (unpublished), which took into consideration the effects of ewe breed composition, proportion of retained heterosis, weaning system, age of ewe, and number of lambs born on lactation traits and the same effects, except number of lambs born, for the trait of litter size. Performance data were available on 1,068 ewes with 2,554 lactation records. Using the regression of ewe performance for each trait on proportion of EF or LA breeding, the predicted performance of purebred EF and LA ewes was estimated. Presented in Table 3 is the predicted performance of purebred EF and LA ewes at 3 yr of age

Table 2 Performance of lambs and ewes sired by purebred East Friesian (n = 14) or purebred Lacaune (n = 6) rams[1]

Trait	Breed of sire of lamb or ewe	
	East Friesian	Lacaune
Lamb growth: (n = 1,794 lambs born)		
Birth wt., kg	5.04 ± 0.09^{a}	4.64 ± 0.09^{b}
30-d wt., kg	14.3 ± 0.2^{a}	13.3 ± 0.3^{b}
150-d wt., kg	48.4 ± 1.1^{a}	48.9 ± 1.2^{a}
Ewe reproduction: (n = 942 exposures)		
Fertility (ewes lambing/ewes exposed),%	96.7 ± 1.4^{a}	94.6 ± 1.4^{a}
Prolificacy (lambs born/ewes lambing), no.	1.85 ± 0.06^{a}	1.69 ± 0.07^{b}
Ewe lactation: (n = 796 lactations)		
Lactation length, d	161.4 ± 3.8^{a}	155.2 ± 4.0^{a}
Milk yield, kg	209.4 ± 9.8^{a}	194.8 ± 11.5^{a}
Fat yield, kg	12.3 ± 0.6^{a}	12.5 ± 0.7^{b}
Fat,%	5.75 ± 0.09^{b}	6.31 ± 0.11^{a}
Protein yield, kg	10.3 ± 0.5^{a}	10.1 ± 0.6^{b}
Protein,%	4.81 ± 0.06^{b}	5.15 ± 0.06^{a}

[1]Performance data were collected from 1999 through 2004 from lambs and ewes born between 1999 and 2004.
[a,b]Means within a row with no superscript in common are different ($P < 0.05$).

Table 3 Predicted performance of pure East Friesian and Lacaune 3-year-old ewes from performance records of crossbred ewes of various percentages of East Friesian and/or Lacaune breeding[1,2]

Trait	Breed	
	East Friesian	Lacaune
Lactation length, d	188.6^{a}	180.3^{a}
Milk yield, kg	359.3^{a}	345.1^{a}
Fat yield, kg	20.9^{a}	22.1^{a}
Fat,%	6.3^{a}	6.5^{b}
Protein yield, kg	18.0^{a}	18.2^{a}
Protein,%	5.2^{a}	5.3^{b}
Litter size, no.	1.97^{a}	1.84^{b}

[1]Previously unpublished data.
[2]Performance records collected from1996 through 2005 on 1,068 individual ewes with 2,554 lactation records.
[a,b]Means within a row with no superscript in common are different ($P < 0.05$).

when they are milked starting from 1 or 2 d after lambing. The conclusions from Tables 2 and 3 are the same, i.e., LA breeding results in milk with a higher ($P < 0.05$) percentage of fat and protein, but the yield of milk, fat and protein is similar between the two breeds, and EF breeding results in more ($P < 0.05$) lambs born per ewe lambing than does LA breeding.

Since there have been no new importations of LA genetics and only limited new importations of EF genetics into the U.S. since the late 1990's and no national or regional programs for genetic improvement of dairy sheep in the U.S., the comparative performance of EF and LA genetics obtained from these studies are expected to be accurate predictions of breed differences that will be observed by commercial dairy sheep producers in the U.S. However, our estimates of breed differences may not be accurate for the current world population of EF and LA sheep. First, our samples of rams from each of the breeds, especially from the LA breed, were small and may not have been good representations of the EF and LA breeds found in Europe in the late 1990's. Second, rates of genetic improvement in these two breeds over the past 15 to 20 yr are probably different. Evidence suggests that the LA breed in France has most likely made greater genetic improvement in lactation traits than has the EF breed in Germany. From 2005 to 2011, an average of 68 flocks and 855 ewes of the EF breed were milk recorded each year in Germany whereas an average of 384 flocks and 172,946 ewes of the LA breed were milk recorded each year over the same period in France [6]. The French LA genetic improvement program is the most sophisticated and effective program among dairy sheep breeds in the world. In 2006, it was estimated that the recorded LA flocks in France had a rate of genetic improvement of +6 liters of milk per ewe per year [7]. While there are no similar estimates of the genetic trend for milk yield in EF sheep in Germany, it is most likely very much lower than for the LA in France due to a less aggressive genetic improvement program for EF compared to LA. Therefore, a sampling of European EF and LA rams today might result in different conclusions than those arrived at from our samples from the late 1990's.

Lamb survival of East Friesian

There are reports in the literature of poor survival of lambs of high percentage EF breeding in France [8] and Greece [9] compared to other breeds with the primary cause of the increased mortality due to respiratory disease. We observed a similar effect in the early years of our work with EF. As we graded-up our non-dairy sheep to higher percentages of EF breeding by top-crossing with EF rams, our lamb mortality increased. Presented in Table 4 is survival of 483 lambs of different proportions of EF breeding born in 1999 (an early year in our grading-up program from non-dairy to dairy sheep) that we reported in 1999 [10] and 2000 [11]. Lambs over 50% EF breeding had significantly lower ($P < 0.05$) survival rates than lambs with 50% or lower EF breeding during all time periods, and lambs of 50% EF breeding had numerically lower, but not significantly lower, survival rates than lambs with less than 50% EF breeding.

Table 4 Least squares means for lamb survival by percentage of East Friesian breeding of the lamb (1999 lamb crop)

Lamb's % EF breeding	No. lambs born alive	Survival rate,% Birth to weaning	Weaning to 7/1/99	Birth to 7/1/99
0	56	96.4 ± 3.5^{a}	100.0 ± 2.9^{a}	96.4 ± 4.2^{a}
>0 to <25	146	96.6 ± 2.2^{a}	99.3 ± 1.8^{a}	95.9 ± 2.6^{a}
≥25 to <50	70	97.1 ± 3.1^{a}	98.5 ± 2.6^{a}	95.7 ± 3.8^{a}
50	60	95.0 ± 3.4^{a}	$93.0 \pm 2.8^{a,b}$	88.3 ± 4.1^{a}
>50	151	83.4 ± 2.1^{b}	86.5 ± 1.9^{b}	72.2 ± 2.6^{b}

[a,b]Within a column, means with a different superscript are different ($P < .05$).

However, a more detailed analysis of lamb survival in this flock for 7,990 lambs born from 1996 through 2011 was reported in an abstract in 2012 [12]. The analysis determined direct and maternal breed effects, individual and maternal heterosis effects, and the effects of the non-genetic factors of lamb type of birth, sex of lamb, month of birth, and age of dam on lamb survival over several age intervals from birth up to 120 d of age. The EF maternal breed effect was significantly negative for lamb survival whereas the EF direct breed effect was generally not significantly different from zero. These results suggest that it is the EF breeding of the ewe and not the EF breeding of the lamb that is responsible for the increased lamb mortality reported in EF breeding programs. It is obvious that more studies are needed to determine the biology of lamb survival in EF populations.

Weaning systems

In 1998, 99 EF-cross ewes in second and third parity were utilized to compare milk production and lamb growth under three weaning systems [13]. One group of ewes (DY1, n = 31) was weaned from their lambs between 24 and 36 h postpartum and machine milked twice daily for the entire lactation. Their lambs were raised on milk replacer until approximately 30 d of age. Another group of ewes (MIX, n = 35) were separated from their lambs at 1700 h. each day and milked once daily each morning at 0600 h from 24 to 36 h after parturition. After the morning milking, ewes were returned to their lambs. MIX ewes were milked twice daily following permanent weaning of their lambs at approximately 30 d of age. The final group of ewes (DY30, n = 33) were left to raise their lambs and not initially milked. Approximately 30 d postpartum, ewes were weaned from their lambs and milked twice daily for the remainder of their lactation. Milk yields of ewes on the three systems were milk obtained during machine milking only and did not estimate milk consumed by the lambs.

Milk yield differed ($P < 0.05$) among the three weaning systems (DY1 = 260 kg, MIX = 236 kg, and DY30 = 172 kg) (Table 5). Lamb weights at 30 d of age when they were weaned from milk replacer or their dams were not significantly different among the weaning treatment groups (averaged 15 kg across treatments). However, lamb weights at 120 d of age tended to be lighter ($P < 0.10$) for lambs from DY1 ewes (44 kg) than for lambs from DY30 ewes (47 kg). Lambs from MIX ewes had intermediate 120 d weights (46 kg). The lamb growth data suggest that artificial rearing of lambs has a slight negative effect on lamb postweaning gain relative to the effects of either limited suckling or *ad libitum* suckling preweaning. Relative to the DY30 system, income from milk and lamb over additional expenses was + $30.66 per ewe for the MIX system and + $5.05 per ewe for the DY1 system. Of course, these economic returns depend on the value of milk and lambs and their relative values, which do change over time. Any one of the three weaning systems could be the most profitable under different sets of prices for milk and lambs.

The economic returns assumed no price differentials for milk composition. Milk from MIX ewes had a lower ($P < 0.05$) fat percentage (4.53%) than milk from either DY1 (5.06%) or DY30 (4.81%) ewes. The lower milk fat percentage of MIX ewes was most dramatic during the first 30 d when they were suckling their lambs during the day. During this period, MIX ewes had a milk fat percentage of 2.80% while DY1 ewes at the same stage of lactation had a milk fat percentage of 4.82%. If price discounts were in place for low-fat milk, the MIX system would have less of an economic advantage than projected in this study. Even so, the MIX system is attractive over the other two systems because the ewes raise their lambs and still produce 85% as much milk as DY1 ewes.

Subsequently, a more detailed study conducted on the UW-Madison campus determined that the low milk fat from MIX ewes while they are nursing their lambs is due to failure of milk ejection from the udder alveoli due to failure of oxytocin release in these ewes during milking [14]. Oxytocin is released as a result of teat and udder stimulation, usually at the time of suckling. Oxytocin is an integral part of milk ejection (the contraction of the alveoli within the udder that causes secreted milk to flow down a system of ducts and canals into the

Table 5 Ewe lactation, lamb growth, and economics of three weaning systems[1]

Trait	Weaning system		
	DY1	MIX	DY30
Ewe lactation traits (n):	(31)	(35)	(33)
Lactation length, d	183.4 ± 5.4	179.2 ± 5.1	182.9 ± 5.5
Machine milking period, d	182.4 ± 5.4[a]	178.2 ± 5.1[a]	152.3 ± 5.5[b]
Commercial milk yield, kg	260.1 ± 9.7[a]	235.8 ± 9.1[b]	171.7 ± 9.9[c]
Fat yield, kg	13.2 ± 0.6[a]	10.9 ± 0.5[b]	8.4 ± 0.6[c]
Fat,%	5.1 ± 0.1[a]	4.5 ± 0.1[b]	4.8 ± 0.1[a,b]
30-d fat,%	4.8 ± 0.2[a]	2.8 ± 0.2[b]	-
Protein yield, kg	13.7 ± 0.5	12.1 ± 0.5	9.0 ± 0.5
Protein,%	5.3 ± 0.1	5.1 ± 0.1	5.2 ± 0.1
Lamb growth traits (n at 120 d):	(64)	(71)	(73)
30-d weight, kg	15.4 ± 0.4	14.5 ± 0.6	15.0 ± 0.5
120-d weight, kg	43.7 ± 1.2[f]	45.9 ± 1.8[e]	47.3 ± 1.6[d]
Economics:			
Total lamb & milk receipts[2], $	506.52 ± 18.07[a]	458.23 ± 17.05[b]	415.25 ± 18.53[b]
Additional expenses[3], $	87.16 ± 2.98[a]	14.40 ± 3.04[b]	-
Receipts – additional expenses, $	420.86 ± 16.87	446.47 ± 15.91	415.81 ± 17.30

[1]Study conducted in 1998.
[2]Total lamb weight at 120 d of age per ewe and commercial milk yield per ewe were valued at $1.87 and $1.32/kg, respectively.
[3]Additional labor and supply expenses per ewe relative to the DY30 system during the first 30 d of lactation.
[a,b,c]Means within a row with no superscript in common are different ($P < 0.05$).
[d,e,f]Means within a row with no superscript in common are different ($P < 0.10$).

storage part of the udder known as the cistern). During machine milking, if there is no release of oxytocin, secreted milk remains in the alveoli along with large quantities of milk fat. The MIX ewes experienced impairment of oxytocin release and the milk ejection reflex because they knew that after milking they would be reunited with their lambs. The milking machine captured their cisternal milk but not their alveolar milk where most of the fat is found.

Fat supplementation

Previous studies with ewes had shown an increased milk fat content when ewes were fed calcium salts of fatty acids (CSFA) during lactation [15-18]. Megalac Rumen Bypass Fat (Church and Dwight Co., Inc.), a CSFA, was added to the diets of EF-cross dairy ewes in early lactation in a study we conducted in 1999 to determine if the treatment could counteract the negative effects of the MIX weaning system on milk fat content [19]. The CSFA was mixed in a diet of whole shelled corn and a protein pellet and fed in the milking parlor to provide 0.10 kg of CSFA per ewe per day.

Ewes (n = 274) lambed over a 6-wk period starting on February 10, 1999 and were randomly allocated to a DY1 or MIX system as they lambed. The CSFA supplementation started on March 3, 1999 and ran for 8 wk. During the first and third 2-wk periods, all ewes received the unsupplemented diet, and during the second and fourth 2-wk periods, all ewes received the CSFA supplemented diet.

The CSFA supplementation had no effect on milk yield but tended to depress milk protein percentage in both DY1 and MIX ewes. The CSFA supplementation resulted in a large increase ($P < 0.05$) in milk fat percentage of approximately +1.19 percentage units in DY1 ewes but had no effect on milk fat percentage of MIX ewes during the milking-suckling period. Therefore, fat supplementation may be a method to increase fat percentage in ewes that have weaned their lambs, but it is not a solution to the low fat percentage of milk from ewes that are still suckling their lambs. Milk fat synthesis was probably not impaired in CSFA-supplemented MIX ewes, but the milk fat was retained within the udder until it was removed by the suckling lamb.

Milking intervals

Three-times-a-day milking

During 2000, 125 multiparous EF crossbred ewes were utilized to compare traditional twice-a-d milking (2×, n = 72) with three-times-a-d milking (3×, n = 53) during the first 30 d of lactation [20]. The 2× ewes were milked at 0630h and 1730h, and the 3× ewes were milked at 0600h, noon, and 1800h each day. After d 30 of lactation, all ewes were milked twice-a-d. All lambs were weaned from their dams within 24 h after parturition, and ewes were immediately assigned to a milking treatment. During the 30-d treatment period, 3× ewes produced a total of 12.6 kg more (+15.2%, $P < 0.05$) milk than 2× ewes (95.2 versus 82.6 kg, respectively). During the 30-day treatment period, the 3× ewes had increased net income of $10.00 per ewe over the 2× ewes. There were no carry-over effects of the treatments in later lactation with 3× and 2× ewes having the same milk yield in wk 7 after both groups had been on 2-times-a-d milking starting in wk 5.

16-Hour milking interval

A trial was conducted in 2001 to determine if the milking interval could be extended from 12 to 16 h starting in mid-lactation without a significant drop in milk yield [21]. Forty-eight third lactation EF crossbred ewes were utilized. Twenty-four ewes were kept on the 12 h milking interval (12H, milked daily at 0600h and 1800h) and 24 ewes were switched from the 12H interval on approximately d 90 of lactation to a 16 h milking interval (16H, milked at 0600h and 2200h one day and at 1400h the following day and then repeating). Lactation performance was measured through d 180 of lactation.

During the 90-d treatment period, 16H ewes produced about 28% more ($P < 0.05$) milk at each 0600h milking than 12H ewes, but there was no difference between treatments in the total amount of milk produced during the entire treatment period (Table 6). The percentage of fat and protein and somatic cell count were not different between the two treatments. From mid- to late lactation, it appears that the number of milkings can be reduced by 25% without a decrease in milk production. This is possible because it has been shown that a larger proportion of the milk yield of dairy goats and dairy sheep is cisternal milk [22] compared to the milk yield of cows [23], i.e, small dairy ruminants have a greater capacity to store milk between milkings than do cows.

Table 6 Lactation performance of ewes[a] milked at 12 or 16 hour intervals from day 90 to 180 of lactation

	Milking interval	
Trait	**12 hour**	**16 hour**
Total number of milkings	180	135
6 a.m. milk yield, kg	0.65 ± 0.03[c]	0.83 ± 0.03[b]
Average 24-hour milk yield, kg	1.34 ± 0.06	1.35 ± 0.06
Total milk yield, kg	119.1 ± 5.3	118.0 ± 5.3
Total parlor time for 24 ewes, h	38.1	27.9

[a]Twenty-four third lactation East Friesian crossbred ewes were used on each treatment.
[b,c]Means within a row with no superscript in common are different ($P < 0.05$). No test of statistical significance was possible for total number of milkings or total parlor time.

Machine stripping

Due to the large cisternal storage capacity and non-vertical teat placement in most dairy ewes, machine stripping is commonly performed to remove milk not obtained by the machine. However, stripping requires individual manual intervention, lengthens the milking routine, and could inadvertently lead to overmilking of other ewes in the parlor [24]. A study was conducted in 2000 to estimate the effect of omission of machine stripping on milk production and parlor throughput of dairy ewes [25].

Forty-eight multiparous ewes of 50 to 75% EF breeding that had been machine milked and stripped twice daily from d 0 to 79 post-partum, were randomly assigned to two stripping treatments for the remainder of lactation: normal stripping (S, n = 24), or no stripping (NS, n = 24). Ewes were milked in a 2 × 12 high-line Casse system milking parlor at 0600h and 1800h each d.

Ewes that were not stripped yielded 14% less ($P < 0.05$) commercial milk during the experiment (NS = 122.7 kg, S = 105.6 kg), but had a similar lactation length (104 d), milk composition, and somatic cell count compared to S ewes. Average machine-on time for S ewes was 10.4 seconds per ewe longer ($P < 0.10$) than for NS ewes because of stripping, which may have resulted in overmilking of some ewes in the S group.

A milking simulation in a double-12 parlor with one or two milkers and stripping or no stripping was conducted. With one milker, elimination of stripping increased the number of ewes milked per hour by 49% (from 103 to 153 ewes per h), and the number of ewes overmilked per side decreased from 11 out of 12 to 0 out of 12. With two milkers, elimination of stripping increased the number of ewes milked per hour by 20% (from 138 to 166 ewes per h), and the number of ewes overmilked per side decreased from 4 out of 12 to 0 out of 12.

These results collectively indicate that elimination of machine stripping will reduce milk yield per ewe, but the loss in milk yield may be somewhat or completely compensated for by increased parlor throughput and the udder health advantages resulting from not overmilking ewes.

Effect of number of corpora lutea on milk yield

The effects of corpora lutea on milk production were examined in 24 second lactation EF crossbred ewes in a study conducted in 1999 [26]. Ewes were synchronized using intravaginal progesterone (controlled intravaginal drug-releasing (CIDR®, Zoetis United States) device), PGF2α, and gonadotropins. After ovulation, corpora lutea (CL) were counted via laparoscopy on d 4 and 11. On d 5, ewes received either saline (CLYES, n = 12) or PGF2α (CLNO, n = 12) to allow CL persistence (2.4 ± 0.3 CL on d 11) or regression (0 CL on d 11), respectively. Each ewe received two CIDR from d 5 to 18 to maintain high concentrations of plasma progesterone (P4) and to suppress estradiol (E2). Each ewe received PGF2α on d 18 to regress all CL. Data were collected during three periods (pre-treatment: d 0 to 5; treatment: d 6 to 18; post-treatment: d 19 to 25). Milk yield and milking time were recorded daily, milk samples were obtained for analyses of fat and protein, and blood samples were collected for P4 and E2 immunoassay.

During treatment, CLYES ewes had higher ($P < 0.05$) milk yield (1.56 vs. 1.44 kg/d), milk fat (92.2 vs. 81.1 g/d), and milk protein (83.7 vs. 77.5 g/d) compared with CLNO ewes, respectively. Differences were maintained post-treatment, despite luteolysis in CLYES ewes. Serum P4 concentrations were greater ($P < 0.05$) during the treatment period for CLYES compared to CLNO ewes (5.3 vs. 2.9 ng/mL, respectively). Estradiol concentrations did not differ between treatments and were low after d 5.

While this study showed an increase in milk yield from ewes with active CL, the biological mechanism was not definitively determined. The increase in milk yield was not due to estradiol, but it may have been a direct effect of the increased P4 concentrations in the CLYES ewes. However, P4 concentrations were not related to milk yield in an older study [27]. While oxytocin levels were not measured in this study, others have reported higher oxytocin levels in ewes with CL compared to ewes without CL [28], and the higher levels of oxytocin may result in greater transfer of milk from the alveoli to the cistern between milkings in CLYES ewes compared to CLNO, resulting in their greater milk yield.

Prepartum photoperiod and milk production

Previous studies have shown that long photoperiods during established lactation increase milk production in dairy cattle [29] and dairy sheep [30], and later studies in dairy cattle [31] and dairy goats [32] suggested that short photoperiods prepartum increased milk yield in the subsequent lactation. The proposed mechanism of function was the level and role of circulating prolactin in mammary development. We conducted a study in 2005-2006 to evaluate the effect of prepartum photoperiod on milk production, milk composition, and prolactin concentration of dairy ewes [33].

Twenty two multiparous crossbred dairy ewes of EF and LA breeding were exposed to short day prepartum photoperiod (SDPP; 8 h of light: 16 h of dark) or long day prepartum photoperiod (LDPP; 16 h of light: 8 h of dark) starting on December 5, 2005 for at least 6 wk prepartum in adjacent environmentally-controlled rooms. Blood samples were collected from each ewe twice weekly during the prepartum period, at the time of lambing, and 1 wk after lambing and analyzed for prolactin concentrations. After lambing, lambs were removed and raised on milk replacer, and all ewes were relocated to a common room in the same facility with 12 h of light: 12 h of dark and milked twice per day in this facility for 53 ± 3 d (trial period). On April 6, 2006 all ewes were returned to the main flock of dairy ewes and maintained under ambient light (post-trial period) under commercial management conditions. The total milking period (trial period + post-trial period) was 180 ± 6 d. Table 7 presents the results from the trial.

During the first 8 wk of lactation, SDPP ewes tended to produce more ($P = 0.053$) milk than LDPP ewes (2.43 vs. 2.29 kg/d, respectively), and the milk of SDPP ewes had a greater ($P < 0.01$) fat percentage than that of LDPP ewes (6.04 vs. 5.51%, respectively). For the entire lactation period of 180 d, SDPP ewes produced more ($P < 0.05$) test day milk than LDPP ewes (1.76 vs. 1.60 ± 0.05 kg/d, respectively), but there were no differences in milk fat or protein percentages.

Ewes in both treatments experienced a prolactin (PRL) surge at lambing (Figure 1), but SDPP ewes had lower ($P < 0.05$) circulating prolactin concentration than LDPP ewes from 4 to 0.5 wk before lambing (14.7 vs. 51.3 ± 4.2 mg/dL, respectively). Similar results have been reported in dairy cows and dairy goats, and some of these studies have conducted more extensive hormonal studies. A possible cascade of events to describe the effect of prepartum photoperiod on milk production may be as follows: 1) short photoperiod in prepartum ruminants results in decreased circulating PRL concentrations; 2) decreased PRL stimulates the increased expression of mRNA for PRL receptors, resulting in a greater number of PRL receptors on mammary secretory epithelial cells; and 3) the natural increase in circulating PRL at parturition stimulates more extensive differentiation and commitment of mammary secretory epithelial cells to produce lactose, thereby increasing milk production in SDPP-treated animals.

These data suggest that decreased prepartum photoperiod may be important for increasing milk production in dairy ewes and may provide a management strategy for dairy sheep producers to increase milk yield. Ewes in late gestation in early winter when day length is short may be expected to produce more milk than ewes in late gestation in late winter or spring when daylength is increasing.

Grazing and supplementation on pasture

Pasture compared to drylot for lactating ewes

In 1998, 97 EF crossbred ewes had been maintained in drylot from early to mid-lactation where they received grain twice per day in the parlor at milking and alfalfa hay during the d in drylot. From mid-lactation to the end of lactation, 48 ewes remained in the drylot and the remaining 49 ewes were grazed during the day on a kura clover-orchard grass pasture. The pastured ewes had 10.5% greater ($P < 0.05$) lactation milk yields than the ewes in drylot (184 versus 167 kg) (unpublished data). As a result of this trial, grazing lactating ewes during the grass-growing season has become a general management practice at the Spooner Agricultural Research Station.

Supplementation on pasture

Trials were conducted in 2005 and 2006 to determine the efficacy of supplementation of lactating ewes while grazing high quality kura clover-orchard grass pastures [34].

In trial 1 conducted in 2005, 56 three-yr-old grazing crossbred dairy ewes of EF and LA breeding in early (21 d in milk, n = 10) or late (136 d in milk, n = 46) lactation were fed 0 (n = 28) or 0.82 (n = 28) kg DM/d per ewe of supplement (16.5% crude protein mixture of corn and a soybean meal-based high-protein pellet) in a 2 × 2 factorial arrangement of treatments. Individual ewe test d milk yield was collected every 14 d for 84 d of the grazing season. Supplementation had similar effects in both early and late lactation ewes. Supplemented ewes had higher ($P < 0.01$) milk production (1.59 vs. 1.36 kg/d, respectively), lower ($P < 0.10$) milk fat percentage (5.75 vs. 6.00%, respectively), lower ($P < 0.01$) milk protein percentage (4.84 vs. 5.04%, respectively) but more ($P < 0.01$) fat and protein corrected milk production (1.09 vs. 0.95 kg/d, respectively) than unsupplemented ewes (Table 8).

Protein has a high nitrogen content. If protein intake is in excess of the needs of the rumen microflora, high levels of protein nitrogen in the form of urea are excreted in the urine, feces, and milk. Therefore, milk urea nitrogen (MUN) is a good indicator of the efficiency of protein utilization by the microflora of the rumen. MUN levels were similar between supplemented and unsupplemented ewes (average of 25.0 mg/dL) but were above recommended levels for dairy ewes (14 to 22 mg/dL,

Table 7 Average daily milk production and milk composition of multiparous crossbred dairy ewes exposed to a short or long photoperiod during the prepartum period

	Trial period (53 d)			Trial + post-trial period (180 d)		
Trait	SDPP[1]	LDPP[2]	*P*	SDPP[1]	LDPP[2]	*P*
Milk, kg/d	2.43 ± 0.051	2.29 ± 0.051	0.053	1.76 ± 0.051	1.60 ± 0.050	0.031
Fat,%	6.04 ± 0.102	5.51 ± 0.104	0.000	6.28 ± 0.110	6.48 ± 0.107	0.216
Protein,%	4.61 ± 0.057	4.54 ± 0.057	0.448	5.23 ± 0.064	5.12 ± 0.062	0.241

[1]Short day prepartum photoperiod (8 h light: 16 h dark), n = 11.
[2]Long day prepartum photoperiod (16 h light: 8 h dark), n = 11.

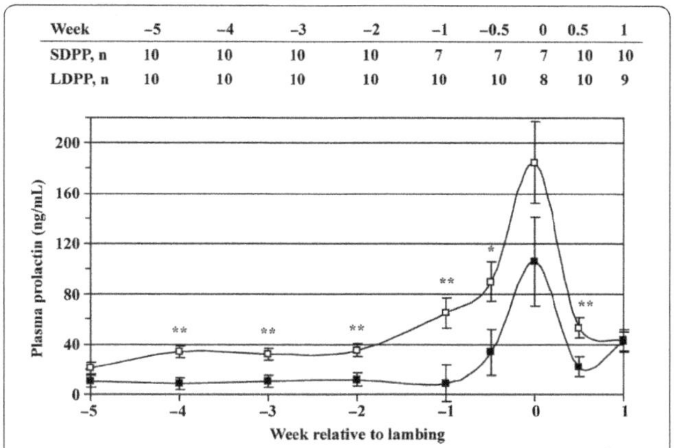

Figure 1 Mean concentration of plasma prolactin of ewes exposed to short day photoperiod (SDPP; 8 h of light: 16 h of dark) or long day prepartum photoperiod (LDPP; 16 h of light: 8 h of dark). The table presents the number of ewes per treatment (n) on each week relative to lambing. Error bars indicate SEM. *Least squares means within a test day are different ($P < 0.05$). **Least squares means within a test day are different ($P < 0.01$).

[35]), indicating excess protein (N) or insufficient energy intake. It appeared that the ewes benefited from the increased energy in the supplement. However, the protein in the pasture, which varied from 16 to 30% during the grazing season, may have been adequate for the level of milk production of these ewes, and the additional protein in the 16.5% crude protein supplement may not have been necessary. This observation resulted in a second supplementation trial the following grazing season.

In trial 2 in 2006, 96 two-, three-, and four-year-old grazing crossbred dairy ewes of EF and LA breeding in mid-lactation (112 d in milk) were randomly assigned to 4 treatments of 0.00, 0.41, 0.82, or 1.24 kg DM/d per ewe of whole shelled corn to determine if a high energy supplement, but of lower protein content than the supplement used in trial 1, may provide a better complement to the high protein content of the pasture [34].

Table 8 Lactation performance of supplemented[1] or unsupplemented crossbred dairy ewes[2] grazing a legume-grass pasture

Trait	Unsupplemented	Supplemented
Test day milk yield, kg	1.36 ± 0.04^{b}	1.59 ± 0.04^{a}
Test day FPCM[3] yield, kg	0.95 ± 0.04^{b}	1.09 ± 0.04^{a}
Milk fat,%	6.00 ± 0.09^{c}	5.75 ± 0.09^{d}
Milk protein,%	5.04 ± 0.04^{a}	4.84 ± 0.04^{b}
Milk urea nitrogen, mg/dL	24.9 ± 1.58	25.1 ± 1.51

[1]0.82 kg DM/d of a 16.5% crude protein mixture of corn and a soybean meal-based high-protein pellet.
[2]Multiparous crossbred ewes of East Friesian and Lacaune breeding, n = 28 ewes per treatment.
[3]6.5% fat-corrected and 5.8% protein-corrected milk yield.
[a,b]Means within a row with no superscript in common are different ($P < 0.01$).
[c,d]Means within a row with no superscript in common are different ($P < 0.10$).

There was a linear increase ($P < 0.01$) in test-day milk yield and test day fat- and protein-corrected milk yield, a linear decrease ($P < 0.01$) in milk fat percentage, and no significant change in milk protein percentage with increasing amounts of corn supplementation (Table 9). Even though the quadratic effects of supplementation levels on milk yields were not statistically significant, the supplementation means indicate that there was no advantage of supplementation above 0.82 kg DM/ewe/d. MUN levels for all four groups were within the range suggested for dairy sheep and decreased ($P < 0.001$) with increasing amounts of corn supplementation. This suggested that protein levels in the high quality legume-grass pasture were adequate for milk production in these ewes and utilization of pasture protein improved with increasing dietary energy intake from whole shelled corn.

Level of protein and rumen undegraded (by-pass) protein

Dietary protein is provided to ruminants in the form of rumen degraded protein (RDP) and rumen undegraded (by-pass) protein (RUP). RDP is utilized by the microflora of the rumen, and the microbial protein is then utilized by the animal. RDP fed in excess of the needs of the rumen microflora is excreted in the feces or as urea in the urine or milk. Rumen undegraded protein cannot be utilized by the rumen microflora but is utilized directly by the animal. High-producing ruminants, like lactating dairy ewes, may increase their productivity if RUP is added to rations already adequate in RDP. A study was conducted in 2008 to test this hypothesis [36].

Three diets were formulated to provide similar energy concentrations and varying concentrations of RDP and RUP: 12% RDP and 4% RUP (12-4) included basal levels of RDP and RUP, 12% RDP and 6% RUP (12-6) included additional RUP, and 14% RDP and 4% RUP (14-4) included additional RDP. Diets were composed of alfalfa-timothy cubes, whole and ground corn, whole oats,

Table 9 Lactation performance of crossbred dairy ewes[1] unsupplemented or supplemented with corn during the grazing season

Trait	Whole shelled corn supplementation, kg DM/ewe/d				SEM	Linear contrast, *P*-value
	0.00	0.41	0.82	1.24		
Test day milk yield, kg	1.30^{a}	1.32^{a}	1.41^{b}	1.44^{b}	0.03	0.001
Test day FPCM[2] yield, kg	1.26^{ab}	1.25^{a}	1.35^{c}	1.33^{bc}	0.03	0.006
Milk fat,%	6.26^{b}	6.40^{b}	6.09^{b}	5.89^{a}	0.11	0.001
Milk protein,%	5.29	5.41	5.37	5.39	0.04	0.093
Milk urea nitrogen, mg/dL	18.9^{a}	17.1^{b}	13.6^{c}	13.6^{c}	0.3	0.001

[1]Multiparous crossbred ewes of East Friesian and Lacaune breeding, n = 24 ewes per treatment.
[2]6.5% fat-corrected and 5.8% protein-corrected milk yield.
[a,b,c]Means within a row with no superscript in common are different ($P < 0.05$).

Table 10 Lactation performance of dairy ewes[1] fed diets with varying levels of rumen degraded (RDP) and undegraded (RUP) protein

Trait	% RDP:%RUP 12:6	14:4	12:4	SEM
Test day milk yield, kg	2.05[a]	1.80[b]	1.79[b]	0.07
Milk fat,%	6.13	6.37	6.18	0.25
Milk protein,%	4.74	4.95	4.80	0.14
Milk urea nitrogen, mg/dL	26.3[a,b]	27.4[a]	23.4[b]	1.4

[1]Third lactation crossbred ewes of East Friesian and Lacaune breeding, n = 6 pens of 3 ewes each per treatment in a Latin square design.
[a,b]Means within a row with no superscript in common are different ($P < 0.05$).

dehulled soybean meal, and expeller soybean meal (SoyPlus, West Central, Ralston, IA). Eighteen third lactation crossbred dairy ewes of EF and LA breeding in mid-lactation were divided by milk yield (low and high) into 2 blocks of 9 ewes each and were randomly assigned within block (low and high) to 3 pens of 3 ewes each. Dietary treatments were arranged in a 3 × 3 Latin square within each block and applied to pens for 14-d periods. Milk yield and composition was determined during the last 4 d of each treatment period.

There was no effect of dietary treatment on dry matter intake. The 18% crude protein diet with the high level of RDP (14-4) resulted in no more milk production than obtained with the 16% crude protein diet with a lower level of RDP and the same level of RUP (12-4). However, the 18% crude protein diet with the high level of RUP (12-6) increased ($P < 0.01$) milk yield over both the 14-4 and 12-4 diets (Table 10). This is strong evidence for the inclusion of RUP in diets of lactating ewes.

Milk urea N concentration was greater ($P < 0.05$) in the 14–4 diet and tended to be greater ($P < 0.10$) in the 12–6 diet compared with the 12–4 diet, indicating that the excretion of urea N in this study was more closely related to dietary crude protein concentration than to protein degradability (Table 10).

Legume content of forage

Our previous trials with dairy ewes fed stored feeds indicated a positive effect of RUP supplementation on milk yield. However, dairy sheep production in the United States is primarily based on grazing mixed grass-legume pastures, which contain a high proportion of RDP. Two trials were conducted in 2008 and 2009 to evaluate the effects of RUP protein supplementation and varying levels of the legume, alfalfa, in legume-grass forages on lactation performance [37].

In a cut-and-carry trial conducted in 2008, 16 multiparous crossbred dairy ewes of EF and LA breeding in mid-lactation were randomly assigned one of two protein supplementation treatments, receiving either 0.0 or 300 g of a high-RUP protein supplement (Soy Pass, LignoTech USA Inc., Rothschild, WI) per d. Within supplementation treatment, ewes were full-fed freshly cut forage of varying percentages of alfalfa:orchardgrass dry matter: 0:100, 25:75, 50:50, or 75:25. Supplementation with a high-RUP source tended to increase ($P < 0.10$) milk yield by 9%. Milk yield ($P < 0.10$), milk protein yield ($P < 0.05$), and MUN ($P < 0.05$) increased with increased percentage of alfalfa (Table 11).

In a grazing trial conducted in 2009, 12 multiparous crossbred dairy ewes of EF and LA breeding in mid-

Table 11 Lactation performance of dairy ewes[1] supplemented with rumen undegraded protein (RUP) and fed or grazing forage of varying proportions of alfalfa

Trait	RUP Supplement			% alfalfa in forage				
	No	Yes	SEM	0	25	50	75	SEM
Cut-and-carry trial:[2]								
Milk yield, kg/d	1.79[f]	1.95[e]	0.06	1.74[g]	1.85[f]	1.94[e]	1.95[e]	0.06
Fat yield, g/d	122	123	3	117	122	124	127	4
Protein yield, g/d	90	95	2	85[b]	90[b]	96[a]	98[a]	3
Milk urea nitrogen, mg/dL	12.3[b]	15.1[a]	0.8	10.9[d]	12.7[c]	14.3[b]	16.8[a]	0.7
Grazing trial:[3]								
Milk yield, kg/d	1.65	1.82	0.16	1.55[g]	1.78[f]	1.87[e]		0.11
Fat yield, g/d	105	115	11	102	113	116		8
Protein yield, g/d	84	94	7	78[a]	90[a,b]	98[b]		5
Milk urea nitrogen, mg/dL	18.2	19.8	1.0	15.0[b]	19.8[a]	22.1[a]		0.8

[1]Multiparous crossbred ewes of East Friesian and Lacaune breeding.
[2]n = 16 ewes; 8 pens of 2 ewes each. 4 pens randomly assigned to each supplement treatment. Within supplementation treatment, pens were randomly assigned to 1 of 4 forage treatments, which were applied in a 4 × 4 Latin square design for 10-d periods.
[3]n = 12 ewes; 3 groups of 4 ewes each. 2 ewes in each group randomly assigned to each supplement treatment. Grazing treatments were arranged in a 3 × 3 Latin square design and randomly applied to groups for 10-d periods.
[a,b,c,d]Means within a row and treatment with no superscript in common are different ($P < 0.05$).
[e,f,g]Means within a row and treatment with no superscript in common are different ($P < 0.10$).

lactation were randomly assigned to receive either 0.0 or 300 g of a high-RUP protein supplement (SoyPlus, West Central Cooperative, Ralston, IA) per day. Within supplementation treatments, ewes grazed paddocks that contained the following percentages of surface area of pure stands of alfalfa:orchardgrass: 0:100, 25:75, or 50:50. Milk yield ($P < 0.10$), milk protein yield ($P < 0.05$), and MUN ($P < 0.05$) increased with increased percentage of alfalfa in the paddock (Table 11).

In conclusion, supplementing with high-RUP protein tended to increase milk yield, and increasing the proportion of alfalfa in the diet increased dry matter intake, milk yield, and protein yield of lactating dairy ewes fed or grazing fresh forage.

Conclusions

The Spooner Agricultural Research Station of the University of Wisconsin-Madison has made significant research contributions to the dairy sheep industry of North America since its establishment in 1993 in several areas; especially in new breed introduction and evaluation, lactational physiology, nutrition of the lactating ewe, and general dairy sheep management. Both the EF and LA dairy breeds were introduced to the industry and evaluated by the Spooner Station. Both breeds produced significantly more milk than traditional meat-wool breeds found in the U.S. East Friesian ewes produced more lambs and slightly more milk than LA ewes whereas LA ewes produced milk with a higher percentage of fat and protein than EF ewes. Studies in lactational physiology showed that ewes with active corpora lutea had increased milk yields, oxytocin release during milking was required to obtain normal fat percentages in milk, large udder cisterns of dairy ewes allowed for increased milking intervals, and short daylengths during late pregnancy resulted in increased milk yield. In the nutrition area, legume-grass pastures and forages with a higher percentage of legumes resulted in increased milk production. Grazing ewes responded to additional supplementation with increased milk yield, but it was important to match the supplement to the quality of the grazing. Ewes on high quality legume-grass pastures that are high in rumen degradable protein responded with increased milk production to supplements high in energy and/or high in rumen undegraded protein.

Competing interest

The authors declare that they have no competing interests.

Authors' contribution

All experiments were conducted under the direction of DLT and YMB. Specific experiments were designed and conducted by either BCM or CMM as part of their M.S. and Ph.D programs, and these experiments can be identified by BCM or CMM being the first author of the reference to the experiment. All authors read and approved the final manuscript.

Author details

[1]University of Wisconsin-Madison, Department of Animal Sciences, Madison, Wisconsin, USA. [2]Spooner Agricultural Research Station, Spooner, Wisconsin, USA. [3]Current address: Merial Ltd, Duluth, Georgia, USA. [4]Current address: Midwest Organic Services Association, Viroqua, Wisconsin, USA.

References

1. Sakul H, Boylan WJ, Shrestha JNB: **Animal model evaluation of dairy traits in US sheep breeds, their crosses and three synthetic populations.** *Small Ruminant Res* 1999, **34**:1–9.
2. Shrestha JNB, Heaney DP: **Review of Canadian, Outaouais and Rideau Arcott breeds of sheep. 1. Development and characterization.** *Small Ruminant Res* 2003, **49**:79–96.
3. Thomas DL, Berger YM, McKusick BC: *Milk and lamb production of East Friesian-cross ewes in northwestern Wisconsin. Proc. 4th Great lakes Dairy Sheep Symp.* Spooner, Wisconsin: University of Wisconsin-Madison, Dept. of Anim. Sci; 1998:11–17.
4. Othmane MH, De La Fuente LF, Carriedo JA, San Primitivo F: **Heritability and genetic correlations of test day milk yield and composition, individual laboratory cheese yield, and somatic cell count for dairy ewes.** *J Dairy Sci* 2002, **85**:2692–2698.
5. Thomas DL, Berger YM, Gottfredson RG, Taylor TA: *Comparison of East Friesian and Lacaune breeds for dairy sheep production in North America. Proc. 10th Great lakes Dairy Sheep Symp.* Hudson, Wisconsin: University of Wisconsin-Madison, Dept. of Anim. Sci; 2004:115–123.
6. ICAR (International Committee for Animal Recording): *Sheep Milk Enquiry.* 2014. Online: http://www.survey-icar.org/sheep_survey4/. Accessed March 28, 2014.
7. Barillet F: **Genetic improvement for dairy production in sheep and goats.** *Small Ruminant Res* 2007, **70**:60–75.
8. Ricordeau G, Flamant JC: **Croisements entre les races ovines Prealpes du Sud et Frisonne (Ostfriesisches Milchschaf). II. Reproduction, viabilite, croissance, conformation.** *Ann Zootech* 1969, **18**:131–149.
9. Katsaounis N, Zygoyiannia D: **The East Friesland sheep in Greece.** *Res Develop Agric* 1986, **3**:19–30.
10. Thomas DL, Berger YM, McKusick BC: *Preliminary results: Survival of high-percentage East Friesian lambs. Proc. 5th Great lakes Dairy Sheep Symp.* Brattleboro, Vermont: University of Wisconsin-Madison: Dept. of Anim. Sci. and University of Vermont, Center for Sustainable Agriculture; 1999:64–66.
11. Thomas DL, Berger YM, McKusick BC: **2000. East Friesian germplasm: effects on milk production, lamb growth, and lamb survival. Proc. American Soc. Anim. Sci.** *J Anim Sci* 1999, **77**:1–6. http://journalofanimalscience.org/content/77/E-Suppl/1.7.
12. Ferreira VC, Thomas DL, Rosa GJ: **Survival in crossbred lambs: individual, maternal, heterosis, and breed effects.** *J Anim Sci* 2012, **90**(3):3. and J. Dairy Sci. Vol. 95, Suppl. 2. Abstr. No. M42:20. (Online: http://www.jtmtg.org/2012/abstracts/15.pdf).
13. McKusick BC, Thomas DL, Berger YM: **Effects of weaning systems on commercial milk production and lamb growth of East Friesian dairy sheep.** *J Dairy Sci* 2001, **84**:1660–1668.
14. McKusick BC, Thomas DL, Romero JE, Marnet PG: **Effect of weaning system on milk composition and distribution of milk fat within the udder of East Friesian dairy ewes.** *J Dairy Sci* 2002, **85**:2521–2528.
15. Hernandez MP, Robinson JJ, Aitken RP, Fraser C: **The effect of dietary supplementation of protected fat on the yield and fat concentration of ewe's milk and on lamb growth.** *Anim Prod* 1986, **42**:455.
16. Sklan D: **A note on production responses of lactating ewes to calcium soaps of fatty acids.** *Anim Prod* 1992, **55**:288–291.
17. Caja G, Bocquier F: *Effects of nutrition on ewe's milk quality. Cooperative FAO – CIHEAM Network on Sheep and Goats.* Grignon, France: Nutrition Subnetwork; 1998:3–5. September. 1-16.
18. Casals R, Caja G, Such X, Torre C, Calsamiglia S: **Effects of calcium soaps and level of protein supplementation on lactational performance of dairy ewes.** *J Dairy Res* 1999, **66**:177–191.
19. McKusick BC, Berger YM, Thomas DL: *Rumen-protected bypass fat for dairy ewe commercial milk production. Proc. 5th Great lakes Dairy Sheep Symp.* Brattleboro, Vermont: University of Wisconsin-Madison: Dept. of Anim. Sci. and University of Vermont, Center for Sustainable Agriculture; 1999:69–80.

20. De Bie LY, Berge M, Thomas DL: *The effect of three times a day milking at the beginning of lactation on the milk production of East Friesian crossbred ewes. Proc. 6th Great lakes Dairy Sheep Symp.* Guelph, Ontario, Canada: University of Wisconsin-Madison, Dept. of Anim. Sci; 2000:1–9.
21. McKusick BC, Thomas DL, Berger YM, Marnet PG: **Effect of milking interval on alveolar versus cisternal milk accumulation, and milk production and composition in dairy ewes.** *J Dairy Sci* 2002, **85:**2197–2206.
22. Marnet PG, McKusick BC: **Regulation of milk ejection and milkability in small ruminants.** *Livest Prod Sci* 2001, **70:**125–133.
23. Pfeilsticker HU, Bruckmaier RM, Blum JW: **Cisternal milk in the dairy cow during lactation and after preceding teat stimulation.** *J Dairy Res* 1996, **63:**509–515.
24. Caja G, Such X, Rovai M: *Udder morphology and machine milking ability in dairy sheep. Proc. 6th Great lakes Dairy Sheep Symp.* Guelph, Ontario, Canada: University of Wisconsin-Madison, Dept. of Anim. Sci; 2000:17–40.
25. McKusick BC, Thomas DL, Berger YM: **Effect of omission of machine stripping on milk production and parlor throughput in East Friesian dairy ewes.** *J Dairy Sci* 2003, **86:**680–687.
26. McKusick BC, Wiltbank MC, Sartori R, Marnet PG, Thomas DL: **Effect of presence or absence of corpora lutea on milk production and composition of mid-lactation dairy ewes.** *J Dairy Sci* 2002, **85:**790–796.
27. Smith LW, Inskeep EK: **Effect of progestins on lactation in the ewe.** *J Anim Sci* 1970, **30:**957–959.
28. Marnet PG, Negrão JA, Labussie're J: **Oxytocin release and milk ejection parameters during milking of dairy ewes in and out of natural season of lactation.** *Small Ruminant Res* 1998, **28:**183–191.
29. Dahl GE, Buchanan BA, Tucker HA: **Photoperiodic effects on dairy cattle: a review.** *J Dairy Sci* 2000, **83:**885–893.
30. Bocquier F, Ligios S, Molle G, Casu S: **Effect of photoperiod on milk, milk composition and voluntary food intake in lactating dairy ewes.** *Ann Zootech (Paris)* 1997, **46:**427–438.
31. Auchtung TL, Rius AG, Kendall PE, McFadden TB, Dahl GE: **Effects of photoperiod during the dry period on prolactin, prolactin receptor, and milk production of dairy cows.** *J Dairy Sci* 2005, **88:**121–127.
32. Mabjeesh SJ, Gal-Garber O, Shamay A: **Effect of photoperiod in the third trimester of gestation on milk production and circulating hormones in dairy goats.** *J Dairy Sci* 2007, **90:**699–705.
33. Mikolayunas CM, Thomas DL, Dahl GE, Gressley TF, Berger YM: **Effect of prepartum photoperiod on milk production and prolactin concentration of dairy ewes.** *J Dairy Sci* 2008, **91:**85–90.
34. Mikolayunas CM, Thomas DL, Albrecht KA, Combs DK, Berger YM, Eckerman SR: **Effects of supplementation and stage of lactation on performance of grazing dairy ewes.** *J Dairy Sci* 2008, **91:**1477–1485.
35. Cannas A: **Feeding of lactating ewes.** In *Pages 123–166 in Dairy Sheep Feeding and Nutrition.* Edited by Pulina G. Avenue Media, Bologna, Italy: 2002.
36. Mikolayunas-Sandrock C, Armentano LE, Thomas DL, Berger YM: **Effect of protein degradability on milk production of dairy ewes.** *J Dairy Sci* 2009, **92:**4507–4513.
37. Mikolayunas C, Thomas DL, Armentano LE, Berger YM: **Effect of rumen undegradable protein supplementation and fresh forage composition on nitrogen utilization of dairy ewes.** *J Dairy Sci* 2011, **94:**416–425.

Amino acid fortified diets for weanling pigs replacing fish meal and whey protein concentrate: Effects on growth, immune status and gut health

Yan Zhao[1], Alexandra C Weaver[1], Vivek Fellner[1], Robert L Payne[2] and Sung Woo Kim[1*]

Abstract

Background: Limited availability of fish meal and whey protein concentrate increases overall feed costs. Availability of increased number of supplemental amino acids including Lys, Met, Thr, Trp, Val, and Ile allows replacing expensive protein supplements to reduce feed costs. This study was to evaluate the effect of replacing fish meal and/or whey protein concentrate in nursery diets with 6 supplemental amino acids on growth performance and gut health of post-weaning pigs. Treatments were 1) FM-WPC: diet with fish meal (FM) and whey protein concentrate (WPC); 2) FM-AA: diet with FM and crystalline amino acids (L-Lys, L-Thr, L-Trp, DL-Met, L-Val, and L-Ile); 3) WPC-AA: diet with WPC and crystalline amino acid; and 4) AA: diet with crystalline amino acid.

Results: Pigs in FM-AA, WPC-AA, and AA had greater ($P < 0.05$) ADG and gain:feed than pigs in FM-WPC during wk 1 (phase 1). Plasma insulin concentration of pigs in AA tended to be greater ($P = 0.064$) than that of FM-WPC at the end of wk 1(phase 1). Plasma concentrations of IgG in AA was lower ($P < 0.05$) compared with WPC-AA and FW, and FM-AA had lower ($P < 0.05$) IgG concentration than WPC-AA at the end of wk 1 (phase 1). Concentration of acetate in cecum digesta in FM-AA tended to be greater ($P = 0.054$) than that of FM-WPC and WPC-AA. Concentration of isovalerate in cecum digesta of pigs in FM-AA was greater ($P < 0.05$) than that of FW and WPC-AA.

Conclusions: This study indicates that use of 6 supplemental amino acids can replace fish meal and/or whey protein concentrate without adverse effects on growth performance, immune status, and gut health of pigs at d 21 to 49 of age. Positive response with the use of 6 supplemental amino acids in growth during the first week of post-weaning may due to increased plasma insulin potentially improving uptake of nutrients for protein synthesis and energy utilization. The replacement of fish meal and/or whey protein concentrate with 6 supplemental amino acids could decrease the crude protein level in nursery diets, and potentially lead to substantial cost savings in expensive nursery diets.

Keywords: Amino acid, Fish meal, Growth performance, Pig, Whey protein concentrate

Background

Extremely high price of fish meal and whey protein concentrate increases overall cost of pork production. Because of advances in biotechnology, the use of supplemental amino acids in animal feeding are extended to Lys, Met, Thr, Trp, Val, Gln, and possibly Ile. Utilization of these essential amino acids as feed additives benefits swine industry by allowing nutritionists to use high levels of supplemental amino acids to replace high cost protein sources [1]; reduce crude protein levels in the diets [2-5]; formulate diets based on ideal protein concept [6-11]; reduce environmental excretion of nitrogen [3,12,13]; and reduce metabolic stress of pigs to detoxify excess ammonia [4,14,15].

Fish meal and whey protein concentrate are 2 of the most widely used animal protein sources in diets for

* Correspondence: sungwoo_kim@ncsu.edu
[1]Department of Animal Science, North Carolina State University, Raleigh, NC 27695, USA
Full list of author information is available at the end of the article

nursery pigs. They are often included in these diets due to their high amino acid content and digestibility, which many believe eases the transition from sow's milk to solid food for nursery pigs [16,17]. However, the prices and availability for these ingredients have become extremely variable in recent years which increases overall cost of pork production and puts pressure on finding alternatives in case of limited availability.

Considering that the objective of a nursery diet is to provide a highly digestible and palatable diet to these young pigs in hopes of getting them to begin consuming feed sooner, perhaps there are opportunities to explore the use of lower crude protein diets supplemented with amino acids. The use of supplemental amino acids would offset or minimize the need to use some of expensive animal proteins, which could reduce the cost of these diets. Furthermore, extensive use of supplemental amino acids would allow to more precisely meet the pigs's dietary requirements while reducing dietary crude protein. This change in formulation can positively impact pig health and the environment by reduction of environmental excretion of nitrogen and reduce metabolic stress of pigs to detoxify excess ammonia. Therefore, the objective of this study was to evaluate the effect of replacing fish meal and/or whey protein concentrate in nursery diets with 6 supplemental amino acids on growth performance, immune status, and gut health of post-weaning pigs.

Methods

Experimental protocols were approved by the Institutional Animal Care and Use Committee at North Carolina State University (Raleigh, USA).

Animals and experimental design

A total of 160 newly weaned pigs (6.49 ± 0.57 kg) at 21 d of age were used in a randomized complete block design with 4 dietary treatments, 8 replicates (4 barrow pens and 4 gilt pens) per treatment, and 5 pigs per pen. Treatments were 1) FM-WPC: diet with fish meal (**FM**) and whey protein concentrate (**WPC**); 2) FM-AA: diet with FM and supplemental amino acids (L-Lys, L-Thr, L-Trp, DL-Met, L-Val, and L-Ile); 3) WPC-AA: diet with WPC and supplemental amino acid; and 4) AA: diet with supplemental amino acid. All diets were formulated according to the nutrient requirements of NRC [18] (Tables 1 and 2). Pigs were fed the assigned experimental diets for a 4-wk period based on a 2-phase-feeding (phase 1: wk 1 post-weaning, Table 1; phase 2: wk 2 to 4 post-weaning, Table 2). During the 4-wk feeding period, all pigs were housed in a temperature-controlled nursery room (25 to 28°C). Pigs had access to diets and water *ad libitum* during the entire experimental period. Fresh feed was provided everyday. At 0800 h, any feed left from the

Table 1 Experimental diets for the study (phase 1)

Items, %	FM-WPC[1]	FM-AA[2]	WPC-AA[3]	AA[4]
Corn grain	40.75	47.66	43.44	50.20
Soybean meal, without hulls	11.00	11.00	11.00	11.00
Whey powder	25.00	25.00	25.00	25.00
Plasma protein[5]	5.00	5.00	5.00	5.00
Whey protein concentrate	9.35	0.00	9.35	0.00
Fish meal, Menhaden	5.00	5.00	0.00	0.00
DL-Met	0.00	0.11	0.08	0.19
L-Thr	0.00	0.18	0.10	0.28
L-Val	0.00	0.15	0.11	0.26
L-Trp	0.00	0.05	0.02	0.08
L-Ile	0.00	0.15	0.10	0.26
L-Lys[6]	0.00	0.50	0.40	0.90
Aureomycin100	0.05	0.05	0.05	0.05
Zinc oxide	0.25	0.25	0.25	0.25
Vitamin-mineral premix[7]	0.50	0.50	0.50	0.50
Dical P	0.00	0.40	0.80	1.20
Ground limestone	0.70	0.80	0.90	1.00
Salt	0.40	0.40	0.40	0.40
Poultry fat	2.00	2.80	2.50	3.43
Total	100.00	100.00	100.00	100.00
Chemical composition:				
DM %	91.2	90.8	91.2	90.7
ME, kcal/kg	3411	3410	3409	3412
CP, %	22.56	20.76	20.17	18.45
SID[8] Lys, %	1.34	1.34	1.34	1.34
SID Met, %	0.68	0.68	0.68	0.68
SID Trp, %	0.26	0.26	0.26	0.26
SID Thr, %	0.93	0.93	0.93	0.93
SID Val, %	0.99	0.99	0.99	0.99
SID Ile, %	0.78	0.78	0.78	0.78
Ca, %	0.91	0.92	0.90	0.91
Available P, %	0.56	0.56	0.56	0.55
Total P, %	0.70	0.72	0.70	0.71
Analyzed value:				
CP, DM basis, %	24.5	21.5	22.2	20.3

[1]FM-WPC: a control diet with fish meal and whey protein concentrate.
[2]FM-AA: a diet with fish meal and amino acids replacing whey protein concentrate.
[3]WPC-AA: a diet with whey protein concentrate and amino acids replacing fish meal.
[4]AA: a diet with amino acids replacing both fish meal and whey protein concentrate.
[5]APC-920 (APC Inc).
[6]Biolys (Evonik-Degussa).
[7]Vitamin-mineral premix provided the following per kilogram of complete phase 1 diet: 62.2 mg of manganese as manganese oxide, 100 mg of iron as iron sulfate, 138.4 mg of zinc as zinc sulfate, 12.7 mg of copper as copper oxide, 0.96 mg of iodide as ethylenediamine dihydroiodide, 0.3 mg of selenium as sodium selenite, 10,072 IU of vitamin A as vitamin A acetate, 1,100 IU of vitamin D3, 82.5 IU of vitamin E, 5.9 IU of vitamin K as menadione sodium bisulfite, 73.2 μg of vitamin B12, 18.3 mg of riboflavin, 58.5 mg of d-pantothenic acid as calcium pantothenate, 73.2 mg of niacin, and 6.4 g of choline as choline chloride.
[8]SID: Standardized ileal digestible.

Table 2 Experimental diets for the study (phase 2)

Items, %	FM-WPC[1]	FM-AA[2]	WPC-AA[3]	AA[4]
Corn grain	50.00	54.15	51.54	55.65
Soybean meal, without hulls	20.00	20.00	20.00	20.00
Whey powder	15.00	15.00	15.00	15.00
Plasma protein[5]	3.00	3.00	3.00	3.00
Whey protein concentrate	5.60	0.00	5.60	0.00
Fish meal, Menhaden	3.00	3.00	0.00	0.00
DL-Met	0.00	0.07	0.05	0.12
L-Thr	0.00	0.11	0.06	0.17
L-Val	0.00	0.09	0.07	0.16
L-Trp	0.00	0.04	0.02	0.05
L-Ile	0.00	0.09	0.06	0.15
L-Lys[6]	0.00	0.30	0.25	0.55
Aureomycin100	0.05	0.05	0.05	0.05
Vitamin-mineral premix[7]	0.50	0.50	0.50	0.50
Dical P	0.20	0.45	0.65	0.90
Ground limestone	0.90	0.95	1.05	1.10
Salt	0.30	0.30	0.30	0.30
Poultry fat	1.45	1.90	1.80	2.30
Total	100.00	100.00	100.00	100.00
Chemical composition:				
DM %	90.3	90.0	90.3	90.1
ME, kcal/kg	3,394	3,392	3,394	3,394
CP, %	22.0	21.0	20.7	19.6
SID[8] Lys, %	1.19	1.19	1.19	1.19
SID Met, %	0.64	0.64	0.64	0.64
SID Trp, %	0.24	0.24	0.24	0.24
SID Thr, %	0.80	0.80	0.80	0.80
SID Val, %	0.91	0.91	0.91	0.91
SID Ile, %	0.76	0.76	0.76	0.76
Ca, %	0.82	0.82	0.82	0.82
Available P, %	0.41	0.41	0.41	0.41
Total P, %	0.61	0.62	0.61	0.62
Analyzed value:				
CP, DM basis, %	23.6	19.9	19.2	18.8

[1]FM-WPC: a control diet with fish meal and whey protein concentrate.
[2]FM-AA: a diet with fish meal and amino acids replacing whey protein concentrate.
[3]WPC-AA: a diet with whey protein concentrate and amino acids replacing fish meal.
[4]AA: a diet with amino acids replacing both fish meal and whey protein concentrate.
[5]APC-920 (APC Inc).
[6]Biolys (Evonik-Degussa).
[7]Vitamin-mineral premix provided the following per kilogram of complete phase 1 diet: 62.2 mg of manganese as manganese oxide, 100 mg of iron as iron sulfate, 138.4 mg of zinc as zinc oxide, 12.7 mg of copper as copper oxide, 0.96 mg of iodide as ethylenediamine dihydroiodide, 0.3 mg of selenium as sodium selenite, 10,072 IU of vitamin A as vitamin A acetate, 1,100 IU of vitamin D3, 82.5 IU of vitamin E, 5.9 IU of vitamin K as menadione sodium bisulfite, 73.2 μg of vitamin B12, 18.3 mg of riboflavin, 58.5 mg of d-pantothenic acid as calcium pantothenate, 73.2 mg of niacin, and 6.4 g of choline as choline chloride.
[8]SID: Standardized ileal digestible.

previous day was removed and weighed. Feed intake and body weight were measured on d 7, 14, 21, and 28 post-weaning for determination of ADG, ADFI, and gain:feed.

Experimental diets

The phase 1 and 2 diets are shown in Tables 1 and 2. All diets contained soybean meal and plasma protein, and the inclusions of these ingredients were held constant across all diets in each growth phase. Soybean meal contents were 11.0 and 20.0% for phase 1 and 2, and plasma protein contents for phase 1 and 2 were 5.0 and 3.0%, respectively. Crude protein content was various among treatment diets. Fish meal (5.0 and 3.0% for phase 1 and 2, respectively) and whey protein concentrate (9.35 and 5.60% for phase 1 and 2, respectively) were used depending on treatment diets. Amino acid contents were measured in triplicate for soybean meal, fish meal, whey protein concentrate, and plasma protein from the same batch to be used for this study (12 samples). These data were used to formulate the diets. All diets were formulated to contain the same standardized ileal digestible levels of Lys, Thr, Trp, Met, Val, and Ile for each respective phase, and all nutrients met or exceeded the NRC [18] requirements. After the diet formulation, amino acid contents in each diet were measured in triplicate to check if diets were formulated to meet the target levels (12 samples). The results showed that all diets had the same content (standard ileal digestible amount basis) of Lys, Met, Thr, Trp, Val, and Ile (Table 3).

Blood cell counts

At the end of each growth phase, blood sample from one pig per pen (same pig was used in both phases) was collected from the jugular vein using 10-mL heparin-coated blood tube and disposable 22-gauge × 0.025 mm Vacutainer needle (BD, Franklin Lakes, NJ). Hematocrit, hemoglobin, mean corpuscular hemoglobin, mean corpuscular hemoglobin concentration, mean corpuscular volume, mean platelet volume (**MPV**), platelet number, red blood cell count, red blood cell distribution width, white blood cell (**WBC**) count, neutrophils, lymphocytes, monocytes, eosinophils, and basophils in whole blood samples were used for hematological evaluation using an automated hematological analyzer (Cell-Dyn 6300, Abbott Diagnostics, Abbott Park, IL). Plasma samples were then collected by centrifuging (5810 R, Eppendorf AG, Hamburg, Germany) at 3,000 g, 15 min, 4°C, allocated into 1.5 mL microcentrifuge tubes, and stored at -80°C for analysis. Plasma from blood was later analyzed for the contents of immunoglobulin G (**IgG**), immunoglobulin A (**IgA**), tumor necrosis factor α (**TNF-α**) at the end of phase 1, and insulin and growth hormone at the end of phase 1 and 2.

Table 3 Growth performance of pigs fed experimental diets

Items	FM-WPC[1]	FM-AA[2]	WPC-AA[3]	AA[4]	SEM	*P* value
n	8	8	8	8		
Age, d	22.1	22.7	23.3	22.5	0.6	NS[5]
Body weight, kg						
Initial	6.50	6.53	6.54	6.38	0.57	NS
Wk 1	7.30	7.74	7.96	7.65	0.58	NS
Wk 2	8.70	9.46	9.52	9.20	0.66	NS
Wk 3	11.07	11.83	11.87	11.50	0.66	NS
Wk 4	15.11	15.85	15.69	16.04	0.87	NS
ADG, g						
Wk 1	114.3[b]	172.6[a]	202.7[a]	182.0[a]	15.2	0.031
Wk 2	200.1	246.5	223.3	221.4	19.9	NS
Wk 3	338.8	338.3	336.3	328.1	26.1	NS
Wk 4	576.1	574.6	545.4	648.0	44.1	NS
Wk 1 to 4	307.3	333.0	326.9	344.9	14.1	NS
ADFI, g						
Wk 1	198.5	235.0	240.4	237.2	15.4	NS
Wk 2	303.0	318.5	326.4	325.0	16.7	NS
Wk 3	539.9	538.7	566.1	524.4	22.9	NS
Wk 4	1,106.1	1,106.6	1,109.1	1,168.2	33.5	NS
Wk 1 to 4	525.6	539.5	556.0	563.7	11.5	NS
Gain:feed						
Wk 1	0.562[b]	0.730[a]	0.836[a]	0.774[a]	0.045	0.008
Wk 2	0.664	0.766	0.681	0.684	0.044	NS
Wk 3	0.634	0.625	0.595	0.625	0.044	NS
Wk 4	0.566	0.551	0.514	0.565	0.055	NS
Wk 1 to 4	0.592	0.618	0.590	0.614	0.026	NS

[1]FM-WPC: a control diet with fish meal and whey protein concentrate.
[2]FM-AA: a diet with fish meal and amino acids replacing whey protein concentrate.
[3]WPC-AA: a diet with whey protein concentrate and amino acids replacing fish meal.
[4]AA: a diet with amino acids replacing both fish meal and whey protein concentrate.
[5]NS: not significant ($P > 0.10$).
[a-b]Means lacking a common superscript are differ within a row ($P < 0.05$).

Immune responses

At the end of phase 1, plasma samples were used to measure IgG, IgA, and TNF-α. Total concentrations of IgG and IgA in plasma of pigs were measured according to the method described by Chaytor et al. [19] using ELISA kits (Bethyl, Montgomery, TX). Firstly, goat anti-pig IgG or IgA were used as capture antibodies to coat wells. Then plasma samples were diluted to 1:100,000 for the measurement. Horseradish peroxidase goat anti-pig IgG or IgA was used as the detection. The plate was read at 450 nm by an ELISA plate reader (Synergy HT, BioTek Instruments, Winooski, VT) and software (KC4 Data Analysis Software, BioTek Tnstrument). Sample concentration was quantified against the standard curve. Detective limits were 7.8 ng/mL for IgG, and 15.6 ng/mL for IgA, respectively.

Determination of TNF-α concentrations in plasma was completed according to the method described by Chaytor et al. [19] using ELISA kits (Pierce Biotechnology, Rockford, IL). Briefly, 50 μL of standard plus dilute or 100 μL of sample was added to microplate wells coated with capture antibody in conjunction with biotinylated antibody reagent. Horseradish peroxidase, TMB substrate, and a stop solution of 2 mol/L H_2SO_4 were used for the detection. Absorbance was read at 450 nm by an ELISA plate reader (Synergy HT, BioTek Instruments, Winooski, VT) and software (KC4 Data Analysis Software, BioTek Tnstrument). Detection limit for TNF-α was 5 pg/mL.

Growth hormone

Plasma growth hormone concentration was determined according to the method described by Atinmo et al. [20]

using the porcine/canine growth hormone radioimmunoassay kit (Millipore, Charles, MO). Briefly, label total count tubes (tube 1-2), non-specific binding (**NSB**) tubes (tube 3-4), reference tubes (tube 5-6), and sample tubes (from tube 7). Assay buffer was first added into the NSB tubes, reference tubes, and sample tubes followed adding standards, or quality controls, or plasma samples. Then ^{125}I-porcine/canine growth hormone was added to all tubes. Guinea pig porcine/canine antiserum was added to all tubes except total count tubes and NSB tubes. After incubation overnight at room temperature, cold (4°C) precipitating reagent was added to all tubes (except total count tubes). After incubation and centrifugal separation, immediately decant the supernate of all tubes except total count tubes, and drain tubes for at least 15 to 60 s. An automatic gamma counter (WIZARD, PerkinElmer, Waltham, MA) was used to count all tubes and calculate the concentration of porcine/canine growth hormone in plasma samples. The limit of sensitivity for the porcine/canine growth hormone assay was 1 ng/mL.

Insulin

Plasma insulin was determined using a commercial ELISA kit (Mercodia AB, Uppsala, Sweden), which allowed for direct measurement of plasma using an ELISA plate reader (Synergy HT, BioTek Instruments, Winooski, VT) and software (KC4 Data Analysis Software, BioTek Tnstrument). All samples were tested in duplicate. Detection limit for insulin was 0.007 ng/mL.

Gut health

At the end of phase 1, one pig from each pen (from 4 barrow pens and 4 gilt pens per treatment, n = 8) were selected and euthanized at 10 am to collect intestinal tissues and digesta samples. After killing, gastrointestinal tract with digesta were removed. The representative digesta from cecum and colon were gently squeezed into a tube and then stored at -80°C until analysis for volatile fatty acids (**VFA**) concentrations [21,22]. Intestinal tissue from middle section of jejunum was isolated, flushed with 0.9% salt solution, fixed in 10% formaldehyde-phosphate buffer, and stored at 4°C for microscopic assessment of mucosal morphology. Briefly, the fixed intestinal segments were embedded in paraffin wax. Cross sections of 6 μm thickness of each intestinal segments were obtained, and stained with hematoxylin and eosin. Villus height was determined at 40× magnification as described by Shen et al. [22]. Villous height from the tip to the crypt-villus junction on 10 well-oriented villi were measured in triplicate for each intestinal segment of each pig [22,23].

Statistical analysis

All the data were analyzed by the GLM procedure (SAS Inst., Inc., Cary, NC) following a randomized complete block design with dietary treatment as the main effect and sex as a block. The pen was the experimental unit. Separation of means was done using the PDIFF option of SAS. Treatment effects were considered statistically significant when probability values were less than 0.05, and probability less than 0.1 and equal or greater than 0.05 was considered as a trend.

Results

As there was no effect ($P > 0.10$) of sex in all measurement, sex effect was excluded in the result. Age and initial body weight of pigs did not differ among treatments (Table 3). Body weights of pigs did not differ among treatments at the end of wk 1, 2, 3, and 4. Average daily gain of pigs did not differ among treatments during wk 2, 3, 4, and the entire 4-wk period. However, pigs in FM-AA, WPC-AA, and AA had greater ($P < 0.05$) ADG than pigs in FM-WPC during wk 1 (phase 1). Average daily feed intake did not differ among treatments during wk 1, 2, 3, 4, and the entire 4-wk period. Gain:feed did not differ among treatments during wk 2, 3, 4, and the entire 4-wk period. However, pigs in FM-AA, WPC-AA, and AA had greater ($P < 0.05$) gain:feed than pigs in FM-WPC during wk 1 (Table 3).

Plasma concentrations of growth hormone did not differ among treatments when measured at the end of wk 1 and 4 (Table 4). Plasma insulin concentration of pigs in AA tended to be greater ($P = 0.064$) than that of FM-WPC at the end of wk 1. However, plasma insulin concentration of pigs did not differ among treatments at the end of wk 4. Plasma concentrations of IgM were not different among treatments at the end of wk 1. Plasma concentrations of IgG in AA was similar to FM-AA, but was lower ($P < 0.05$) compared with WPC-AA and FM-WPC, and FM-AA had lower ($P < 0.05$) IgG concentration than WPC-AA at the end of wk 1. Plasma concentration of TNF-α did not differ among treatments at the end of wk 1 (Table 4). There was no treatment × sex interaction for all plasma measurements.

Numbers of blood cells and hematological characteristics were not different among treatments at the end of wk 1 and wk 4 (Table 5). Height of villi in jejunum of pigs did not differ among treatments at the end of wk 4 (Table 6). Concentration and percentage of propionate, isobutyrate, butyrate, and valerate in digesta did not differ among treatments at the end of wk 4 (Table 6). However, acetate concentration in cecum digesta of pigs in FM-AA tended to be greater ($P = 0.054$) than that of FM-WPC and WPC-AA (Table 6). Concentration of isovalerate in cecum digesta of pigs in FM-AA was greater ($P < 0.05$) than that of FM-WPC and WPC-AA. Percentage of isovalerate in total volatile fatty acids in colon digesta of pigs in FM-AA tended to be greater ($P = 0.057$) than that of WPC-AA (Table 6).

Table 4 Plasma mearurements of pigs fed experimental diets

Items	FM-WPC[1]	FM-AA[2]	WPC-AA[3]	AA[4]	SEM	*P* value
n	8	8	8	8		
Phase 1, d 7						
Growth hormone, ng/mL	9.205	5.277	7.232	8.225	2.170	NS[5]
Insulin, μg/L	0.026^{A}	0.034^{AB}	0.028^{AB}	0.058^{B}	0.012	0.064
Immunoglobulin G, ug/mL	0.325^{ab}	0.290^{bc}	0.362^{a}	0.275^{c}	0.028	0.024
Immunoglobulin M, ug/mL	0.058	0.060	0.066	0.060	0.006	NS
TNF-α, pg/mL	47.0	47.3	35.0	47.3	8.9	NS
Phase 2, d 28						
Growth hormone, ng/mL	4.163	4.140	6.251	7.298	1.686	NS
Insulin, μg/L	0.026	0.049	0.034	0.024	0.011	NS

[1]FM-WPC: a control diet with fish meal and whey protein concentrate.
[2]FM-AA: a diet with fish meal and amino acids replacing whey protein concentrate.
[3]WPC-AA: a diet with whey protein concentrate and amino acids replacing fish meal.
[4]AA: a diet with amino acids replacing both fish meal and whey protein concentrate.
[5]NS: not significant ($P > 0.10$).
[a-b]Means lacking a common superscript are differ within a row ($P < 0.05$).
[A-B]Means lacking a common superscript tend to differ within a row ($0.05 \leq P < 0.1$).

Discussion

The reduction in growth rate during the period immediately after weaning is a major problem in nursery pig nutrition and management. It was hypothesized that supplementing highly digestible crystalline amino acids to nursery diets could support growth performance, minimize the need for the use some of expensive animal proteins such as fish meal and/or WPC, and reduce the crude protein level in diets to reduce the cost [4]. Our data demonstrate that supplementing amino acids to replace fish meal and/or WPC in nursery diet increased ADG and gain:feed than pigs in FM-WPC during wk 1 (phase 1). This is in agreement with the result of previous study [12]. In this study, we did not observe changes in voluntary feed intake, indicating that the amount of energy and standardized amino acids provided for utilization did not differ among treatments, therefore feed intake was not the main reason for the increased ADG and feed:gain in phase 1 but rather it could be contributed from enhanced amino acid availability and reduced energy costs for digestion of crude protein. True digestibilities of supplemental amino acids were often found to be 100% digestible [24]. The average apparent ileal digestibilities of fish meal was reported to be a approximately 73.6% [6]. In this study, although all diets were formulated to contain the same standardized ileal digestible levels of Lys, Thr, Trp, Met, Val, and Ile, the different digestibilities of amino acid from supplemented amino acid source or fish meal source could result in the difference of amino acid digestion and absorption between FM-WPC and amino acid supplemented treatments in phase 1. In addition, plasma insulin concentration of pigs in AA in this study tended to be greater than that of FM-WPC at the end of wk 1 (phase 1). Amino acids and insulin are critical regulators of muscle protein synthesis in neonatal pigs [25]. Thus, the improved growth performance was possibly induced by enhanced insulin sensitivity due to increased plasma free amino acid contents.

When supplementing amino acid to replace fish meal and/or WPC in phase 2 nursery diets, although there was numerically increase in ADG and gain:feed in amino acid supplemented treatments compared with FM-WPC, statistically there was no difference for growth performance among treatments. This is in agreement with previous studies [2,26,27]. Nyachoti et al. [4] showed that daily feed intake, ADG, and G:F of post-weaning pigs were not affected by reducing CP level from 23% to 21%, however, they also reported that piglet performance may suffer when dietary CP level was reduced by 4 or more percentage units from 23%. Results in this study indicate that replacing fish meal and/or WPC with crystalline amino acid may improve growth performance of pigs immedically after weaning. When CP level was reduced to 20% by supplementing amino acid during the nursery phase, the growth performance was not affected.

This study also examed the plasma concentrations of growth hormone which did not differ among treatments when measured at the end of wk 1 and 4. Variations among values of growth hormones were rather large which may have contributed to no significant differences among means. The result indicates that supplemented amino acid to replace fish meal and/or WPC in nursery diets did not affect growth hormone changes in plasma of pigs.

In this study, supplemental amino acid to fish meal and/ or WPC improve ADG and gain:feed in phase 1 diet without adversely affecting immune status of pigs in phase1, as showing by plasma concentrations of IgM and TNF-α which were not different among treatments at the end of

Table 5 Hematological evaluation of pigs fed experimental diets

Items	FM-WPC[1]	FM-AA[2]	WPC-AA[3]	AA[4]	SEM	*P* value
n	8	8	8	8		
Phase 1, d 7						
White blood cell, 10^3 cell/μL	13.09	12.78	13.81	13.83	1.16	NS[5]
Red blood cell, 10^6 cell/μL	6.18	6.14	5.93	6.23	0.28	NS
Hemoglobin, g/dL	10.60	10.64	9.99	10.76	0.64	NS
Hematocrit, %	33.60	34.23	32.56	34.61	1.77	NS
Mean corpuscular volume, fL	54.25	56.11	55.11	55.54	1.79	NS
Mean corpuscular hemoglobin, pg/cell	17.09	17.46	16.93	17.21	0.73	NS
Mean corpuscular hemoglobin concentration, %	31.40	31.03	30.64	30.95	0.40	NS
Red blood cell distribution width, μm	18.13	19.71	20.73	20.80	1.78	NS
Platelet count, /μL	407.9	393.8	346.6	439.6	57.0	NS
Mean platelet volume, fL	10.24	9.84	10.28	10.61	0.44	NS
Neutrophil, 10^3 cell/μL	4.86	5.78	6.05	6.16	0.65	NS
Lymphocyte, 10^3 cell/μL	7.13	5.72	6.41	6.50	0.65	NS
Monocyte, 10^3 cell/μL	0.59^{A}	0.62^{A}	0.91^{B}	0.56^{A}	0.09	0.057
Eosinophil, 10^3 cell/μL	0.23	0.33	0.41	0.29	0.08	NS
Basophil, 10^3 cell/μL	0.05	0.08	0.06	0.06	0.04	NS
Phase 2, d 28						
White blood cell, 10^3 cell/μL	17.99	14.51	18.74	18.46	1.863	NS
Red blood cell, 10^6 cell/μL	6.01^{ab}	5.93^{a}	5.96^{a}	6.15^{b}	0.09	0.005
Hemoglobin, g/dL	10.05^{ab}	10.27^{a}	9.88^{b}	10.30^{a}	0.08	0.002
Hematocrit, %	31.65	32.11	30.54	32.59	0.73	NS
Mean corpuscular volume, fL	52.96	54.53	51.38	53.04	1.45	NS
Mean corpuscular hemoglobin, pg/cell	16.84	17.43	16.63	16.79	0.50	NS
Mean corpuscular hemoglobin concentration, %	31.75	31.96	32.30	31.63	0.23	NS
Red blood cell distribution width, μm	18.50	19.27	19.90	19.31	1.11	NS
Platelet count, /μL	467.0	453.1	357.6	506.9	60.2	NS
Mean platelet volume, fL	9.75	9.49	9.56	9.81	0.46	NS
Neutrophil, 10^3 cell/μL	5.79^{A}	5.33^{A}	8.02^{B}	5.95^{A}	1.52	0.082
Lymphocyte, 10^3 cell/μL	10.15	7.33	8.61	10.42	0.91	NS
Monocyte, 10^3 cell/μL	1.34	1.02	1.02	1.16	0.18	NS
Eosinophil, 10^3 cell/μL	0.43	0.37	0.42	0.38	0.06	NS
Basophil, 10^3 cell/μL	0.11	0.10	0.13	0.12	0.02	NS

[1]FM-WPC: a control diet with fish meal and whey protein concentrate.
[2]FM-AA: a diet with fish meal and amino acids replacing whey protein concentrate.
[3]WPC-AA: a diet with whey protein concentrate and amino acids replacing fish meal.
[4]AA: a diet with amino acids replacing both fish meal and whey protein concentrate.
[5]NS: not significant ($P > 0.10$).
[A-B]Means lacking a common superscript tend to differ within a row ($0.05 \leq P < 0.1$).
[a-d]Means lacking a common superscript are differ within a row ($P < 0.05$).

wk 1. In addition, plasma concentrations of IgG was lower in AA compared with FM-WPC at the end of wk 1. The reduced immune challenge could be another reasons for the positive effect of dietary supplementation of the amino acid on pig's growth performance in wk 1.

To obtain further knowledge about how supplemental amino acids affect the post-weaning pigs, the blood cell counts and plasma composition were measured during 4-wk study. Results indicate that supplemented amino acids did not affect numbers of blood cells and hematological characteristics among treatments at the end of wk 1 and wk 4. Numbers of white blood cells and red blood cells, neutrophils, lymphocytes, monocytes, eosinophils, and basophils in pigs are shown as normal compared with reference values suggested by Friendship et al. [28], Dubreuil and Lapierre [29], and Klem et al. [30].

Table 6 Villus height and concentrations of volatile fatty acids in digesta of pigs fed experimental diets (d 28)

Items	FM-WPC[1]	FM-AA[2]	WPC-AA[3]	AA[4]	SEM	*P* value
n	8	8	8	8		
Villus height, μm	376.8	477.8	505.4	472.8	42.2	0.208
Cecum digesta						
Acetate, mmol/L	66.04^{A}	94.63^{B}	65.24^{A}	90.86^{AB}	10.74	0.054
Propionate, mmol/L	40.94	50.15	39.25	47.51	8.37	NS[5]
Isobutyrate, mmol/L	0.54	0.98	0.40	0.83	0.22	NS
Butyrate, mmol/L	18.16	22.30	14.41	25.63	4.45	NS
Isovalerate, mmol/L	0.41^{a}	0.82^{b}	0.30^{a}	0.58^{ab}	0.12	0.032
Valerate, mmol/L	2.99	2.70	2.22	4.38	0.67	NS
Total, mmol/L	129.06	171.57	121.79	169.78	21.78	NS
Acetate, % of total	53.21	56.15	54.45	52.55	3.40	NS
Propionate, % of total	29.05	28.67	32.27	28.81	3.51	NS
Isobutyrate, % of total	0.31	0.61	0.27	0.45	0.13	NS
Butyrate, % of total	13.95	12.38	11.03	15.35	1.81	NS
Isovalerate, % of total	0.24	0.55	0.22	0.32	0.10	NS
Valerate, % of total	3.24	1.65	1.77	2.53	0.73	NS
Colon digesta						
Acetate, mmol/L	92.90	110.52	93.78	117.08	16.81	NS
Propionate, mmol/L	50.58	54.17	52.10	51.78	10.32	NS
Isobutyrate, mmol/L	1.76	1.18	1.69	1.81	0.36	NS
Butyrate, mmol/L	22.67	25.16	20.98	28.60	5.43	NS
Isovalerate, mmol/L	1.99	1.93	1.11	1.44	0.44	NS
Valerate, mmol/L	5.34	3.80	3.70	4.00	1.20	NS
Total, mmol/L	175.22	196.75	173.34	204.68	31.82	NS
Acetate, % of total	54.64	57.20	53.56	57.23	2.51	NS
Propionate, % of total	26.45	26.82	31.14	25.35	2.88	NS
Isobutyrate, % of total	0.77	0.64	0.94	0.88	0.17	NS
Butyrate, % of total	13.36	12.30	11.71	13.90	1.68	NS
Isovalerate, % of total	0.85^{AB}	1.05^{B}	0.59^{A}	0.69^{AB}	0.18	0.057
Valerate, % of total	3.95	2.00	2.07	1.95	0.73	NS

[1]FM-WPC: a control diet with fish meal and whey protein concentrate.
[2]FM-AA: a diet with fish meal and amino acids replacing whey protein concentrate.
[3]WPC-AA: a diet with whey protein concentrate and amino acids replacing fish meal.
[4]AA: a diet with amino acids replacing both fish meal and whey protein concentrate.
[5]NS: not significant ($P > 0.10$).
[a-b]Means lacking a common superscript are differ within a row ($P < 0.05$).
[A-B]Means lacking a common superscript tend to differ within a row ($0.05 \leq P < 0.1$).

Weaning is one of the most critical developmental stages of the digestive tract of pigs when food is changed from liquid milk to a solid diet. The process of weaning could be divided into the acute phase and adaptive phase. During the acute phase, there is a significant reduction in villus height in the small intestine, causing severe damages to intestinal health and function [31,32]. Nyachoti et al. [4] reported that supplemented amino acid in nursery diet did not affect gut morphology of post-weaning pigs. In this study, we also did not observe an improvement in the height of villi in jejunum of pigs in amino acid supplemented treatments compared with FM-WPC at the end of wk 4.

In this study, VFA concentrations were used as an index of the changes in microbial fermentation in colon and cecal digest samples. The VFA, also known as short-chain fatty acids, are produced in pigs by microbial fermentation of undigested nutrients such as protein and carbohydrates in the gastrointestinal tract [4,33]. The degree of fermentation depends on the presence of nitrogen, minerals, vitamins, and primarily upon the source of dietary fiber which are essential for the microbial populations residing in the hindgut [33]. Diets for early-weaned pigs usually contain high levels of protein, which may enhance proliferation of pathogenic bacteria in the gastrointestinal tract [4]. It was shown that ammonia N in ileal digesta was reduced linearly as dietary CP was decreased, indicating that low CP diets help maintain enteric health in pigs by lowering toxic microbial metabolites such as ammonia [4]. Therefore, in the current study, it was anticipated that feeding post-weaning pigs with supplemented amino acid diets to reduce CP level would minimize microbial proliferation in the hindgut and also limit their byproducts such as VFA, thus giving pigs an opportunity to better utilize dietary nutrients for growth. However, results in this study showed that concentration of acetate in cecum digesta of pigs in FM-AA was the same to WPC-AA and AA, but tended to be greater than that of FM-WPC. Concentration of isovalerate in cecum digesta of pigs in FM-AA was the same to AA, but was greater than that of FM-WPC and WPC-AA, indicating increased fermentation of hind gut in FM-AA group. This was opposite to our previous hypothesis. In this study, we did not observe changes in voluntary feed intake, indicating that the amount of energy and std AA were not different among treatment. However, it is important to note that the levels of dietary fiber in the current experimental diets changed with differences in CP content. Treatment FM-AA in phase 1 and 2 contained higher corn content than FM-WPC and WF, which means dietary fiber content could be higher than FM-WPC and WF. Because feed intake was not different among treatment, and dietary fiber affects microbial activities in the gut [4], especially in the cecum and colon, it could speculated that the higher fiber content supplied in FM-WPC could be the reason to cause increased fermentation in FM-AA treatment.

The principal VFA produced in either the rumen or large intestine are acetate, propionate, and butyrate, which are produced in a ratio varying from approximately 75:15:10 to 40:40:20 [33-35]. Franklin et al. [36] showed a ratio of 71 acetate: 28 propionate: 10 butyrate in cecum digesta of pigs. Grieshop et al. [33] reproted a ratio of 60 acetate: 20 propionate: 15 butyrate in colon digesta of pigs. These VFA compositions were comparable to our current measurements with an average ratio of 54 acetate: 30 propionate: 13 butyrate. These results demonstrate that supplementing amino acid to replace the use of fish meal and/or whey protein concentrate did not adversely affect gut morphology and health of pigs.

Conclusions

This study demonstrates that supplemental L-Lys, L-Thr, L-Trp, DL-Met, L-Val, and L-Ile successfully replaced the use of fish meal and/or whey protein concentrate without adverse effects on growth performance, immune status, and gut health of pigs at d 21 to 49 of age. Positive response with supplemental amino acid in growth during the first week of post-weaning may be partially due to increased plasma insulin concentration potentially improving uptake of nutrients for protein synthesis and energy utilization. The replacement of fish meal and/or whey protein concentrate with supplemental amino acids could decrease the crude protein level in nursery diets, and potentially lead to substantial cost savings in expensive nursery diets.

Competing interests

The authors declare that they have no competing interests.

Authors' contributions

YZ, ACW carried out the animal studies. YZ, ACW, VF, SWK carried out sample analysis. YZ, RLP, SWK participated in the sequence alignment and drafted the manuscript. All authors read and approved the final manuscript.

Author details

[1]Department of Animal Science, North Carolina State University, Raleigh, NC 27695, USA. [2]Evonik Corp, Kennesaw, GA 30144, USA.

References

1. Junghans P, BeyeR M, Derno M, Petzke KJ, Küchenmeister U, Hennig U, Jentsch W, Schwerin M: **Studies on persisting effects of soy-based compared with amino acid-supplemented casein-based diet on protein metabolism and oxidative stress in juvenile pigs.** *Arch Anim Nutr* 2007, **61**:75–89.
2. Hansen JA, Knabe DA, Burgoon KG: **Amino acid supplementation of low-protein sorghum-soybean meal diets for 5- to 20-kilogram swine.** *J Anim Sci* 1993, **71**:452–458.
3. Shriver JA, Carter SD, Sutton AL, Richert BT, Senne BW, Pettey LA: **Effects of adding fiber sources to reduced-crude protein, amino acidsupplemented diets on nitrogen excretion, growth performance, and carcass traits of finishing pigs.** *J Anim Sci* 2003, **81**:492–502.
4. Nyachoti CM, Omogbenigun FO, Rademacher M, Blank G: **Performance responses and indicators of gastrointestinal health in early weaned pigs fed low-protein amino acid-supplemented diets.** *J Anim Sci* 2006, **84**:125–134.
5. Eklund M, Mosenthin R, Piepho HP, Rademacher M: **Effect of dietary crude protein level on basal ileal endogenous losses and standardized ileal digestibilities of crude protein and amino acids in newly weaned pigs.** *J Anim Physol Anim Nutr* 2008, **92**:578–590.
6. Sauer WC, Ozimek L: **Digestibility of amino acids in swine: Results and their practical applications.** *A review. Livest Prod Sci* 1986, **15**:367–388.
7. Baker DH, Hahn JD, Chung TK: **Nutrition and growth: The concept and application of an ideal protein for swine growth.** In *Growth of the Pig.* Edited by Hollis GR. Wallingford, Oxon, UK: CAB Intl; 1993:133.
8. Mosenthina R, Sauerb WC, Blankc R, Huismand J, Fan MZ: **The concept of digestible amino acids in diet formulation for pigs.** *Livest Prod Sci* 2000, **64**:265–280.
9. Kim SW, Baker DH, Easter RA: **Dynamic ideal protein and limiting amino acids for lactating sows: Impact of amino acid mobilization.** *J Anim Sci* 2001, **79**:2356–2366.
10. Kim SW, Hurley WL, Wu G, Ji F: **Ideal amino acid balance for sows during gestation and lactation.** *J Anim Sci* 2009, **87**:E123–E132.
11. Eklunda M, Mosenthina R, Piephob HP, Rademacher M: **Estimates of dietary threshold levels for crude protein and amino acids to obtain plateau values of apparent ileal crude protein and amino acid digestibilities in newly weaned pigs.** *Arch Anim Nutr* 2010, **64**:357–372.
12. Jin CF, Kim IH, Han K, Bae SH: **Effects of supplemental synthetic amino acids to the low protein diets on the performance of growing pigs.** *Asian-Aus J Anim Sci* 1998, **11**:1–7.
13. Dai ZL, Zhang J, Wu GY, Zhu WY: **Utilization of amino acids by bacteria from the pig small intestine.** *Amino Acids* 2010, **39**:1201–1215.

14. Figueroa JL, Lewis AJ, Miller PS, Fischer RL, Gomez RS, Diedrichsen RM: **Nitrogen metabolism and growth performance of gilts fed standard corn-soybean meal diets or low-crude protein, amino acid supplemented diets.** *J Anim Sci* 2002, **80:**2911–2919.
15. Columbus D, Cant JP, de Lange CFM: **Estimating fermentative amino acid losses in the upper gut of pigs.** *Livest Sci* 2010, **133:**124–127.
16. Stoner GR, Allee GL, Nelssen JL, Johnston ME, Goodband RD: **Effect of select menhaden fish meal in starter diets for pigs.** *J Anim Sci* 1990, **68:**2729–2735.
17. Grinstead GS, Goodband RD, Dritz SS, Tokach MD, Nelssen JL, Woodworth JC, Molitor M: **Effects of a whey protein product and spray-dried animal plasma on growth performance of weanling pigs.** *J Anim Sci* 2000, **78:**647–657.
18. NRC: *Nutrient Requirements of Swine.* 10th edition. Washington, DC: Natl Acad Press; 1998.
19. Chaytor AC, See MT, Hansen JA, de Souza ALP, Middleton TF, Kim SW: **Effects of chronic exposure of diets with low levels of aflatoxin and deoxynivalenol on growth and immune status of pigs.** *J Anim Sci* 2011, **89:**124–135.
20. Atinmo T, Baldijão C, Pond WG, Barnes RH: **Decreased dietary protein or energy intake and plasma growth hormone levels of the pregnant pig, its fetuses and developing progeny.** *J Nutr* 1976, **106:**940–946.
21. Mathew AG, Chattin SE, Robbins CM, Golden DA: **Effects of a direct-fed yeast culture on enteric microbial populations, fermentation acids, and performance of weaning pigs.** *J Anim Sci* 1998, **76:**2138–2145.
22. Shen YB, Piao XS, Kim SW, Wang L, Liu P, Yoon I, Zhen YG: **Effects of yeast culture supplementation on growth performance, intestinal health, and immune response of nursery pigs.** *J Anim Sci* 2009, **87:**2614–2624.
23. Touchette KJ, Carroll JA, Allee GL, Matteri RL, Dyer CJ, Beausang LA, Zannelli ME: *Effect of spraydried plasma and lipopolysaccharide exposure on weaned pig diets.* Columbia: PhD Diss. Univ. Missouri; 2002.
24. Chung TK, Baker DH: **Apparent and true amino acid digestibility of a crystalline amino acid mixture and of casein: comparison of values obtained with ileal-cannulated pigs and cecectomized cockerels.** *J Anim Sci* 1992, **70:**3781–3790.
25. Wu GY: **Amino acids: metabolism, functions, and nutrition.** *Amino Acids* 2009, **37:**1–17.
26. Southern LL, Baker DH: **Performance and concentration of amino acids in plasma and urine of young pigs fed diets with excesses of either arginine or lysine.** *J Anim Sci* 1982, **55:**857–866.
27. Le Bellego L, Noblet J: **Performance and utilization of dietary energy and amino acids in piglets fed low protein diets.** *Livest Prod Sci* 2002, **76:**45–58.
28. Friendship RM, Lumsden JH, McMillan I, Wilson MR: **Hematology and biochemistry reference values for Ontario swine.** *Can J Comp Med* 1984, **48:**390–393.
29. Dubreuil P, Lapierre H: **Biochemistry reference values for Quebec lactating dairy cows, nursing sows, growing pigs and calves.** *Can J Vet Res* 1997, **61:**235–239.
30. Klem TB, Bleken E, Morberg H, Thoresen SI, Framstad T: **Hematologic and biochemical reference intervals for Norwegian crossbreed grower pigs.** *Vet Clin Patho* 2009, **39:**221–226.
31. Pluske JR, Williams IH, Aherne FX: **Nutrition of the neonatal pig.** In *The Neonatal Pig: Development and Survival.* Edited by Varley MA. Wallingford, UK: CAB Int; 1996:187–235.
32. Hou YQ, Wang L, Ding BY, Liu YL, Zhu HL, Liu J, Li YT, Wu X, Yin YL, Wu GY: **Dietary a-ketoglutarate supplementation ameliorates intestinal injury in lipopolysaccharide-challenged piglets.** *Amino Acids* 2010, **39:**555–564.
33. Grieshop CM, Reese DE, Fahey GC: **Nonstarch polysaccharides and oligosaccharides in swine nutrition**. In *Swine Nutrition.* Edited by Lewis AJ, Southern LL. Boca Raton, FL: CRC Press; 2000.
34. Rao SO: *Effects of dietary supplementation of Lactobacillus based probiotics on growth and gut environment of nursery pigs.* Texas Tech University: M.S. Thesis; 2007.
35. Vasquez CM: *Effects of yeast culture supplementation to gestation and lactation diets on performance of sows and their offspring.* Texas Tech University: M.S. Thesis; 2008.
36. Franklin MA, Mathew AG, Vickers JR, Clift RA: **Characterization of microbial populations and volatile fatty acid compositions in the jejunum, ileum, and cecum of pigs weaned at 17 vs. 24 days of age.** *J Anim Sci* 2002, **80:**2904–2910.

28

Poor maternal nutrition inhibits muscle development in ovine offspring

Sarah A Reed*, Joseline S Raja†, Maria L Hoffman†, Steven A Zinn and Kristen E Govoni

Abstract

Background: Maternal over and restricted nutrition has negative consequences on the muscle of offspring by reducing muscle fiber number and altering regulators of muscle growth. To determine if over and restricted maternal nutrition affected muscle growth and gene and protein expression in offspring, 36 pregnant ewes were fed 60%, 100% or 140% of National Research Council requirements from d 31 ± 1.3 of gestation until parturition. Lambs from control-fed (CON), restricted-fed (RES) or over-fed (OVER) ewes were necropsied within 1 d of birth (n = 18) or maintained on a control diet for 3 mo (n = 15). Semitendinosus muscle was collected for immunohistochemistry, and protein and gene expression analysis.

Results: Compared with CON, muscle fiber cross-sectional area (CSA) increased in RES (58%) and OVER (47%) lambs at 1 d of age ($P < 0.01$); however at 3 mo, CSA decreased 15% and 17% compared with CON, respectively ($P < 0.01$). Compared with CON, muscle lipid content was increased in OVER (212.4%) and RES (92.5%) at d 1 ($P < 0.0001$). Muscle lipid content was increased 36.1% in OVER and decreased 23.6% in RES compared with CON at 3 mo ($P < 0.0001$). At d 1, *myostatin* mRNA abundance in whole muscle tended to be greater in OVER ($P = 0.07$) than CON. *Follistatin* mRNA abundance increased in OVER ($P = 0.04$) and tended to increase in RES ($P = 0.06$) compared with CON at d 1. However, there was no difference in myostatin or follistatin protein expression ($P > 0.3$). Phosphorylated Akt (ser473) was increased in RES at 3 mo compared with CON ($P = 0.006$).

Conclusions: In conclusion, maternal over and restricted nutrient intake alters muscle lipid content and growth of offspring, possibly through altered gene and protein expression.

Keywords: Growth, Maternal nutrition, Muscle development, Sheep

Background

Poor maternal nutrition can be defined as the overall nutritional excess or restriction during gestation as well as excess or lack of specific nutrient classes, such as proteins or minerals. Restricted- or over-feeding during gestation has immediate and long-term impacts on growth and development of offspring [1-4]. The type and extent of phenotypic alterations vary depending on the duration, type, and form of inappropriate maternal nutrition. Maternal over or restricted nutrition during gestation can affect pre- and post-natal muscle growth, altering skeletal muscle fiber number, intramuscular adiposity, and fiber type composition [5-7]. The fetal period is crucial to muscle development. In sheep, muscle fiber formation begins around d 20 and is completed between d 80 and 125 of gestation [8,9], after which fiber number is fixed. Postnatal muscle growth occurs through hypertrophy with increases in protein accretion and contributions from muscle satellite cells. Satellite cell activity can be evaluated by the expression of Pax7 and the myogenic regulatory factors (MRF) MyoD, Myf5, myogenin and MRF4 as markers of specific stages in myogenesis [10,11]. A previous study demonstrated that muscle from 75-d old fetuses from ewes fed an obesogenic diet exhibited reduced gene expression of *MyoD* and *mgn* compared with fetuses from control ewes [12].

Although the various phenotypic effects of maternal restricted- and over-nutrition on offspring growth and development are known, the mechanisms through which poor maternal nutrition alters animal body composition, particularly muscle development, are not well understood. We hypothesized that over and restricted nutrition (total nutrient intake) in gestating ewes would negatively affect

* Correspondence: sarah.reed@uconn.edu
†Equal contributors
Department of Animal Science, University of Connecticut, Storrs, CT 06269, USA

muscle growth in the offspring by reducing muscle fiber size, and altering gene and protein expression via the Akt signaling pathway. The objective of this study was to determine the effects of maternal over or restricted nutrition during gestation on muscle development in the offspring.

Methods

Animals

All animal experiments were reviewed and approved by the University of Connecticut Institutional Animal Care and Use Committee.

Thirty-six multiparous ewes (25 Dorsets, 7 Shropshires, and 4 Southdowns) were selected from the University of Connecticut sheep flock and bred within breed by live cover to 1 of 5 different rams (3 Dorsets, 1 Southdown and 1 Shropshire) following estrous synchronization with progesterone controlled intravaginal drug release devices (Pfizer Animal Health, New York, NY) and Lutalyse (Pfizer Animal Health [13,14]). Date of breeding was considered the day that ewes were marked by the ram. At approximately d 30 of gestation, ewes were moved into individual pens. Pregnancy was confirmed using ultrasonography. Ewes determined to be pregnant were balanced for breed and ram exposure, and randomly assigned to 1 of 3 diets: 100, 60, or 140% National Resource Council (NRC) requirements for total digestible nutrients (TDN) for ewes pregnant with twins [15]. Ewes were fed a complete pelleted feed (Central Connecticut Farmer's Co-Op, Manchester, CT) which contained 12.8% crude protein, 31.10% acid digestible fiber, 42.10% neutral digestible fiber, and 74.00% total digestible nutrients (TDN). Rations were calculated weekly on an individual BW basis and fed daily to provide 100, 60, or 140% of NRC requirements for TDN for gestating ewes. Ewes were transitioned onto diets at d 31 ± 1.3 of gestation and remained on their respective diets until parturition. Individual housing ensured proper feed intake and any refusals were measured. No refusals were present from any ewe in any group on any day. Upon parturition, lambs from control-fed ewes (CON; n = 12), lambs from over-fed ewes (OVER, n = 12) and lambs from restricted-fed ewes (RES; n = 12) were allowed to nurse their mother for colostrum for up to 24 h. Within 1 d of parturition, one lamb was removed from the ewe for use on study and any remaining lambs and the ewe were returned to the flock. A total of 62 lambs were born to all ewes (Dorset = 44; Southdown = 7; Shropshire = 11). Actual distribution of singletons, twins and triplets were singleton (control-fed = 4; restricted-fed = 0; over-fed = 4); twins (control-fed = 7; restricted-fed = 10; over-fed = 7); triplets (control-fed = 1; restricted-fed = 2; over-fed = 1). The largest of the lambs was chosen when lambs were the same gender. If lambs were different genders, the largest male lamb was chosen for use on the study since males are often slaughtered for market and females kept for breeding. One-half of the lambs from each diet group were slaughtered within 1 d of birth. The remaining lambs were maintained on the same control diet until 3 mo of age regardless of maternal dietary treatment. These lambs were fed milk replacer (1.7% of BW; Land O'Lakes Animal Milk Product Company; Shoreview, MN) from a bottle until weaning at 60 d of age and allowed ad libitum access to water, creep feed (Lamb BT, Blue Seal Feeds; Litchfield, CT), and second cutting hay for the entire 3 mo period. Feed intake was not measured for lambs. Three animals in the 3 mo group died due to causes unrelated to the study. At time of slaughter and sampling, animals were considered fed. The final distribution for gender across treatments was as follows: CON (d 1 = 3 rams, 3 ewes; 3 mo = 3 rams, 3 ewes), RES (d 1 = 3 rams, 3 ewes; 3 mo = 5 rams) and OVER (d 1 = 4 rams, 2 ewes; 3 mo = 3 rams, 3 ewes).

Sample collection and processing

Animals were euthanized with an intravenous injection of Beuthanasia-D Special (Merck Animal Health; Summit, NJ) containing 390 mg/mL sodium pentobarbital and 50 mg/mL phentoin based on BW, followed by exsanguination. Muscle samples were collected from the midpoint of the left semitendinosus immediately after euthanasia. Samples for RNA or protein extraction were immediately snap frozen in liquid nitrogen. Muscle samples for histology were embedded in Tissue-Tek OCT (Fisher Scientific, Pittsburgh, PA) and frozen in dry ice-cooled isopentane. Samples were stored at −80°C until further use.

Immunohistochemistry

To determine muscle fiber cross-sectional area (CSA), 10 μm muscle sections were collected using a Leica CM 3050S cryostat (Wetzlar, Germany). Sections were rehydrated in PBS with 0.1% TritonX-100 (PBS-T) for 5 min and fixed with 4% paraformaldehyde (PF) for 10 min. To determine myonuclear number, Hoescht 33342 (1:1,000, Invitrogen, Carlsbad, CA) was used to visualize nuclei and Alexafluor 568 conjugated wheat germ agglutinin (WGA, 1:50, Invitrogen, Carlsbad, CA) was used to visualize the sarcolemma [16]. The number of myonuclei per fiber was determined by dividing the number of myonuclei by the number of muscle fibers in a 20X image. To identify muscle fiber types, cross sections were incubated for 30 min in 5% horse serum in PBS-T to block non-specific antigen sites. Sections were incubated for 1 h with anti-bovine type IIB myosin heavy chain (1:10, BF-F3, Developmental Studies Hybridoma Bank), anti-bovine type IIA myosin heavy chain (1:10, SC-71, Developmental Studies Hybridoma Bank), anti-bovine myosin heavy chain slow (1:10, BAD5, Developmental Studies Hybridoma Bank), and anti-dystrophin (1:50, Pierce, Rockford, IL). Following extensive washing with PBS, sections were incubated with the appropriate AlexaFluor antibodies (1:1,000) and Hoescht 33342 for 30 min. Sections were rinsed with PBS. To quantify intramuscular adiposity, sections were blocked with 5% horse serum in PBS-T for 1 h followed by three 5 min washes with PBS. The sections were incubated with 4,4-Difluoro-1,3,5,7,8-Pentamethyl-4-Bora-3a,4a-Diaza-s-Indacene (BODIPY 493/503, 1:400, Invitrogen D-3922), WGA (1:400), and Hoescht 33342 (1:1,000) in PBS for 1 h in the dark in a humidified box. Following extensive washing with PBS, sections were cover-slipped with 9:1 glycerol/PBS solution. Images for all immunohistochemistry procedures were captured using an AxioCam camera (Zeiss, Jena, Germany) mounted to an AxioObserver microscope (Zeiss) or an Orca ER camera (Hamamatsu, Boston, MA) mounted to an Axiovert Wide-field microscope (Zeiss), false colored and merged using ImageJ (NIH). Cross-sectional area was measured as the region within the fiber boundary using the area measurement tool in ImageJ. At least 10 images were obtained from 4 different muscle sections, resulting in the analysis of a minimum of 500 fibers per muscle. Intramuscular adiposity was quantified as a percent of the total area stained in 25 images per animal [17].

RNA extraction

Tissue was homogenized using the Qiagen Tissuelyser system with 1 mL TriReagent (Sigma Aldrich). A Qiagen Mini Kit was used to extract RNA according to the manufacturer's protocol (Qiagen; Valencia, CA). Genomic DNA was removed from samples using a Turbo DNA Free kit (Ambion, Foster City, CA). The quality of RNA was determined using a Bioanalyzer analysis system (Agilent Technologies, Santa Clara, CA).

Real-time reverse transcription (RT)-PCR

Reverse transcription was performed using 300 ng total RNA with OligodT primer (Ambion) and master mix containing 5.5 μL of 5X Buffer (Invitrogen), 1.0 μL dNTP (Promega, Madison, WI), 2.0 μL DTT and 0.5 μL Superscript II (Invitrogen) for a total reaction volume of 20 μL. The samples and master mix underwent a standard RT protocol starting at 70°C for 10 min, 4°C for 20 min, 37°C for 3 min, 42°C for 1 h, 4°C for 3 min, 90°C for 2.5 min. Real-time RT-PCR primers were designed using Primer3 and NCBI BLAST, validated as previously described [18] and synthesized by Integrated DNA Technologies (Coralville, IA). Primer sequences are presented in Table 1. Real-time RT-PCR was performed using Power SybrGreen Master Mix (Invitrogen) and the ABI 7900 HT Fast Real-time PCR machine (Applied Biosystems, Foster City, CA) as previously described [19]. The total volume of the reaction mixture was 25 μL (5 μL cDNA, 3 μL nuclease free water, 1 μL each 10 nmol/L forward and reverse primer, and 10 μL SybrGreen). Real-time RT-PCR was performed using standard cycling conditions (Stage 1: 50°C for 2 min and 95°C for 10 min, Stage 2: 95°C for 15 s and 60°C for 1 min for 40 cycles, Stage 3: 95°C for 15 s and 60°C for 15 s with a 2% ramp to 95°C for 5 min). $^{\Delta}$Ct values were obtained and used to calculate the $^{\Delta\Delta}$Ct values to determine relative gene expression [20]. Glyceraldehyde 3-phosphate dehydrogenase (GAPDH) mRNA expression was used as the internal control and did not differ between treatment groups ($P \geq 0.25$).

Western blotting and analyses

Whole muscle protein was extracted by a standard bead beating method (TissueLyser II, Qiagen) in SDS-PAGE sample buffer (250 mmol/L Tris, pH 6.8, 8% SDS, 40% glycerol, and 0.4% β-mercaptoethanol). Protein content was measured by DC Protein Assay (BioRad, Hercules, CA) and an equal amount of protein was separated through 12% polyacrylamide gels. Proteins were transferred to polyvinylidene difluoride (PVDF) membranes and blocked with blocking buffer (Licor Biosciences, Lincoln, NE) for 1 h with gentle agitation. Blots were incubated with primary antibodies overnight at 4°C with agitation. Following extensive washing, blots were incubated with the appropriate infrared secondary antibody. Blots were washed with PBS-T, scanned with the Licor Odyssey Scanning system, and analyzed with Image Studio (Licor Biosciences). After analysis, blots were converted to black and white, inverted, and cropped for presentation (Adobe Photoshop CS5, San Jose, CA). All analysis was done before any adjustment. Antibodies used were: anti-phospho Akt (ser473; 1:1,000, Cell Signaling Technologies, Beverly, MA), anti-phospho Akt (thr308; 1:1,000, Cell Signaling Technologies), anti-Akt (1:1,000, Cell Signaling Technologies), anti-myostatin (N-19; 1:200, Santa Cruz Biotechnology, Dallas, TX), anti-follistatin (H-114; 1:200, Santa Cruz Biotechnology), anti-tubulin (loading control; 0.5 μg/mL, Thermo Scientific, Waltham, MA), goat anti-mouse IRDye 800CW (1:2,000, Licor Biosciences), and goat anti-rabbit IRDye 680RD (1:2,000, Licor Biosciences).

Statistical analysis

Data were sorted by age (1 d or 3 mo of age) and analyzed using the PROC MIXED procedure (SAS Inst. Inc., Cary, NC) using dietary treatment as the fixed effect. Treatment mean comparisons were performed using LSMEANS statement and PDIFF option. Differences were determined to be significant at $P \leq 0.05$ or a tendency at $0.05 < P \leq 0.10$. Gender and breed differences were not evaluated due to the limited number of females, Shropshire and Southdown lambs in each treatment at each time point. For CSA, fiber type composition, and lipid accumulation, 3 lambs per treatment group were used at d 1 (n = 9) and 3 mo of age (n = 9). For protein expression analysis, 4 lambs per

Table 1 Primer sequences

Gene	Primer sequences (5′ → 3′)	Amplicon length	Reference
Pax7			
Forward	GAGACCGACTGCTGAAGGAC	167	XM_002685738
Reverse	ATGCTGTGCTTGGCTTTCTT		
Myf5			
Forward	AGACGCCTGAAGAAGGTGAA	134	XM_004006219.1
Reverse	AGCAGCTCCTGCAGACTCTC		
Myogenin			
Forward	TGGGCGTGTAAGGTGTGTAA	169	Tong et al., 2009 [12]
Reverse	TGCACAGGATCTCCACTTTG		
MyoD			
Forward	CCCTGGTGACTTCAGCTGTT	239	Tong et al., 2009 [12]
Reverse	CCTGCCTGCCGTATAAACAT		
Myostatin			
Forward	CACAGAAGGTCTTCCCCTCA	147	NM_001001525.2
Reverse	GGTTAAATGCCAACCATTGC		
Follistatin			
Forward	AAAACCTACCGCAACGAATG	120	NM_001257093.1
Reverse	GAGCTGCCTGGACAGAAAAC		
Foxo1			
Forward	TGACTTGGACGGCATGTTTA	157	XM_004012275.1
Reverse	CCAGCTGTGTGTTGTCGTCT		
Foxo3a			
Forward	GGGGAGTTTGGTCAATCAGA	170	NM_001267889.1
Reverse	TTTGCATAGACTGGCTGACG		
GAPDH[1]			
Forward	GGCGTGAACCACGAGAAGTATAA	118	Buza et al., 2009 [21]
Reverse	CCCTCCACGATGCCAAAGT		

[1]glyceraldehyde 3-phosphate dehydrogenase.

treatment group were used at each time point. For gene expression analysis, 15 lambs were used at 1 d (CON: n = 4, RES: n = 6, and OVER: n = 5) and 12 lambs were used at 3 mo (CON: n = 5, RES: n = 4, and OVER: n = 3).

Results

By the end of gestation, restricted and overfed diets effectively reduced (−18.9%; 18.3 ± 3.6 kg; $P < 0.01$) or increased (6.6%; 119.7 ± 3.6 kg; $P < 0.10$) ewe body weight (BW), respectively compared with control-fed ewes (112.9 ± 3.6 kg). Overall, OVER lambs weighed 13% more than CON lambs from 1 d to 3 months of age [$P \leq 0.05$; [22]]. However, there were no differences in BW between RES and CON lambs from 1 d to 3 months of age ($P = 0.70$).

The effects of maternal nutrition during gestation on muscle development were examined in the semitendinosus (**STN**) muscle of a subset of lambs at 1 d and 3 mo of age. At d 1, the CSA of muscle fibers from the STN muscle of OVER and RES lambs was 47% and 57% greater than CON lambs, respectively ($P < 0.0001$, Figure 1A). At 3 mo of age the CSA of OVER and RES lambs was 17% and 15% less, respectively, than the CSA of CON lambs ($P < 0.0001$, Figure 1B). Differences in CSA could have resulted from alterations in myonuclear accretion, thus the number of myonuclei per fiber was calculated. The number of myonuclei per fiber was not different between treatment groups at 1 d ($P = 0.16$) or 3 mo ($P = 0.32$) of age (Table 2). The number of Type IIa fibers was 13.3% greater in OVER lambs and 14.6% greater in RES lambs compared with CON lambs at d 1 ($P < 0.01$, Figure 1C). The percent of Type I fibers was decreased by 35.4% and 40.5% in OVER and RES, respectively, compared with CON ($P < 0.01$, Figure 1C). There was no difference in the muscle fiber composition of the OVER or RES lambs compared with CON lambs at 3 mo of age ($P > 0.1$, Figure 1D). No Type IIb fibers were identified in the STN muscle of any of

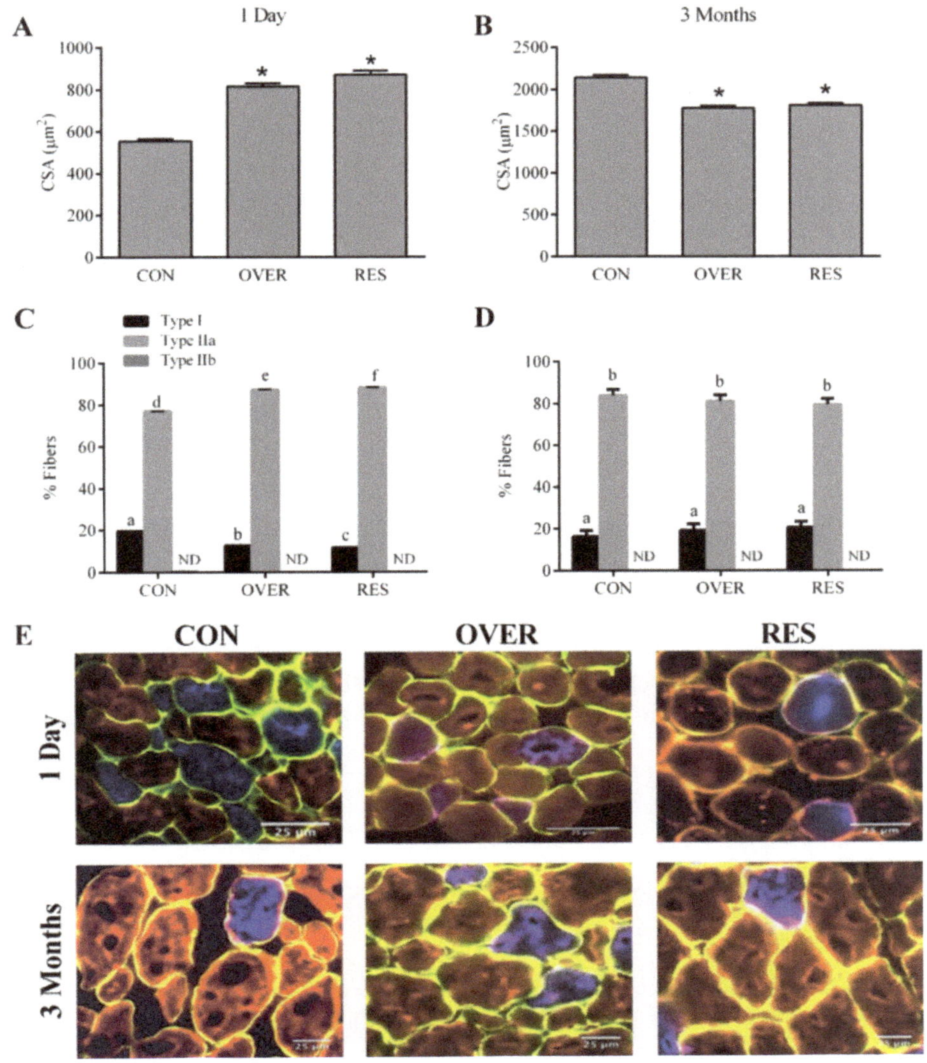

Figure 1 Poor maternal nutrition alters cross sectional area and fiber type composition of offspring. Muscle cross sections (10 µm) were immunostained with α-dystrophin to mark muscle fiber membranes (yellow), α-myosin heavy chain (MyHC) I (red), α-MyHC IIA (blue), or α-MyHC IIB (green) to identify the different muscle fiber types. Cross sectional area was measured at 1 d **(A)** and 3 mo **(B)**. Fiber type distribution was determined at 1 d **(C)** and 3 mo **(D)**. Representative photographs are shown **(E)**. $^{*}P \leq 0.05$ compared with CON, different letters indicate $P \leq 0.05$ compared with CON. CON = lambs from control-fed ewes (100% NRC), OVER = lambs from overfed ewes (140% NRC), RES = lambs from restricted-fed ewes (60% NRC), n = 3 lambs/group.

the lambs. Changes in CSA were not specific to Type I or Type IIa fibers at either age (Table 3). At d 1 lipid content in STN muscle from OVER and RES lambs was 212.4% and 92.5% greater ($P < 0.0001$, Figure 2B) compared with CON lambs. At 3 mo of age, OVER lambs had a 36.1% greater ($P = 0.003$) while RES lambs had 23.6% less ($P = 0.03$) lipid accumulation compared with CON lambs (Figure 2C).

The regulation of postnatal muscle growth is complex, involving many different signaling pathways and cellular functions. The lack of postnatal muscle fiber growth in OVER and RES lambs could be due to alterations in

Table 2 Number of myonuclei per muscle fiber[1] in the semitendinosus muscle

Treatment[2]	Day 1	3 Months
CON	0.57 ± 0.13	0.72 ± 0.08
OVER	0.37 ± 0.06	0.62 ± 0.13
RES	0.30 ± 0.02	0.46 ± 0.02

[1]Mean ± SEM.
[2]CON = lambs from control-fed ewes (100% NRC), OVER = lambs from overfed ewes (140% NRC), RES = lambs from restricted-fed ewes (60% NRC), n = 3 lambs/group.

Table 3 Cross-sectional area (CSA)[1] of Type I and Type IIa fibers in the semitendinosus muscle

	Day 1		3 Months	
Treatment[2]	Type I, CSA[3]	Type IIa, CSA	Type I, CSA	Type IIa, CSA
CON	306.9 ± 92.5	337.3 ± 42.3	1,791.1 ± 417.2	1,934.3 ± 119.6
OVER	570.7 ± 80.5	570.2 ± 75.0	1,099.6 ± 219.1	1,109.7 ± 20.4
RES	515.2 ± 55.7	622.0 ± 14.7	1,028.2 ± 46.7	1,151.3 ± 37.5

[1]Mean μm^2 ± SEM.
[2]CON = lambs from control-fed ewes (100% NRC), OVER = lambs from overfed ewes (140% NRC), RES = lambs from restricted-fed ewes (60% NRC), n = 3 lambs/group.
[3]Cross sectional area, mean ± SEM.

satellite cell function or changes in anabolic/catabolic signaling pathways. Over or restricted nutrient consumption during gestation did not alter expression of *Pax7*, *MyoD*, or *myogenin* in whole muscle tissue of lambs at 1 d ($P \geq 0.57$) or 3 mo ($P \geq 0.24$; Table 4) of age.

Reduced CSA at 3 mo of age may be a result of altered expression of myostatin or follistatin, which influence accretion of muscle mass. At d 1, OVER lambs exhibited a tendency toward increased *myostatin* gene expression ($P = 0.06$; Table 4); however no change was observed in RES lambs ($P = 0.78$). At d 1, *follistatin* gene expression was greater in OVER lambs ($P = 0.04$) and tended to be increased in RES lambs ($P = 0.06$, Table 4). No change in *myostatin* or *follistatin* gene expression was observed at the 3 mo time point ($P = 0.36$, Table 4). Further, over or restricted maternal nutrition did not affect the expression of the precursor ($P = 0.5$), active dimeric ($P = 0.7$) or the active monomeric ($P \geq 0.7$) forms of myostatin protein at d 1 or 3 mo of age (Figure 3D, E). The active monomeric isoform of myostatin was not detectable at d 1. There was also no effect of poor maternal nutrition on the expression of follistatin protein in lambs at 1 d ($P = 0.5$) or 3 mo of age ($P = 0.7$, Figure 3F).

The balance of protein accretion and degradation is crucial to the maintenance of muscle mass and muscle hypertrophy. Increased transcription of forkhead box O1 (*FoxO1*) and *FoxO3a* is associated with catabolic states in muscle [23-26]. At 3 mo of age, reduced *FoxO1* mRNA

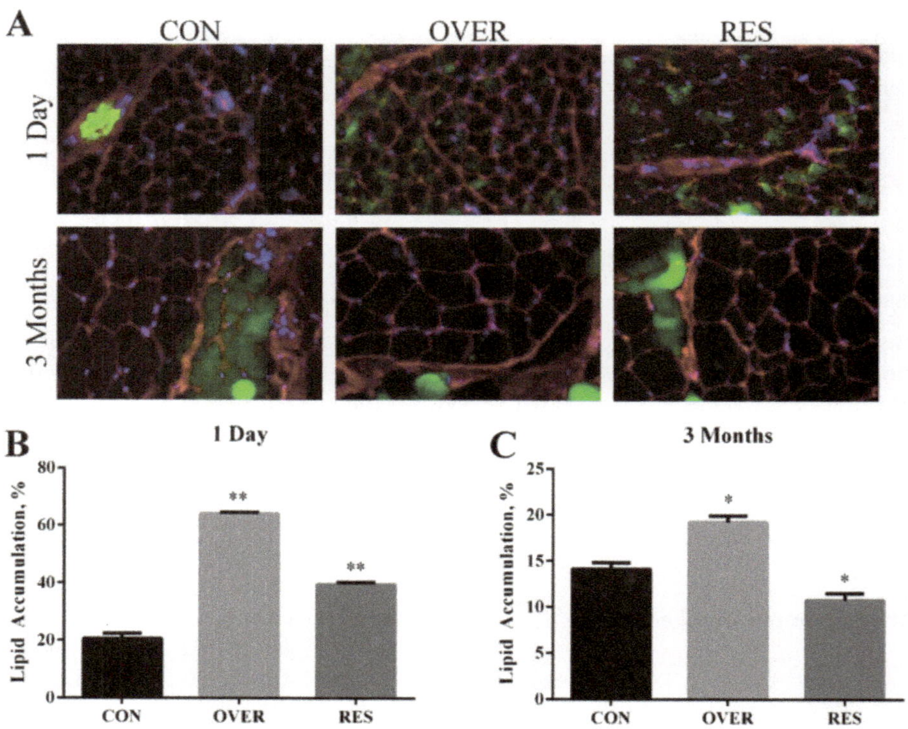

Figure 2 Poor maternal nutrition alters lipid accumulation in the semitendinosus muscle of offspring. Cross sections (10 μm) of the semitendinosus muscle were stained with wheat germ agluttinin (WGA; red) to delineate the muscle fiber membrane, BODIPY 493/503 (green) which stains lipids and Hoescht 33342 (blue) to identify nuclei. Representative photographs are shown **(A)**. Data from 1 d **(B)** and 3 mo of age **(C)** are presented as mean ± SEM. $**P < 0.0001$ vs. CON; $*P < 0.05$ vs. CON. CON = lambs from control-fed ewes (100% NRC), OVER = lambs from overfed ewes (140% NRC), RES = lambs from restricted-fed ewes (60% NRC), n = 3 lambs/group.

Table 4 Gene expression[1] in semitendinosus muscle of CON[2], OVER or RES lambs

	Day 1				3 Months			
Gene	CON	OVER	RES	*P*-value	CON	OVER	RES	*P*-value
Pax7	1.38 ± 0.67	1.20 ± 0.27	1.86 ± 0.51	0.70	1.14 ± 0.31	1.42 ± 0.24	1.94 ± 0.77	0.46
MyoD	2.16 ± 1.36	3.51 ± 0.90	2.70 ± 0.98	0.88	1.11 ± 0.29	1.01 ± 0.01	1.27 ± 0.29	0.83
Myogenin	1.53 ± 0.74	2.13 ± 0.74	2.65 ± 0.71	0.57	1.21 ± 0.35	1.24 ± 0.47	2.41 ± 0.66	0.24
Myostatin	1.17 ± 0.36	2.94 ± 0.50**	1.63 ± 0.51	0.11	1.02 ± 0.10	1.89 ± 0.26	1.51 ± 0.58	0.36
Follistatin	1.03 ± 0.16	1.92 ± 0.39*	1.73 ± 0.28**	0.09	1.10 ± 0.21	1.56 ± 0.21	1.51 ± 0.34	0.36
Foxo1	1.12 ± 0.30	0.64 ± 0.29	0.74 ± 0.29	0.37	1.06 ± 0.16	0.91 ± 0.36	0.32 ± 0.07*	0.04
Foxo3a	1.60 ± 0.73	0.72 ± 0.26	1.12 ± 0.56	0.70	1.09 ± 0.24	1.42 ± 0.43	0.77 ± 0.14	0.21

[1]Relative to CON, mean ± SEM.
[2]CON = lambs from control-fed ewes (100% NRC, d 1, n = 4; 3 mo, n = 5), OVER = lambs from overfed ewes (140% NRC, d 1, n = 5; 3 mo, n = 3), RES = lambs from restricted-fed ewes (60% NRC, d 1, n = 6; 3 mo, n = 4).
*$P < 0.05$ compared with CON.
**$P < 0.01$ compared with CON.

expression was observed in RES lambs ($P = 0.04$, Table 4). No change was observed in *FoxO1* mRNA expression at d 1 ($P = 0.37$) or *FoxO3a* mRNA expression at d 1 or 3 mo ($P \geq 0.20$, Table 4). Increased signaling through the insulin receptor substrate/phosphoinositide-3-kinase (IRS/PI3K)-Akt axis is associated with muscle growth [27]. There were no significant effects of poor maternal nutrition on the expression of pAkt (ser473), pAkt (thr308) or total Akt protein at d 1 ($P \geq 0.30$; Figure 3B). There was no effect of maternal diet composition on the amount of phosphorylated Akt (thr308) or total Akt protein expression at d 1 ($P \geq 0.3$; Figure 3B) or 3 mo of age ($P \geq 0.7$; Figure 3C). However, at 3 mo of age, pAkt (ser473) increased 4.5-fold in RES lambs compared with CON (CON: 1.0 ± 0.3; RES: 4.5 ± 1.2; $P = 0.006$; Figure 3C), whereas OVER lambs had similar expression of pAkt (ser473) to CON ($P = 0.6$).

Discussion

In the present study, over or restricted maternal nutrition during gestation altered muscle growth and lipid accumulation postnatally, muscle fiber type at d 1, postnatal muscle gene expression, and postnatal phosphorylation of Akt in the offspring at 3 mo of age. Thus, poor maternal nutrition during gestation, whether restricted- or overfeeding, impairs muscle development and initiates lasting negative effects on muscle growth of the offspring.

The total number of muscle fibers in mammals is predetermined at birth [28-30]; thus the prenatal development of muscle fibers is crucial for normal muscle growth and function [28,29,31,32]. Importantly, muscle growth is permanently impaired by poor maternal nutrition, both under- and over-feeding [29]. The thrifty phenotype hypothesis suggests that a poor nutrient environment prenatally programs the animal for a poor environment after birth [33]. Given the timing of the poor nutrition (starting at d 31 ± 1.3 d of gestation), it is likely that secondary and tertiary myogenesis were affected resulting in the development of fewer, larger muscle fibers in OVER and RES lambs. This is consistent with other work that suggests that secondary myofibers are preferentially affected by environmental conditions, including nutrition [34]. The poor postnatal muscle growth observed in OVER lambs may be due to prenatally programmed changes in nutrient partitioning, resulting in decreased muscle mass and increased adiposity at 3 mo of age. Although not measured in the current study, changes in fibrosis may also result in decreased muscle mass. Indeed, both maternal obesity and nutrient restriction result in increased collagen content and crosslinking in the muscle [35,36]. However, RES lambs exhibited increased lipid accumulation in the muscle at 1 d of age but decreased lipid content at 3 mo compared with CON lambs, indicating that the mechanisms by which poor maternal nutrition affects both muscle growth and fat deposition in the muscle is specific to the nutritional insult during gestation. In other studies, maternal obesity during gestation increased lipid content in fetal muscle and offspring at 22 mo of age [6,17] whereas nutrient restriction from 28 to 78 d of gestation increased intramuscular triglyceride content of the longissimus muscle in offspring at 8 mo of age [5]. Differences between the current work and that of others may be due to differences in the timing and length of poor nutrition, the severity of over- or under-feeding, or strategies for postnatal feeding. In our study, lambs were removed from the ewe and bottle fed until weaning to remove the impact of changes in milk quantity and composition due to differences in maternal dietary intake during gestation [37]. Importantly, the changes in muscle growth and fat deposition in the current experiment occurred in a flock representative of a production flock, which are heterogeneous for number and gender of the offspring.

Muscle fiber type is related to oxidative capacity [38], insulin sensitivity [39], and meat quality [40]. Alterations in fiber type composition due to poor maternal diet have been conflicting. Under-nutrition during gestation has been shown to increase [5,41] or decrease

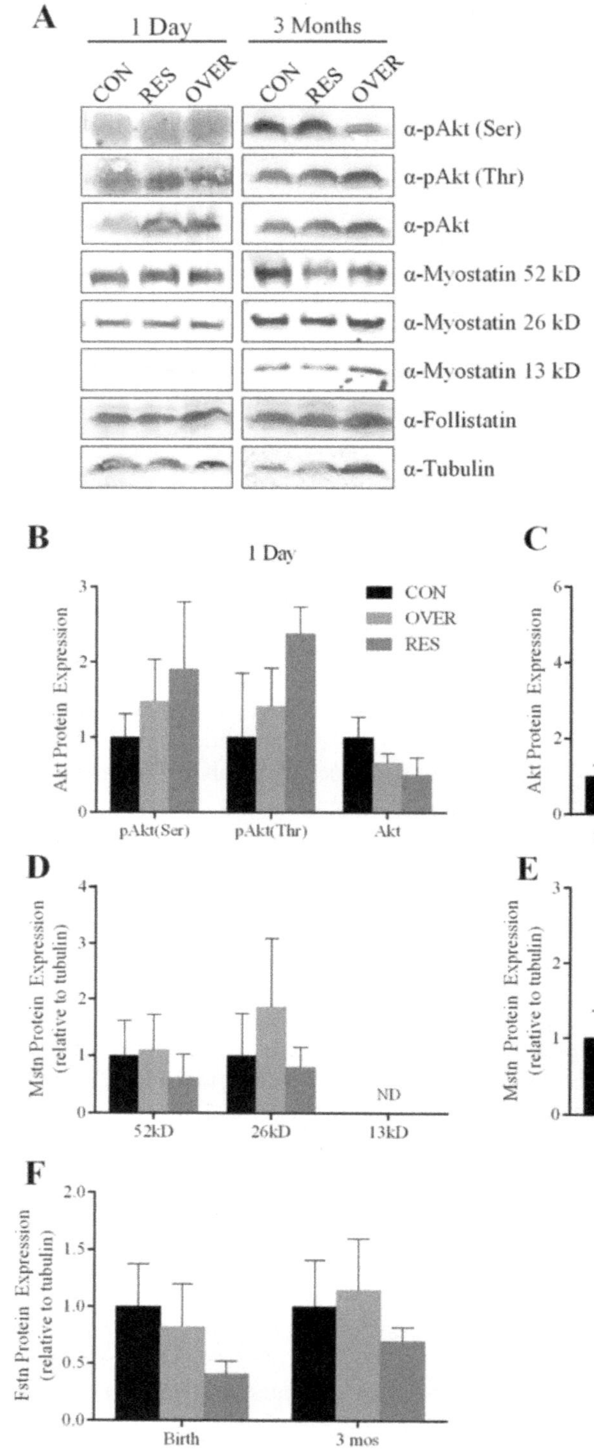

Figure 3 Restricted maternal nutrition increases Akt phosphorylation postnatally but does not affect myostatin or follistatin protein expression. Semitendinosus muscle was analyzed for Akt (**B & C**), myostatin (mstn, **D & E**) and follistatin (fstn, **F**) by western blot. Tubulin was used as a loading control. Representative blots are shown **(A)**. Data are presented as mean ± SEM. *$P = 0.006$. CON = lambs from control-fed ewes (100% NRC), OVER = lambs from overfed ewes (140% NRC), RES = lambs from restricted-fed ewes (60% NRC), n = 4 lambs/group. ND = not detectable.

[42] the number of Type II fibers. Likewise, overnutrition during gestation decreased [43,44] or had no effect [29] on the percent of Type II fibers. In the current study, the percent of Type II fibers is increased due to restricted- or over-feeding at d 1. The contrasting results are likely due to differences in species, developmental stages, and the timing and/or duration of maternal dietary changes. The availability of an optimal diet postnatally may have allowed for fiber type switching [45] as there was no difference in the fiber type composition of the STN muscle at 3 mo of age. Interestingly, there was no effect of poor maternal nutrition on the CSA of Type I and Type IIa fibers, suggesting that regulation of fiber size by maternal nutrition is not fiber type specific, consistent with previous reports [28,46].

In an effort to understand the factors that may be involved in the phenotypic changes resulting from a poor maternal diet during gestation, we investigated the expression of several factors that regulate muscle development and growth. The lack of fiber growth in OVER or RES lambs could result from failure of a number of mechanisms, including alterations in satellite cell function or changes in anabolic/catabolic mechanisms. Calves from restricted-fed dams possess larger muscle fibers with fewer Pax7 immunopositive cells at d 85 of gestation [36]. The lack of change in *Pax7, MyoD,* and *myogenin* mRNA expression in lambs at d 1 and 3 mo in the current study suggests that changes in satellite cell number and/or function may be more evident in the prenatal animal. The protein kinase Akt has a pivotal role in many cell processes including proliferation, glucose metabolism and angiogenesis [47]. Activated (phosphorylated) Akt enhances skeletal muscle growth through increased protein synthesis and inhibition of protein degradation [27]. Akt is fully activated when phosphorylated at Thr308 and Ser473 [48]. The threonine residue is phosphorylated by pyruvate dehydrogenase kinase (PDK)-1 which leads to phosphorylation of the serine residue. The mechanisms which result in Ser473 phosphorylation are currently under investigation, but may be due to autophosphorylation [48] or phosphorylation by several other candidate kinases, including mTORC2 [49-51]. Constitutively active Akt induces hypertrophy of transfected muscle fibers [52,53] and in muscle-specific transgenic models [27,54]. Further, in response to the activation of IRS/PI3K pathway, Akt phosphorylates FoxO1, blocking its nuclear localization and transcriptional activity [55]. FoxO1 regulates muscle metabolism and protein breakdown and increased mRNA expression is associated with catabolic states in muscle [23-26,56,57]. In this study, Akt phosphorylation at Ser473 was increased and *FoxO1* mRNA expression was reduced in 3-mo old offspring born to restricted-fed mothers, suggesting alterations to the signaling pathways that control protein synthesis/degradation. The upstream pathways controlling increased Akt activity in RES animals at this time point are unclear, but may be related to changes in circulating growth factors. It is unlikely due to changes in circulating insulin like growth factor (IGF) or IGF binding proteins (IGFBP) as there was no difference in the concentrations of these proteins between RES and CON animals [22]. However, the reduced CSA in RES lambs at 3 mo suggests that changes in Akt phosphorylation are not sufficient to overcome the deficit in muscle growth at this time. Indeed, at 8 mo of age, lambs subjected to maternal nutrient restriction during gestation exhibited greater muscle fiber diameter than control animals [5]. Increased signaling through the Akt pathway at 3 mo may provide a mechanism for growth later in life. Interestingly, whereas similar effects were observed in CSA changes in OVER and RES lambs, overfeeding during gestation did not affect activation of Akt or mRNA expression of *FoxO1* in the semitendinosus of lambs at d 1 or 3 mo of age, suggesting that poor postnatal growth in these animals may occur through a separate mechanism.

Myostatin is a member of the transforming growth factor (TGF)-β family which negatively regulates myogenesis at multiple levels [58,59]. Myostatin inhibits muscle growth and development via Smad dependent and independent pathways resulting in decreased protein synthesis and satellite cell proliferation [60,61]. No change in myostatin protein was observed at d 1 or 3 mo in the offspring of poorly nourished ewes. Similarly, no change in follistatin protein, an antagonist of myostatin activity, was observed. However, *myostatin* mRNA expression tended to be increased at d 1 in lambs from poorly nourished ewes, which may be a result of post-translational regulatory mechanisms. Nutritional status can effect microRNA expression resulting in changes in myostatin protein expression in rats and humans [62,63]. In sheep, microRNA can regulate myostatin protein expression [61]. The microRNA regulation of translation may explain the lack of increase in myostatin protein expression observed despite the tendency for an increase in *myostatin* gene expression. Although the regulatory mechanisms are not fully understood, our observations suggest that post-translational modifications may be involved and justify additional studies.

Conclusions

In summary, over or restricted maternal nutrition in ewes during gestation altered CSA, lipid accumulation, muscle fiber type, and gene and protein expression in the semitendinosus muscle tissue of offspring, resulting in poor postnatal muscle growth. Thus, maternal nutrition during gestation is a critical factor in determining the growth potential of the offspring. The mechanisms by which these changes occur are currently unknown and warrant significant further investigation.

Competing interests
The authors declare that they have no competing interests.

Authors' contributions
SAR – conceived of study, participated in design and coordination, drafted the manuscript. JSR – carried out histochemical analysis and western blots, assisted with statistical analysis. MLH – carried out gene analysis, performed statistical analysis. SAZ – conceived of study, participated in design and coordination, critically evaluated the manuscript. KEG - conceived of study, participated in design and coordination, critically evaluated the manuscript. All authors read and approved the final manuscript.

Acknowledgements
The authors thank Pfizer, Inc. for the donation of CIDRs, the staff of the University of Connecticut Livestock Unit, Dave Schreiber, Victor Delaire, and the numerous students that assisted with this project.

References

1. Ford SP, Hess BW, Schwope MM, Nijland MJ, Gilbert JS, Vonnahme KA, Means WJ, Han H, Nathanielsz PW: **Maternal undernutrition during early to mid-gestation in the ewe results in altered growth, adiposity, and glucose tolerance in male offspring.** *J Anim Sci* 2007, **85:**1285–1294.
2. Ford SP, Zhang L, Zhu M, Miller MM, Smith DT, Hess BW, Moss GE, Nathanielsz PW, Nijland MJ: **Maternal obesity accelerates fetal pancreatic beta-cell but not alpha-cell development in sheep: Prenatal consequences.** *Am J Physiol Regul Integr Comp Physiol* 2009, **297:**R835–R843.
3. Long NM, George LA, Uthlaut AB, Smith DT, Nijland MJ, Nathanielsz PW, Ford SP: **Maternal obesity and increased nutrient intake before and during gestation in the ewe results in altered growth, adiposity, and glucose tolerance in adult offspring.** *J Anim Sci* 2010, **88:**3546–3553.
4. Meyer AM, Neville TL, Reed JJ, Taylor JB, Reynolds LP, Redmer DA, Hammer CJ, Vonnahme KA, Caton JS: **Maternal nutritional plane and selenium supply during gestation impact visceral organ mass and intestinal growth and vascularity of neonatal lamb offspring.** *J Anim Sci* 2013, **91:**2628–2639.
5. Zhu MJ, Ford SP, Means WJ, Hess BW, Nathanielsz PW, Du M: **Maternal nutrient restriction affects properties of skeletal muscle in offspring.** *J Physiol* 2006, **575:**241–250.
6. Yan X, Huang Y, Zhao JX, Long NM, Uthlaut AB, Zhu MJ, Ford SP, Nathanielsz PW, Du M: **Maternal obesity-impaired insulin signaling in sheep and induced lipid accumulation and fibrosis in skeletal muscle of offspring.** *Biol Reprod* 2011, **85:**172–178.
7. Huang Y, Yan X, Zhu MJ, McCormick RJ, Ford SP, Nathanielsz PW, Du M: **Enhanced transforming growth factor-beta signaling and fibrogenesis in ovine fetal skeletal muscle of obese dams at late gestation.** *Am J Physiol Endocrinol Metab* 2010, **298:**E1254–E1260.
8. Ashmore CR, Robinson DW, Rattray P, Doerr L: **Biphasic development of muscle fibers in the fetal lamb.** *Exp Neurol* 1972, **37:**241–255.
9. Maier A, McEwan JC, Dodds KG, Fischman DA, Fitzsimons RB, Harris AJ: **Myosin heavy chain composition of single fibres and their origins and distribution in developing fascicles of sheep tibialis cranialis muscles.** *J Muscle Res Cell Motil* 1992, **13:**551–572.
10. Yablonka-Reuveni Z, Rivera AJ: **Temporal expression of regulatory and structural muscle proteins during myogenesis of satellite cells on isolated adult rat fibers.** *Dev Biol* 1994, **164:**588–603.
11. Cornelison DD, Wold BJ: **Single-cell analysis of regulatory gene expression in quiescent and activated mouse skeletal muscle satellite cells.** *Dev Biol* 1997, **191:**270–283.
12. Tong JF, Yan X, Zhu MJ, Ford SP, Nathanielsz PW, Du M: **Maternal obesity downregulates myogenesis and beta-catenin signaling in fetal skeletal muscle.** *Am J Physiol Endocrinol Metab* 2009, **296:**E917–E924.
13. Knights M, Hoehn T, Lewis PE, Inskeep EK: **Effectiveness of intravaginal progesterone inserts and FSH for inducing synchronized estrus and increasing lambing rate in anestrous ewes.** *J Anim Sci* 2001, **79:**1120–1131.
14. Knights M, Maze TD, Bridges PJ, Lewis PE, Inskeep EK: **Short-term treatment with a controlled internal drug releasing (CIDR) device and FSH to induce fertile estrus and increase prolificacy in anestrous ewes.** *Theriogenology* 2001, **55:**1181–1191.
15. National Research Council: *Nutrient Requirements of Small Ruminants: Sheep, Goats, Cervids, and New World Camelids.* Washington, DC: The National Academies Press; 2007.
16. Pena SD, Gordon BB, Karpati G, Carpenter S: **Lectin histochemistry of human skeletal muscle.** *J Histochem Cytochem* 1981, **29:**542–546.
17. Tong J, Zhu MJ, Underwood KR, Hess BW, Ford SP, Du M: **AMP-activated protein kinase and adipogenesis in sheep fetal skeletal muscle and 3 T3-L1 cells.** *J Anim Sci* 2008, **86:**1296–1305.
18. Govoni KE, Lee SK, Chadwick RB, Yu H, Kasukawa Y, Baylink DJ, Mohan S: **Whole genome microarray analysis of growth hormone-induced gene expression in bone: T-box3, a novel transcription factor, regulates osteoblast proliferation.** *Am J Physiol Endocrinol Metab* 2006, **291:**E128–E136.
19. Glynn ER, Londono AS, Zinn SA, Hoagland TA, Govoni KE: **Culture conditions for equine bone marrow mesenchymal stem cells and expression of key transcription factors during their differentiation into osteoblasts.** *J Anim Sci Biotechnol* 2013, **4:**40–1891. 4-40.
20. Livak KJ, Schmittgen TD: **Analysis of relative gene expression data using real-time quantitative PCR and the 2(−delta delta C(T)) method.** *Methods* 2001, **25:**402–408.
21. Buza JJ, Mori Y, Bari AM, Hikono A, Hirayama HHS, Shu Y, Momotani E: **Mycobacterium avium subsp. paratuberculosis infection causes suppression of RANTES, monocyte chemoattractant protein 1, and tumor necrosis factor alpha expression in peripheral blood of experimentally infected cattle.** *Infect Immun* 2003, **71:**7223–7227.
22. Peck KN, Hoffman ML, Forella ME, Fox AR, Govoni KE, Zinn SA: **Poor maternal nutrition affects postnatal growth and development of lambs.** *J Anim Sci* 2013, **91**(E-Suppl. 2):424. Abstr.
23. Lecker SH, Jagoe RT, Gilbert A, Gomes M, Baracos V, Bailey J, Price SR, Mitch WE, Goldberg AL: **Multiple types of skeletal muscle atrophy involve a common program of changes in gene expression.** *FASEB J* 2004, **18:**39–51.
24. Crossland H, Constantin-Teodosiu D, Gardiner SM, Constantin D, Greenhaff PL: **A potential role for Akt/FOXO signalling in both protein loss and the impairment of muscle carbohydrate oxidation during sepsis in rodent skeletal muscle.** *J Physiol* 2008, **586:**5589–5600.
25. Sandri M, Sandri C, Gilbert A, Skurk C, Calabria E, Picard A, Walsh K, Schiaffino S, Lecker SH, Goldberg AL: **Foxo transcription factors induce the atrophy-related ubiquitin ligase atrogin-1 and cause skeletal muscle atrophy.** *Cell* 2004, **117:**399–412.
26. Smith IJ, Alamdari N, O'Neal P, Gonnella P, Aversa Z, Hasselgren PO: **Sepsis increases the expression and activity of the transcription factor forkhead box O 1 (FOXO1) in skeletal muscle by a glucocorticoid-dependent mechanism.** *Int J Biochem Cell Biol* 2010, **42:**701–711.
27. Lai KM, Gonzalez M, Poueymirou WT, Kline WO, Na E, Zlotchenko E, Stitt TN, Economides AN, Yancopoulos GD, Glass DJ: **Conditional activation of akt in adult skeletal muscle induces rapid hypertrophy.** *Mol Cell Biol* 2004, **24:**9295–9304.
28. Nissen PM, Danielsen VO, Jorgensen PF, Oksbjerg N: **Increased maternal nutrition of sows has no beneficial effects on muscle fiber number or postnatal growth and has no impact on the meat quality of the offspring.** *J Anim Sci* 2003, **81:**3018–3027.
29. Greenwood PL, Hunt AS, Hermanson JW, Bell AW: **Effects of birth weight and postnatal nutrition on neonatal sheep: II. skeletal muscle growth and development.** *J Anim Sci* 2000, **78:**50–61.
30. Glore SR, Layman DK: **Cellular development of skeletal muscle during early periods of nutritional restriction and subsequent rehabilitation.** *Pediatr Res* 1983, **17:**602–605.
31. Jue T, Rothman DL, Shulman GI, Tavitian BA, DeFronzo RA, Shulman RG: **Direct observation of glycogen synthesis in human muscle with 13C NMR.** *Proc Natl Acad Sci U S A* 1989, **86:**4489–4491.
32. Meyer C, Dostou JM, Welle SL, Gerich JE: **Role of human liver, kidney, and skeletal muscle in postprandial glucose homeostasis.** *Am J Physiol Endocrinol Metab* 2002, **282:**E419–E427.
33. Hales CN, Barker DJ: **Type 2 (non-insulin-dependent) diabetes mellitus: The thrifty phenotype hypothesis.** *Diabetologia* 1992, **35:**595–601.
34. Wilson SJ, McEwan JC, Sheard PW, Harris AJ: **Early stages of myogenesis in a large mammal: Formation of successive generations of myotubes in sheep tibialis cranialis muscle.** *J Muscle Res Cell Motil* 1992, **13:**534–550.
35. Huang Y, Zhao JX, Yan X, Zhu MJ, Long NM, McCormick RJ, Ford SP, Nathanielsz PW, Du M: **Maternal obesity enhances collagen accumulation and cross-linking in skeletal muscle of ovine offspring.** *PLoS One* 2012, **7:**e31691.

36. Gonzalez JM, Camacho LE, Ebarb SM, Swanson KC, Vonnahme KA, Stelzleni AM, Johnson SE: **Realimentation of nutrient restricted pregnant beef cows supports compensatory fetal muscle growth.** *J Anim Sci* 2013, **91**:4797–4806.
37. Meyer AM, Reed JJ, Neville TL, Thorson JF, Maddock-Carlin KR, Taylor JB, Reynolds LP, Redmer DA, Luther JS, Hammer CJ, Vonnahme KA, Caton JS: **Nutritional plane and selenium supply during gestation affect yield and nutrient composition of colostrum and milk in primiparous ewes.** *J Anim Sci* 2011, **89**:1627–1639.
38. Quiroz-Rothe E, Rivero JL: **Coordinated expression of myosin heavy chains, metabolic enzymes, and morphological features of porcine skeletal muscle fiber types.** *Microsc Res Tech* 2004, **65**:43–61.
39. James DE, Jenkins AB, Kraegen EW: **Heterogeneity of insulin action in individual muscles in vivo: Euglycemic clamp studies in rats.** *Am J Physiol* 1985, **248**:E567–E574.
40. Franck M, Figwer P, Godfraind C, Poirel MT, Khazzaha A, Ruchoux MM: **Could the pale, soft, and exudative condition be explained by distinctive histological characteristics?** *J Anim Sci* 2007, **85**:746–753.
41. Daniel ZC, Brameld JM, Craigon J, Scollan ND, Buttery PJ: **Effect of maternal dietary restriction during pregnancy on lamb carcass characteristics and muscle fiber composition.** *J Anim Sci* 2007, **85**:1565–1576.
42. Fahey AJ, Brameld JM, Parr T, Buttery PJ: **The effect of maternal undernutrition before muscle differentiation on the muscle fiber development of the newborn lamb.** *J Anim Sci* 2005, **83**:2564–2571.
43. Cerisuelo A, Baucells MD, Gasa J, Coma J, Carrion D, Chapinal N, Sala R: **Increased sow nutrition during midgestation affects muscle fiber development and meat quality, with no consequences on growth performance.** *J Anim Sci* 2009, **87**:729–739.
44. Bee G: **Effect of early gestation feeding, birth weight, and gender of progeny on muscle fiber characteristics of pigs at slaughter.** *J Anim Sci* 2004, **82**:826–836.
45. Khan M, Couturier A, Kubens JF, Most E, Mooren FC, Kruger K, Ringseis R, Eder K: **Niacin supplementation induces type II to type I muscle fiber transition in skeletal muscle of sheep.** *Acta Vet Scand* 2013, **55**:85–0147. 55-85.
46. Mallinson JE, Sculley DV, Craigon J, Plant R, Langley-Evans SC, Brameld JM: **Fetal exposure to a maternal low-protein diet during mid-gestation results in muscle-specific effects on fibre type composition in young rats.** *Br J Nutr* 2007, **98**:292–299.
47. Fayard E, Tintignac LA, Baudry A, Hemmings BA: **Protein kinase B/Akt at a glance.** *J Cell Sci* 2005, **118**:5675–5678.
48. Toker A, Newton AC: **Akt/protein kinase B is regulated by autophosphorylation at the hypothetical PDK-2 site.** *J Biol Chem* 2000, **275**:8271–8274.
49. Sarbassov DD, Guertin DA, Ali SM, Sabatini DM: **Phosphorylation and regulation of Akt/PKB by the rictor-mTOR complex.** *Science* 2005, **307**:1098–1101.
50. Partovian C, Simons M: **Regulation of protein kinase B/Akt activity and Ser473 phosphorylation by protein kinase calpha in endothelial cells.** *Cell Signal* 2004, **16**:951–957.
51. Lu D, Huang J, Basu A: **Protein kinase cepsilon activates protein kinase B/Akt via DNA-PK to protect against tumor necrosis factor-alpha-induced cell death.** *J Biol Chem* 2006, **281**:22799–22807.
52. Pallafacchina G, Calabria E, Serrano AL, Kalhovde JM, Schiaffino S: **A protein kinase B-dependent and rapamycin-sensitive pathway controls skeletal muscle growth but not fiber type specification.** *Proc Natl Acad Sci U S A* 2002, **99**:9213–9218.
53. Bodine SC, Stitt TN, Gonzalez M, Kline WO, Stover GL, Bauerlein R, Zlotchenko E, Scrimgeour A, Lawrence JC, Glass DJ, Yancopoulos GD: **Akt/mTOR pathway is a crucial regulator of skeletal muscle hypertrophy and can prevent muscle atrophy in vivo.** *Nat Cell Biol* 2001, **3**:1014–1019.
54. Blaauw B, Canato M, Agatea L, Toniolo L, Mammucari C, Masiero E, Abraham R, Sandri M, Schiaffino S, Reggiani C: **Inducible activation of akt increases skeletal muscle mass and force without satellite cell activation.** *FASEB J* 2009, **23**:3896–3905.
55. Brunet A, Bonni A, Zigmond MJ, Lin MZ, Juo P, Hu LS, Anderson MJ, Arden KC, Blenis J, Greenberg ME: **Akt promotes cell survival by phosphorylating and inhibiting a forkhead transcription factor.** *Cell* 1999, **96**:857–868.
56. Kamei Y, Miura S, Suzuki M, Kai Y, Mizukami J, Taniguchi T, Mochida K, Hata T, Matsuda J, Aburatani H, Nishino I, Ezaki O: **Skeletal muscle FOXO1 (FKHR) transgenic mice have less skeletal muscle mass, down-regulated type I (slow twitch/red muscle) fiber genes, and impaired glycemic control.** *J Biol Chem* 2004, **279**:41114–41123.
57. Zhao J, Brault JJ, Schild A, Cao P, Sandri M, Schiaffino S, Lecker SH, Goldberg AL: **FoxO3 coordinately activates protein degradation by the autophagic/lysosomal and proteasomal pathways in atrophying muscle cells.** *Cell Metab* 2007, **6**:472–483.
58. Grobet L, Martin LJ, Poncelet D, Pirottin D, Brouwers B, Riquet J, Schoeberlein A, Dunner S, Menissier F, Massabanda J, Fries R, Hanset R, Georges M: **A deletion in the bovine myostatin gene causes the double-muscled phenotype in cattle.** *Nat Genet* 1997, **17**:71–74.
59. McPherron AC, Lawler AM, Lee SJ: **Regulation of skeletal muscle mass in mice by a new TGF-beta superfamily member.** *Nature* 1997, **387**:83–90.
60. Argiles JM, Figueras M, Ametller E, Fuster G, Olivan M, de Oliveira CC, Lopez-Soriano FJ, Isfort RJ, Busquets S: **Effects of CRF2R agonist on tumor growth and cachexia in mice implanted with lewis lung carcinoma cells.** *Muscle Nerve* 2008, **37**:190–195.
61. Huang Z, Chen X, Chen D: **Myostatin: A novel insight into its role in metabolism, signal pathways, and expression regulation.** *Cell Signal* 2011, **23**:1441–1446.
62. Davidson LA, Wang N, Shah MS, Lupton JR, Ivanov I, Chapkin RS: **n-3 polyunsaturated fatty acids modulate carcinogen-directed non-coding microRNA signatures in rat colon.** *Carcinogenesis* 2009, **30**:2077–2084.
63. Drummond MJ, Glynn EL, Fry CS, Dhanani S, Volpi E, Rasmussen BB: **Essential amino acids increase MicroRNA-499, −208b, and -23a and downregulate myostatin and myocyte enhancer factor 2C mRNA expression in human skeletal muscle.** *J Nutr* 2009, **139**:2279–2284.

Quantitative analysis of ruminal methanogenic microbial populations in beef cattle divergent in phenotypic residual feed intake (RFI) offered contrasting diets

Ciara A Carberry[1,2], David A Kenny[1], Alan K Kelly[2] and Sinéad M Waters[1*]

Abstract

Background: Methane (CH_4) emissions in cattle are an undesirable end product of rumen methanogenic fermentative activity as they are associated not only with negative environmental impacts but also with reduced host feed efficiency. The aim of this study was to quantify total and specific rumen microbial methanogenic populations in beef cattle divergently selected for residual feed intake (RFI) while offered (i) a low energy high forage (HF) diet followed by (ii) a high energy low forage (LF) diet. Ruminal fluid was collected from 14 high (H) and 14 low (L) RFI animals across both dietary periods. Quantitative real time PCR (qRT-PCR) analysis was conducted to quantify the abundance of total and specific rumen methanogenic microbes. Spearman correlation analysis was used to investigate the association between the relative abundance of methanogens and animal performance, rumen fermentation variables and diet digestibility.

Results: Abundance of methanogens, did not differ between RFI phenotypes. However, relative abundance of total and specific methanogen species was affected ($P < 0.05$) by diet type, with greater abundance observed while animals were offered the LF compared to the HF diet.

Conclusions: These findings suggest that differences in abundance of specific rumen methanogen species may not contribute to variation in CH_4 emissions between efficient and inefficient animals, however dietary manipulation can influence the abundance of total and specific methanogen species.

Keywords: Bovine, qRT-PCR, Residual feed intake, Rumen methaongens

Background

Enteric methane (CH_4) emissions from ruminant livestock accounts for approximately 80% of emissions from the agricultural-sector [1]. In the rumen, CH_4 production (methanogenesis) occurs during microbial fermentation of feed by a specific group of microbes known as archaea called methanogens. The majority of CH_4 produced in the rumen is eructed and thus released into the atmosphere. Enteric CH_4 production also represents a significant (2-15%) loss in energy from dietary gross energy intake [2]. Therefore, any reduction in enteric CH_4 emissions would potentially represent both environmental and economic benefits. In beef cattle production, feed efficiency is defined as the efficiency with which dietary substrate is converted into product (e.g. carcass) and is an important determinant of the profitability of meat production [3]. In recent times residual feed intake (RFI) has become the index of choice for measuring feed efficiency [4]. A number of research groups [5,6] including our own [7], have shown that CH_4 produced in the rumen has been shown to be negatively associated with host feed efficiency.

Enteric CH_4 emissions are not only influenced by the quantity of feed consumed by ruminants, but also by its chemical composition [8]. For example forage based diets such as grass are composed of structural carbohydrates

* Correspondence: sinead.waters@teagasc.ie
[1]Animal and Bioscience Research Department, Animal and Grassland Research and Innovation Centre, Teagasc, Grange, Dunsany, Co. Meath, Ireland
Full list of author information is available at the end of the article

such as cellulose and hemicellulose, which produce predominantly acetate and butyrate as fermentation end-products compared to the propionate dominant fermentation patterns from a cereal based diets. Recently, using clone-sequencing and next generation methanogen-specific tag-encoded pyrosequencing, we showed that specific species of archaea, *Methanobrevibacter* spp. are the dominant methanogens in the rumen across contrasting diets, with *Methanobrevibacter smithii* being the most abundant species followed by *Methanobrevibacter ruminantium* and *Methanosphaera stadtmanae* [9]. However, particularly when species are present in low abundance, these technologies are not sufficiently accurate to reliably quantify specific methanogens. Quantitative real-time PCR (qRT-PCR) has become a popular method for estimation of methanogen abundance in the ruminant digestive tract [10]. Therefore the objectives of the present study were to quantify the relative abundance of total methanogens and key species *viz Methanobrevibacter smithii, M. ruminantium* and *Methanosphaera stadtmanae* in the ruminal fluid of cattle divergent for RFI offered two contrasting diets: a low energy, high forage diet (HF) and a high energy, low forage diet (LF) respectively. Additionally, correlation analysis was used to assess the association between methanogen abundance and animal performance, diet digestibility and rumen fermentation variables.

Materials and methods

All procedures involving animals were approved for the use of live animals in experiments by the Animal Research Ethics Committee, University College Dublin, Belfield, Dublin, Ireland, and were licensed by the Irish government Department of Health and Children, in accordance with the Cruelty to Animals Act (Ireland 1897) and European Community Directive 86/609/EC (http://ec.europa.eu/food/fs/aw/aw_legislation/scientific/86-609-eec_en.pdf).

Animal management

This experiment was conducted as part of a larger study designed to examine the physiological control of energetic efficiency in growing beef heifers [11]. Details of the animal experiment used in this study are as previously described [12]. Dietary ingredients and chemical composition have been previously described in detail [11] and both diets were offered *ad libitum*. Briefly, individual dry matter intake (DMI) and growth were recorded for 86 yearling Limousin × Friesian heifers offered *ad libitum* access to a high energy low forage (LF) concentrate based diet over 112 d. The LF diet consisted of 30% corn silage and 70% pelleted concentrate. All animals were subsequently ranked retrospectively on phenotypic RFI. The 14 heifers with the highest (inefficient; high RFI) and 14 heifers with the lowest (efficient; low RFI) RFI coefficients during the study of Kelly et al. [11] were selected for use in the current study. Following initial selection, all 28 animals were re-allocated to a low energy high forage (HF) grass silage diet and individual feed intake recorded for a 44 d period. Following this 44 d period (Period 1) all animals were turned out to pasture for a 56 d dietary "washout" period. Subsequently, all 28 animals were re-housed and re-allocated to a LF concentrate based diet and individual feed intake was recorded for 35 d (Period 2). Individual feed intake and body weight gain were recorded for a further 84 d and RFI was re-calculated. All 28 animals remained within their respective RFI groups [13]. The experiment was therefore designed to have two factors (i) RFI phenotype and (ii) diet type.

Rumen sampling and methane measurements

Rumen sampling was performed at the end of both dietary periods. Samples of ruminal fluid were collected using a transesophageal sampling device (FLORA rumen scoop; Guelph, Ontario, Canada) according to manufacturer's instructions. Subsequently, a 20 mL aliquot was transferred using a pipette and sterilized tip into a separate labeled sterilized container, immediately frozen in liquid nitrogen, and stored at −80°C until processing [12].

Methane measurements recorded during the study of Kelly et al. [11] were utilised in the current study (Additional file 1: Table S1). In brief, daily CH_4 emissions were determined using a calibrated tracer (sulphur hexafluoride (SF_6)) technique [14] during the last 5 d of each dietary period.

DNA extraction from ruminal fluid

A detailed description of the DNA extraction method has been described by Carberry et al. [12]. Total microbial DNA was extracted from the 28 ruminal fluid samples by using a repeated bead beating and column purification method which provides efficient recovery of PCR-quality microbial DNA [12].

Quantitative polymerase chain reaction (qRT-PCR) assays

The differences between SYBR green and TaqMan qRT-PCR chemistries have been reviewed [15]. As previously published primers were available for quantification of total methanogens [16] and *M. stadtmanae* [17] via SYBR green chemistry, we chose to utilise these in our study (Table 1).

The primers were commercially synthesized (Sigma-Aldrich Ireland Ltd. Dublin, Ireland) and end point PCR was conducted for the validation of the specificity of the primers against target species. For the quantification of *M smithii* and *Methanobrevibcater ruminantium*, it was necessary to utilise TaqMan chemistry for these qRT-

Table 1 PCR primers used for SYBR green qRT-PCR analysis

Target Taxon	Primer[1] Forward	Primer[1] Reverse	E[2]	Product size, bp	Reference
16 s V3[3]	CCTACGGGAGGCAGCAG	ATTACCGCGGCTGCTGG	2.00	194	Muyzer et al., 1993 [18]
Total methanogens	GGATTAGATACCCSGGTAGT	GTTGARTCCAATTAAACCGCA	1.96	173	Hook et al., 2009 [16]
Methanosphaera stadtmanae	CTTAACTATAAGAATTGCTGGAG	TTCGTTACTCACCGTCAAGATC	2.01	150	Zhou et al., 2009 [17]

[1]5′→3′.
[2]Efficiency.
[3]Primers used for qRT-PCR normalisation.

PCR assays to allow for species specificity. Details of TaqMan primer and probe sets used in this study, to amplify *M smithii* and *M ruminantium* species are listed in Table 2. For the detection of *M smithii*, primers and a specific FAM labelled probe F were utilised which was designed based on previously published sequences [19]. The same primer set was used in conjunction with a FAM labelled probe designed in the current study to target *M ruminantium* species. The probe was of the same length as the *M smithi* probe with the exception of 3 nucleotide changes in the sequence. All primers and probe sets for the TaqMan qRT-PCR assays were generated using the Primer Express algorithm v1.0 from published sequences (National Centre For Biotechnology Information, NCBI: http://www.ncbi.nlm.nih.gov/) and commercially synthesized (Applied Biosystems, Warrington, UK).

DNA was extracted from individual ruminal fluid samples and diluted to a concentration of 100 ng/μL and 1 μL of diluted DNA was used as a template in all PCR reactions. All SYBR PCR amplifications were initially optimised and performed in 0.5 mL tubes in a DNA thermal cycler (Eppendorf Mastercycler®) using the following program: an initial denaturation step of 95°C for 2 min; 30 cycles of 95°C for 30 s, 60°C for 30 s, 72°C for 30 s and a final elongation step of 72°C for 7 min. The PCR reaction solution (50 μL) consisted of 1 × PCR buffer (final concentrations, 16 mmol/L $(NH_4)_2$ SO_4, 67 mmol/L Tris–HCl [pH 8.8], 0.01% Tween-20 and 1.5 mmol/L $MgCl_2$), 10 pmol of each primer, a 200 μmol/L concentration of each deoxynucleoside triphosphate, 100 ng of template DNA, and 5.0 U of *Taq* DNA polymerse (Bioron, Ludwigshafen, Germany) and 40 μL of molecular grade H_2O. Aliquots of 10 μL PCR products were analysed by electrophoresis on a 2% (w/v) agarose gel to verify the presence and size of the amplicons. Furthermore amplicons were sequenced to confirm their identity. Using BLAST analysis on the NCBI website, all amplicons were confirmed 100% homologous to their target species. Negative controls without template DNA were included in parallel. Specificity of TaqMan assays for the quantification of *M smithii* and *M ruminantium* were verified before quantification of ruminal DNA. Each probe was validated by running a non-target clone standard as a negative control. In addition, a 10 fold dilution series starting from 25 ng/μL of clones obtained from our own library [9] and identified as either *M smithii* or *M ruminantium* were used as standards and run in triplicate to verify the reproducibility of the TaqMan probe assay. These dilutions were used to calculate PCR efficiencies for each assay. DNA from rumen samples was also diluted and used to calculate PCR efficiencies. For all primer sets, PCR efficiencies were equivalent for both clone libraries and DNA from rumen samples and within the acceptable range to be used to perform qRT-PCR analysis.

The qRT-PCR assays were performed on an ABI 7500 Fast Real-Time PCR system using Fast SYBR® Master Mix (Applied Biosystems, Warrington, UK). Optimization of assay conditions was performed for template DNA concentrations. To reduce inhibition total microbial DNA was diluted to 10 ng/uL. All amplified qRT-PCR reactions were carried out in a 96 well plate format. Non-template controls were included on every plate for each assay to allow screening for possible contamination and primer dimer formation. To minimise variation, all samples included in each analysis were derived from the same extracted DNA batch, prepared under the same conditions and samples were run in triplicate.

Table 2 PCR primers and probes used in this study for TaqManqRT-PCR analysis

Primer/Probe	Sequence (5′→3′)	Efficiency	Product size, bp	Reference
MsmithiiF	CCGGGTATCTAATCCGGTTC			Dridi et al., 2009 [19]
MsmithiiR	CTCCCAGGGTAGAGGTGAAA	1.98	123	Dridi et al., 2009 [19]
Msmithii probe	CCGTCAGAATCGTTCCAGTCAG			Dridi et al., 2009 [19]
Mruminantium F	CCGGGTATCTAATCCGGTTC			Dridi et al., 2009 [19]
Mruminantium R	CTCCCAGGGTAGAGGTGAAA	2.02	123	Dridi et al., 2009 [19]
Mruminantium probe	CCGTCAGGTTCGTTCCAGTTAG			Current study

All SYBR green reaction mixtures were prepared in a total volume of 20 μL. The reaction consisted of 1 μL DNA, 10 μL Fast SYBR Green master mix, 1 μL of forward and reverse primers (10 ng of each) and 8 μL nuclease-free H_2O. Thermal cycling conditions applied to each assay consisted of an initial *Taq* activation step at 95°C for 15 min followed by 40 cycles of 95°C for 15 s, 60°C for 60 s, followed by an amplicon dissociation stage (95°C for 15 s, 60°C for 1 min, increasing 0.5°C/cycle until 95°C was reached) which confirmed specificity via dissociation curve analysis of PCR end products. Fluorescence detection was also performed at the end of each denaturation and extension step. TaqMan qRT-PCR assays were performed using TaqMan Fast Universal Master Mix (2×) (Applied Biosystems, Warrington, UK) and prepared in a total volume of 20 μL with 1 μL DNA (10 ng/μL), 1 μL 20× TaqMan assay (forward and reverse primer plus probe), 10 μL TaqMan Fast Universal Master Mix (2×) and 8 μL nuclease-free H_2O. Thermal cycling conditions applied to each TaqMan assay consisted of an initial hold step at 50°C for 2 min and 95°C for 20 s followed by 40 cycles of 95°C for 3 s and 60°C for 30 s.

qRT-PCR data analysis

Real time PCR amplification efficiencies (*e*) were estimated for all assays using a linear regression of the threshold cycle (Ct) for each dilution versus the log dilution using the formula: $e = 5^{-1/\text{slope}}$ [20] where '5' is the corresponding fold dilution. Efficiencies of the species specific TaqMan probes and primers sets are presented in Tables 1 and 2. Efficiencies for both SYBR green and TaqMan assays ranged from 1.96 to 2.02, close to the theoretical value of 2.0 which is representative of the doubling effect of the target sequence during the qRT-PCR cycle. All primers had PCR efficiencies between 90% and 110% and were therefore acceptable for use in this study.

Inter-plate calibration based on a calibrator sample included on all plates, efficiency correction of the raw cycle threshold (**Ct**) values and results from triplicate PCR reactions for each target species were averaged and the means calculated using the software package GenEx 5.2.1.3 (MultiD Analyses AB, Gothenburg, Sweden). Within GENEX, the stability of the reference bacterial or methanogen 16S rRNA gene was assessed and assured stable across diet and RFI. Abundance of total methanogens were expressed as a proportion of total estimated rumen bacterial 16S rRNA gene as described previously [21,22] according to the equation: relative quantification = $2^{-(\text{Ct target-Ct total bacteria})}$, where Ct represents threshold cycle. Abundance of specific methanogen species were expressed as a proportion of total estimated methanogen 16S rRNA gene according to the equation: relative quantification = $2^{-(\text{Ct target-Ct total methanogens})}$.

Statistical analysis

All data were analysed using Statistical Analysis Systems v9.1 2002 (SAS Institute, Cary, NC, USA). Data were examined for normality and homogeneity of variance by histograms, qqplots and formal statistical tests as part of the UNIVARIATE procedure of SAS. Data that were not normally distributed were transformed by raising the variable to the power of lambda. The appropriate lambda value was obtained by conducting a Box-Cox transformation analysis using the TRANSREG procedure in SAS. Delta Ct values for total Methanogens were transformed using a lambda value of –0.5 while *M. stadtmanae, M smithii* and *M ruminantium* were transformed using a lambda value of 0.25. The transformed data were used to calculate *P*-values. However, the corresponding least squares means and standard errors of the non-transformed data are presented in the results for clarity.

A mixed model ANOVA (PROC MIXED) was conducted to determine the effect of RFI phenotype and diet type on the relative abundance of each species of interest. Fixed effects included RFI phenotype (H or l), diet type (LF or HF), and their interaction. The interaction term, if not statistically significant ($P > 0.10$), was subsequently excluded from the final model. In all analyses the individual animal was denoted as the experimental unit and animal was included as a random effect.

The statistical model used: $Y = \mu + Ri + Dj + (Rj \times Dj) + Ak + \varepsilon ijk$, where μ was the overall mean, R_i was the fixed effect of RFI phenotype ($_I$ = H to L), D_j the fixed effect of diet ($_j$ = LF to HK), $R_i \times D_j$ is the interaction between RFI phenotype and diet type, A_k the random effect of animal, and εijk is the associated error. Differences between treatments were determined by F-tests using Type III sums of squares. The PDIFF command incorporating the Tukey test was applied to evaluate pairwise comparisons between treatment means.

Spearman partial correlation analysis (PROC CORR, SAS), was calculated to examine associations amongst physiological data, rumen fermentation variables [23] and measured relative methanogen microbial abundance values using analysis with RFI and dietary treatment included as fixed effects in the analysis. Differences were considered significant where $P < 0.05$, while tendencies were considered where $P < 0.10$.

Results

qRT-PCR quantification of methanogens

The relative abundance of total methanogens, *Methanosphaera stadtmanae, M smithii* and *M ruminantium* are presented in Table 3. No RFI phenotype × diet interactions ($P > 0.05$) were observed for any of the methanogens measured, however a tendency ($P = 0.08$) towards an RFI × diet interaction was observed for the total methanogen population. Additionally, no effect of RFI

Table 3 Effect of phenotypic RFI and diet on ruminalmethanogen populations[1]

Items	RFI			Diet[3]			Significance[4]		
	H	L	SED	HF	LF	SED	RFI	Diet	RxD
Methanogens[1]	0.07	0.08	0.010	0.07	0.09	0.010	0.71	0.01	0.08
Methanosphaera stadtmanae[2]	0.36	0.37	0.069	0.09	0.64	0.069	0.44	<0.0001	0.68
M smithii[2]	0.32	0.25	0.057	0.14	0.43	0.057	0.29	<0.0001	0.33
M ruminantium[2]	0.18	0.29	0.089	0.04	0.43	0.089	0.65	<0.0001	0.92

[1]Methanogens measured as a proportion of total estimated rumen bacterial 16S rDNA, relative quantification = $2^{-(Ct\ target-Ct\ total\ bacteria)}$.
[2]Methanogen spp. measured as a proportion of total estimated rumen methanogen 16S rDNA, relative quantification = $2^{-(Ct\ target-Ct\ total\ methanogens)}$.
[3]Diet = HF = high forage (grass silage), LF = Low forage (maize silage (30):concentrate (70)).
[4]Significance values for transformed data. Back transformed means presented for clarity. R = RFI, D = DIET.

phenotype was detected for the relative abundance of any of the methanogens analysed ($P > 0.05$). However, diet type affected ($P = 0.03$) the relative abundance of the overall methanogen population with a greater relative abundance of total methanogens in cattle offered the LF compared to the HF diet (Table 3). Furthermore, the relative abundance of *M. stadtmanae* and *M. ruminantium* was higher ($P < 0.001$) when animals were offered the LF compared to the HF diet.

Correlation analysis

Correlation coefficients for the association between animal performance, rumen fermentation variables and relative methanogen abundance are presented in Table 4. When animals were offered the LF diet, abundance of total methanogens was positively correlated with acetate (r = 0.44, $P = 0.02$) and acetate: propionate (r = 0.42, $P = 0.03$) and negatively correlated with propionate (r = −0.41, $P = 0.03$) and rumen pH (r = −0.49, $P = 0.01$). *Methanosphaera stadtmanae* was negatively correlated with CH_4 (r = −0.47, $P < 0.02$). In addition, although not statistically significant a tendency towards a positive relationship between total methanogens and dry matter digestibility (DMD) (r = 0.35, $P = 0.08$), organic matter digestibility (OMD) (r = −0.34, $P = 0.09$) and crude protein digestibility (CPD) (r = −0.38, $P = 0.05$) was observed when animals were offered the LF diet. When animals were offered the HF diet a tendency towards a negative relationship between both *M. smithii* and *M. ruminantium* and CH_4 (r = −0.36, $P = 0.07$) (r = −0.38, $P = 0.05$) was detected.

Discussion

Efficiency of feed utilisation is a key economically relevant trait to the beef cattle industry worldwide given that feed typically accounts for the greatest single input cost [24]. Improved feed efficiency is not only linked to increased profitability, but also reduces the environmental burden, with efficient animals producing less nutrient excretion [25] and reduced CH_4 emissions [5,6]. As CH_4 is a terminal product of methanogen mediated feed fermentation, recent research has focused on characterising the methanogen population in animals selected for divergent feed efficiency. Studies [17,26,27] have identified both a diet effect and a correlation between host feed efficiency and rumen microbial composition. *Methanobrevibacter* spp. and *Methanosphaera* sp. are consistently identified as dominant methanogenic archaea in the rumen irrespective of geographical location or dietary feeding regime [17,27-31]. Rumen methanogens were previously characterised by our group [9] in cattle divergent for phenotypic RFI and it was reported that *Methanosphaera* and *Methanobrevibacter* OTUs were identified as the most dominant methanogens. Therefore, in our study total methanogens and specific ruminal methanogen populations of *Methanosphaera, M smithii,* and *M rumantium* were selected for relative qRT-PCR analysis to quantitatively assess whether these populations are associated with variance in feed efficiency and/or dietary energy type in cattle.

There is evidence to suggest that efficient L-RFI animals produce less CH_4 both on a daily basis (g/d) [6,32,33] and as per unit of body weight (g/kg) [23] than their inefficient H-RFI counterparts. However work conducted from our own group examining the effect of phenotypic RFI on CH_4 emissions from the animals utilised in the present study showed no difference in CH_4 between animals ranked as either high or low RFI across the two dietary periods [23]. In our study quantitative analysis showed that the total and specific methanogen communities of animals ranked as either H or L-RFI was not different. This is consistent with previous reports when a low energy feed lot diet was offered [17] and also when the diet was switched from a low to a high energy diet [27]. While there is a sparsity of information in the literature with regard to studies exploring the links between enteric CH_4 emissions and rumen methanogen abundance there is evidence to suggest that CH_4 emissions may be consistent with the population size of methanogens [34,35]. However, since the microbial inhabitants of the rumen function in symbiosis, the metabolic activity and substrate specificity of both the methanogens and the wider rumen microbial community must also be considered. Indeed a growing number of studies have reported

Table 4 Association between physiological and rumen fermentation variables and relative methanogen abundance in beef heifers divergent for residual feed intake (RFI)[1]

Items	HF Diet				LF Diet			
	Methanogens	*M. stadtmanae*	*M. smithii*	*M. ruminantium*	Methanogens	*M. stadtmanae*	*M. smithii*	*M. ruminantium*
DMI[2]	−0.30	−0.05	0.14	0.06	0.13	−0.14	0.22	0.08
CH_4[3]	−0.27	−0.36	−0.36[a]	−0.38*	0.13	−0.47**	−0.07	0.03
MLW[4]	−0.19	−0.07	−0.17	−0.15	0.22	0.14	−0.20	−0.06
tVFA[5]	0.13	0.11	0.08	−0.00	0.01	0.00	−0.00	−0.18
acetate	−0.01	0.06	0.11	−0.04	0.44**	0.07	−0.18	−0.14
propionate	0.01	0.00	−0.13	−0.08	−0.41**	0.30	−0.23	0.09
isobutyrate	0.07	−0.05	−0.12	0.06	0.20	−0.15	−0.21	0.08
butyrate	−0.01	0.06	−0.24	−0.16	0.15	0.03	−0.08	−0.19
isovalerate	−0.01	−0.05	0.08	−0.30	−0.12	−0.13	0.17	0.12
valerate	−0.02	−0.06	0.11	−0.05	−0.31	−0.10	0.14	0.18
A:P[6]	−0.02	0.05	0.14	0.07	0.42**	−0.14	0.22	0.08
pH	−0.11	−0.05	0.14	0.07	−0.49**	0.14	0.22	0.08
CH_4GEI[7]	−0.01	−0.03	−0.15	−0.15	0.03	−0.23	−0.24	−0.05
DMD[8]	0.21	0.16	−0.21	−0.25	0.35*	0.14	−0.21	−0.16
OMD[9]	0.22	0.15	−0.20	−0.14	0.34*	0.14	−0.22	−0.15
CPD[10]	0.14	0.20	−0.23	−0.24	0.38*	0.17	−0.22	−0.17
NDFD[11]	0.05	0.09	−0.18	−0.14	0.20	0.13	−0.24	−0.16
ADFD[12]	0.05	0.09	−0.17	−0.13	0.26	0.13	−0.24	−0.23
GED[13]	0.17	0.14	−0.18	−0.13	0.35	0.16	−0.25	−0.16

[1]Spearman correlation coefficient in boldface are different from zero ($P < 0.10$).
[2]Dry matter intake.
[3]Methane.
[4]Mean live weight.
[5]Total volatile fatty acids.
[6]Acetate:propionate ratio.
[7]Methane energy from gross energy intake.
[8]Dry matter digestibility.
[9]Organic matter digestibility.
[10]Crude protein digestibility.
[11]Neutral detergent fiber digestibility.
[12]Acid detergent fiber digestibility.
[13]Gross energy digestibility.
*$P < 0.10$, **$P < 0.05$.

that methanogen abundance does not reflect the amount of enteric CH_4 emissions [35-37]. Furthermore, no significant relationship between methanogen abundance and CH_4 production potential in other ecosystems has been reported [38]. Our previous work [9] reported that the rumen appears to harbour a core group of methanogens regardless of host classification for feed efficiency. Therefore, as suggested by the results obtained for total methanogen abundance in our study, the quantity of the total methanogen population may not be indicative of divergence in CH_4 yield between RFI phenotypes.

While work from our own group has shown moderate within animal repeatability of RFI while maintained on a constant diet type [13] other recent work has shown that the relative ranking of animals for this trait may change when moved from a low to a high energy diet [39]. This highlights the necessity to investigate both RFI and the rumen microbial community of animals divergent for RFI across different diet types. Previous research has shown that the chemical composition of the diet can have a great effect on the overall rumen bacterial composition in animals divergent for RFI [12]. Despite this, previous reports have suggested that no change in total methanogen abundance of either feed efficient or inefficient cattle divergent for RFI occurs when the diet is changed from a low to high energy diet [27]. Furthermore, it has been suggested that diet has a greater effect on the diversity of the methanogen community rather than total methanogen abundance [17,27]. However, in the current study the abundance of total methanogens was found to be affected by the change in dietary substrate offered. While it was surprising that the relative total methanogen abundance was greater when animals were offered the LF compared to the HF diet, this may have arisen due to the nature of the LF diet (30:70 maize silage:concentrate). With the inclusion of 30% forage in this diet hydrogen fermenting bacteria would still have the ability to proliferate, albeit to a lesser extent, thus providing hydrogen for methanogen growth. In addition, on the LF diet, intakes were higher and there was more easily digestible substrate available to support CH_4 emissions and the growth of methanogens. Indeed, overall the CH_4 emissions were higher from animals on the LF diet [23]. Furthermore, decreased rumen pH is often associated with high concentrate diets due to a lowering of the acetate:propionate ratio [40] Decreased ruminal pH can indirectly affect CH_4 synthesis due to its influence on VFA production [5] and directly through inhibition of methanogen activity [41]. However, data from our own laboratory generated from the same pool of animals as the current study showed that although mean ruminal pH for both high and low RFI phenotypes was lower on the LF diet, the pH did not drop below the optimum range for rumen methanogen growth (pH 6.0-7.5) [23]. Therefore, it is hypothesized that methanogen proliferation would not have been inhibited while animals were offered the LF diet in our study. Differences in our findings to that of others could also have arisen due to differences in qRT-PCR quantification methods. Our qRT-PCR results represent the relative abundance of total methanogens to the total bacterial population (relative quantification) while DNA copy number was used to quantify the methanogen population (absolute quantification) in the study by Zhou et al., [27]. While both these methods have been extensively utilised throughout the literature for the quantification of rumen microbiota, it is important to acknowledge that neither method is without its drawbacks [15]. In the current study, quantification using the relative method was favoured due to the implications of quantification of the target in comparison to a standard curve while using absolute quantification [15]. Additionally, host breed genotype has also been shown to influence methanogen diversity in the rumen and therefore could be a contributing factor to the variation across studies [29].

In the present study when animals were offered the LF diet overall methanogen abundance was positively associated with acetate and negatively associated with propionate. It is widely accepted that both acetate and butyrate promote CH_4 production, while propionate is considered a rival pathway to methanogenesis [42,43], directly competing with methanogens for available hydrogen during formation [5]. In addition, when total methanogens was correlated with protozoa abundance from our previous study [12] there was strong a positive relationship (data not shown). This positive relationship is compounded by the fact that among H_2 producers, protozoa have a prominent position, which is strengthened by their close physical association with methanogens, which favours H_2 transfer from one to the other [37].

An effect of diet type on the relative abundance of *M. stadtmanae* was observed in the present study. This result is consistent with the pyrosequencing analysis conducted in our previous study [9] and may be attributed to the difference in the chemical compostion of the diets and also abundance of other rumen microbes. Energy metabolism of *M. stadtmanae* is more restricted than other methanogenic species, growing only on methanol and H_2 [44]. In the rumen large quantities of methanol are produced via the degradation of pectin by protozoa and other anaerobic bacteria [45]. A HF grass silage based diet is a relatively poor source of dietary pectin, while some constituents of concentrate feeds such as citrus and beet pulp are abundant. We previously reported a tendency towards greater relative abundance of rumen protozoa between the LF and HF dietary periods [12]. Due to the restricted energy metabolism of *Methanosphaera*, the dietary effect observed on this methanogenic species may have arisen due to the fermentation products produced by other rumen microbial populations. In addition, the fastidious energy requirements of *Methanosphaera stadtmanae* compared to other methanogens was most likely the causative reason why when animals were offered the LF diet this methanogenic species was negatively correlated with CH_4.

The effect of diet on the relative abundance of *M smithii* is most likely due to substrate specificity preference for methanogenesis by this methanogen species. Energy metabolism of *M smithii* is less restricted than *M. stadtmanae* in that it can produce CH_4 via CO_2-H_2 and or formate [45]. In the rumen, formate and hydrogen are the fermentative end products of several types of rumen bacteria and protozoa [46]. We have previously reported a greater abundance of *Fibrobacter succinogenes* and *Prevotella*, both of which produce formate as an end product of fermentation [46], in animals offered LF compared with HF diets [12]. This, coupled with the tendency towards greater relative abundance of hydrogen producing rumen protozoa between the LF and HF dietary periods, suggests that the appropriate substrates for *M smithii* were likely to be in abundance while animals were offered the LF diet. In addition, *M smithii* also possesses enzymes which have been shown in other methanogens to facilitate utilisation of methanol and ethanol [44,47], which are products of bacterial fermentation. Therefore abundance of these methanogens is most likely affected by both chemical composition of the diet and the availability of appropriate end products produced from rumen microbial fermentation.

The relative abundance of *M ruminantium* was higher when animals were offered the LF compared to the HF diet. This is in agreement with Zhou et al. [27] who reported a marked shift in the number of *M ruminantium* when the diet was changed from a low energy to a high energy diet, using PCR-DGGE microbial community analysis. Similar to *M smithii*, it is believed that formate may also be an important substrate for methanogenesis by *M ruminantium* [48]. In addition, the genome sequence of this methanogen has revealed an abundance of encoding adhesion proteins [48]. Initial experiments show that some are involved in mediating close assimilations with hydrogen producing bacteria while others synergise association between protozoa and fungi [48]. Although the LF diet offered in the current study would be expected to yield less CH_4 emissions compared to a HF diet, the genomic niche adaptation of *M ruminantium* may explain the increase in abundance through mutual associations with other rumen microbial populations.

Conclusion

In conclusion, this study is the first to quantify both total and a panel of key candidate rumen methanogens in cattle divergent for phenotypic RFI across two contrasting diets. Quantification of the methanogen community showed that feed efficiency alone had no significant effect on the abundance of total or specific methanogens at the species level. However, our results extend the findings of others by demonstrating that the type of dietary substrate offered greatly influences the abundance of specific methanogen species and in addition the density of total methanogens may also be affected by changes in the diet. It is concluded that diet alone has a greater influence on the relative abundance of specific methanogens at a species level. Future work focused on designing specific qRT-PCR assays which target these rarer, less studied methanogens will provide further insight into the effect of host feed efficiency and diet on the rumen methanogen population. Furthermore, studies investigating the association between CH_4 emissions, rumen methanogen abundance and diet digestibility are required to allow the development of more specific nutritional management strategies to reduce CH_4 emissions without impacting on host health, growth or productivity.

Additional file

Additional file 1: Table S1. Information of tested cattle.

Competing interests

The authors declare that they have no competing interests.

Authors' contributions

DAK and AKK performed the animal study and collected data relating to animal performance and the rumen samples. SMW and CAC conceptualised the study and were responsible for experimental design. SMW and CAC designed the primers and CAC performed all laboratory and real time PCR analysis. DAK and AKK performed statistical analysis. CAC, SMW and DAK participated in the data collection, data analysis and interpretation. CAC drafted the manuscript. All authors approved the final version of the manuscript for publication.

Acknowledgements

Funding for the development and main work of this research was provided under the National Development Plan, through the Research Stimulus Fund, administered by the Department of Agriculture, Fisheries & Food, Ireland RSF 05 224.

Author details

[1]Animal and Bioscience Research Department, Animal and Grassland Research and Innovation Centre, Teagasc, Grange, Dunsany, Co. Meath, Ireland. [2]UCD School of Agriculture, Food Science and Veterinary Medicine, College of Life Sciences, University College Dublin, Belfield, Dublin 4, Ireland.

References

1. Steinfeld H, Gerber P, Wassenaar T, Castel V, de Haan C: **Livestock's long shadow: issues and options. Food and agriculture organization of the united nations.** 2006. http://www.Fao.Org/docrep/010/a0701e/a0701e00.Htm. Accessed December 21, 2011.
2. Van Nevel CJ, Demeyer DI: **Control of rumen methanogenesis.** *Environ Mon Assess* 1996, **42:**73–97.
3. Reynolds CK, Crompton LA, Mills JAN: **Improving the efficiency of energy utilisation in cattle.** *Anim Prod Sci* 2011, **51:**6–12.
4. Crews DH: **Genetics of feed utilization and national cattle evaluation: a review.** *Genet Mol Res* 2005, **4:**152–165.
5. Hegarty RS: **Reducing rumen methane emissions through elimination of rumen protozoa.** *Austral J Agri Res* 1999, **50:**1321–1327.
6. Nkrumah JD, Okine EK, Mathison GW, Schmid K, Li C, Basarab JA, Price MA, Wang Z, Moore SS: **Relationships of feedlot feed efficiency, performance, and feeding behavior with metabolic rate, methane production, and energy partitioning in beef cattle.** *J Anim Sci* 2006, **84:**145–153.
7. Fitzsimons C, Kenny DA, Deighton MH, Fahey AG, McGee M: **Methane emissions, body composition, and rumen fermentation traits of beef heifers differing in residual feed intake.** *J Anim Sci* 2013, **91:**5789–5800.
8. Beauchemin KA, Kreuzer M, O' Mara F, McAllister TA: **Nutritional management for enteric methane abatement: a review.** *Austral J Exp Agri* 2008, **48:**21–27.
9. Carberry CA, Waters SM, Kenny DA, Creevey CJ: **Rumen methanogenic genotypes differ in abundance according to host residual feed intake phenotype and diet type.** *Appl Environ Microbiol* 2014, **80:**586–594.
10. McCartney CA, Bull ID, Waters SM, Dewhurst RJ: **Technical note: comparison of biomarker and molecular biological methods for estimating methanogen abundance.** *J Anim Sci* 2013, **91**(12):5724–5728.
11. Kelly AK, McGee M, Crews DH Jr, Fahey AG, Wylie AR, Kenny DA: **Effect of divergence in residual feed intake on feeding behavior, blood metabolic variables, and body composition traits in growing beef heifers.** *J Anim Sci* 2010, **88:**109–123.
12. Carberry CA, Kenny DA, Han S, McCabe MS, Waters SM: **Effect of phenotypic residual feed intake and dietary forage content on the rumen microbial community of beef cattle.** *Appl Environ Microbiol* 2012, **78**(14):4949–4958.
13. Kelly AK, McGee M, Crews DH Jr, Sweeney T, Boland TM, Kenny DA: **Repeatability of feed efficiency, carcass ultrasound, feeding behavior, and blood metabolic variables in finishing heifers divergently selected for residual feed intake.** *J Anim Sci* 2010, **88:**3214–3225.
14. Johnson K, Huyler M, Westberg H, Lamb B, Zimmerman P: **Measurement of methane emissions from ruminant livestock using a sulfur hexafluoride tracer technique.** *Environ Sci Technol* 1994, **28:**359–362.
15. Smith CJ, Osborn AM: **Advantages and limitations of quantitative PCR (q-PCR)-based approaches in microbial ecology.** *FEMS Microbiol Ecol* 2009, **67:**6–20.

16. Hook SE, Wright A, Denis G, McBride BW: **Methanogens: methane producers of the rumen and mitigation strategies.** *Archaea* 2010, **2010:**945785.
17. Zhou M, Hernandez-Sanabria E, Guan LL: **Assessment of the microbial ecology of ruminal methanogens in cattle with different feed efficiencies.** *Appl Environ Microbiol* 2009, **75:**6524–6533.
18. Muyzer G, de Waal EC, Uitterlinden AG: **Profiling of complex microbial populations by denaturing gradient gel electrophoresis analysis of polymerase chain reaction-amplified genes coding for 16S rRNA.** *Appl Environ Microbiol* 1993, **59:**695–700.
19. Dridi B, Henry M, El Khechine A, Raoult D, Drancourt M: **High prevalence of *Methanobrevibacter smithii* and *Methanosphaera stadtmanae* detected in the human gut using an improved DNA detection protocol.** *PLoS ONE* 2009, **4:**e7063.
20. Pfaffl MW: **A new mathematical model for relative quantification in real-time RT-PCR.** *Nucleic Acids Res* 2001, **29**(9):e45.
21. Chen XL, Wang JK, Wu YM, Liu JX: **Effects of chemical treatments of rice straw on rumen fermentation characteristics, fibrolytic enzyme activities and populations of liquid- and solid-associated ruminal microbes in vitro.** *Anim Feed Sci Technol* 2008, **141:**1–14.
22. Guo YQ, Liu JX, Lu Y, Zhu WY, Denman SE, Mc Sweeney CS: **Effect of tea saponin on methanogenesis, microbial community structure and expression of *mcra* gene, in cultures of rumen micro-organisms.** *Lett Appl Microbiol* 2008, **47:**421–426.
23. McDonnell RP: *An examination of the effects of divergent phenotypic selection for residual on enteric methane emissions in beef cattle.* Ireland: University College Dublin; 2008. Masters thesis.
24. Arthur PF, Herd RM: **Efficiency of feed utilization by livestock – implications and benefits of genetic improvement.** *Can J Anim Sci* 2005, **85:**281–290.
25. Herd RM, Hegarty RS, Dicker RW, Archer JA, Arthur PF: **Selection for residual feed intake improves feed efficiency in steers on pasture.** *Anim Prod Austral* 2002, **24:**85–88.
26. Guan LL, Nkrumah JD, Basarab JA, Moore SS: **Linkage of microbial ecology to phenotype: correlation of rumen microbial ecology to cattle's feed efficiency.** *FEMS Microbiol Lett* 2008, **288:**85–91.
27. Zhou M, Hernandez-Sanabria E, Guan LL: **Characterization of variation in rumen methanogenic communities under different dietary and host feed efficiency conditions, as determined by PCR-denaturing gradient gel electrophoresis analysis.** *Appl Environ Microbiol* 2010, **76:**3776–3786.
28. Jeyanathan J, Kirs M, Ronimus RS, Hoskin SO, Janssen PH: **Methanogen community structure in the rumens of farmed sheep, cattle and red deer fed different diets.** *FEMS Microbiol Ecol* 2011, **76:**311–326.
29. King EE, Smith RP, St-Pierre B, Wright ADG: **Differences in the rumen methanogen populations of lactating jersey and holstein dairy cows under the same diet regimen.** *Appl Environ Microbiol* 2011, **77:**5682–5687.
30. Skillman LC, Evans PN, Strompl C, Joblin KN: **16 s rDNA directed pcr primers and detection of methanogens in the bovine rumen.** *Lett Appl Microbiol* 2006, **42:**222–228.
31. Whitford MF, Teather RM, Forster RJ: **Phylogenetic analysis of methanogens from the bovine rumen.** *BMC Microbiol* 2001, **1:**5.
32. Muro-Reyes A, Gutierrez-Banuelos H, Diaz-Garcia LH, Gutierrez-Pina FJ, Escareno-Sanchez LM, Banuelos-Valenzuela R, Medina-Flores CA, Corral Luna A: **Potential environmental benefits of residual feed intake as strategy to mitigate methane emissions in sheep.** *J Anim Vet Adv* 2011, **10:**1551–1556.
33. Hegarty RS, Goopy JP, Herd RM, McCorkell B: **Cattle selected for lower residual feed intake have reduced daily methane production.** *J Anim Sci* 2007, **85:**1479–1486.
34. Lillis L, Boots B, Kenny DA, Petrie K, Boland TM, Clipson N, Doyle EM: **The effect of dietary concentrate and soya oil inclusion on microbial diversity in the rumen of cattle.** *J Appl Microbiol* 2011, **111:**1426–1435.
35. Liu C, Zhu Z, Liu Y, Guo T, Dong H: **Diversity and abundance of the rumen and fecal methanogens in altay sheep native to xinjiang and the influence of diversity on methane emissions.** *Arch Microbiol* 2012, **194:**353–361.
36. Machmüller A, Soliva CR, Kreuzer M: **Effect of coconut oil and defaunation treatment on methanogenesis in sheep.** *Reprod Nut Dev* 2003, **43:**41–55.
37. Morgavi DP, Martin C, Jouany J-P, Ranilla MJ: **Rumen protozoa and methanogenesis: not a simple cause-effect relationship.** *Brit J Nut* 2012, **107:**388–397.
38. Liu DY, Ding WX, Jia ZJ, Cai ZC: **Relation between methanogenic archaea and methane production potential in selected natural wetland ecosystems across china.** *Biogeosciences* 2011, **8:**329–338.
39. Durunna ON, Mujibi FD, Goonewardene L, Okine EK, Basarab JA, Wang Z, Moore SS: **Feed efficiency differences and reranking in beef steers fed grower and finisher diets.** *J Anim Sci* 2011, **89:**158–167.
40. Slyter LL: **Influence of Acidosis on rumen function.** *J Anim Sci* 1976, **43:**910–929.
41. Pitt RE, Van Kessel JS, Fox DG, Pell AN, Barry MC, Van Soest PJ: **Prediction of ruminal volatile fatty acids and pH within the net carbohydrate and protein system.** *J Anim Sci* 1996, **74:**226–244.
42. Whitelaw FG, Eadie JM, Bruce LA, Shand WJ: **Methane formation in faunated and ciliate-free cattle and its relationship with rumen volatile fatty acid proportions.** *Brit J Nut* 1984, **52:**261–275.
43. Moss AR, Jouany JP, Newbold J: **Methane production by ruminants: its contribution to global warming.** *Ann Zootech* 2000, **49:**231–253.
44. Fricke WF, Seedorf H, Henne A, Krüer M, Liesegang H, Hedderich R, Gottschalk G, Thauer RK: **The genome sequence of methanosphaera stadtmanae reveals why this human intestinal archaeon is restricted to methanol and H_2 for methane formation and ATP synthesis.** *J Bacteriol* 2006, **188:**642–658.
45. Samuel BS, Hansen EE, Manchester JK, Coutinho PM, Henrissat B, Fulton R, Latreille P, Kim K, Wilson RK, Gordan JI: **Genomic and metabolic adaptations of *Methanobrevibacter smithii* to the human gut.** *Proc Natl Acad Sci U S A* 2007, **104:**10643–10648.
46. Hobson PN, Stewart CS: *The rumen microbial ecosystem.* London, UK: Blackie academic and professional; 1997:467–491.
47. Berk H, Thauer RK: **Function of coenzyme F_{420}-dependent NADP reductase in methanogenic archaea containing an NADP-dependent alcohol dehydrogenase.** *Arch Microbiol* 1997, **168:**396–402.
48. Leahy SC, Kelly WJ, Altermann E, Ronimus RS, Yeoman CJ, Pacheco DM, Li D, Kong Z, Mc Tavish S, Sang C, Lambie SC, Janseen PH, Dey D, Attwood GT: **The genome sequence of the rumen methanogen Methanobrevibacter ruminantium reveals new possibilities for controlling ruminant methane emissions.** *PLoS ONE* 2010, **5:**e8926.

Occurrence of mycotoxins in feed ingredients and complete feeds obtained from the Beijing region of China

Xiaoying Li[1†], Lihong Zhao[1†], Yu Fan[1], Yaxiong Jia[2], Lei Sun[3], Shanshan Ma[1], Cheng Ji[1], Qiugang Ma[1] and Jianyun Zhang[1*]

Abstract

Background: The current study was carried out to provide a reference for the control of mycotoxin contamination in feed ingredients and complete feeds for swine.

Methods: A total of 55 feed ingredients, including 14 corn, 13 wheat bran, 11 soybean meal and 17 dried distillers grains with solubles (DDGS) as well as 76 complete swine feeds including 7 creep feeds, 14 starter feeds, 14 grower feeds, 18 grower-finisher feeds, 10 gestating sow feeds, and 13 lactating sow feeds were randomly collected from 15 swine farms located in the Beijing region of China from July to August 2011. Immunoaffinity clean-up, using High-Performance Liquid Chromatography (HPLC) in combination with UV or Fluorescence Detection, was used for quantitative analysis of aflatoxin B_1 (AFB_1), deoxynivalenol (DON), zearalenone (ZEA) and ochratoxin A (OTA) in the ingredients and complete feeds.

Results: DON and ZEA were the most prevalent mycotoxins found. DON was detected at percentages of 93, 92, 54, 100 and 97% with a mean level of 1.01, 0.44, 0.05, 1.36 and 0.65 ppm in the samples of corn, wheat bran, soybean meal, DDGS and complete feeds, respectively. The detected percentages of ZEA were 100, 100, 54, 100 and 100 with mean levels of 109.1, 14.9, 9.2, 882.7 and 58.9 ppb in the same samples. In the wheat bran and soybean meal samples, the content of all four mycotoxins were below the maximum limits set by Chinese regulations while the percentage of samples that exceeded regulatory limits were 7, 57 and 7% for corn, and 7, 14 and 3% for the complete feeds for AFB_1, DON and OTA respectively. DDGS showed the most serious mycotoxin contamination and the percentage of samples that exceeded regulatory limits were 6, 88 and 41%, for AFB_1, DON and ZEA, respectively.

Conclusions: This paper is the first to present data on the natural occurrence of AFB_1, DON, ZEA and OTA in ingredients and complete feeds obtained from swine farms in China's Beijing region. The data shows that feed ingredients and complete swine feeds obtained from these farms are most often contaminated with DON followed by contamination with AFB_1 and ZEA.

Keywords: AflatoxinB_1, Beijing region, Deoxynivalenol, Ochratoxin A, Swine feeds, Zearalenone

* Correspondence: jyzhang@cau.edu.cn
†Equal contributors
[1]State Key Laboratory of Animal Nutrition, China Agricultural University, No. 2. West Road Yuanming Yuan, Beijing 100193, P.R. China
Full list of author information is available at the end of the article

Background

Mycotoxins are a large group of fungal metabolites found worldwide in cereal grains and other feed products. Mycotoxins pose a severe threat to humans and as well as causing huge economic losses in the feed and food industries [1]. Aflatoxin, deoxynivalenol (DON), zearalenone (ZEA) and ochratoxin A (OTA) are considered the most economically important mycotoxins in terms of their prevalence and their negative effects on animal performance [2,3]. Pigs are particularly susceptible to mycotoxins, suffering a variety of chronic or acute syndromes depending on the amount of contaminants they consume [4].

Aflatoxins are the most potent of the known mutagenic and carcinogenic natural substances produced by *Aspergillus* spp. [5]. There are six types of aflatoxins that frequently contaminate feeds and foods, including B_1, B_2, G_1 and G_2. Aflatoxin B_1 (AFB_1) has long been considered one of the most poisonous carcinogens [6]. A potent hepatocarcinogenic and hepatotoxic mycotoxin [7], it has been placed in Group IA by the International Agency for Research on Cancer IARC: [8].

DON is one of the most frequently detected trichothecene contaminants in cereals [9]. The ingestion of feeds contaminated with DON leads to feed refusal and weight loss, decreases nutritional efficiency and causes lesions in the gastrointestinal tract, vomiting, bloody diarrhea and severe dermatitis accompanied by hemorrhaging and immune dysregulation. Swine are typically the most sensitive of the susceptible species in that they exhibit the most severe symptoms of acute DON toxicity [10,11].

ZEA is a non-steroidal estrogenic toxin produced by certain *Fusarium* species [12]. ZEA can competitively bind to estrogen receptors leading to reproductive disorders and estrogenic dysfunction in humans and animals (especially in breeding animals), impairing fertility and increasing the frequency of stillbirths along with reducing sperm quality [12,13]. Swine are the animal species most severely affected by ZEA [14,15].

OTA, produced by the *Penicillium verrucosum* fungus and various species of *Aspergillus* [5] is a nephrotoxic mycotoxin that causes renal toxicity and possesses carcinogenic, teratogenic immunotoxic and possibly neurotoxic properties [6]. This toxin has been classified by the IARC as a possible human carcinogen [8]. Research on the effects of this mycotoxin in swine has revealed that it causes changes in renal function [16].

China faces a feed shortage. The feed that is available is frequently contaminated with mycotoxins [17]. Furthermore, the high prices and scarcity of protein sources for animal feeds have led to the use of alternative protein sources such as distillers dried grains with solubles (DDGS). However, the supplementation of animal diets with DDGS must be carefully monitored for the presence of toxic compounds such as mycotoxins, especially AFB_1, DON, ZEA and OTA [18].

Although Beijing is one of the largest cities in China, there has been little research conducted on the presence of the mycotoxins AFB_1, DON, ZEA and OTA in the Beijing region. As the surveys that have been done have been limited and because more thorough research is very much necessary, the current study investigated the occurrence of four major mycotoxins in feed ingredients (corn, wheat bran, soybean meal and DDGS) and complete swine feeds for various stages of growth on a variety of swine farms in China's Beijing region.

Methods

Sample collection and preparation

The protocol for the experiment was reviewed and approved by the Institutional Animal Care and Use Committee at China Agricultural University.

A total of 131 samples including 55 feed ingredients and 76 complete swine feeds were randomly sampled from 15 swine farms in the Beijing region from July through August 2011. The samples included 16 samples of corn, 13 samples of wheat bran, 12 samples of soybean meal, and 17 DDGS samples. The swine complete feeds included creep feeds (n = 7), starter feeds (n = 14), grower feeds (n = 14), grower-finisher feeds (n = 18), gestating sow feeds (n = 10) and lactating sow feeds (n = 13). The sampling was undertaken in accordance with European Regulation No. 401/2006 [19]. All samples were ground, mixed and stored at 4°C prior to analysis.

Methods of analysis for mycotoxins

AFB_1 analysis

Samples were analyzed according to the methods of the AOAC [20]. Briefly, 25 g of ground feed was extracted with 100 mL of methanol:water (80:20, v/v), blended at high speed for 3 min and then filtered through a sheet of Waterman Filter Paper No. 4. The extract was diluted with a phosphate-buffered saline solution (PBS, pH 7.40), mixed well and filtered through microfiber filter paper. The immunoaffinity column (AokinImmunoClean CF AFLA, Aokin AG, Berlin, Germany) was conditioned with 1 mL of sodium azide, and 10 mL of the diluted filtrate was passed through the column at a flow rate of 1 mL/min. The column was then washed with 10 mL of a methanol: water solution (10:90, v/v) at a flow rate of 3 mL/min. The retained chemicals were then eluted with 1 mL of methanol at a flow rate of 1 mL/min. Subsequently, 20 μL of the clear eluate was injected directly into an HPLC system. If the eluate was found not to be clear, it was passed through an organic filter unit (0.45 μm) before injection. The mobile phase utilized a methanol:water solution (50:50, v/v) with the flow rate set at 1 mL/min. Postcolumn derivatization was performed with a photochemical reactor (AURA, Los Angeles, CA). A C18 column (4.6 mm × 150 mm, 5 μm, Dikma, Beijing, China) was employed with the LOD set at 0.03 ppb and the LOQ at 0.1 ppb. A fluorescence detector (SHIMADZU, Kyoto, Japan) was set for excitation and emission wavelengths of 360 and 440 nm, respectively. The retention time was 16.5 min.

DON analysis

Samples were analyzed according to the method of GB/T 23503–2009. Briefly, 25 g of ground feed was extracted with 100 mL of a solution of methanol:water (60:40, v/v), blended at high speed for 3 min and then filtered through a sheet of Waterman Filter Paper No. 4. The extract was diluted with a phosphate-buffered saline solution (PBS, pH 7.4), mixed well and filtered through microfiber filter paper. The immunoaffinity column (AokinImmunoClean CF DON, Aokin AG, Berlin. Germany) was conditioned with 1 mL of sodium azide, and 10 mL of the diluted filtrate was passed through the column by gravity at a flow rate of 1 mL/min. The column was then washed with a 10 mL solution of methanol:water (10:90, v/v) at a flow rate of 3 mL/min. The bound compounds were then eluted with 3 mL of methanol at a flow rate of 1 mL/min. The purified samples were dried under a stream of nitrogen gas at 50°C. For the mobile phase, the residue was dissolved in 200 μL after evaporation. Subsequently, 20 μL of the eluate was injected into an HPLC system. The mobile phase utilized a methanol: water solution (30: 70, v/v)with the flow rate set at 1 mL/min. A C18 column (4.6 mm × 150 mm, 5 μm, Agilent, Santa Clara, CA) was employed with the LOD set at 0.02 ppm and the LOQ at 0.06 ppm. The absorption UV wavelength (SHIMADZU, Kyoto, Japan) was set at 218 nm. The retention time was 5.6 min.

ZEA analysis

Samples were analyzed according to the certified Chinese GB/T 23504–2009 method. Briefly, 25 g of feed was extracted with 100 mL of a methanol:water solution (60:40, v/v), blended at high speed for 3 min and then filtered through a sheet of Waterman Filter Paper No. 4. The extract was diluted with a phosphate-buffered saline solution (PBS, pH 7.40), mixed well and filtered through microfiber filter paper. The immunoaffinity column (AokinImmunoClean CF ZEA, Aokin AG, Berlin, Germany) was conditioned with 1 mL of sodium azide, and 10 mL of the diluted filtrate were passed through the column by gravity at a flow rate of 1 mL/min. The column was then washed with 10 mL of a methanol:water solution (10:90, v/v) at a flow rate of 3 mL/min. The bound chemicals were then eluted with 3 mL of methanol at a flow rate of 1 mL/min. Subsequently, 20 μL of the clear eluate was injected directly into an HPLC system. If the eluate was found not to be clear, it was passed through an organic filter unit (0.45 μm) before injection. The mobile phase utilized an acetonitrile:water:methanol solution (46: 46: 8, v/v/v with the flow rate set at 1 mL/min. A C18 column (4.6 mm × 150 mm, 5 μm, Dikma, Beijing, China) was employed with the LOD set at 1.5 ppb and the LOQ at 4 ppb. A fluorescence detector (SHIMADZU, Kyoto, Japan) was set for excitation and emission wavelengths of 274 and 440 nm, respectively. The retention time was 7.3 min.

OTA analysis

Samples were analyzed according to the methods of the AOAC [21]. Briefly, 25 g of feed was extracted with 100 mL of a methanol:water solution (60:40, v/v), blended at high speed for 3 min and then filtered through a sheet of Waterman Filter Paper No. 4. The extract was diluted with a phosphate-buffered saline solution (PBS, pH 7.4), mixed well and filtered through microfiber filter paper. The immunoaffinity column (AokinImmunoClean CF OTA, Aokin AG, Berlin, Germany) was conditioned with 1 mL of sodium azide, and 10 mL of the diluted filtrate was passed through the column by gravity at a flow rate of 1 mL/min. The column was then washed with 10 mL of a methanol:water solution (10:90, v/v) at a flow rate of 3 mL/min. The bound compounds were then eluted with 3 mL of methanol at a flow rate of 1 mL/min, after which 20 μL of the clear eluate was injected directly into an HPLC system. If the eluate was found not to be clear, it was passed through an organic filter unit (0.45 μm) before injection. The mobile phase utilized an acetonitrile:water: glacial acetic acid solution (99: 99: 2, v/v/v) with the flow rate set at 0.9 mL/min. A C18 column (4.6 mm × 150 mm, 5 μm, Dikma, Beijing, China) was employed with the LOD set at 0.035 ppb and the LOQ at 0.117 ppb. A fluorescence detector was set for excitation and emission wavelengths of 333 and 477 nm, respectively. The retention time was 11.4 min.

Validation of analytical methods

The analytical methods used were evaluated for linearity, recovery and reproducibility. Standard curves were generated by comparing the linear regressions of peak regions against concentrations. The recovery (precision) of mycotoxins was determined by spiking blank samples (corn, wheat bran, soybean meal, DDGS and complete feeds) with known levels of AFB_1, DON, ZEA or OTA standards dissolved during the mobile phase. The spiked concentrations of AFB_1, DON, ZEA and OTA were evaluated at high, intermediate and low levels for each standard mycotoxin concentration. Each sample was analyzed after being allowed to stand for 2 h to allow solvent equilibration between the analytes and the matrix. The reproducibility of the method was established by injecting five replicates of the same standard solution on the same day and over five different days.

Statistical analysis

The data were analyzed using MS Excel, expressed as average contents, detection rates and the percentage of samples that exceeded regulatory limits.

Results and discussion

Performance of the analytical methods

The standard curves for AFB_1, DON, ZEN and OTA produced from three determinations of five concentration levels were linear with correlation coefficients (R^2) ranging from 0.9990 to 0.9999. Both the intra-day and inter-day precision measurements showed relative standard deviations below 15%. The determination of recoveries in the blank samples spiked with AFB_1, DON, ZEN and OTA were 80.3 to 85.0, 84.1 to 89.8, 86.6 to 90.4 and 87.3 to 90.8, respectively. Therefore, the performances of the employed analytical methods were shown to be satisfactory for the study's purposes. The analytical methods were validated for all feed samples and met the performance criteria set by European Regulation No. 401/2006 [14] for mycotoxin analysis.

The occurrence of mycotoxins in feed ingredients and complete feeds

Corn

The occurrence of mycotoxin contamination in corn is summarized in Table 1. AFB_1 was detected in 50% of the corn samples and 7% of the corn samples showed concentrations exceeding regulatory limits. The highest concentration of AFB_1 in corn was 58.9 ppb while the average value was 6.0 ppb. The corn samples were also contaminated with DON (93% detection rate). The percentage of corn samples exceeding regulatory limits for DON was very high, with 57% of samples exceeding the maximum limits set by Chinese regulations. The average content of ZEA in corn was 109.1 ppb and the average value of OTA was 22.1 ppb. Additionally, the percentage of corn samples exceeding regulatory limits for OTA was 7%.

Corn is an energy ingredient frequently used in animal feeds in China. The results showed that corn is frequently contaminated with AFB_1, DON, ZEA and OTA. The percentages of samples in which AFB_1 exceeded regulatory limits were similar to those in corn samples surveyed by Rodrigues and Naehrer in North, South-East and South Asia and in Oceania [22]. The results also showed that corn samples were contaminated with DON at an average value of 1.01 ppm while the highest value detected was 2.13 ppm which agrees with the maximum value in corn reported by Rodriguez and Naehrer [22]. The average content of ZEA in corn was similar to the 121 ppb reported by the two researchers in China [22]. It was similar to the concentrations of DON (93.20%) and ZEA (96.60%) in corn as reported by Guan et al. [17]. The average values and concentrations of OTA in the samples that exceeded regulatory limits was in agreement with survey results that have shown lower contamination levels of OTA than other mycotoxins in corn [5,23].

DDGS

The occurrence of mycotoxin contamination with DDGS is summarized in Table 1. The results showed that 6% of the DDGS contained concentrations of AFB_1 that exceeded regulatory limits, with an average content of 9.8 ppb. Approximately 88% of the samples contained concentrations of DON that exceeded regulatory limits with an average content of 1.4 ppm. The ZEA contamination levels in the DDGS samples ranged from 49 to 2,958 ppb. We found that 41% of the samples contained concentrations of ZEA that exceeded regulatory limits, with an average content of 883 ppb. However, the OTA content in the DDGS samples did not exceed the standard limits. These samples had an average OTA content of 23 ppb.

The occurrence of mycotoxins in DDGS used as an ingredient in pig diets has rarely been studied. Four mycotoxins were found to be prevalent in the DDGS samples used in our study. Contamination with DON and ZEN was particularly serious. The percentage of samples containing concentrations of DON that exceeded regulatory limits and the average content of DON detected in these samples were similar to the concentrations of DON that have been found in distillers dried grains with solubles sourced worldwide [18]. These results may be explained by the fact that mycotoxins in DDGS (constituting the remaining portions within the final by-product) are up to three times more concentrated than in corn grain [24]. Furthermore, if improperly stored, DDGS are easily contaminated with more mycotoxins due to their high moisture content. In addition, grains are damaged during the process of DDGS production, which can easily cause the production of more mycotoxins. Moreover, in our study, the DDGS were obtained from the regions lying along the middle and lower reaches of the Yangtze River (in Henan, Sichuan and Anhui provinces) which are characterized by a moist climate which may contribute to the contamination of corn with mycotoxins [25,26]. What is more, this study has shown that multiple mycotoxins coexist in most feeds and feedstuffs [27,28]. About 29% (5/17) of the DDGS samples used in this study were co-contaminated with DON and ZEN at levels of 1 ppm and 500 ppb, respectively, which exceeds Chinese regulatory limits for feedstuffs. The simultaneous occurrence of contamination by various kinds of mycotoxins leads not only

Table 1 Occurrence of mycotoxins in feed ingredients and complete swine feeds[1,2]

Item	Total samples, n	Average contents				Content range				Detection rates, %				Percentage of samples over the maximum limits, %[1,2]			
		AFB, ppb	DON, ppm	ZEA, ppb	OTA, ppb	AFB_1, ppb	DON, ppm	ZEA,ppb	OTA,ppb	AFB_1	DON	ZEA	OTA	AFB_1	DON	ZEA	OTA
Corn	14	5.98	1.01	109	22	<LOQ-58.9	<LOQ-2.13	6.25-321	<LOQ-136	50	93	100	93	7	57	0	7
Bran	13	0.25	0.44	159	8	<LOQ-1.1	<LOQ-0.89	<LOQ-44.4	<LOQ-36	46	92	100	77	0	0	0	0
Soybean meal	11	1	0.05	9	11	<LOQ-10.7	<LOQ-0.21	<LOQ-35.4	<LOQ-42	36	54	54	64	0	0	0	0
DDGS	17	9.83	1.36	883	23	<LOQ-64.1	0.85-1.72	48.89-2958	1.79-89	94	100	100	100	6	88	41	0
Creep feed	7	0.54	0.28	39	5	<LOQ-3.7	<LOQ-0.53	4.7-107	<LOQ −22	57	86	100	100	0	0	0	0
Starter feed	14	0.51	0.85	54	16	0.1-3.9	0.16-2.31	<LOQ-115	<LOQ-84	71	100	100	86	0	14	0	0
Grower feed	14	0.31	0.62	67	12	<LOQ-3.9	0.12-1.14	7.59-132	<LOQ-94	57	100	100	86	0	14	0	0
Grower-finisher feed	18	4.33	0.58	59	13	<LOQ-38.3	<LOQ-1.44	7.1-150	<LOQ-92	67	94	100	94	11	22	0	0
Gestating sow feed	10	0.25	0.82	63	48	<LOQ-1.64	0.13-1.45	9.2-149	<LOQ-212	70	100	100	90	0	10	0	20
Lactation sow feed	13	9.55	0.62	76	15	<LOQ-41.0	0.07-1.52	7.4-231	1.67-92	69	100	100	100	23	23	0	0
Total	76	2.89	0.65	59	18	<LOQ-41.0	<LOQ-2.31	<LOQ-232	<LOQ-212	70	97	100	92	7	14	0	3

[1] Samples exceeding regulatory limits of four mycotoxins: higher than the maximum limits set by the Chinese regulation (Table 2).
[2] Maximum limit of mycotoxins in other feed ingredients is in accordance with the regulations on mycotoxins in corn.

to immune suppression in animals, but also lowered efficiency in animal production [29].

Wheat bran and soybean meal

The occurrence of mycotoxin contamination in wheat bran and soybean meal is summarized in Table 1. The detection rate of AFB_1, DON, ZEA and OTA was 46, 92, 100 and 77% in wheat bran, and 36, 54, 54 and 64% in soybean meal, respectively.

The samples of wheat bran and soybean meal were analyzed for the presence of four mycotoxins. The contamination levels of these mycotoxins were lower than those found in the corn and DDGS samples and did not exceed the maximum limits set in China (Table 2). In general, feed ingredients such as soybean meal and wheat bran were less contaminated with mycotoxins, a finding in agreement with the results of a survey on the worldwide occurrence of mycotoxins in feedstuffs and feed conducted by Rodrigues and Naehrer [22]. This may be due to the low moisture content and the composition (high protein/carbohydrate ratio) of soybean meal, which inhibits the growth of mold. Another reason is that producers and farmers pay more attention to storage conditions because of the high price of soybean meal.

Complete feeds

The occurrence of mycotoxin contamination in complete swine feed is summarized in Table 1. The incidences of the four investigated mycotoxins ranged from 57 to 100% in complete feeds obtained from swine farms. The detection rate of AFB_1 in the samples of creep feed, starter pig feed, grower pig feed, grower-finisher pig feed, gestating sow feed and lactating sow feed were 57, 71, 57, 67, 70 and 69%, respectively, with mean concentrations of 0.54, 0.51, 0.31, 4.33, 0.25 and 9.55 ppb, respectively. In addition, the percentage of grower-finisher pig feed and lactating sow feed that contained concentrations of AFB_1 that exceeded regulatory limits was 11 and 23%, respectively.

The detection rate of DON was also very high, with percentages in creep feed, starter pig feed, grower pig feed, grower-finisher pig feed, gestating sow feed and lactating sow feed of 86, 100, 100, 100, 94, 100 and 100%, with average concentrations of 0.28, 0.85, 0.62, 0.58, 0.82 and 0.62 ppm, respectively. The percentage of samples with DON concentrations exceeding regulatory limits in complete feeds for the various stages of pig growth was 0, 14, 14, 22, 10 and 23%, respectively. The highest concentration of DON was detected in starter pig feeds (2.31 ppm). Moreover, although high incidences of ZEA and OTA (86 to 100%) were found in swine feeds, the mean concentrations were low (that is, they did not exceed the maximum limits set in China), except for the concentrations of OTA found in gestating sow feed.

Although the levels of ZEA and OTA detected in our study did not seem to have any clear clinical effects, there may well be chronic accumulated carry-over effects on animal products over a long period of feed consumption [2]. In contrast, AFB_1 and DON were found in a high percentage of samples in concentrations that exceeded regulatory limits for swine feeds, which could have clinical effects on gilt and sow health. The fact that AFB_1 was found at concentrations exceeding regulations in feeds for lactating sows could be of importance in terms of animal's health. Indeed, it could lead to the excretion of AFM_1 and to an exposure of piglets to that toxin, since young animals are more sensitive to AFB_1

Table 2 Maximum limit regulations for mycotoxins in feedstuffs in China[1,2]

Mycotoxin	Feedstuff	Maximum limit, ppb	Reference
AFB_1	Corn	50	GB1 13078-2001
	Soybean meal	30	GB 13078-2001
	Starter pig feed	10	GB 13078-2001
	Grower pig feed	20	GB 13078-2001
	DDGS	50	NY/T2 1968-2010
DON	Swine complete feed	1,000	GB 13078.3-2007
	Corn	1,000	GB 2761–2011.
	DDGS	1,000	NY/T 1968-2010
ZEA	Corn	500	GB 13078.2-2006
	Complete feed	500	GB 13078.2-2006
OTA	Corn	100	GB 13078.2-2006
	Swine complete feed	100	GB 13078.2-2006
	DDGS	100	NY/T 1968-2010

[1] *GB*: General Administration of Quality Supervision, Inspection and Quarantine of the People's Republic of China. National Standard of the People's Republic of China.

[2] NY/T: Ministry of Agriculture of the People's Republic of China. Agricultural Standard of the People's Republic of China.

than adults. Concentrations as low as 1 ppm may cause loss of appetite, vomiting, low feed efficiency and a reduction in body weight gain [23,30]. Therefore, the high incidence of mycotoxins found in complete feed samples obtained from swine farms highlights the need for surveillance and prevention.

The natural occurrence of mycotoxins has been reported in a variety of foods and feeds in many countries [31]. The maximum admissible levels of these four mycotoxins in feed ingredients and swine complete feeds vary from one country to another with legislation prescribing mandatory upper limits that often take the form of product standards [32]. The United States' Food and Drug Administration has set a maximum permissible level of 20 ppb for total aflatoxins in all foodstuffs, while in China, the legal limit for AFB_1 contamination in corn is 50 ppb. The wide variations in standards have resulted in conflicting perceptions regarding safe levels among the various national agencies. The regulatory limits of the four mycotoxins in this study is higher than the maximum limits set by the Chinese regulation (Table 2).

Pigs are among the most sensitive species to all of these toxins [4]. This is also one of the most frequently exposed species due to its diet based mainly on cereals. The regulation is presented as a consequence of pigs sensitivity (in EU, limits for pigs feeds are often lower than those for other species).

The toxicology of mycotoxins and the incidence of mycotoxin contamination have aroused public awareness. Health authorities need to take action to mitigate the serious negative effects they can have on humans and animals in China. The reduction of mycotoxin contamination will require an integrated understanding of agronomy, fungal ecology, harvesting methods, storage conditions, feed processing and detoxification strategies [33,34]. Effective control strategies have been implemented based on the establishment of the Hazard Analysis and Critical Control Point (HACCP). This approach can minimize mycotoxin contamination in the food chain through appropriate management of products [35].

Because mycotoxins are stable chemicals, physical and chemical degradation methods are limited in their effectiveness and include disadvantages such as loss of product nutrition, organoleptic qualities, undesirable health effects and high equipment cost [36]. These disadvantages have encouraged the development of promising biodegradation methods by which mycotoxins may be detoxified [37,38]. Further periodic investigations of feed ingredients and complete feed samples from various regions in China taken during all seasons are ongoing.

This paper is the first to present data on the natural occurrence of AFB_1, DON, ZEA and OTA in feed ingredients and swine complete feeds obtained from livestock farms in China's Beijing region. The results of this study should serve as a significant reference for feed manufacturers, animal farms and Chinese regulatory authorities for the future consideration of feed and food safety issues. Periodic surveillance and monitoring of the occurrence of mycotoxins in feed ingredients and in animals are very important for minimizing animal and human health risks.

Conclusion

The content of all four mycotoxins were below the maximum limits set by Chinese regulations in the wheat bran and soybean meal samples while the percentage of samples that exceeded regulatory limits were 7, 57 and 7% for corn, and 7, 14 and 3% for the complete feeds for AFB_1, DON and OTA respectively. DDGS exhibited the most serious mycotoxin contamination and the percentage of samples that exceeded regulatory limits were 6, 88 and 41%, for AFB_1, DON and ZEA, respectively.

Competing interests

The authors declare that they have no competing interests.

Authors' contributions

XYL carried out the statistical analysis and drafted the manuscript. LHZ participated in the chemical analysis. YF, YXJ, LS, SSM, CJ, QGM and JYZ conceived the study, participated in its design and coordination, and helped draft the manuscript. All authors read and approved the final manuscript.

Acknowledgements

This study was supported by the Introduction of International Advanced Agriculture Science and Technology Project (948 Project) (Grant No. 2013-Z20), the Program of Beijing Municipal Bureau of Agriculture (Grant No. 20120633), and National Key Technology Research and Development Program of the Ministry of Science and Technology of China (Grant No. 2011BAD26B04).

Author details

[1]State Key Laboratory of Animal Nutrition, China Agricultural University, No. 2. West Road Yuanming Yuan, Beijing 100193, P.R. China. [2]Beijing Municipal General Station for Animal Husbandry and Veterinary Service, Beijing, China. [3]Institute of Plant Protection, Chinese Academy of Agricultural Sciences, Beijing, China.

References

1. Sundstol EG, Pettersson H, Lundh T: **Comparative cytotoxicity of deoxynivalenol, nivalenol, their acetylated derivatives and de-epoxy metabolites.** *Food Chem Toxicol* 2004, **42**:619–624.
2. Lawlor PG, Lynch PB: **Mycotoxins in pig feeds. 1: Source of toxins, prevention and management of mycotoxicosis.** *Irish Vet J* 2001, **54**:117–120.
3. Miller JD: **Fungi and mycotoxins in grain: implications for stored product research.** *J Stored Prod Res* 1995, **31**:1–16.
4. Chayto AC, Hansen JA, Eric Van Heugten M, Todd S, Sung Woo K: **Occurrence and decontamination of mycotoxins in swine feed.** *Asian-Aust J Anim Sci* 2011, **5**:723–738.
5. Kumagai S, Nakajima M, Tabata S, Ishikuro E, Tanaka T, Norizuki H, Itoh Y, Aoyama K, Fujita K, Kai S, Sato T, Saito S, Yoshiike N, Sugita-Konishi Y: **Aflatoxin and ochratoxin A contamination of retail foods and intake of these mycotoxins in Japan.** *Food Add Contam Part A Chem Anal Control Expo Risk Assess* 2008, **25**:1101–1106.
6. EFSA: *Definition and description of emerging risks within the EFSA's mandate.* Italy: European Food Safety Authority; 2007.

7. Tulayakul P, Sakuda S, Dong KS, Kumagai S: **Comparative activities of glutathione-S-transferase and dialdehyde reductase toward aflatoxin B1 in livers of experimental and farm animals.** *Toxicology* 2005, **46:**204–209.
8. IARC: ***Monograph on the evaluation of carcinogenic risk to humans: Some traditional herbal medicines some mycotoxins, naphthalene and styrene***, Volume 82. Lyon, France: 2002:1–556.
9. Jelinek CF, Pohland AE, Wood GE: **Worldwide occurrence of mycotoxins in foods and feeds: An update.** *J Assoc Offic Anal Chem* 1989, **72:**223–230.
10. Pestka JJ: **Deoxynivalenol: toxicity, mechanisms and animal health risks.** *Anim Feed Sci Technol* 2007, **137:**283–298.
11. Wu F: **Measuring the economic impacts of Fusarium toxins in animal feeds.** *Anim Feed Sci Technol* 2007, **137:**363–374.
12. Zinedine A, Soriano JM, Molto JC, Manes J: **Review on the toxicity, occurrence, metabolism, detoxification, regulations and intake of zearalenone: An oestrogenic mycotoxin.** *Food Chem Toxicol* 2007, **45:**1–18.
13. Borutova R, Aragon YA, Hrer NAK, Berthiller F: **Co-occurrence and statistical correlations between mycotoxins in feedstuffs collected in the Asia–Oceania in 2010.** *Anim Feed Sci Technol* 2012, **178:**190–197.
14. CAST: *Mycotoxins: Risks in plant, animal, and human systems.* Ames, IA: Council for Agric Sci Technol Task Force. Report No. 139; 2003.
15. Richard JL: **Some major mycotoxins and their mycotoxicoses: An overview.** *Int J Food Microbiol* 2007, **119:**3–10.
16. Krogh P: **Ochratoxins.** In *Mycotoxins in Human and Animal Health.* Edited by Rodricks JV, Hesseltine CW, Mehlman MA. Park Forest South, IL: Pathotox Publishers, Inc; 1977:489–498.
17. Guan S, Gong M, Yin Y, Huang R, Ruan Z, Zhou T, Xie M: **Occurrence of mycotoxins in feeds and feed ingredients in China.** *J Food Agric Environ* 2011, **9:**163–167.
18. Rodrigues I, Chin LJ: **A comprehensive survey on the occurrence of mycotoxins in maize dried distillers' grain and solubles sourced worldwide.** *World Mycotoxin J* 2012, **5:**83–88.
19. European Commission: **Commission regulation (EC) no. 401/2006 of 23 February 2006 laying down the methods of sampling and analysis for the official control of the levels of mycotoxins in foodstuffs.** *Offic J European Union* 2006, **L70:**12–34.
20. AOAC: *Official Methods of Analysis. 18th ed. AOAC Official Method 999.07 – Aflatoxin B1 and total aflatoxins in peanut butter, pistachio paste, fig paste and paprika powder – immunoaffinity column liquid chromatography with post-column derivatisation first action 2008.* Arlington, Virginia: Assoc Offic Anat Chem; 2008.
21. AOAC: *Official Methods of Analysis. Determination of ochratoxin A in wine and beer.* Arlington, Virginia: Assoc Offic Anal Chem; 2001.
22. Rodrigues I, Naehrer K: **A three-year survey on the worldwide occurrence of mycotoxins in feedstuffs and feed.** *Toxins (Basel)* 2012, **4:**663–675.
23. Van Egmond HP, Schothorst RC, Jonker MA: **Regulations relating to mycotoxins in food: Perspectives in a global and European context.** *Anal Bioanal Chem* 2007, **389:**147–157.
24. Wu F, Munkvold GP: **Mycotoxins in ethanol co-products: Modeling economic impacts on the livestock industry and management strategies.** *J Agric Food Chem* 2008, **56:**3900–3911.
25. Cui L, Selvaraj JN, Xing F, Zhao Y, Zhou L, Liu Y: **A minor survey of deoxynivalenol in Fusarium infected wheat from Yangtze–Huaihe river basin region in China.** *Food Control* 2013, **30:**469–473.
26. Wang X, Liu X, Liu J, Wang G, Wan K: **Contamination level of T-2 and HT-2 toxin in cereal crops from Aba area in Sichuan Province, China.** *Bull Environ Contam Toxicol* 2012, **88:**396–400.
27. Binder EM, Tan LM, Chin LJ, Handl J, Richard J: **Worldwide occurrence of mycotoxins in commodities, feeds and feed ingredients.** *Anim Feed Sci Technol* 2007, **137:**265–282.
28. Monbaliu S, Van Poucke C, Detavernier C, Dumoulin F, Van De Velde M, Schoeters E, Van Dyck S, Averkieva O, Van Peteghem C, De Saeger S: **Occurrence of mycotoxins in feed as analyzed by a multi-mycotoxin LC-MS/MS method.** *J Agric Food Chem* 2009, **58:**66–71.
29. Speijers GJA, Speijers MHM: **Combined toxic effects of mycotoxins.** *Toxicol Letters* 2004, **153:**91–98.
30. Eriksen GS, Pettersson H: **Toxicological evaluation of trichothecenes in animal feed.** *Anim Feed Sci Technol* 2004, **114:**205–239.
31. Gutema T, Munimbazi C, Bullerman LB: **Occurrence of fumonisins and moniliformin in corn and corn-based food products of U.S. origin.** *J Food Prot* 2000, **63:**1732–1737.
32. Afsah-Hejri L, Jinap S, Arzandeh S, Mirhosseini H: **Optimization of HPLC conditions for quantitative analysis of aflatoxins in contaminated peanut.** *Food Control* 2011, **22:**381–388.
33. Bryden WL: **Mycotoxin contamination of the feed supply chain: Implications for animal productivity and feed security.** *Anim Feed Sci Techno* 2012, **173:**134–158.
34. Ma QG, Gao X, Zhou T, Zhao LH, Fan Y, Li XY, Lei YP, Ji C, Zhang JY: **Protective effect of Bacillus subtilis ANSB060 on egg quality, biochemical and histopathological changes in layers exposed to aflatoxin B1.** *Poult Sci* 2012, **91:**2852–2857.
35. Scudamore KA: **Control of mycotoxins: Secondary processing.** In *Mycotoxins in Food: Detection and Control.* Edited by Magan N, Olsen M. Cambridge CB1 6AH, England: Woodhead Publishing Ltd; 2004:174–189. ISBN 1 85573 733 7.
36. Teniola OD, Addo PA, Brost IM, Rber PFA, Jany KD, Alberts JF, Van Zyl WH, Steyn PS, Holzapfel WH: **Degradation of aflatoxin B1 by cell-free extracts of Rhodococcus erythropolis and Mycobacterium fluoranthenivorans sp. nov. DSM44556 T.** *Intern J Food Microbiol* 2005, **105:**111–117.
37. Guan S, He J, Young JC, Zhu H, Li XZ, Ji C, Zhou T: **Transformation of trichothecene mycotoxins by microorganisms from fish digesta.** *Aquaculture* 2009, **290:**290–295.
38. Zhao LH, Guan S, Gao X, Ma QG, Lei YP, Bai XM, Ji C: **Preparation, purification and characteristics of an aflatoxin degradation enzyme from Myxococcus fulvus ANSM068.** *J Appl Microbiol* 2011, **110:**147–155.

Permissions

All chapters in this book were first published in JASB, by BioMed Central; hereby published with permission under the Creative Commons Attribution License or equivalent. Every chapter published in this book has been scrutinized by our experts. Their significance has been extensively debated. The topics covered herein carry significant findings which will fuel the growth of the discipline. They may even be implemented as practical applications or may be referred to as a beginning point for another development.

The contributors of this book come from diverse backgrounds, making this book a truly international effort. This book will bring forth new frontiers with its revolutionizing research information and detailed analysis of the nascent developments around the world.

We would like to thank all the contributing authors for lending their expertise to make the book truly unique. They have played a crucial role in the development of this book. Without their invaluable contributions this book wouldn't have been possible. They have made vital efforts to compile up to date information on the varied aspects of this subject to make this book a valuable addition to the collection of many professionals and students.

This book was conceptualized with the vision of imparting up-to-date information and advanced data in this field. To ensure the same, a matchless editorial board was set up. Every individual on the board went through rigorous rounds of assessment to prove their worth. After which they invested a large part of their time researching and compiling the most relevant data for our readers.

The editorial board has been involved in producing this book since its inception. They have spent rigorous hours researching and exploring the diverse topics which have resulted in the successful publishing of this book. They have passed on their knowledge of decades through this book. To expedite this challenging task, the publisher supported the team at every step. A small team of assistant editors was also appointed to further simplify the editing procedure and attain best results for the readers.

Apart from the editorial board, the designing team has also invested a significant amount of their time in understanding the subject and creating the most relevant covers. They scrutinized every image to scout for the most suitable representation of the subject and create an appropriate cover for the book.

The publishing team has been an ardent support to the editorial, designing and production team. Their endless efforts to recruit the best for this project, has resulted in the accomplishment of this book. They are a veteran in the field of academics and their pool of knowledge is as vast as their experience in printing. Their expertise and guidance has proved useful at every step. Their uncompromising quality standards have made this book an exceptional effort. Their encouragement from time to time has been an inspiration for everyone.

The publisher and the editorial board hope that this book will prove to be a valuable piece of knowledge for researchers, students, practitioners and scholars across the globe.

List of Contributors

Sung Woo Kim
Department of Animal Science, North Carolina State University, North Carolina, Raleigh 27695, USA

Alexandra C Weaver
Department of Animal Science, North Carolina State University, North Carolina, Raleigh 27695, USA

Yan Bin Shen
Department of Animal Science, North Carolina State University, North Carolina, Raleigh 27695, USA

Yan Zhao
Department of Animal Science, North Carolina State University, North Carolina, Raleigh 27695, USA

Cletos Mapiye
Agriculture and Agri-Food Canada, Lacombe Research Centre, 6000 C & E Trail, Lacombe, Alberta T4L 1 W1, Canada

Tyler D Turner
Agriculture and Agri-Food Canada, Lacombe Research Centre, 6000 C & E Trail, Lacombe, Alberta T4L 1 W1, Canada

John A Basarab
Alberta Agriculture and Rural Development, Lacombe Research Centre, 6000 C & E Trail, Lacombe, Alberta T4L 1 W1, Canada

Vern S Baron
Agriculture and Agri-Food Canada, Lacombe Research Centre, 6000 C & E Trail, Lacombe, Alberta T4L 1 W1, Canada

Jennifer L Aalhus
Agriculture and Agri-Food Canada, Lacombe Research Centre, 6000 C & E Trail, Lacombe, Alberta T4L 1 W1, Canada

Michael ER Dugan
Agriculture and Agri-Food Canada, Lacombe Research Centre, 6000 C & E Trail, Lacombe, Alberta T4L 1 W1, Canada

Xi Liang
Faculty of Agriculture, Iwate University, 3-18-8 Ueda, Morioka 020-8550, Iwate, Japan

Kyota Yamazaki
Faculty of Agriculture, Iwate University, 3-18-8 Ueda, Morioka 020-8550, Iwate, Japan

Mohammad Kamruzzaman
Faculty of Agriculture, Iwate University, 3-18-8 Ueda, Morioka 020-8550, Iwate, Japan

Xue Bi
Faculty of Agriculture, Iwate University, 3-18-8 Ueda, Morioka 020-8550, Iwate, Japan

Arvinda Panthee
Faculty of Agriculture, Iwate University, 3-18-8 Ueda, Morioka 020-8550, Iwate, Japan

Hiroaki Sano
Faculty of Agriculture, Iwate University, 3-18-8 Ueda, Morioka 020-8550, Iwate, Japan

Tércia Cesária Reis de Souza
Facultad de Ciencias Naturales, Universidad Autónoma de Querétaro, Santiago de Querétaro, Av. De las Ciencias s/n Juriquilla 76000, Querétaro, Querétaro, México

Araceli Aguilera Barreyro
Facultad de Ciencias Naturales, Universidad Autónoma de Querétaro, Santiago de Querétaro, Av. De las Ciencias s/n Juriquilla 76000, Querétaro, Querétaro, México

Gerardo Mariscal-Landín
Centro Nacional de Investigación en Fisiología Animal, Instituto Nacional de Investigaciones Forestales Agrícolas y Pecuarias, Km 1, Carretera a Colón, Ajuchitlán Colón Querétaro, 76280 México

Bo Wang
State Key Laboratory of Animal Nutrition, College of Animal Science and Technology, China Agricultural University, Beijing 100193, China

Zhizhi Min
State Key Laboratory of Animal Nutrition, College of Animal Science and Technology, China Agricultural University, Beijing 100193, China

Jianmin Yuan
State Key Laboratory of Animal Nutrition, College of Animal Science and Technology, China Agricultural University, Beijing 100193, China

Bingkun Zhang
State Key Laboratory of Animal Nutrition, College of Animal Science and Technology, China Agricultural University, Beijing 100193, China

Yuming Guo
State Key Laboratory of Animal Nutrition, College of Animal Science and Technology, China Agricultural University, Beijing 100193, China

Jeffrey L Vallet
U.S. Department of Agriculture, Agricultural Research Service, U.S. Meat Animal Research Center, State Spur 18D, Clay Center, NE 68933, USA

Anthony K McNeel
U.S. Department of Agriculture, Agricultural Research Service, U.S. Meat Animal Research Center, State Spur 18D, Clay Center, NE 68933, USA

Jeremy R Miles
U.S. Department of Agriculture, Agricultural Research Service, U.S. Meat Animal Research Center, State Spur 18D, Clay Center, NE 68933, USA

Bradley A Freking
U.S. Department of Agriculture, Agricultural Research Service, U.S. Meat Animal Research Center, State Spur 18D, Clay Center, NE 68933, USA

Krystal K Lum
Department of Animal Science, Cornell University, Ithaca, NY 14850, USA

Jonggun Kim
Department of Animal Science, Cornell University, Ithaca, NY 14850, USA

Xin Gen Lei
Department of Animal Science, Cornell University, Ithaca, NY 14850, USA

Jan Erik Lindberg
Department of Animal Nutrition and Management, Swedish University of Agricultural Sciences, P.O. Box 7024, Uppsala SE-75007, Sweden

Shailendra Raghuwanshi
Department of Microbiology, University of Delhi South Campus, Benito Juarez Road, New Delhi 110 021, India

Swati Misra
Department of Microbiology, University of Delhi South Campus, Benito Juarez Road, New Delhi 110 021, India

Rajendra Kumar Saxena
Department of Microbiology, University of Delhi South Campus, Benito Juarez Road, New Delhi 110 021, India

Haoyu Liu
Department of Animal Nutrition and Management, Swedish University of Agricultural Sciences, P.O. Box 7024, Uppsala SE-750 07, Sweden

Emma Ivarsson
Department of Animal Nutrition and Management, Swedish University of Agricultural Sciences, P.O. Box 7024, Uppsala SE-750 07, Sweden

Torbjörn Lundh
Department of Animal Nutrition and Management, Swedish University of Agricultural Sciences, P.O. Box 7024, Uppsala SE-750 07, Sweden

Jan Erik Lindberg
Department of Animal Nutrition and Management, Swedish University of Agricultural Sciences, P.O. Box 7024, Uppsala SE-750 07, Sweden

Guoyao Wu
Department of Animal Science, Texas A&M University, College Station, Texas 77843, USA

Wenming Huang
State Key Laboratory of Animal Nutrition, College of Animal Science and Technology, China Agricultural University, Beijing 100193, China

Yujia Tian
State Key Laboratory of Animal Nutrition, College of Animal Science and Technology, China Agricultural University, Beijing 100193, China

Yajing Wang
State Key Laboratory of Animal Nutrition, College of Animal Science and Technology, China Agricultural University, Beijing 100193, China

Aminamu Simayi
State Key Laboratory of Animal Nutrition, College of Animal Science and Technology, China Agricultural University, Beijing 100193, China

Amingguli Yasheng
State Key Laboratory of Animal Nutrition, College of Animal Science and Technology, China Agricultural University, Beijing 100193, China

Zhaohai Wu
State Key Laboratory of Animal Nutrition, College of Animal Science and Technology, China Agricultural University, Beijing 100193, China

Shengli Li
State Key Laboratory of Animal Nutrition, College of Animal Science and Technology, China Agricultural University, Beijing 100193, China

Zhijun Cao
State Key Laboratory of Animal Nutrition, College of Animal Science and Technology, China Agricultural University, Beijing 100193, China

Shelby Marie Curry
Department of Animal Sciences, University of Illinois, Urbana 61801, USA

Diego Mario David Labadan Navarro
Department of Animal Sciences, University of Illinois, Urbana 61801, USA

Ferdinando Nielsen Almeida
Department of Animal Sciences, University of Illinois, Urbana 61801, USA

Juliana Abranches Soares Almeida
Department of Animal Sciences, University of Illinois, Urbana 61801, USA

Hans Henrik Stein
Department of Animal Sciences, University of Illinois, Urbana 61801, USA

Kun Jun Han
Louisiana State University Agricultural Center, School of Plant, Environmental, and Soil Sciences, 104 M.B. Sturgis Hall, Baton Rouge, LA 70803, USA

Michael E McCormick
Louisiana State University Agricultural Center, Southeast Region Office, 21549 Old Covington, Hammond, LA 70403, USA

Rafael A Cabrera
Laboratory of Developmental Nutrition, Department of Animal Science, North Carolina State University, 101 Polk Hall, North Carolina State University, Raleigh, NC 27695, USA
Author current employment: Huvepharma USA, 525 West Park Drive Suite 230, Peachtree City, GA 30269, USA

James L Usry
Ajinomoto Heartland Lysine, Chicago, IL 60631, USA
Author current employment: Micronutrients, 1550 Research Way, Indianapolis, IN 46231, USA

Consuelo Arrellano
Department of Statistics, North Carolina State University, Raleigh, NC 27695-8203, USA

Eduardo T Nogueira
Ajinomoto do Brasil. Ajinomoto Animal Nutrition, São

Marianne Kutschenko
Ajinomoto do Brasil. Ajinomoto Animal Nutrition, São

Adam J Moeser
Department of Population Health and Pathobiology, College of Veterinary Medicine, North Carolina State University, Raleigh, NC 27695, USA

Jack Odle
Laboratory of Developmental Nutrition, Department of Animal Science, North Carolina State University, 101 Polk Hall, North Carolina State University, Raleigh, NC 27695, USA

Ferdinando Nielsen Almeida
Department of Animal Sciences, University of Illinois, 1207 W Gregory Dr, Urbana, IL 61801, USA

John Kyaw Htoo
Evonik Industries AG, Rodenbacher Chaussee 4, Hanau 63457, Germany

John Thomson
Evonik Degussa Corporation, 1701 Barrett Lakes Blvd NW, Kennesaw, GA 30144, USA

Hans Henrik Stein
Department of Animal Sciences, University of Illinois, 1207 W Gregory Dr, Urbana, IL 61801, USA

Ashley Gelderman
Department of Animal Science, South Dakota State University, Brookings 57007SD, USA

Jeffrey Clapper
Department of Animal Science, South Dakota State University, Brookings 57007SD, USA

Huili Wang
College of Engineering, China Agricultural University, Beijing 100083, P. R. China

Chuncheng Xu
College of Engineering, China Agricultural University, Beijing 100083, P. R. China

Christopher D Reinhardt
Department of Animal Sciences and Industry, 232 Weber Hall, Kansas State University, Manhattan, KS 66506, USA

Tiffany L Lee
Department of Clinical Sciences, A-111 Mosier Hall, Kansas State University, Manhattan, KS 66506, USA

Daniel U Thomson
Department of Clinical Sciences, A-111 Mosier Hall, Kansas State University, Manhattan, KS 66506, USA

Laman K Mamedova
Department of Animal Sciences and Industry, 232 Weber Hall, Kansas State University, Manhattan, KS 66506, USA

Barry J Bradford
Department of Animal Sciences and Industry, 232 Weber Hall, Kansas State University, Manhattan, KS 66506, USA

Chong Wang
College of Animal Science and Technology, Zhejiang Agriculture and Institute of Dairy Science, Zhejiang University, Hangzhou, P. R. China

Zhen Liu
Institute of Dairy Science, Zhejiang University, Hangzhou, P. R. China

Diming Wang
Institute of Dairy Science, Zhejiang University, Hangzhou, P. R. China

Jianxin Liu
Institute of Dairy Science, Zhejiang University, Hangzhou, P. R. China

Hongyun Liu
Institute of Dairy Science, Zhejiang University, Hangzhou, P. R. China

Zhiguo Wu
School of Veterinary Medicine, University of Pennsylvania, 382 West Street Road, 19348 Kennett Square, USA

Quanfeng Li
State Key Laboratory of Animal Nutrition, Ministry of Agriculture Feed Industry Centre, China Agricultural University, Beijing 100193, China

Meng Shi
State Key Laboratory of Animal Nutrition, Ministry of Agriculture Feed Industry Centre, China Agricultural University, Beijing 100193, China

Chuanxin Shi
State Key Laboratory of Animal Nutrition, Ministry of Agriculture Feed Industry Centre, China Agricultural University, Beijing 100193, China

Dewen Liu
State Key Laboratory of Animal Nutrition, Ministry of Agriculture Feed Industry Centre, China Agricultural University, Beijing 100193, China

Xiangshu Piao
State Key Laboratory of Animal Nutrition, Ministry of Agriculture Feed Industry Centre, China Agricultural University, Beijing 100193, China

Defa Li
State Key Laboratory of Animal Nutrition, Ministry of Agriculture Feed Industry Centre, China Agricultural University, Beijing 100193, China

Changhua Lai
State Key Laboratory of Animal Nutrition, Ministry of Agriculture Feed Industry Centre, China Agricultural University, Beijing 100193, China

Simin Poorkasegaran
Department of Animal Science, Animal Science and Aquaculture Faculty, Sari Agricultural and Natural Resource University, Sari, Mazandaran, Iran

Asadollah Teimouri Yansari
Department of Animal Science, Animal Science and Aquaculture Faculty, Sari Agricultural and Natural Resource University, Sari, Mazandaran, Iran
Department of Animal Science, Educational Center of Tajan, GhaemShahar, Mazandaran, Iran

Xiangjun Leng
Department of Animal Science, Cornell University, Ithaca, NY 14853, USA
College of Aquaculture and Life Science, Shanghai Ocean University, Shanghai 201306, China

Kun-Nan Hsu
Department of Animal Science, Cornell University, Ithaca, NY 14853, USA

Richard E Austic
Department of Animal Science, Cornell University, Ithaca, NY 14853, USA

Xin' gen Lei
Department of Animal Science, Cornell University, Ithaca, NY 14853, USA

Gerardo Mariscal-Landín
CENID Fisiología, Instituto Nacional de Investigaciones Forestales Agrícolas y Pecuarias, km 1 Carretera a Ajuchitlán, 76280 Querétaro, México

Tércia Cesária Reis de Souza
Facultad de Ciencias Naturales, Universidad Autónoma de Querétaro, México, Av. de las Ciencias s/n, 76230 Querétaro, México

Ericka Ramírez Rodríguez
CENID Fisiología, Instituto Nacional de Investigaciones Forestales Agrícolas y Pecuarias, km 1 Carretera a Ajuchitlán, 76280 Querétaro, México

David L Thomas
University of Wisconsin-Madison, Department of Animal Sciences, Madison, Wisconsin, USA

Yves M Berger
Spooner Agricultural Research Station, Spooner, Wisconsin, USA

Brett C McKusick
University of Wisconsin-Madison, Department of Animal Sciences, Madison, Wisconsin, USA
Merial Ltd, Duluth, Georgia, USA

Claire M Mikolayunas
University of Wisconsin-Madison, Department of Animal Sciences, Madison, Wisconsin, USA
Midwest Organic Services Association, Viroqua, Wisconsin, USA

Yan Zhao
Department of Animal Science, North Carolina State University, Raleigh, NC 27695, USA

Alexandra C Weaver
Department of Animal Science, North Carolina State University, Raleigh, NC 27695, USA

Vivek Fellner
Department of Animal Science, North Carolina State University, Raleigh, NC 27695, USA

Robert L Payne
Evonik Corp, Kennesaw, GA 30144, USA

Sung Woo Kim
Department of Animal Science, North Carolina State University, Raleigh, NC 27695, USA

Sarah A Reed
Department of Animal Science, University of Connecticut, Storrs, CT 06269, USA

Joseline S Raja
Department of Animal Science, University of Connecticut, Storrs, CT 06269, USA

Maria L Hoffman
Department of Animal Science, University of Connecticut, Storrs, CT 06269, USA

Steven A Zinn
Department of Animal Science, University of Connecticut, Storrs, CT 06269, USA

Kristen E Govoni
Department of Animal Science, University of Connecticut, Storrs, CT 06269, USA

Ciara A Carberry
Animal and Bioscience Research Department, Animal and Grassland Research and Innovation Centre, Teagasc, Grange, Dunsany, Co. Meath, Ireland
UCD School of Agriculture, Food Science and Veterinary Medicine, College of Life Sciences, University College Dublin, Belfield, Dublin 4, Ireland

David A Kenny
Animal and Bioscience Research Department, Animal and Grassland Research and Innovation Centre, Teagasc, Grange, Dunsany, Co. Meath, Ireland

Alan K Kelly
UCD School of Agriculture, Food Science and Veterinary Medicine, College of Life Sciences, University College Dublin, Belfield, Dublin 4, Ireland

Sinéad M Waters
Animal and Bioscience Research Department, Animal and Grassland Research and Innovation Centre, Teagasc, Grange, Dunsany, Co. Meath, Ireland

Xiaoying Li
State Key Laboratory of Animal Nutrition, China Agricultural University, No. 2. West Road Yuanming Yuan, Beijing 100193, P.R. China

Lihong Zhao
State Key Laboratory of Animal Nutrition, China Agricultural University, No. 2. West Road Yuanming Yuan, Beijing 100193, P.R. China

Yu Fan
State Key Laboratory of Animal Nutrition, China Agricultural University, No. 2. West Road Yuanming Yuan, Beijing 100193, P.R. China

Yaxiong Jia
Beijing Municipal General Station for Animal Husbandry and Veterinary Service, Beijing, China. 3Institute of Plant Protection, Chinese Academy of Agricultural Sciences, Beijing, China

Lei Sun
Institute of Plant Protection, Chinese Academy of Agricultural Sciences, Beijing, China

Shanshan Ma
State Key Laboratory of Animal Nutrition, China Agricultural University, No. 2. West Road Yuanming Yuan, Beijing 100193, P.R. China

Cheng Ji
State Key Laboratory of Animal Nutrition, China Agricultural University, No. 2. West Road Yuanming Yuan, Beijing 100193, P.R. China

Qiugang Ma
State Key Laboratory of Animal Nutrition, China Agricultural University, No. 2. West Road Yuanming Yuan, Beijing 100193, P.R. China

Jianyun Zhang
State Key Laboratory of Animal Nutrition, China Agricultural University, No. 2. West Road Yuanming Yuan, Beijing 100193, P.R. China